TRANSITION METAL COMPOUNDS

Describing all aspects of the physics of transition metal compounds, this book provides a comprehensive overview of this unique and diverse class of solids.

Beginning with the basic concepts of the physics of strongly correlated electron systems, the structure of transition metal ions, and the behavior of transition metal ions in crystals, it goes on to cover more advanced topics such as metal–insulator transitions, orbital ordering, and novel phenomena such as multiferroics, systems with oxygen holes, and high-T_c superconductivity.

Each chapter concludes with a summary of key facts and concepts, presenting all the most important information in a consistent and concise manner. Set within a modern conceptual framework, and providing a complete treatment of the fundamental factors and mechanisms that determine the properties of transition metal compounds, this is an invaluable resource for graduate students, researchers, and industrial practitioners in solid-state physics and chemistry, materials science, and inorganic chemistry.

DANIEL I. KHOMSKII is a Professor at the University of Köln, Germany, where his research interests focus on metal–insulator transitions, magnetism, orbital ordering, and superconductivity.

TRANSITION METAL COMPOUNDS

DANIEL I. KHOMSKII
University of Köln

CAMBRIDGE
UNIVERSITY PRESS

CAMBRIDGE
UNIVERSITY PRESS

University Printing House, Cambridge CB2 8BS, United Kingdom

Cambridge University Press is part of the University of Cambridge.

It furthers the University's mission by disseminating knowledge in the pursuit of
education, learning and research at the highest international levels of excellence.

www.cambridge.org
Information on this title: www.cambridge.org/9781107020177

© Daniel I. Khomskii 2014

First published 2014

Printed in the United Kingdom by Clays, St Ives plc

A catalogue record for this publication is available from the British Library

Library of Congress Cataloguing in Publication data
Khomskii, Daniel, 1938–
Transition metal compounds / Daniel I. Khomskii, University of Köln.
pages   cm
Includes bibliographical references and index.
ISBN 978-1-107-02017-7
1. Transition metal compounds.   I. Title.
QD411.8.T73K46   2014
546′.6–dc23
2014006192

ISBN 978-1-107-02017-7 Hardback

Contents

Introduction

Transition metal (TM) compounds present a unique class of solids. The physics of these materials is extremely rich. There are among them good metals and strong, large-gap insulators, and also systems with metal–insulator transitions. Their magnetic properties are also very diverse; actually, most strong magnets are transition metal (or rare earth) compounds. They display a lot of interesting phenomena, such as multiferroicity or colossal magnetoresistance. Last but not least, high-T_c superconductors also belong to this class.

Transition metal compounds are manifestly the main area of interest and the basis for a large field of physical phenomena: the physics of systems with strong electron correlations. Many novel ideas, such as Mott insulators, were first suggested and developed in application to transition metal compounds.

From a practical point of view, the magnetic properties of these materials have been considered and used for a long time, but more recently their electronic behavior came to the forefront. The ideas of spintronics, magnetoelectricity and multiferroicity, and high-T_c superconductivity form a very rich and fruitful field of research, promising (and already having) important applications.

There are many aspects of the physics of transition metal compounds. Some of these are of a fundamental nature – the very description of their electronic structure is different from the standard approach based on the conventional band theory and applicable to standard metals such as Na or Al, or insulators or semiconductors such as Ge or Si. Furthermore, transition metal compounds also possess many specific and unusual features, largely related to the details of the structure of the atoms or ions that form these compounds. These features have to be taken into account if we want to make our description realistic. Many interesting and diverse phenomena in transition metal compounds are based directly on the specific features of the corresponding atoms or ions. A book devoted to this class of materials has to deal with both of these aspects; it should describe the conceptual problems in the description of systems in which electron–electron interactions (or strong electron correlations) largely determine the very type of states we are dealing with, but it should also combine this with the specific details of particular types of transition metal ions, etc.

The aim of this book is to give a coherent general description of the main aspects of the physics of transition metal compounds. By these, I have in mind solids made up on the basis of transition metal elements. A large and separate field of molecular systems and chemical compounds containing transition metals belongs rather to inorganic chemistry, and is left out of this book. These topics are covered, for example, in Cotton *et al.* (1995) or Bersuker (2010). Nevertheless, many notions which first appeared and are crucial in this field are also very important for solids, and these are discussed to a certain extent in the present book. However, the main focus is indeed on the aspects of the physics of transition metal solids, both conceptual and general, and those specific for particular classes of materials and phenomena. A more general treatment of the theoretical aspects of solids in general can be found in many books, in particular in Khomskii (2010), which I will refer to relatively often. This and the present book can be considered in tandem.

I hope to cover the main aspects of this whole field. Of course, the field of transition metal compounds is actually enormous, and if one goes into all the details, every particular subfield would require a separate book three times bigger than the present volume. Therefore, I have chosen to concentrate on the basic notions and ideas, trying to explain them as qualitatively as possible, sometimes leaving out many technical details. I hope that this style will make this book accessible and useful to a broad audience: both specialists in the physics of transition metal compounds as well as people working in inorganic chemistry and material sciences. I also hope that it will be useful both for more senior and beginning scientists and for graduate students specializing in these fields. One can find more detailed treatments of some specific problems in the literature cited.

This book originated from a set of lectures, usually in the form of short "crash courses" on the physics of transition metal compounds, which I have given during the last 10 years at several places: Cologne, Grenoble, Loughborough, and Korea. These were usually lectures for both senior and younger researchers and also for graduate students specializing or interested in various aspects of the physics of transition metal compounds; the audiences came from physics, inorganic chemistry, and material science departments. Thus I hope that this book may be useful for such audiences, in particular as an advanced textbook.

The general approach taken in this book determines some of its specific features. Thus I prefer to illustrate the main notions by schematic figures, very rarely using real experimental data – rather, presenting such data in a schematic form and stressing the main conceptual features. Of course, this field of research is largely an experimental one. However, sometimes the real experimental data for a particular compound show not only the generic features, but also some specific details which mask or obscure the main effect we want to discuss.

This also relates to the references. Where possible, I have tried to refer not so much to original publications (although of course there are many such references too) but to books or review articles, whenever they exist. I apologize in advance to many of my colleagues whose important papers are not mentioned in this book.

As the material of different chapters often covers quite different fields and is to some extent independent, and I anticipate that some readers will not read the whole book but may skip some parts. Therefore in several places, to preserve the continuity of discussion, I repeat some of the material discussed in more detail in other chapters. This is done in very few places, but I think it justified as it makes each chapter more self-contained.

One more specific feature of the book is determined by its scope and style. I decided to conclude each chapter with a short summary or "digest," presenting in a few pages the main notions and material discussed. The aim is two- or even threefold. The short summaries should remind readers of the main ideas of the corresponding chapters, and help to "enforce" their understanding so that at least the main ideas settle in. Also, these short digests could be used at a first reading so that the reader can immediately see which problems are discussed in the respective chapters and then decide whether he or she needs to study a particular chapter in more detail, or maybe postpone it until later. Thus, for example, the chapter on multiferroics may be interesting for people working in this and related fields, but less so for those specializing in high-T_c superconductivity. These short summaries would give such people at least the opportunity to understand quickly "what it is all about," even in fields far from their own narrow interests.

In effect, these "digests" collected together would form something like a "book within a book," giving in a small volume a qualitative presentation of this whole field. I hope such a "book within a book" will be useful for a broad audience, both for younger people just starting to study and work in the field of transition metal compounds as well as for more mature scientists who could quickly refresh their memory on some topics of both general and more specific character.

A few more words regarding the general scheme and layout of the presentation. As I already mentioned above, in dealing with transition compounds the main physical effects are connected with strong electron correlations. This gives rise to such fundamental notions as electron localization, Mott insulators, etc. The usual starting point in discussing these questions is the simplest case of electrons in a nondegenerate band, with one electron per site ($n = 1$), and with strong electron interaction $U/t \gg 1$, where U is the on-site Coulomb ("Hubbard") repulsion and t is an electron hopping. With this example one can illustrate some key notions mentioned above. However, to make the description more realistic, one has to include many details such as intra-atomic characteristics, orbital structure, spin–orbit interaction, etc. Then, one can gradually relax the restrictions imposed at the beginning, such as integer electron occupation (e.g., one electron per site) or the condition of strong interaction.

The layout of the book more or less follows this scheme. Chapter 1 discusses the basic phenomena, using the example of the simple nondegenerate Hubbard model. Then, in Chapters 2–8, various specific features are gradually included such as the atomic structure of the corresponding elements, modifications occurring when atoms or ions are in a crystal, effects connected with orbital structure including orbital degeneracy, etc. This treatment still deals predominantly with strongly correlated electrons with integer occupation of d-levels, although already in some places the restrictions are relaxed and we treat, for example, charge ordering which mainly takes place for other filling of d-levels. The main effects connected with lifting this restriction, and the treatment of systems with doping and arbitrary band filling, are considered in Chapter 9. Finally, in Chapter 10 we lift the final "constraint" of strong electron correlations, $U \gg t$, and consider the general case of variable U/t, in particular paying attention to the extremely interesting phenomenon of metal–insulator or Mott transitions, which occur in this situation. Finally, in Chapter 11, which stands somewhat apart, we briefly discuss the main properties of another class of solids with strong electron correlations – those made not on the basis of transition metals, but of the $4f$ and $5f$ compounds (rare earths and actinides). Many interesting phenomena in these materials, such as the Kondo effect, are seen in one form or another also in transition metal compounds, and the main physics of these compounds is also similar, although they also have some special features. I think it makes sense for completeness to discuss these materials and phenomena in the present book, as they have close relations to the phenomena observed in transition metal compounds.

In conclusion, I am grateful to many of my coworkers and colleagues, whose contributions and discussions over many years contributed significantly to my understanding of this very large field. I am especially grateful to L. Bulaevskii, K. Kugel, I. Mazin, T. Mizokawa, M. Mostovoy, G. Sawatzky, S. Streltsov, Hao Tjeng, and Hua Wu.

1

Localized and itinerant electrons in solids

The main topic of this book is the physics of solids containing transition elements: $3d$ – Ti, V, Cr, Mn, . . . ; $4d$ – Nb, Ru, . . . ; $5d$ – Ta, Ir, Pt, . . . These materials show extremely diverse properties. There are among them metals and insulators; some show metal–insulator transitions, sometimes with a jump of conductivity by many orders of magnitude. Many of these materials are magnetic: practically all strong magnets belong to this class (or contain rare earth ions, the physics of which is in many respects similar to that of transition metal compounds). And last but not least, superconductors with the highest critical temperature also belong to this group (high-T_c cuprates, with T_c reaching ~ 150 K, or the recently discovered iron-based (e.g., FeAs-type) superconductors with critical temperature reaching 50–60 K).

The main factor determining the diversity of behavior of these materials is the fact that their electrons may have two conceptually quite different states: they may be either localized at corresponding ions or delocalized, itinerant, similar to those in simple metals such as Na (and, of course, their state may be something in between). When dealing with localized electrons, we have to use all the notions of atomic physics, and for itinerant electrons the conventional band theory may be a good starting point.

This division in fact goes back to the first half of the 20th century. In the early stages of the development of quantum mechanics one used to treat in detail the electrons in atoms, with different aspects of atomic structure, shell model, atomic quantum numbers, etc. All these details are indeed important for transition metal compounds as well, and we will discuss these problems in the main body of the book. However at the beginning, in this introductory chapter, we will treat the simplest case, ignoring these complications and paying most attention to the competition between localized and itinerant states of electrons in solids.

1.1 Itinerant electrons, band theory

The "fate" of atomic electrons when individual atoms form a concentrated system – a solid – was treated in the first half of the last century, and it led to a very successful picture known as band theory; see, for example, Mott and Jones (1958), Ashcroft and Mermin (1976), Kittel (2004a), and many other textbooks on solid-state physics. In this theory

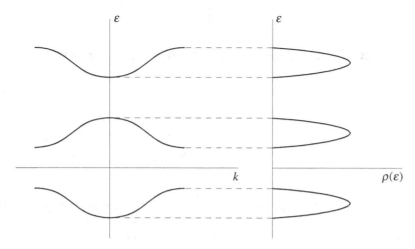

Figure 1.1 Schematic form of the energy spectrum $\varepsilon(k)$ and the density of states $\rho(\varepsilon)$ of free electrons in a crystal.

one considers the motion of noninteracting electrons on the background of a periodic lattice of ions. The spectrum of an electron in the periodic potential consists of allowed energy bands with forbidden states – energy gaps – between them; see the schematic picture of Fig. 1.1, where we show the energy bands $\varepsilon_i(k)$ and the corresponding density of states $\rho(\varepsilon)$.

If there are N atoms in a crystal, each band contains N k-points; for example in the one-dimensional case $k_n = 2\pi n/N$, $n = -\frac{1}{2}N, \ldots, +\frac{1}{2}N$, so that in the continuous limit $-\pi \leqslant k < \pi$ (here and below in most cases we will take the lattice constant $a = 1$). The values $-\pi \leqslant k \leqslant \pi$ form the (first) Brillouin zone. For a system with N sites each band contains N energy levels and, according to the Pauli principle, one can put two electrons with spins $\uparrow$ and $\downarrow$ on each level, so that each band has room for $2N$ electrons.

In this scheme the electrons occupy the lowest energy levels, and if the number of electrons N_{el} is less than $2N$, that is the electron density $n = N_{el}/N < 2$, the electrons would occupy the lowest energy band only partially, up to a certain maximal momentum k_F and energy ε_F (Fig. 1.2) and the system would be a metal. k_F and ε_F are correspondingly the Fermi momentum and Fermi energy.

In the one-dimensional (1d) case we would have two Fermi points $\pm k_F$. In two-dimensional (2d) and three-dimensional (3d) systems the electrons occupy the states $\varepsilon(k) \leqslant \varepsilon_F$, and the boundary of these occupied states forms the Fermi surface. There may exist several energy bands, which may intersect, and the corresponding Fermi surface of metals may in general be very complicated.

If, in the simplest case of one nondegenerate band of Fig. 1.1, we had $N_{el} = 2N$, the electrons would fully occupy this first band and the system would be insulating, with an energy gap separating the completely full valence band and the empty conduction

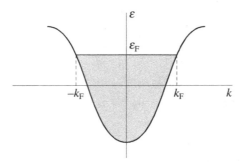

Figure 1.2 Energy spectrum of a metal: the occupied states are shaded. ε_F and k_F are the Fermi energy and Fermi momentum.

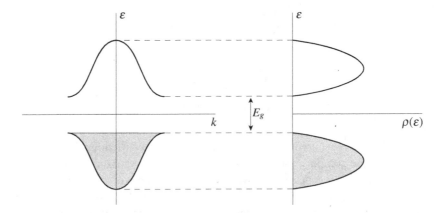

Figure 1.3 Typical energy spectrum of a semiconductor. The occupied states are shaded.

band (Fig. 1.3). This is the standard description of the ordinary band insulators or semiconductors such as Ge and Si.

There are two methods of describing band formation in solids, usually presented at the very beginning of textbooks on solid-state physics, see for example Ashcroft and Mermin (1976) and Kittel (2004a). The first method treats the motion of independent, noninteracting electrons in a periodic lattice potential (Fig. 1.4). One can start from free electrons with the spectrum $\varepsilon(k) = k^2/2m$ in a periodic potential. The corresponding Schrödinger equation for the electron is known in mathematics as the Mathieu equation, and its spectrum, shown in Fig. 1.5, has the form of energy bands separated by energy gaps at wave vectors equal to the Umklapp wave vectors of the given periodic lattice, $\mathcal{K} = 2\pi n/a$, where a is the corresponding lattice parameter. For a weak periodic potential we have the picture shown in Fig. 1.5 which, after we fold the spectrum to the first Brillouin zone, would give the spectrum shown schematically in Figs 1.1 and 1.3. This is the so-called *free electron approximation*. Of course, in contrast to the simple 1d case shown

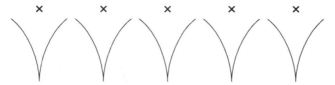

Figure 1.4 Periodic potential for treating the motion of electrons in the band theory.

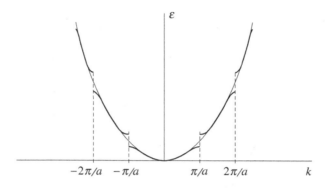

Figure 1.5 The origin of energy bands for free electrons in a periodic potential.

in Figs 1.4 and 1.5, in real crystals with complicated crystal structures the band structure may look much more complicated, with some bands possibly crossing in some directions, but the general rule remains the same: if we have an odd number of electrons per ion, or per unit cell, then in this approximation some bands will necessarily be partially filled and the system should be a metal; and if there is an even number of electrons per unit cell, we may have an insulator of the type shown in Fig. 1.3 (though in the case of overlapping bands we may still obtain a metal or semimetal).

Another approximation often used in band theory is the *tight-binding approximation*; this picture is usually closer to reality for d-electrons in transition metal compounds, and we will mostly use this approximation in what follows. This approach starts by considering isolated atoms with their localized atomic levels, and then treats the tunneling or hopping of electrons from one atom to another, that is from one potential well to the next in the crystal (Fig. 1.6). For two neighboring potential wells this leads to a splitting of energy levels (dashed lines in Fig. 1.6(a)) into bonding and antibonding configurations, $|b\rangle = \frac{1}{\sqrt{2}}(|1\rangle + |2\rangle)$ and $|a\rangle = \frac{1}{\sqrt{2}}(|1\rangle - |2\rangle)$), and in a periodic lattice composed of such centers each atomic level is broadened into a band, Fig. 1.6(b), with the states in the form of a plane wave with momentum $\boldsymbol{k}$,

$$|\boldsymbol{k}\rangle = \frac{1}{\sqrt{N}} \sum e^{i\boldsymbol{k}\cdot\boldsymbol{n}}|\boldsymbol{n}\rangle \, , \qquad (1.1)$$

where $|\boldsymbol{k}\rangle$ is the plane wave wavefunction and $|\boldsymbol{n}\rangle$ is the atomic state at site $\boldsymbol{n}$. As a result we have again a band picture as shown in Fig. 1.1, with each band originating from the

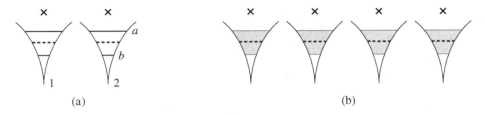

Figure 1.6 The origin of energy bands in the tight-binding approximation.

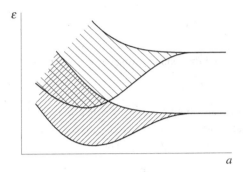

Figure 1.7 Typical dependence of the energy bands on interatomic distance a.

corresponding atomic level (again, these bands can in principle overlap; see the famous picture of Fig. 1.7, showing schematically the broadening of atomic levels into bands when the atoms are moved closer together, i.e. the interatomic distance a is reduced). For large interatomic distances, that is for small overlap of the corresponding wavefunctions of neighboring atoms and hence for small probability of tunneling between neighboring potential wells of Fig. 1.6, the bands will be narrow and one can treat each such band separately, ignoring the others. This is the approximation often used to describe the crossover from the band picture with itinerant electrons to the picture with localized electrons.

1.2 Hubbard model and Mott insulators

Consider the simplest idealized case of a lattice consisting of atoms with nondegenerate electron levels – for example, one can visualize it as a lattice of hydrogen atoms or protons separated by distance a (taken as 1) with one nondegenerate $1s$ level at each site (the dashed line in the potential wells of Fig. 1.6). The hopping of electrons from site to site,

$$\mathcal{H} = -t \sum_{\langle ij \rangle, \sigma} c_{i\sigma}^{\dagger} c_{j\sigma} \, , \tag{1.2}$$

where $c_{i\sigma}^{\dagger}$, $c_{i\sigma}$ are creation and annihilation operators of electrons at site i with spin σ, t is the hopping matrix element, and the summation $\langle ij \rangle$ goes over nearest neighbors,

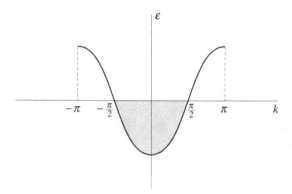

Figure 1.8 Nondegenerate band for noninteracting electrons in the tight-binding approximation for one electron per site, $n = N_{el}/N = 1$.

leads to the formation of an energy band. (We use here and below the formalism and language of second quantization, which is widely used nowadays; a simple introduction to this technique is presented in Appendix B.) The Fourier transform of (1.2) gives for this band the Hamiltonian

$$\mathcal{H}_b = \sum_{k,\sigma} \varepsilon(k)\, c^\dagger_{k\sigma} c_{k\sigma} \qquad (1.3)$$

with the spectrum (with one, two, or three terms for the simple chain, square, or cubic lattice)

$$\varepsilon(k) = -2t(\cos k_x + \cos k_y + \cos k_z)\,, \qquad (1.4)$$

see Fig. 1.8 (we again set the lattice constant $a = 1$). This is the standard tight-binding approximation. As discussed above, for a lattice of N sites there will be N energy levels in this band, which for N (or volume V) going to infinity gives the continuous spectrum (1.4). According to the Pauli principle there will be $2N$ places for electrons in this band. Thus, if there is one electron per site, with electron density $n = N_{el}/N = 1$, the band will be half-filled, as shown in Fig. 1.8, and the system should be a metal.

Note that this conclusion does not depend on the distance between atoms in Fig. 1.6, that is on the value of the hopping matrix element t in (1.2), (1.4), which determines the total bandwidth $W = 2zt$ (where z is the number of nearest neighbors – e.g., $z = 2$ in 1d chain, $z = 4$ in square lattice, etc.). However this hopping t, which actually is proportional to the probability of electron tunneling from site to site, will be exponentially small when we increase the distance between sites. Nevertheless, according to (1.3), (1.4) and Fig. 1.8, such systems should still be metallic for any distance between sites and for arbitrarily small values of t. Thus, for example, we can put our "hydrogen atoms" one meter apart and still formally the system should be metallic!

Of course that is a very unphysical result. Intuitively it is evident that in this case the system would consist of neutral hydrogen atoms, with exactly one electron localized at each site, and it would be insulating. What is wrong then, what is missing in the treatment

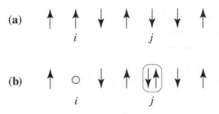

Figure 1.9 Creation of charge-carrying excitations (holes and extra electrons, or doublons) from the state with electrons localized one per site.

which led to the conclusion of Fig. 1.8 that such systems would be metallic for arbitrarily small hopping t and bandwidth $W \sim t$? At least one very important physical effect was missing in this treatment and in the corresponding one-electron Hamiltonian (1.2): it completely ignored the interaction, the Coulomb repulsion between the electrons. Physically we expect that if the distance between atoms is large enough, there will be exactly one electron localized at each site, and because of the repulsion with the first electron, the second electron would not go to the already occupied site.

Let us start from such a state, with one electron per site – see Fig. 1.9(a). To create charge-carrying excitations we should take one electron from a certain site, say site i, and transfer it to another site j – see Fig. 1.9(b). Then the hole left at site i and the extra electron at site j (a doubly occupied site or "doublon") can start to move across the crystal, contributing to the electric current (nobody has told us that the hole should remain at site i; an electron from a neighboring site $i + 1$ can hop to site i, i.e. the hole would move to site $i + 1$, without any energy cost; the same is true for an extra electron or doublon).

However, to create such charge-carrying excitations, a hole and a doublon, we first had to move electrons and put an extra electron at site j, which already had an electron! And this would cost us at least the energy of Coulomb repulsion of two electrons at site j. This energy is traditionally denoted U, and the corresponding term in the Hamiltonian describing this interaction has the form

$$\mathcal{H}_{\text{int}} = U \sum_i n_{i\uparrow} n_{i\downarrow} \qquad (1.5)$$

(for the nondegenerate case, such as our "$1s$" level, one can put only two electrons with opposite spins at each site, and that is why the interaction (1.5) contains the electron densities at each site i, $n_{i\sigma} = c^\dagger_{i\sigma} c_{i\sigma}$, with opposite spins). The resulting full model would then be (combining (1.2) and (1.5))

$$\mathcal{H} = -t \sum_{\langle ij \rangle, \sigma} c^\dagger_{i\sigma} c_{j\sigma} + U \sum_i n_{i\uparrow} n_{i\downarrow} ; \qquad (1.6)$$

this is known as the *Hubbard model* (Hubbard, 1963).

It is clear that when we include this physical effect, the on-site electron repulsion (1.5) to create such an electron–hole (or doublon–hole) pair would cost us energy U. What we can gain here is the kinetic energy of both electron and hole: when they start to move through the crystal, they would form energy bands (1.3), (1.4) and both would occupy the lowest states in these bands, with energy $-\frac{1}{2}W = -zt$. That is, the total energy gain obtained by creating such excitations would be $W = 2zt$, but the energy loss would be U – the on-site electron repulsion. Qualitatively, we expect that if $U > W$ then the electrons would remain at their sites and the system would remain insulating. To create electron and hole excitations which would then be able to carry currents, we need to overcome the energy gap

$$E_g \sim U - W = U - 2zt , \tag{1.7}$$

which plays the same role as the energy gap between the filled valence band and the empty conduction band in ordinary insulators or semiconductors such as Ge or Si (see Fig. 1.3).

Thus for one electron per site, $n = 1$, and for small electron hopping t (or for a narrow band W) the inclusion of the on-site Coulomb repulsion $Un_{i\uparrow}n_{i\downarrow}$ (1.5) can make the system insulating if $U \gtrsim W = 2zt$, despite the fact that this system would have been metallic in the conventional band picture, which is a one-electron picture and ignores electron–electron interactions. Such insulators are called *Mott* or *Mott–Hubbard insulators* (see, e.g., Mott, 1990).[1] The nature of this insulating state is quite different from that of ordinary band insulators; it is caused not by the periodic potential of the lattice, as is the case for band insulators, but is due completely to electron–electron interaction or, as this is frequently called, *strong electron correlations*. Correspondingly, most of the properties of Mott insulators are also very different from those of ordinary insulators, although some of the notions from ordinary insulators (such as energy gaps) can also be used in the description of Mott insulators. Often this analogy is helpful, but one has to be careful in using it and in transferring the notions of band insulators to Mott insulators. We will see many examples of differences between these two types of insulators later on.

The connection with the theory of chemical bonding in molecules should be mentioned here. The simplest description of the formation of a chemical bond, for example in the H_2 molecule, uses the so-called molecular orbital (MO) description: the wavefunction of an electron moving between sites (protons) a and b is written as

$$|\Psi_{\pm}\rangle = \frac{1}{\sqrt{2}}(\Psi_a \pm \Psi_b), \tag{1.8}$$

where the $+$ and $-$ signs refer to bonding and antibonding orbitals. The ground state of the H_2 molecule with two electrons on bonding orbitals (with antiparallel spins) will then be described by the state

$$|\Psi_{MO}\rangle = \frac{1}{2}\big(\Psi_a(r_1) + \Psi_b(r_1)\big)\big(\Psi_a(r_2) + \Psi_b(r_2)\big) \cdot \frac{1}{\sqrt{2}}(1\uparrow 2\downarrow - 1\downarrow 2\uparrow) \tag{1.9}$$

(a singlet state, symmetric in coordinates and antisymmetric in spins). This state is called an MO state, often also an MO LCAO state (molecular orbital – linear combination of atomic orbitals), or Hund–Mulliken state. One sees that in expression (1.9) there are terms corresponding to electrons located at different sites ($\Psi_a(r_1)\Psi_b(r_2)$) – these are nonpolar

---

[1] Some interesting historical notes connected with the origin of the notion of Mott insulators are presented in Appendix A.

or homopolar states. However the MO state also contains, with equal probability, the ionic states of type $\Psi_a(r_1)\Psi_a(r_2)$ in which both electrons reside on the same atom a. When extended to a large periodic crystal, such MO states give rise to the standard band picture of noninteracting electrons (1.3), (1.4).

It is, however, clear that the ionic states such as $\Psi_a(r_1)\Psi_a(r_2)$ cost a large Coulomb on-site energy of electron–electron repulsion. To avoid this cost, one often describes the chemical bond using another state, known as the Heitler–London state:

$$|\Psi_{HL}\rangle = \tfrac{1}{\sqrt{2}}\Big(\Psi_a(r_1)\!\uparrow\ \Psi_b(r_2)\!\downarrow - \Psi_a(r_1)\!\downarrow\ \Psi_b(r_2)\!\uparrow\Big). \qquad (1.10)$$

In this state all ionic configurations with two electrons at the same site are excluded, and the corresponding energy loss is avoided.[2] On the energy diagram the MO state corresponds to two electrons occupying a bonding orbital with energy $-t$, see Fig. 1.10(a). The Heitler–London state can also be shown on a similar diagram, see Fig. 1.10(b), but one has to remember that the basic states in this case are not the single-electron states of noninteracting electrons but rather many-electron (here two-electron) states, with the electron–electron interaction taken into account. As a result, the energy of such a Heitler–London bonding state is not $-t$ as was the case for the MO state, but rather $\sim -t^2/U$, see below, where U is the on-site Hubbard repulsion (1.5). Once again, these are many-electron states, the lower one corresponding to a singlet state and the upper one to a triplet state of our two sites–two electrons problem. In fact, these two approaches are the main ones used in the

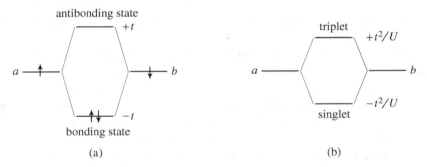

Figure 1.10 (a) Creation of bonding and antibonding states for two centers in the MO LCAO approximation. (b) Creation of singlet bonding and triplet antibonding states in the Heitler–London approximation.

[2] In the second quantization formalism, the wavefunctions (1.9) and (1.10), written using the electron creation and annihilation operators $c^\dagger$, c used for example in eq. (1.6) have the form

$$|\Psi_{MO}\rangle = \tfrac{1}{2}(c_{1\uparrow}^\dagger + c_{2\uparrow}^\dagger)(c_{1\downarrow}^\dagger + c_{2\downarrow}^\dagger)|0\rangle\,,$$

$$|\Psi_{HL}\rangle = \tfrac{1}{\sqrt{2}}(c_{1\uparrow}^\dagger c_{2\downarrow}^\dagger - c_{1\downarrow}^\dagger c_{2\uparrow}^\dagger)|0\rangle\,,$$

where $|0\rangle$ is the vacuum state without any electrons. The required antisymmetry of the total wavefunction is guaranteed by the anticommutativity of the fermion operators c_i, c_j on different sites i, j: $c_{i\sigma}^\dagger c_{j\sigma'}^\dagger = -c_{j\sigma'}^\dagger c_{i\sigma}^\dagger$.

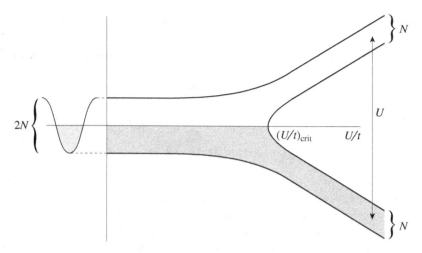

Figure 1.11 Schematic form of the energy band in the Hubbard model as a function of the strength of electron–electron (Hubbard) repulsion. The two bands on the right are the lower (shaded) and the upper (unshaded) Hubbard (sub)bands.

description of chemical bonds in molecules, for example in the hydrogen molecule H_2 (see, e.g., Slater, 1963).

We see that the MO or MO LCAO description of the chemical bond corresponds to the band picture of concentrated solids, while the Heitler–London description corresponds to Mott or Mott–Hubbard insulators, where the electrons avoid each other and are localized at different sites, one per site.

The picture often used to describe Mott insulators, constructed by analogy with band insulators, is that of Hubbard subbands (Fig. 1.11). We can say that the energy band, which for noninteracting electrons ($U = 0$) would be half-filled, for large enough U, $(U/t) > (U/t)_{\mathrm{crit}}$ (or $U > U_{\mathrm{crit}} \sim W = 2zt$), would split into two subbands, with the energy gap between them being $E_g \sim U$ (or, more accurately, $E_g \sim U - W = U - 2zt$). For one electron per site, $n = N_{\mathrm{el}}/N = 1$, each of these subbands will then have "space" for N electrons. The lower band will then be occupied, and the upper one empty. These bands are called *lower and upper Hubbard (sub)bands*. This picture resembles that of ordinary semiconductors, see Fig. 1.3.

However, there is an important difference here. If for band insulators or semiconductors the energy gap was determined by the periodic potential of the crystal lattice, here it is entirely due to electron–electron interactions. Also, each band in Fig. 1.3 contained $2N$ places. In Mott insulators, however, the band for $U = 0$ contains $2N$ places, but when the band is split into two Hubbard subbands for large U (Fig. 1.11), each of these subbands (for one electron per site) would contain only N places; it is precisely because of this that the system with one electron per site, $n = N_{\mathrm{el}}/N = 1$, would completely occupy the lower Hubbard band, leaving the upper one empty.

In fact, the situation is even more complicated. In ordinary band insulators the "capacity" (spectral weight) of each band is constant, $2N$, or two states per site. In contrast, in Hubbard subbands the "capacity" (number of states) in each subband is not constant, but depends on the number of electrons present. When we change the number of electrons, there occurs a redistribution of states between these subbands so that the number of electrons per site becomes, for example, less than 1 in one subband and more than 1 in the other. This *spectral weight redistribution* or *spectral weight transfer*, described in more detail in Section 9.1, is very characteristic of Mott insulators; it is seen experimentally, for example in optics, and it plays a very important role in these systems, putting in particular some restrictions on the use of the semiconductor analogy for them.

One important phenomenon can be deduced directly from the form of Fig. 1.11, and also from the preceding discussion. We have established that for one electron per site and for strong interaction $U \gg t$ the system described by the nondegenerate Hubbard model (1.6) is a Mott insulator. However, it is definitely metallic for $U = 0$. Thus, if we start from noninteracting electrons and increase U (or U/t), there should occur a metal–insulator transition. This transition from a metal to a Mott insulator is called *Mott transition*. There exist many transition metal compounds which undergo such a transition, which is driven either by a change in pressure, temperature, or composition. Usually such metal–insulator transitions are also accompanied by a change in the crystal and/or magnetic structure, which in itself can lead to opening the gap in the spectrum. Thus the question of the detailed nature of a particular metal–insulator transition, in particular the relative role played by electronic correlations and structural distortions in the transition, is usually a rather nontrivial one and requires a separate study. Many examples of metal–insulator transitions, in particular Mott transitions, will be given in Chapter 10. Different theoretical approaches to the description of Mott transitions will also be discussed there.

1.3 Magnetism of Mott insulators

We have seen in the previous section that the state of Mott insulators is characterized by electron localization, one per site. Correspondingly, the appearance of localized electrons in a system implies simultaneously the appearance of *localized spins* or *localized magnetic moments*. In a sense we have come back from the band description of electrons in a crystal, where the electrons were treated as itinerant, delocalized, forming plane waves, to the atomic description with electrons localized at corresponding sites, with all the atomic physics to be taken into account. This will be done in the following chapters; here however we continue to illustrate the basic effects using our simplified model (1.6), ignoring all complications and considering nondegenerate electrons characterized only by site index i and spin σ. (A more detailed theoretical treatment can be found, e.g., in chapter 12 of Khomskii, 2010.) But we do have to include spins, otherwise the system with one electron per site would be insulating even in the standard band theory.

Thus, when we go over to electrons localized at each site, localized spins would appear, which have to be ordered somehow at low temperatures. Without such ordering the state of the system would remain highly degenerate (with degeneracy 2^N – each of N sites with one electron would have two equivalent states, with spins ↑ and ↓). This degeneracy would contradict the Nernst theorem (which states that at zero temperature any system should be ordered, with zero entropy), and a certain spin ordering should occur at $T = 0$.

When we consider the Hubbard model (1.6) with one electron per site, $n = 1$, and with strong interaction $U \gg t$, the description should be in a sense opposite to what we usually do for interacting systems. Normally we first solve the problem of noninteracting electrons (the first term in the Hamiltonian (1.6)) and then take into account interactions (the second term in (1.6)), for example using perturbation theory. Here, however, the situation is just the opposite. The ground state of Mott insulators, with one electron localized at a site, is chosen so as to minimize the interaction energy, the second term in the Hamiltonian (1.6). Thus this term should be taken as the main, or unperturbed Hamiltonian, and the first term – electron hopping – can be treated in the case $t/U \ll 1$ as a perturbation. Indeed, the interaction term in (1.6) leads to electrons localized one at a site, but leaves the spin degeneracy. The first term in (1.6), acting twice, in the second order in small t/U, lifts this degeneracy and leads to (antiferro)magnetic exchange and to magnetic ordering.

One can qualitatively understand this process in the following way, see Fig. 1.12. We have to consider two situations: the spins of electrons at neighboring sites are either parallel, Fig. 1.12(a), or antiparallel, Fig. 1.12(b). Both situations correspond to a minimum in the interaction energy, $E_0 = 0$. The first term in the Hamiltonian (1.6), electron hopping, in principle moves an electron from one site to a neighboring site. To first order it will create states of the type shown in Fig. 1.9(b), with one empty site and another (neighboring) site containing two electrons; the energy of this excited state is $E_{\rm ex} = U$. This state, however, does not belong to the 2^N-degenerate ground state manifold of nonpolar states. However, the second application of the same hopping term, the first term in (1.6), can move an electron back from the doubly occupied site to the empty site and return us to the ground state. Thus, second-order processes are possible in principle.

As always in quantum mechanics, the second-order contribution to the ground state energy,

$$\Delta E_2 = \sum_n \frac{|\langle 0| H'|n\rangle|^2}{E_0 - E_n}, \qquad (1.11)$$

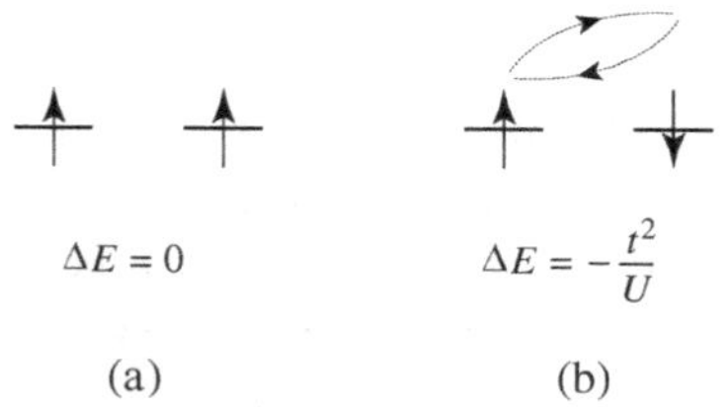

$$\Delta E = 0 \qquad\qquad \Delta E = -\frac{t^2}{U}$$

(a)   (b)

Figure 1.12 The scheme explaining the origin of the antiferromagnetic (super)exchange in Mott insulators (with nondegenerate levels).

where H' is the perturbation and where the summation goes over all excited states with energies $E_n > E_0$, is negative. In our case the role of perturbation H' is played by electron hopping, the first term in (1.6). Thus, if the process of virtual hopping of an electron to a neighboring site and back is allowed, this process will decrease the energy (for the ground state the denominator in (1.11) is always negative).

For parallel spins (Fig. 1.12(a)), however, this process is forbidden by the Pauli exclusion principle, that is, the corresponding energy gain is zero. For antiparallel spins (Fig. 1.12(b)), it is allowed and gives the energy gain $-2t^2/U$. Indeed, this term describes the hopping of an electron occurring twice, therefore we have t^2 in the numerator and in the denominator there appears the energy of the intermediate state with two electrons on one site, equal to U (the factor 2 comes from the fact that there are two such processes: hopping first from the left to the right and back, and from the right to the left). Thus we see that as a result of this process the antiparallel arrangement of neighboring spins is more favorable, and the ground state of our system would be antiferromagnetic. This, in fact, is the origin of the singlet–triplet splitting in the state described by the Heitler–London wavefunction (1.10).

One can indeed carry out this calculation (for details see, e.g., Khomskii, 2010), and as a result we find that the magnetic state of our system may be described, instead of the original electronic Hamiltonian (1.6), by the effective Heisenberg Hamiltonian

$$\mathcal{H}_{\text{eff}} = J \sum_{\langle ij \rangle} S_i \cdot S_j , \qquad J = \frac{2t^2}{U} . \tag{1.12}$$

This Hamiltonian acts on the 2^N-degenerate subspace of nonpolar states, all having one electron per site, and it describes the ground state and the lowest excited states of our system, which are magnetic excitations (spin waves). In deriving (1.12) one uses the connection between electronic and spin operators, which for the subspace of one electron per site has the form

$$c_{i\uparrow}^{\dagger} c_{i\uparrow} \Rightarrow \tfrac{1}{2} + S_i^z , \qquad c_{i\downarrow}^{\dagger} c_{i\downarrow} \Rightarrow \tfrac{1}{2} - S_i^z ,$$
$$c_{i\uparrow}^{\dagger} c_{i\downarrow} \Rightarrow S_i^+ , \qquad c_{i\downarrow}^{\dagger} c_{i\uparrow} \Rightarrow S_i^- , \tag{1.13}$$

where $S_i^+ = S_i^x + i S_i^y$ and $S_i^- = S_i^x - i S_i^y$ are operators reversing the spins from $\downarrow$ to $\uparrow$ (S_i^+) and from $\uparrow$ to $\downarrow$ (S_i^-). Note that in the process of virtual hopping shown in Fig. 1.12(b) not only the "hopping" electron can return to its own original place, but also at the second step the "own" electron from site j can move to site i, that is there will occur *an exchange* of electrons. As a result the effective Hamiltonian (1.12) has a Heisenberg form, that is it contains not only the Ising (classical) terms $S_i^z S_j^z$ but also the exchange (quantum) terms $S_i^x S_j^x + S_i^y S_j^y = \tfrac{1}{2}(S_i^+ S_j^- + S_i^- S_j^+)$.

One general remark is warranted here. In writing the exchange interaction such as (1.12), different conventions are used in the literature. Sometimes one uses the opposite sign, $-J \sum S_i \cdot S_j$, sometimes one writes $2J \sum S_i \cdot S_j$, etc. Also, one has to specify whether the summation $\sum_{\langle ij \rangle}$ is carried out over indices i and j independently (this is the convention used below, and in Khomskii, 2010) – in which case each pair of spins, for example $S_1 \cdot S_2$, is counted twice, $i = 1$, $j = 2$ and $i = 2$, $j = 1$ – or whether each pair is counted only once. In the latter case the exchange constant in (1.12) would not be $2t^2/U$ but $4t^2/U$; this convention can also be found in the literature. Thus, when consulting the values of exchange constants presented in the literature one should always verify which particular form of the exchange integral is being used.

From the form of the effective Hamiltonian for Mott insulators, (1.12), we indeed see that the ground state of the simple nondegenerate Hubbard model (1.6) for $n = 1$ and $U \gg t$ is a state with antiferromagnetic ordering, in accordance with the qualitative considerations illustrated in Fig. 1.12. This mechanism of exchange interaction is called *superexchange* (Anderson, 1959), or sometimes kinetic exchange. As we have seen above, although it has the form of a Heisenberg exchange interaction ($\sim S_i \cdot S_j$), it is in fact not the original Heisenberg exchange interaction (the exchange part of the Coulomb interaction),

$$\int \Psi_1^*(r)\Psi_2^*(r') \frac{e^2}{|r - r'|} \Psi_1(r')\Psi_2(r) , \qquad (1.14)$$

but is caused by effective delocalization of electrons, that is by a decrease of their kinetic energy: for an antiparallel arrangement of neighboring spins an electron from one site can partially delocalize to neighboring sites, due to virtual hopping, and this partial delocalization leads to a decrease in kinetic energy, according to the uncertainty principle $\delta x \, \delta p \gtrsim \hbar$. Such a superexchange is actually the main mechanism of exchange interaction in magnetic insulators; the real Heisenberg exchange interaction $\int \Psi^* \Psi^* (e^2/r) \Psi \Psi$ (1.14) is usually much smaller (and of opposite sign). In real transition metal compounds, with their complicated structure of atomic levels, the resulting form of magnetic exchange can be significantly more complicated and is not necessarily antiferromagnetic; we will discuss all these details in the next chapters. But at least conceptually the mechanism of the formation of Mott insulators and of magnetic exchange in them is captured correctly already in the simple Hubbard model (1.6) (simple in its formulation, which ignores many details; not at all simple in its properties!).

1.4 Interplay of electronic motion and magnetism in Mott insulators

Following our general strategy, and anticipating the phenomena which will be described in much more detail later in this book, we use the example of the nondegenerate Hubbard model (1.6) to describe several other effects which would play an important role in real situations. One of the most important such effects is that for Mott insulators, in contrast to ordinary insulators, there exists a very intricate interplay between electron and spin degrees of freedom. Consider the ground state of a strongly correlated system, which, as we have shown above, for $n = 1$ and $U \gg t$ is an antiferromagnetic Mott insulator, and let us look at the motion of an extra electron or hole doped into the system (or created by electron–hole excitation from the ground state). Such an electron or hole could in principle be at any site i: once created, it could in principle move due to the hopping term in the Hamiltonian (1.6),

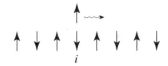

Figure 1.13 The motion of an "extra" electron on the antiferromagnetic background.

without the extra energy cost of the Coulomb repulsion U. This could broaden the state of such carriers into bands: the lower (for holes) and upper (for electrons) Hubbard bands of Fig. 1.11. These electrons or holes, however, would move not in an empty lattice, but on the background of other electrons, with their antiferromagnetic ordering (Fig. 1.13). Thus, if we put an extra electron with spin ↑ at site i in Fig. 1.13, this electron could in principle move to a neighboring site without extra cost in the Hubbard energy U. However, as we can see from Fig. 1.13, for antiferromagnetic ordering this motion would be prevented (due to the Pauli principle) by the presence of electrons with the same spin sitting on neighboring sites. Thus it appears that the background antiferromagnetic ordering would hinder, and maybe totally suppress, the motion of the extra electron. This motion would be allowed if at finite temperatures (or even at $T = 0$ due to zero-point fluctuations) the spins of neighboring sites were reversed, but the stronger the antiferromagnetic correlations, the more difficult such electron motion would be.

The theoretical treatment indeed shows (Bulaevskii *et al.*, 1968; Khomskii, 2010) that due to this effect the effective electron hopping in the mean-field approximation is reduced,

$$t_{ij} \longrightarrow \left(t_{ij}\right)_{\text{eff}} = t_{ij} \frac{\frac{1}{4} + \langle S_i \cdot S_j \rangle}{\sqrt{\frac{1}{4} - \overline{S}^2}} \simeq t\sqrt{\frac{1}{4} - \overline{S}^2}, \qquad (1.15)$$

where $\overline{S} = \langle S_i \rangle$ is the average sublattice magnetization. Thus if $\overline{S} \to \frac{1}{2}$, the electron hopping would be completely suppressed. Finite temperatures and zero-point vibrations, always present in an antiferromagnet, reduce $\overline{S}$ and allow for certain hopping, but with a greatly reduced amplitude.

However, the treatment presented above ignores one important factor, which at first glance completely invalidates these arguments. In treating the electron motion, we have considered the hopping of an *extra* electron in Fig. 1.13. But all electrons are equivalent, and if the extra electron with spin ↑ of Fig. 1.13 cannot move to neighboring sites because of the presence of electrons with the same spin there, an "own" electron with spin ↓ can still hop there! As a result it looks like at each step one "own" electron can move (Fig. 1.14(a)), and consequently the extra negative charge (the doubly occupied site), could move through the antiferromagnetically ordered crystal. This is even more evident if we consider the motion not of one extra electron (doublon) of Figs 1.13, 1.14 but of a hole in the crystal (Fig. 1.15). Indeed, at each step an electron from a neighboring site can hop to the empty site, and as a result the hole would move through the crystal.

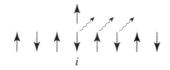

Figure 1.14 Possible motion of an extra charge in an antiferromagnetic Mott insulator.

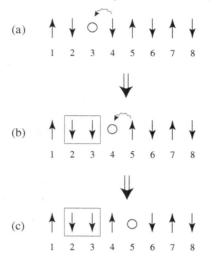

Figure 1.15 The motion of a hole showing spin–charge separation – the formation of a domain wall (two spins ↓ ↓) and a hole which does not disturb the antiferromagnetic background.

But we can notice here one nontrivial effect. Each time an electron moves to an empty site, it interchanges the spins of the sublattices. Thus, if in the original state of Fig. 1.15(a) all odd sites had spin ↑ and all even sites had spin ↓ (and a doped hole somewhere), after two hoppings of the hole (Fig. 1.15(c)), we have wrong, reversed spins at sites 3 and 4: ↓ ↑ instead of the original ↑ ↓ spins in the undoped case. Thus a hole moving by this process through an antiferromagnetic crystal leaves behind a trail of wrong spins (Bulaevskii *et al.*, 1968).[3]

This is not so dangerous in 1d systems: such a process would create one "domain wall" – the wrong ferromagnetic bond (2–3) marked in Fig. 1.15(b, c), but any further

---

[3] Note also that the excitations created after a few steps, for example in Fig. 1.15(c), are rather strange: the hole is surrounded by spins ↑ and ↓, that is it does not carry spin (if we "contract" the lattice by removing such a hole, we would recover the undistorted antiferromagnetic ordering). In contrast, there is a spin excitation – two parallel spins in another place of the lattice, here at sites 2 and 3. This defect (it is actually an excitation, and can move by interchanging e.g. the spins 1 and 2) carries spin $\frac{1}{2}$. This spin-$\frac{1}{2}$ neutral excitation is called a *spinon* and the hole (on site 5 in Fig. 1.15(c)) is a *holon*. As a result, the original (real) hole on site 3 in Fig. 1.15(a), which has both charge $+e$ and spin $\frac{1}{2}$ (or, here, rather $-\frac{1}{2}$, as it is missing an original spin $\frac{1}{2}$ at site 3), is split into *two* excitations, a holon carrying charge but no spin and a spinon, a neutral spin-$\frac{1}{2}$ excitation. This is the *spin–charge separation* (Anderson, 1997) typical for one-dimensional systems, but which is often invoked also in higher-dimensional systems, for example frustrated systems (see Section 5.7.2) or in high-T_c superconductors (see Chapter 9).

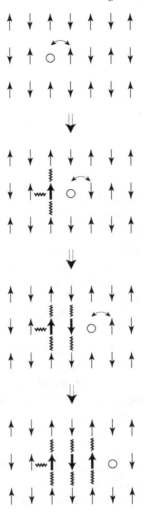

Figure 1.16 Scheme illustrating the formation of a trail of wrong spins left by a hole moving in an antiferromagnetic background in two-dimensional (and three-dimensional) systems.

motion of the hole will not cost any extra energy. The situation is completely different in 2d or 3d systems, see Fig. 1.16. We see that in this case this process – the motion of a hole by interchange with neighboring electrons – will leave a trail of wrong spins, which would have "wrong" ferromagnetic bonds with their neighbors (the wavy bonds in Fig. 1.16). (This trajectory need not be a straight line.) As a result the further the hole travels from its initial position, the more energy it costs; each step increases the energy by $J(z-2)$, where z is the number of nearest neighbors. Consequently, the energy loss will be proportional to the length of the hole trajectory, $\sim Jl$ (Bulaevskii *et al.*, 1968).

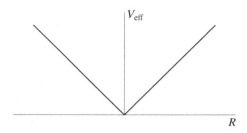

Figure 1.17 Confining potential for the motion of a hole on an antiferromagnetic background in 2d and 3d cases.

In 2d and 3d cases the hole can move along a curved trajectory, with loops, etc. But even the shortest, that is straight, trajectory between the original point $r = 0$ and the point $r = R$ would cost energy $\sim J|R|$. Thus even if we consider the least costly (i.e., straight) trajectory from $r = 0$ to $r = R$, the energy would grow linearly with distance, that is it would appear as if the hole moved in a potential growing linearly with distance, $V_{\text{eff}} \sim J|R|$ (Fig. 1.17) (and for curved trajectories the potential would grow even faster). In other words, there will be a constant force

$$F = -\frac{\partial V_{\text{eff}}}{\partial R} \tag{1.16}$$

pulling the hole back to its origin. This situation is known as *confinement*: effectively, a string with constant tension would bind the hole to its origin. (This situation strongly resembles quark confinement in quantum chromodynamics.) The motion of the hole will have the form qualitatively illustrated in Fig. 1.18: a random wandering around the initial position, constantly returning to the origin, with rare excursions to further distances. Thus, the rapid, unhindered motion of the hole with the hopping amplitude t would indeed delocalize the hole, but only in a certain vicinity of its original position, and it would not produce a net current or charge transfer to large distances. Only the reduced hopping (1.15) would allow the hole to delocalize fully and contribute to the electric current.[4] One can easily see that the situation would be exactly the same not for a hole, but for an extra electron (see Fig. 1.14): the motion of an "own" electron at each step would also produce the same trail of "wrong" spins, with the same confinement, etc.

We thus see that the motion of a hole (or of an extra electron) on an antiferromagnetic background is severely hindered, and we cannot gain as much kinetic energy as we would have liked. However if for example the background spin ordering was ferromagnetic instead of antiferromagnetic, the extra electron or hole could move unimpeded, and we would gain the maximum kinetic energy possible, $\sim W \sim zt$ (the extra electron or

---

[4] Another mechanism giving such real delocalization is the following: due to quantum effects (the terms $S_i^+ S_j^-$ in the exchange interaction (1.12)) the trail of "wrong" spins left by the moving hole can be "healed" – these "wrong" spins can be reversed back to the correct ones by the terms $S^+ S^-$. But these processes occur with the energy scale $J \ll t$, that is the corresponding motion of holes would be very slow and inefficient. The eventual coherent motion of the hole (with its spin-distorted region) due to the "healing" caused by terms $S^+ S^-$ is further strongly suppressed by quantum interference effects, see Weng *et al.* (1997).

Figure 1.18 Schematic trajectory of a hole in an antiferromagnet.

hole would be at the bottom of the corresponding band). Thus for very strong interactions $U \gg t$, if we make the system (or part of it) ferromagnetic instead of antiferromagnetic, we lose the exchange energy $\sim J V_f \sim t^2 V_f / U$, where V_f is the volume of the ferromagnetic part of the sample; but we gain the electron or hole kinetic energy $\sim t\delta$, where δ is the doping concentration. And if we gain more than we lose, it would be favorable to transform a part of the sample, or even the whole sample, to the ferromagnetic state.

This can be shown rigorously to happen if $U = \infty$ and when for a simple (so-called bipartite) lattice we have one extra electron or hole; this is known as Nagaoka ferromagnetism (Nagaoka, 1966). The situation for finite U and finite doping δ is much more complicated. It will be partially discussed below, in Chapters 5 and 9, but here we stress one general conclusion which is also valid for more realistic cases discussed below: there is a general tendency that undoped systems with integer occupation of d levels are often insulating and antiferromagnetic, whereas many doped systems become metallic and ferromagnetic. Thus the general trend is that (Mott) insulating states prefer to be antiferromagnetic, and ferromagnetism usually goes hand in hand with metallic conductivity.

There are of course exceptions from this general rule: some metallic systems may be antiferromagnetic, and there exist some insulating ferromagnets. But these cases are relatively rare, and in each case there are special reasons for such behavior. Thus metallic antiferromagnets typically have rather weak sublattice magnetization, and they are usually described by spin density waves (SDW) – special states of some metals with particular features of the Fermi surface called nesting, see for example chapter 11 in Khomskii (2010). Insulating ferromagnets, in turn, most often have their origin in a particular orbital structure of the corresponding ions. This will be described in detail in Chapters 5, 6, and 9.

1.5 Doped Mott insulators

In the previous section we have already gradually switched from consideration of the conventional Mott insulators with one electron per site to the case of partial filling of d levels, $n \neq 1$. The situation with doped strongly correlated systems with partially filled Hubbard bands is even more complicated and much less clear than that with integer occupation. Again, these problems will be discussed in detail below; here we only give some qualitative arguments and present some general results, deferring detailed discussion till later.

As we have already mentioned above, there is a very strong interplay between the motion of doped carriers, electrons or holes, and the background magnetic structure. Undoped systems with $U \gg t$ are typically antiferromagnetic insulators, whereas in doped systems it may be favorable to change the magnetic ordering to ferromagnetic. This can occur locally, leading to the formation of ferromagnetic microregions – magnetic polarons, or ferrons, trapping extra electrons or holes (Nagaev, 1983). This is one kind of phase separation in strongly correlated systems, to be discussed in more detail in Section 9.7. Such phase separation, however, may be prevented, or at least partially suppressed, by the long-range Coulomb interaction which "does not like" charge segregation and enforces electroneutrality. In this case there may still appear a homogeneous ferromagnetic state in a system, see the qualitative phase diagram of Fig. 1.19 (Penn, 1966; Khomskii, 1970). Here, for one electron per site ($n = 1$) we have an antiferromagnetic state – definitely for $U/t \gg 1$, but for certain special cases (bipartite lattices with nesting) even down to smaller values of U/t. For large U/t and strong doping the ferromagnetic state may appear; the detailed conditions for its appearance and the limits of its existence are still a matter of debate. And between antiferromagnetic and ferromagnetic states there may exist an intermediate, crossover phase – either in the form of a canted antiferromagnet, with the canting angle increasing when approaching the ferromagnetic region; or in the form of a spiral with the wavelength increasing and becoming infinite in the ferromagnetic state; or, as mentioned above, there may occur in this intermediate region a phase separation into ferromagnetic and antiferromagnetic regions, with the volume of the ferromagnetic regions increasing with doping until they occupy the entire sample.

Yet another very exciting possibility is that the doping of Mott insulators can create not just a metallic state, but a superconducting one. Especially after the discovery of high-T_c

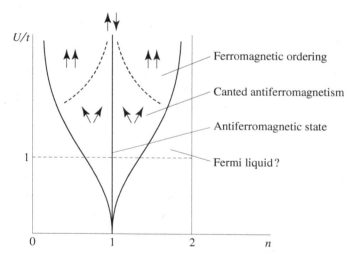

Figure 1.19 Qualitative phase diagram of the Hubbard model for arbitrary band filling and for different interaction strengths, after Penn (1966) and Khomskii (1970).

superconductivity in cuprates this possibility began to be investigated extensively. The outcome is still uncertain. There are some results which indeed point toward the possibility that superconductivity exists in the simple Hubbard model, but there are opposite claims in the literature as well. Again, we will discuss these points in more detail below (see Section 9.6), after we have gained a better understanding of the important details of the structure and properties of the corresponding compounds. In any case, considerable theoretical activity is devoted to these problems today, even using the prototype Hubbard model (1.6).

S.1 Summary of Chapter 1

According to our general plan, we give here a short qualitative summary of the main points of this chapter, so that the reader can get an understanding of the main results without going into too many details. In this chapter we have considered possible states of electrons in solids. If we first treat electrons using a single-electron picture, ignoring the presence of and interaction with other electrons and only taking into account the periodic potential of the crystal lattice, we already obtain a description of the electronic structure and of many properties of solids, which is quite satisfactory in many cases. In this picture the energy spectrum of electrons consists of energy bands with forbidden states – energy gaps – between them. Each of these bands contains $2N$ states, $2N$ places for electrons. If some bands are completely filled and the next, empty bands are separated from the occupied ones by energy gaps, we have the standard band insulators or semiconductors such as Ge or Si. If, however, some of these bands (they may in principle overlap) are partially filled, we obtain a metal. This is always the case if we have odd or noninteger number of electrons per unit cell. We get insulators in this band picture only when the number of electrons per unit cell is even (and even in this case we can have metals or semimetals, if the energy bands overlap).

In any case, for one (or an odd number of) electron(s) per site the system in this standard band picture *should be* metallic. However, we know many compounds, especially those containing transition metals (and rare earths), which are very good insulators even in the case of one (or an odd number of) electron(s) per site. The classical example is MnO, with five *d*-electrons; it is a very good insulator, with energy gap $E_g \sim 4\,\text{eV}$. There are many more systems like that. Why are they insulating?

It turns out that these insulators are of completely different type than the conventional band insulators; they are called *Mott insulators*. In contrast to the conventional band insulators, semiconductors or metals, in which electrons are itinerant or delocalized, in Mott insulators the electrons are *localized*, in the simplest case one at a site. In this sense their state resembles more that of electrons in atoms or ions than that of itinerant electrons in solids. Physically the origin of this state lies in the important role of electron–electron interaction, or strong electron–electron correlations. The simplest model describing this phenomenon is the nondegenerate Hubbard model (1.6) which describes electrons at sites i, hopping from site to site with the hopping matrix element t and with on-site Coulomb repulsion U. Intersite hopping of electrons would lead to their delocalization and to the formation of bands which, for one electron per site, $n = N_{el}/N = 1$, would give a metal independent of the value of hopping t, however small. But for large distances between sites, with small overlap of wavefunctions at neighboring sites and correspondingly small hopping amplitude t (and narrow bandwidth $W \sim t$), the electrons prefer to stay localized one per site: this allows minimization of the repulsion energy U, which would have increased the total energy if there were doubly occupied sites. As for the itinerant (metallic) states such doubly occupied sites necessarily appear, the metallic state would have the energy $\sim U P(2)$ where $P(2)$ is the probability of finding doubly occupied sites. For strong interaction $U \gg t$ this is very unfavorable, and the electrons would stay localized one per site, with $P(2) \simeq 0$. But such a state would be insulating: to create charge-carrying excitations one has to move an electron from its site and put it on another, already occupied one. This, however, would cost us the energy of repulsion from the electron already present on that site, U. And the most we can gain is the kinetic energy of the created electron and hole, $\sim t$. Thus, if $U > t$ (or rather $U > W = 2zt$), this would definitely be unfavorable, and the system would remain insulating with one localized electron per site. This state is a *Mott insulator*. Thus, the nature of such an insulating state is quite different from that of ordinary band insulators such as silicon: whereas in the latter case the insulating state is obtained already in the one-electron picture due to the action of the periodic potential of the crystal, the origin of the Mott insulating state is entirely due to strong electron–electron interaction (strong electron correlations).

The descriptions of electrons in solids (the band picture of itinerant electrons and the picture of Mott insulators with their localized electrons) have a counterpart in the theory of chemical bonds in molecules: the analogy of the band picture is the MO (molecular orbital), or MO LCAO (molecular orbitals – linear combination of atomic orbitals) description, sometimes called the Hund–Mulliken picture. The alternative description of chemical bonds – the Heitler–London picture – excludes the ionic states and is thus equivalent to the picture of Mott insulators with localized electrons. However, whereas for small molecules such as H_2 there is a continuous crossover between the two pictures, in concentrated solids they really describe two different thermodynamic states, with quite different properties, and with a well-defined (phase) transition between them – the *metal–insulator* or *Mott transition*.

As the electrons in Mott insulators are localized, the creation of such a state simultaneously means the creation of localized spins, or *localized magnetic moments*. Consequently, most Mott insulators are strong magnets and, vice versa, practically all strong magnets are either Mott insulators or at least systems with strong electron correlations; the very nature of strong magnetism lies in strong electron–electron interaction.

In the simplest cases (one electron per site, nondegenerate electron levels, strong interaction $U \gg t$), the effective exchange interaction between these localized electrons turns out to be antiferromagnetic, eq. (1.12). It is due to virtual hopping of electrons to neighboring sites, that is caused by a tendency to gain extra kinetic energy due to partial delocalization, which for nondegenerate electrons is only possible when the spins of neighboring sites are antiparallel. Consequently, in undoped Mott insulators with $n = 1$ and $U \gg t$ and with simple lattices, one would have an antiferromagnetic ordering in the ground state, and the lowest excited states would be magnetic excitations (spin waves). The exchange interaction (1.12), with the exchange integral $J \sim 2t^2/U$, is called a *superexchange*; this is the main mechanism of exchange in insulators.

When we dope Mott insulators, the doped electrons and holes can move through the crystal and the resulting state may become metallic. However, the presence of strong electron correlations can make these metals anomalous. Thus the background antiferromagnetic ordering hinders the motion of charge carriers. When the extra electron, or hole, moves through an antiferromagnet, it leaves behind a trail of wrong spins, which in two-dimensional and three-dimensional systems can lead to *confinement* – the electron or hole remains close to its original position. To gain the kinetic (band) energy it may be favorable to change the antiferromagnetic ordering to a ferromagnetic one. In this we lose the superexchange interaction (1.12), but can gain more in the kinetic energy of electrons and holes. As a result we may get a ferromagnetic metal (and for small doping this can lead to more complicated types of magnetic ordering, such as canted antiferromagnetism or a magnetic spiral, or to phase separation, with the formation of ferromagnetic droplets in an antiferromagnetic matrix). This is a general tendency, observed also in more complicated cases: the Mott insulating states are most often antiferromagnetic, but ferromagnetism typically coexists with metallic states (although there are interesting exceptions to this rule, which will be discussed later in this book).

Furthermore, the ferromagnetic metallic state in doped Mott insulators is not the only option; the resulting metallic state may still be a conventional paramagnet, but it can even be superconducting. There are ideas that the electron correlation and the corresponding magnetic degrees of freedom actually determine the nature of superconductivity in high-T_c cuprates, and possibly in the new class of iron-based superconductors, although many details in this picture are still missing.

2

Isolated transition metal ions

After having presented briefly in Chapter 1 the general approach to the description of correlated electrons in solids using a simplified model – the nondegenerate Hubbard model (1.6) – from this chapter on we turn toward a more detailed treatment of the physics of transition metal compounds, which will take into account the specific features of d-electrons. The well-known saying is that "the devil is in the details." Thus if we want to make our description realistic, we have to include all the main features of the d states, the most important interactions of d-electrons, etc. We begin by summarizing briefly in this chapter the basic notions of atomic physics, with specific applications to d-electrons in isolated transition metal ions. For more details, see the many books on atomic physics; specifically for application to transition metals, see Ballhausen (1962), Abragam and Bleaney (1970), Griffith (1971), Cox (1992), and Bersuker (2010).

2.1 Elements of atomic physics

Here we recall some basic facts from atomic physics, which will be important later on. We give here only a very sketchy presentation; one can find the details in many specialized books on atomic physics, for example the works cited above and Slater (1960, 1968).

The state of an electron in an atom is characterized by several quantum numbers. These are the principal quantum number n, the orbital moment $l \leqslant n - 1$, and the spin S. The shells corresponding to different values of l are usually denoted as follows: s-shell for $l = 0$, p-shell for $l = 1$, d-shell for $l = 2$, f-shell for $l = 3$. The values of l^z can be $\{l, l - 1, \ldots, -l\}$ – altogether $(2l + 1)$ states. On each of them we can put, according to the Pauli principle, two electrons with spins $\uparrow$ and $\downarrow$. In effect there are $2 \times (2l + 1)$ states in each shell with orbital quantum number l: 2 for s-electrons ($l = 0$), 6 for p-electrons ($l = 1$), 10 for d-electrons ($l = 2$), 14 for f-electrons ($l = 3$).

In an isolated atom the shells are initially filled one after another: $n = 1$ ($l = 0$); $n = 2$ ($l = 0$; $l = 1$); $n = 3$ ($l = 0$; $l = 1$). But starting from $n = 3$, the situation changes. After filling the $n = 3$ s- and p-shell the electrons, instead of filling the $3d$ levels, first occupy the $4s$ states ($4s^1$ and $4s^2$), and only after that the $3d$ levels start to be filled. This produces the $3d$ transition metal series, starting from Sc, see the Periodic Table at the end of this book. Only after all $3d$ levels are filled do the $4p$ states start to be occupied. A similar situation is repeated for the $4d$ and $5d$ series of transition metals (with some small irregularities – see below).

Why do we have such, at first glance, strange behavior? To understand that we have to go back to the quantum-mechanical description of atomic structure. The motion of an electron in a Coulomb potential of a nucleus with charge Z is described by the standard Schrödinger equation

$$\nabla^2 \Psi + \frac{2m}{\hbar^2}\left(E - V(r)\right)\Psi = 0, \qquad (2.1)$$

where $V(r) = -Ze^2/r$. In this spherically symmetric potential the general solution is

$$\Psi_{lm} = f(r)\, Y_{lm}(\theta, \phi), \qquad (2.2)$$

with spherical harmonics $Y_{lm} = P_l^{|m|}(\cos\theta)\, e^{im\phi}$, where $P_l^{|m|}$ are the Legendre functions. For given l, m the problem reduces to a one-dimensional equation for the function $f(r) = \chi(r)/r$ (defined in the interval $(0, \infty)$):

$$\frac{d^2\chi}{dr^2} + \left[\frac{2m}{\hbar^2}(E - V) - \frac{l(l+1)}{r^2}\right]\chi = 0 . \qquad (2.3)$$

This has the form of a one-dimensional Schrödinger equation of the same type as eq. (2.1), with V replaced by an effective potential for the electron with given angular momentum l:

$$V(r) \Longrightarrow \widetilde{V}(r) = V(r) + \frac{\hbar^2}{2m}\frac{l(l+1)}{r^2} , \qquad V(r) = -\frac{Ze^2}{r} . \qquad (2.4)$$

Because of the strong increase in the centrifugal term $\sim l(l+1)/r^2$ in $\widetilde{V}$, the effective potential for d shells ($l = 2$) and even more so for f shells ($l = 3$) develops a rather narrow potential well, which pushes the corresponding energy levels up. As a result the $3d$ levels lie higher in energy than the $4s$ ones and, in moving along the Periodic Table, after filling the $2s$ and $2p$ levels we first fill the $4s$ levels, and only after reaching the configuration $\cdots 3s^2 3p^6 4s^2$ (the element Ca) do we start to occupy the $3d$ levels.[1] This gives rise to the transition metal series, in this case the $3d$ series (from Sc, which has configuration $4s^2 3d^1$, to Cu – which has configuration $4s^1 3d^{10}$ rather than $4s^2 3d^9$, but for which both one $4s$-electron and one d-electron often go to establish chemical bonds, leaving $Cu^{2+}(3d^9)$ – the state one would expect if the configuration of neutral Cu was indeed $4s^2 3d^9$). Such redistribution between $3d$ and $4s$ states, which lie close in energy, is met also in neutral transition metal atoms in a few other cases, see the Periodic Table. Thus, for example, the neutral Cr atom could have been $4s^2 3d^4$, but is $4s^1 3d^5$ instead. This effect is usually connected with the extra stability of completely full shells ($3d^{10}$ in Cu) and half-filled shells ($3d^5$ in Cr) compared with their fractional occupation.[2] For our purposes, however, in concentrated transition metal compounds these effects do not play a very important role: typically, the valence states of transition metal ions in solids can be

---

[1] Further on we will omit the filled (deep) levels and mark in the configurations only those levels ($4s$, $3d$) which are relevant for our discussion.

[2] This effect is especially important for rare earth compounds with intermediate valence and heavy fermions, see Chapter 11.

understood as if the original state was always $4s^2 3d^n$. As mentioned above, this happens for Cu which easily accepts the valence $Cu^{2+}(4s^0 3d^9)$, although the state $Cu^{1+}(4s^0 3d^{10})$ is often more stable. This is also the case for Cr, which typically exists in the valence state $Cr^{3+}(4s^0 3d^3)$, or sometimes as $Cr^{2+}(4s^0 3d^4)$ or $Cr^{4+}(4s^0 3d^2)$, but almost never as $Cr^+(4s^0 3d^5)$, which one might have expected based on the configuration of the neutral atom $Cr(4s^1 3d^5)$. A similar situation occurs in the $4d$ and $5d$ transition metal series.

One very important point is worth mentioning here. For the solutions of the single-particle Schrödinger equation both the energy levels and the radii of the corresponding wavefunctions increase when the principal quantum number n increases (typically, the radius of the maximum electron density is $\sim n^2 a_0$, where a_0 is the Bohr radius). For real atoms, however, with the electrons moving in the potential of the nuclei and in the self-consistent potential of other electrons, the situation may be more complicated. Thus, whereas for TM atoms (e.g., of the $3d$ series) the energies of the $4s$ and $3d$ states are comparable, the "sizes" of their corresponding orbitals are very different. Thus, for Ti the radius of the $4s$ orbital is $\sim 1.48\,\text{Å}$, but the $3d$ orbitals are much more localized; the typical $3d$ radius in Ti is $\sim 0.49\,\text{Å}$ (see table 7-2 of Slater, 1968). The same is also true for other $3d$ elements. Thus, although $3d$-electrons belong energetically to valence electrons and can participate in the formation of chemical bonds (going e.g. to anions such as F^- or O^{2-} in ionic crystals, or participating in covalent bonds), they are, in contrast, rather strongly localized close to the ionic core, with typical radius of d states $\sim 0.5\,\text{Å}$. The same is true, although to a lesser extent, also for $4d$-electrons, and even less so for $5d$-electrons: the effective radius of d-electrons increases in going from $3d$ to $4d$ to $5d$. In effect, $5d$-electrons more often behave as itinerant ones, even in compounds such as oxides.

We can learn yet another lesson when looking at the systematics of the behavior of d-electrons in the $3d$ (or $4d$, or $5d$) series. Since the charge of the nucleus Z increases when we move from left to right in a row in the Periodic Table, we may expect some shrinking of the size of the corresponding orbitals. And indeed this is what happens. Thus if (looking at neutral atoms) the $3d$ radius in Ti is $\sim 0.49\,\text{Å}$, it reduces to $\sim 0.39\,\text{Å}$ in Mn, and to $0.324\,\text{Å}$ in Ni (table 7-2 of Slater, 1968). The same trend is observable in $4d$ and $5d$ series, and also in rare earths ($4f$ series), where it is known as a *lanthanoid contraction*. This tendency is very important for the formation of corresponding solids; thus, for example, the ionic radii of TM ions determining the interatomic distances and, often, the very type of crystal structure follow the same trend (Shannon, 1976). We will see many examples of this later.

When transition metals form chemical compounds such as oxides, usually the s-electrons go to form valence bonds (or go to the anions in the case of ionic bonds). Some d-electrons can also become valence electrons and go, for example, to the oxygens. Thus, the typical valence of Sc is 3+ (all three outer electrons $4s^2 3d^1$ "go away"); Ti may have valence 2+, 3+, or 4+, etc. In effect, after the compounds are formed, the transition metal ions in many cases "donate" some of their d-electrons to the chemical bonds, but there may still be several d-electrons which in the first approximation remain on the

d-shells of the corresponding ions. As we will see later, these electrons may still have certain covalent bonding with, for example, oxygens in oxides, but in the first approximation they may be treated as localized on the ionic cores. And when we form a solid out of these ions, according to the treatment of the previous chapter these electrons can either become itinerant (largely losing their atomic character), thus contributing to the *metallic bonding*, or they can remain localized, so that the corresponding materials should be treated as Mott insulators. In the latter case these localized d-electrons of partially filled d-shells can have localized magnetic moments and can give rise to some kind of magnetic ordering. However, in contrast to the nondegenerate case of Chapter 1, where every localized state with one electron per site meant the existence of a localized spin $S = \frac{1}{2}$ at a site, for real d-electrons there may be not one, but two, three, ..., up to 10 d-electrons per ion. Correspondingly, the total spin can be different. The orbital moment $l = 2$ for d-electrons also contributes to the total magnetic moment.

2.2 Hund's rules

The classification of possible resulting states of many-electron atoms and ions is an important aspect of atomic physics. Without going into all the details, we formulate here only the most important conclusions. We first consider the case of isolated atoms or ions, with full rotational symmetry, and then discuss what modifications would appear for ions in solids.

For atoms with several electrons on the same atomic shell, there exist certain rules determining the filling of corresponding atomic levels. The first and most important of these is the *first Hund's rule* (or just simply Hund's rule). It states that, first of all, the electrons fill the levels so as to make the largest possible total spin. For d-electrons with $l = 2$ there are $2l + 1 = 5$ different orbital states, with $l^z = 2, 1, 0, -1, -2$ (see Fig. 2.1). In an isolated atom or ion these levels are degenerate. In that case, according to the first Hund's rule, the electron will first fill different levels with parallel spins, for example with four d-electrons the orbital occupation will be that shown in Fig. 2.1(a). Correspondingly, the total spin in this case will be $S = 2$ – the maximal possible spin of four d-electrons. The filling of different levels by electrons with parallel spins, according to the Pauli principle, is possible up to $n_d = 5$, after which we have to start filling the d levels with electrons having opposite spins, see for example Fig. 2.1(b) for the case of 7 d-electrons. Thus for less-than-half-filled d-shells with n_d d-electrons the total spin is $S = \frac{1}{2}n_d$, and for more-than-half-filled shells the total spin will start to decrease, $S = \frac{1}{2}(10 - n_d)$.

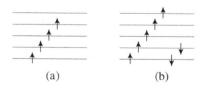

(a)   (b)

Figure 2.1 Filling of d levels according to the first Hund's rule.

The second part of the first Hund's rule states that among different possible configurations with maximal total spin S, or maximal multiplicity $(2S + 1)$, the ground state is the state with maximal possible orbital angular momentum.[3] The point is that d-electrons have orbital moment $l = 2$, and thus, for example, two d-electrons would have $\boldsymbol{L} = \boldsymbol{l}_1 + \boldsymbol{l}_2$; according to the quantum-mechanical rules of the addition of angular momenta, the total moment L can take values from 4 to 0. But, according to the Pauli principle, not all possible combinations with total spin S and total orbital moment L are allowed. Thus, for example, suppose we indeed have two d-electrons. The state of the first one, characterized by spin $S_1^z = \pm\frac{1}{2}$ and orbital moment $m_{\text{orb}} = l^z = (2, 1, 0, -1, -2)$ could thus be $|l^z, S^z\rangle = |2, +\frac{1}{2}\rangle, |2, -\frac{1}{2}\rangle, |1, +\frac{1}{2}\rangle, |1, -\frac{1}{2}\rangle, |0, +\frac{1}{2}\rangle, |0, -\frac{1}{2}\rangle$ (the states with negative l^z do not give anything new). Now consider the second d-electron. According to the Pauli principle, two d-electrons can be in the states $|(2, +\frac{1}{2})_1(1, \pm\frac{1}{2})_2\rangle$, $|(2, +\frac{1}{2})_1(0, \pm\frac{1}{2})_2\rangle$, etc. – but not, for example, in the state $|(2, +\frac{1}{2})_1(2, +\frac{1}{2})_2\rangle$. The first part of the Hund's rule tells us that the ground state should be the state with maximum spin, $S = 1$, for example the state $(S_1^z = +\frac{1}{2}, S_2^z = +\frac{1}{2})$. Then the second part tells us that among all such states the one with maximum L should be the ground state. But this could only be a state with $L = 3$, for example $|(2, +\frac{1}{2})_1(1, +\frac{1}{2})_2\rangle$, but not with $L = 4$, because that would contain $|L^z = 4, S^z = 1\rangle = |(2, +\frac{1}{2})_1(2, +\frac{1}{2})_1\rangle$ which is forbidden by the Pauli principle. In other words, the state of two d-electrons with total spin $S = 1$ has a wavefunction symmetric in spins (e.g., $|S_1^z = +\frac{1}{2}, S_2^z = +\frac{1}{2}\rangle$). This however implies that the coordinate part of the wavefunction has to be antisymmetric, which excludes the state with $L = 4$ (which, in particular, contains the symmetric combination $|l_1^z = +2, l_2^z = +2 >$). Thus we conclude that the ground term of two d-electrons should be the term $(S = 2, L = 3)$, that is in the atomic nomenclature the term ^5F (the standard notation of atomic terms is $^{2S+1}L_J$, where L is the total orbital moment, $2S + 1$ is the spin multiplicity, and J is the total angular momentum $\boldsymbol{J} = \boldsymbol{L} + \boldsymbol{S}$, which is determined by spin–orbit coupling, discussed below; according to the standard convention, the state with $L = 0$ is labeled the S state, the state with $L = 1$ the P state, and the state with $L = 2$ the D state, etc.). These arguments give the systematics of the ground-state terms of transition metal atoms and ions. For isolated ions with full spherical symmetry, these ground-state terms are:

d^1	d^2	d^3	d^4	d^5	d^6	d^7	d^8	d^9
^2D	^3F	^4F	^5D	^6S	^5D	^4F	^3F	^2D

One may notice that the terms with n and $(10 - n)$ d-electrons are the same. This is easy to understand: the filled shell with 10 d-electrons has $S = 0$, $L = 0$ and one can treat, for example, the state with occupation d^9 or d^8 as having respectively one or two d-*holes*, instead of one or two d-electrons. Indeed, the quantum numbers of, for example, the state

---

[3] The terminology used in different books varies. Sometimes one speaks of the first and second Hund's rules, the first being the statement about the maximal total spin and the second being the statement about the maximal possible orbital moment consistent with maximal spin. The next rule, dealing with the total moment, which we here call the second Hund's rule, would then be referred to as the third Hund's rule.

with two holes on top of the fully occupied shell (d^{10}) are the same as for two electrons in an empty shell; this electron–hole symmetry leads to the symmetry of the ground states presented in the table above.

The interaction leading to the first Hund's rule is often called the intra-atomic Hund's exchange, which should then be ferromagnetic; it is usually written in the form

$$\mathcal{H}_{\text{H}} \sim -J_{\text{H}} \sum_{\alpha \neq \beta} (\tfrac{1}{2} + 2\mathbf{S}_{i\alpha} \cdot \mathbf{S}_{i\beta}) + \text{const.} \tag{2.5}$$

where J_{H} is called Hund's exchange constant, and $\mathbf{S}_{i\alpha}$ is the spin of an electron at site i with orbital quantum number $\alpha = l^z$. Its origin is the standard Coulomb interaction: according to the Pauli principle the electrons with parallel spins avoid each other, as a result of which the average Coulomb repulsion is reduced, which decreases the total energy and stabilizes the states with parallel spins (or with maximum possible spin for many electrons). The same applies to orbitals as well. This is the physical mechanism of the first Hund's rule.

Another factor leading to the same conclusion is that the average radius of a doubly occupied orbital is somewhat larger than that of a singly occupied one due to more efficient screening. As a result, an electron in a singly occupied orbital is closer to the nucleus and experiences stronger Coulomb attraction to it, which stabilizes this occupation. In any case, we see that both these factors, which lead to the first Hund's rule, are in fact largely due to the direct part of the Coulomb interaction.

The resulting contribution to the total energy can still be written in the form of eq. (2.5), which consequently can be used to describe partially filled shells. This expression is also useful because it allows us to describe a purely quantum-mechanical effect: it is not just that the *parallel* spins $S_1{\uparrow}$, $S_2{\uparrow}$ are stabilized by the Hund's rule interaction, but it is rather important that they form a *triplet* state with $S_{\text{tot}} = 1$. It is in such a triplet state that the electrons "avoid" one another and gain Coulomb energy. But the triplet state, besides the states $|S_{\text{tot}}^z = +1\rangle = |S_1{\uparrow}, S_2{\uparrow}\rangle$ and $|S_{\text{tot}}^z = -1\rangle = |S_1{\downarrow}, S_2{\downarrow}\rangle$, contains also a third state $|S_{\text{tot}}^z = 0\rangle = \frac{1}{\sqrt{2}}\{|S_1{\uparrow}, S_2{\downarrow}\rangle + |S_1{\downarrow}, S_2{\uparrow}\rangle\}$ – a symmetric state with total $S_{\text{tot}}^z = 0$. Indeed, according to quantum mechanics, a symmetric spin state in a triplet implies an antisymmetric coordinate wavefunction, in which, consequently, the probability of having two electrons with close coordinates r_1, r_2 is reduced (it is zero for $r_1 = r_2$). Similarly, for more than two d-electrons the state with maximal total spin S has a "maximally anti-symmetric" coordinate wavefunction, which minimizes the Coulomb repulsion; this is the essence of the first Hund's rule.

Nevertheless, in a mean-field approximation, one can often neglect states such as the triplet state with $S_{\text{tot}}^z = 0$ and consider only states of the type shown in Fig. 2.1(a, b). The Hund's exchange (2.5) can then be rewritten in such a mean-field approximation as

$$\mathcal{H}_{\text{H}}^{\text{MF}} = -J_{\text{H}} \sum (\tfrac{1}{2} + 2S_{i\alpha}^z S_{i\beta}^z). \tag{2.6}$$

Then one can calculate the total contribution of Hund's interaction (2.5) to the energy of an atom or ion by counting the number of *pairs of parallel spins*: each such pair gives a contribution $-J_H$ to the total energy. Thus, the Hund's rule energy of the state of Fig. 2.1(a) would be $-6J_H$ and that of Fig. 2.1(b) would be $-11J_H$ ($-10J_H$ for the contribution from spins ↑, and $-J_H$ from spins ↓).

We will often use below this rule of counting the contribution of the Hund's rule interaction to the total energy, and in most cases this is sufficient, though one has to be aware that in principle there may exist quantum effects connected with the $S_{tot}^z = 0$ states discussed above. An example of their possible role will be given in Section 5.2.

The Hund's coupling is introduced here in a rather simplified way. The accurate treatment should use all the machinery of atomic physics, operating with such notions as Slater–Koster integrals or Racah parameters, see e.g. Slater (1960), Griffith (1971), Sugano *et al.* (1970). The simplified treatment presented in this book is usually sufficient for the purposes of qualitative (but not quantitative) discussion of most phenomena considered below; however one has to be aware of its limitations. In particular the numerical coefficients with which the Hund's rule coupling J_H, introduced in (2.5) and (2.6), enters different expressions, may differ somewhat from those following from the simple treatment formulated here – see for example Section 3.3. For most practical purposes, however, and to elucidate the qualitative trends in various situations, the treatment of this book is usually sufficient, and we will often use it below – also mentioning its possible limitations.

Typical values of the Hund's rule energy for transition metals are $J_H \sim 0.8$–0.9 eV for $3d$, $J_H \sim 0.6$–0.7 eV for $4d$, and $J_H \sim 0.5$ eV for $5d$ elements. Interestingly enough, these values remain practically unchanged when transition metal ions are in crystals: in contrast to the direct Coulomb repulsion U of Chapter 1, J_H is not screened and not modified when we form solids (even metallic) from isolated ions. This is connected with the fact that this energy is actually the *difference* of the energies of electrons with different spins or orbitals on the same atomic shell, and whereas the separate terms do change in going from atoms or ions to a solid, their differences usually remain practically unchanged.

One can show that for an isolated atom or ion the following relation holds:

$$U_{mn} = U_{mm} - 2J_H \,, \tag{2.7}$$

where U_{mn} is the direct Coulomb repulsion of electrons on different orbitals, $m \neq n$, and U_{mm} is the Coulomb (Hubbard) repulsion on the same orbital. And whereas in solids both U_{mm} and U_{mn} are screened, their difference remains practically unchanged. Note however that this relation is only valid for spherically symmetric states; it does not in general hold for ions in a crystal, see Chapter 3.

2.3 Spin–orbit interaction

Yet another important factor determining the electronic terms of atoms or ions is the relativistic spin–orbit interaction,

$$\mathcal{H}_{SO} = \sum_i \zeta_i \, l_i \cdot S_i, \tag{2.8}$$

where the sum goes over all electrons.

The spin–orbit coupling leads to the splitting of electronic terms into multiplets with definite total angular momentum $J = L + S$. In atomic physics one discriminates between two schemes for many-electron states: the LS or Russell–Saunders coupling scheme and the jj coupling scheme, see for example Slater (1968) or Abragam and Bleaney (1970).

In the jj scheme, used for very strong spin–orbit coupling, applicable for example to $4f$ and $5f$ elements, one first adds the spin and orbital moment for each electron, $j_i = l_i + s_i$, and then these total angular momenta j_i are summed up to the total angular momentum of the atom, $J = \sum j_i$. However, we will mostly be dealing with the Russell–Saunders (LS) case.

The LS scheme is applicable to most TM elements, in particular to all $3d$ elements (the situation in the $4d$ and especially in the $5d$ series is somewhat more complicated). In this scheme one first sums the electron spins to obtain the total spin $S = \sum S_i$, and all orbital moments of separate electrons to obtain the total orbital moment $L = \sum l_i$, and as the last step one includes the spin–orbit coupling, which couples S and L and produces the total angular momentum J.

The full spin–orbit coupling in the LS scheme can be written in the form

$$\mathcal{H} = \lambda L \cdot S . \tag{2.9}$$

The coupling constant λ is composed of the partial spin–orbit coupling constants ζ_i (2.8). For one shell all ζ_i are the same, $\zeta_i = \zeta$, and then

$$\lambda = \pm \frac{\zeta}{2S} , \tag{2.10}$$

where the "+" sign is used for less-than-half-filled shells and "−" for more-than-half-filled shells. From the relativistic treatment one can show that the ζ_i are positive, and consequently $\lambda > 0$ for less-than-half-filled shells and $\lambda < 0$ for more-than-half-filled shells. Note right away that because of the presence of the "normalization factor" $2S$ in the denominator of eq. (2.10) the values of the spin-orbital coupling λ in general differ from the "atomic" values ζ, and, moreover, depend on the valence and spin state of the corresponding ions, so that for example the values of λ for $V^{3+}(d^2, S = 1)$ and $V^{4+}(d^1, S = \frac{1}{2})$ would be different.

The interaction (2.9) couples L and S and produces the total angular momentum $J = L + S$, with different energies for terms with different J. By the usual rules of quantum mechanics in the state with given J ($= L + S$) the contribution to the energy of the corresponding term, according to (2.9), is determined by the scalar product

$$\langle L \cdot S \rangle = \frac{J(J + 1) - L(L + 1) - S(S + 1)}{2} . \tag{2.11}$$

The normalization coefficient $1/2S$ in eq. (2.10) is chosen in such a way that the difference in energies for terms with different J is of the order of J.

According to quantum-mechanical rules, the quantum number J can take values $J = L + S, L + S - 1, \ldots, |L - S|$. For less-than-half-filled shells, with coupling constant $\lambda > 0$, the lowest level is the one with the smallest possible value of J, and the energies of multiplet levels increase with increasing J; this is the regular multiplet structure. For more-than-half-filled shells one can go over to the hole representation, taking $\lambda < 0$, and the ordering of multiplet levels is inverted; the lowest level is that with the maximum value of J. Sometimes this rule is referred to as the *second Hund's rule*.[4] Thus the second Hund's rule tells us that for less-than-half-filled shells we have a *normal multiplet structure*, with the energy of multiplet levels increasing with the total angular momentum J, and for more-than-half-filled shells we have the *inverted multiplet structure*: the terms with larger J are lower.

As mentioned above, the standard notation of the corresponding terms is $^{2S+1}L_J$, where for the orbital moment L one uses the conventional labeling: S for $L = 0$, P for $L = 1$, D for $L = 2$, etc. Thus the lowest terms of a free atom or ion, taking into account the spin–orbit interaction and the second (or third) Hund's rule, are:

ion	d^1	d^2	d^3	d^4	d^5	d^6	d^7	d^8	d^9
term	$^2D_{3/2}$	$^3F_2$	$^4F_{3/2}$	$^5D_0$	$^6S_{5/2}$	$^5D_4$	$^4F_{9/2}$	$^3F_4$	$^2D_{5/2}$
g_J	$\frac{4}{5}$	$\frac{2}{3}$	$\frac{2}{5}$	–	2	$\frac{3}{2}$	$\frac{4}{3}$	$\frac{5}{4}$	$\frac{6}{5}$

(This table "upgrades" the table presented above on p. 29.) The last row of this table shows the values of the effective g-factor for the full moment J, see eq. (2.12) below. Note that the electron–hole symmetry, present in the nonrelativistic atomic structure, is lost when we include the spin–orbit interaction, cf. the ground state multiplets in this table and in the table on p. 29.

According to the standard quantum-mechanical treatment, see for example Landau and Lifshitz (1965), the effective moment of a free atom with the ground-state multiplet described by the orbital moment L, spin S, and total moment J is determined by the effective g-factor (Landé factor)

$$g_J = 1 + \frac{J(J+1) + S(S+1) - L(L+1)}{2J(J+1)} = \frac{3}{2} + \frac{S(S+1) - L(L+1)}{2J(J+1)}, \quad (2.12)$$

so that the magnetic moment M entering the Zeeman interaction with the external field, $-HM$, is $M = g_J\mu_B J$ and the effective moment μ_{eff} entering the Curie (or Curie–Weiss) magnetic susceptibility, $\chi = C/T$, $C = \frac{1}{3}\mu_{\text{eff}}^2$, is $\mu_{\text{eff}} = g_J\mu_B\sqrt{J(J+1)}$. The values of g_J are presented in the table above.

The effects of spin–orbit coupling take a somewhat different form for transition metal ions in crystals; we will discuss this in detail in Chapter 3. One generalization of the Landé formula can be presented here: if the spin and orbital moment g-factors are g_S and g_L, respectively, then the g-factor for the total angular momentum $\boldsymbol{J} = \boldsymbol{L} + \boldsymbol{S}$ is

$$g_J = \frac{g_L + g_S}{2} + \frac{L(L+1) - S(S+1)}{2J(J+1)}(g_L - g_S). \quad (2.13)$$

---

[4] Or the third Hund's rule, see the footnote on p. 29.

For the usual values $g_L = 1$ and $g_S = 2$ we obtain the standard expression (2.12); but this generalized expression will be useful later on, when we consider transition metal ions in crystals.

Spin–orbit coupling becomes more important for heavier elements: $4d$-elements and especially $5d$-elements (for which the situation may actually be intermediate between that of LS and jj coupling schemes, although still closer to the LS one). This is due to the fact that the spin–orbit coupling constant strongly increases with the atomic number Z, $\lambda \sim Z^4$ – or, rather, the "atomic" coefficient ζ in (2.8) is $\zeta \sim Z^4$. As mentioned after eq. (2.10), the coefficient λ also depends on the valence and spin state of an ion.[5] Thus it becomes more and more important for heavier elements. Whereas for light $3d$ elements such as Ti $\zeta \sim 20\,\text{meV}$, already for heavier $3d$ metals it is much stronger: thus for Co $\zeta \sim 70\,\text{meV}$, see for example appendix 6 in Griffith (1971). And it is even stronger for $4d$ and $5d$ elements; thus for $5d$ elements ζ or λ can reach values $\sim 0.5\,\text{eV}$ – already comparable with the relevant electron hopping t and the on-site Hubbard repulsion U, which is ~ 3–$6\,\text{eV}$ for $3d$ elements but reduced to ~ 1.5–$2\,\text{eV}$ for $5d$ elements.

---

[5] The dependence $\zeta \sim Z^4$ is usually presented in all the literature on spin-orbit interaction. However, there are also alternative estimates: thus, for example, another treatment (see Landau and Lifshitz, 1965), leads to the result $\zeta \sim Z^2$. Both these dependences are of course only rough guides and cannot serve for obtaining real numerical values of spin-orbit coupling constants, which have to be determined experimentally.

S.2 Summary of Chapter 2

In this chapter we have briefly discussed the basic aspects of atomic physics, paying most attention to the effects and factors which will be important for us further on, when considering the properties of TM compounds. The "classical" quantum theory of atomic structure describes first the state of electrons moving in the Coulomb potential of a nucleon. The electronic states form shells with the principal quantum number $n = 1, 2, 3, \ldots$, and for each n the electrons can have orbital moment $l < n$: for $n = 1$ we have $l = 0$ (s-shells); for $n = 2$ we have $l = 0, 1$ (s- and p-shells); for $n = 3$ we have $l = 0, 1, 2$ (s-, p- and d-shells). For each l there are $2l + 1$ states with the magnetic quantum number $m = l^z = \{l, l - 1, \ldots, -l\}$, and on each such level one can put, according to the Pauli principle, two electrons with spins up and down. Thus, d-shells ($l = 2$) for an isolated atom or ion are fivefold degenerate and may contain up to 10 d-electrons.

In the Periodic Table (see the end of the book) different shells are filled first in order of increasing n ($1s$; $2s$, $2p$) but starting from $n = 3$, the filling scheme changes: after filling the $3s$ and $3p$ shells, the $4s$ shells start to be filled, and only after that the inner $3d$ shells. The same happens in $4d$ and $5d$ elements. Thus we have the $3d$ series (Sc, Ti, V, $\ldots$, Cu) and similarly the $4d$ and $5d$ series with a gradual filling of d shells. Both the outer s-electrons and some of the d-electrons can participate in ionic or covalent bonds, so that in effect the valence of TM ions in compounds may vary (e.g., Ti^{2+}, Ti^{3+}, Ti^{4+}; V^{2+}, $\ldots$, V^{5+}, etc.).

Despite the fact that the binding energy of d-electrons is relatively small, and in this case they have to be considered as valence electrons – the electrons of the outer shells, participating in chemical bonds together with $4s$ (or $5s$, $6s$ in the $4d$ and $5d$ series) – the actual radii or d states are rather small, ~ 0.5 Å for $3d$-electrons. Thus, if these d-electrons do not go to anions such as F^- or O^{2-}, they can in a first approximation be treated as localized, with the correlation effects of Chapter 1 playing potentially a very important role. This leads to all the variety of the properties of transition metal compounds. In particular, we have here a good chance to have a Mott insulator state, with diverse magnetic properties.

When consecutive d-electrons start to fill the degenerate d levels, the interaction between these electrons determines which levels will be occupied and which states will be formed. The requirement that the Coulomb repulsion between electrons should be minimized implies that the most favorable states would be those with the "most antisymmetric" coordinate wavefunctions, where the electrons avoid one another. These states are the states with maximal possible total spin, and among these, with maximal orbital moment. This is the first (or first and second) Hund's rule. The requirement to have the maximal possible spin plays an especially important role in TM compounds (in crystals the orbital moment is no longer a valid quantum number, although the atomic systematics is still very useful in solids as well). A simple way to find the Hund's energy of an atom or an ion is to count the number of pairs of parallel spin: in a mean-field approximation each such pair adds the stabilization energy $-J_H$ (where $J_H \sim 0.8$–0.9 eV in $3d$ series, ~ 0.6–0.7 eV in $4d$ series, and ~ 0.5 eV in $5d$ series). Interestingly enough, the atomic values of the Hund's rule energy J_H are not screened, and they remain in solids practically equal to those of isolated atoms or ions.

The remaining important interaction is the relativistic spin–orbit coupling $\sim l \cdot s$. In most light elements, including $3d$ and $4d$ transition metals, the so-called LS (or Russell–Saunders) coupling scheme is applicable: first the spins of electrons combine to form the total spin $S = \sum s_i$ and the orbital moments of each electron form the total orbital moment $L = \sum l_i$ (consistent with the Pauli principle for given S), and then the spin–orbit coupling $\lambda L \cdot S$ couples orbital and spin moments to form the total angular momentum $J = L + S$.[6] This determines the multiplet terms, denoted as $^{2S+1}L_J$ (where for L one uses the standard notation: S for $L = 0$, P for $L = 1$, D for $L = 2$, etc.). The second (or third) Hund's rule claims that for less-than-half-filled shells the order of multiplets is normal, that is the energy of a multiplet increases with J, while for more-than-half-filled shells the multiplet scheme is inverted (the states with larger J have lower energy). These rules are also very important for transition metals in solids, although the surrounding crystal will modify their specific form.

When going from the $3d$ to $4d$ and $5d$ series, the general trend is that d-electrons become less correlated. $4d$ and especially $5d$-electrons have larger radii than $3d$, and correspondingly larger covalency with surrounding ions and larger bandwidth; simultaneously the effective Hubbard repulsion energy decreases, as does the Hund's rule energy. In contrast, for these heavier elements the relativistic spin–orbit interaction becomes more and more important, and it can determine many properties of $4d$ and especially $5d$ systems.

---

[6] The situation in $5d$ elements may be more complicated, and it may be intermediate between the LS and jj schemes (typical for $4f$ and $5f$ elements). Nevertheless, even here one can often use the LS scheme in a first approximation.

3

Transition metal ions in crystals

3.1 Crystal field splitting

When we put a transition metal ion in a crystal, the systematics of the corresponding electron states changes. For isolated atoms or ions we have spherical symmetry, and the corresponding states are characterized by the principal quantum number n, by orbital moment l and, with spin–orbit coupling included, by the total angular momentum J. When the atom or ion is in a crystal, the spherical symmetry is violated; the resulting symmetry is the local (point) symmetry determined by the structure of the crystal. Thus, if a transition metal ion is surrounded by a regular octahedron of anions such as O^{2-} (Fig. 3.1) (this is a typical situation in many TM compounds, e.g. in oxides such as NiO or LaMnO$_3$), the d levels which were fivefold degenerate in the isolated ion ($l = 2$; $l^z = 2, 1, 0, -1, -2$) are split into a lower triplet, t_{2g}, and an upper doublet, e_g (Fig. 3.2).[1] The corresponding splitting is caused by the interaction of d-electrons with the surrounding ions in the crystal, and is called crystal field (CF) splitting. The type of splitting and the character of the corresponding levels is determined by the corresponding symmetry. The detailed study of such splittings is a major field in itself, and is mostly treated using group-theoretical methods. We will not discuss such methods here; suffice it to say that the splitting shown for example in Fig. 3.2, with the corresponding degeneracies of the resulting levels, follows from just such a treatment. Instead of providing a mathematical description of the corresponding phenomena, which is usually done in terms of the representations of the respective symmetry groups, we will give a qualitative explanation of the resulting physical effects (though sometimes we will use symmetry terminology and the corresponding notation). One can find relevant mathematical details in many specialized books (Ballhausen, 1962; Jorgensen, 1962; Griffith, 1971).

As shown in Fig. 3.2, five degenerate wavefunctions $|l = 2, l^z\rangle$ (we will henceforward simply denote them by $|l^z\rangle$, e.g. $|l^z = +2\rangle$, $|+1\rangle$, $|0\rangle$, etc.) are split in a cubic crystal field of a regular octahedron into an upper doublet e_g and a lower triplet t_{2g}.

---

[1] If the total crystal symmetry is lower than cubic, there may be an additional splitting of d levels due to interaction with further neighbors, even for regular MO_6 octahedra; for a while we ignore these effects.

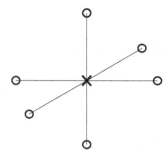

Figure 3.1 Transition metal ion (cross) in a regular ligand (e.g., oxygen) octahedron (circles).

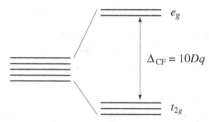

Figure 3.2 Crystal field splitting of *d* levels for a transition metal ion in a regular octahedron.

This notation comes from group theory: the doubly degenerate representation of the cubic group is denoted E_g, and the triply degenerate representation is denoted T_{2g} (Mulliken notation). The subscript *g* here, from the German *gerade* (even), shows that the corresponding base functions are even with respect to spatial inversion $r \to -r$. Sometimes one employs a different notation for these terms, introduced by H. Bethe: Γ_3 for E_g (double orbitally degenerate states) and Γ_5 for T_{2g} (triple orbitally degenerate states). The symmetric one-dimensional representation, A_{1g} in the Mulliken notation, becomes Γ_1 in the Bethe notation. There exists yet another important representation of the cubic group, also triply degenerate: the Γ_3 or T_{1u} representation, which is realized in the ground state of some ions, for example Co^{2+}. (Here the subscript *u*, from the German *ungerade* (odd), shows that the corresponding wavefunctions change sign under spatial inversions, like components of a vector.)

The corresponding splitting is denoted Δ_{CF}; quite often it is also denoted $10Dq$.[2] Keeping the center of gravity of these levels unchanged (e.g., taken as zero), we will then have the energy of the lower triplet $E_{t_{2g}} = -\frac{2}{5}\Delta_{CF} = -4Dq$, and that of the upper doublet $E_{e_g} = +\frac{3}{5}\Delta_{CF} = +6Dq$. The corresponding wavefunctions are linear superpositions of the five basis functions $|l^z\rangle$ (Ballhausen, 1962; Griffith, 1971):

---

[2] The latter notation, widely used nowadays, especially in the chemistry literature, apparently originated in the paper of Schlapp and Penney (1932). In that paper *D* was the constant specifying the strength of the main component of the cubic crystal field $\sim D(x^2 + y^2 + z^2)$, and *q* was the ratio of certain matrix elements used in calculating the crystal field splitting. Nowadays, however, "*Dq*" or "$10Dq$" is always used as a single symbol.

$$e_g : \begin{cases} |z^2\rangle = |3z^2 - r^2\rangle = |l^z = 0\rangle \sim \tfrac{1}{2}(3z^2 - r^2) = \tfrac{1}{2}(2z^2 - x^2 - y^2)\,, \\[2mm] |x^2 - y^2\rangle = \tfrac{1}{\sqrt{2}}(|2\rangle + |-2\rangle) \sim \tfrac{\sqrt{3}}{2}(x^2 - y^2)\,, \end{cases} \tag{3.1}$$

$$t_{2g} : \begin{cases} |xy\rangle = -\tfrac{i}{\sqrt{2}}(|2\rangle - |-2\rangle) \sim \sqrt{3}\,xy\,, \\[2mm] |xz\rangle = -\tfrac{1}{\sqrt{2}}(|1\rangle - |-1\rangle) \sim \sqrt{3}\,xz\,, \\[2mm] |yz\rangle = \tfrac{i}{\sqrt{2}}(|1\rangle + |-1\rangle) \sim \sqrt{3}\,yz\,. \end{cases} \tag{3.2}$$

Here we have chosen real combinations of the basic spherical harmonics $|l^z\rangle$.[3]

Often one uses a slightly different normalization for these wavefunctions, especially e_g ones; instead of (3.1), one can take these wavefunctions in the form

$$\begin{aligned} |z^2\rangle &= \tfrac{1}{\sqrt{6}}(3z^2 - r^2) = \tfrac{1}{\sqrt{6}}(2z^2 - x^2 - y^2)\,, \\[2mm] |x^2 - y^2\rangle &= \tfrac{1}{\sqrt{2}}(x^2 - y^2) \end{aligned} \tag{3.3}$$

(of course, the ratio of the coefficients in (3.1) and (3.3) is the same). We will often use this normalization in what follows. This normalization will be very convenient when we discuss hopping matrix elements between different e_g orbitals in Chapter 6.

Instead of the real basis of eqs (3.1) or (3.3) and (3.2), one can also take other linear combinations. This may be especially useful for t_{2g} levels, where one can choose the basis states as

$$\begin{aligned} |t_{2g}^0\rangle &\sim -\tfrac{i}{\sqrt{2}}(|2\rangle - |-2\rangle) \sim |xy\rangle\,, \\[2mm] |t_{2g}^1\rangle &= |1\rangle \sim -\tfrac{1}{\sqrt{2}}(|xz\rangle + i|yz\rangle)\,, \\[2mm] |t_{2g}^{-1}\rangle &= |-1\rangle \sim \tfrac{1}{\sqrt{2}}(|xz\rangle - i|yz\rangle)\,. \end{aligned} \tag{3.4}$$

Again, one often uses a different normalization for the t_{2g} functions (3.2), (3.4); this does not change the physical results.

One sees right away that the orbital angular momentum $l = 2$ for isolated ions is quenched for some states. Thus in the doublet e_g all matrix elements of l, both diagonal and nondiagonal, are zero: from (3.1) it is clear that $\langle e_g | l^z | e_g \rangle = 0$, and the nondiagonal matrix elements of l^x or l^y (or of $l^\pm = l^x \pm i l^y$) are zero because the operators $l^\pm$ can only mix states with $\delta l^z = \pm 1$, but e_g states are composed of the states $|l^z = 0\rangle$ and $|l^z = \pm 2\rangle$.

For t_{2g} levels that is not the case. These states form a triplet, and one can easily see that, in the basis (3.4), the matrix elements of l^z for the corresponding states have values $\langle l^z \rangle = 0, \pm 1$, and the nondiagonal elements are also in principle non-zero.

---

[3] In this book we use the short notation $|z^2\rangle$ for $|3z^2 - r^2\rangle = |2z^2 - x^2 - y^2\rangle$, and the same for similar orbitals oriented in x- and y-directions.

The fact that t_{2g} levels form a triplet leads to the possibility of mapping this triplet onto the triplet of states with an effective orbital moment $\tilde{l} = 1$. This mapping is extremely convenient, and we will often use it below. One important point is that this effective moment should rather be taken negative (and the corresponding spin–orbit coupling can be multiplied by some constant, see below). Alternatively one takes $\tilde{l} = +1$, but then the sign of the effective spin–orbit coupling constant $\tilde{\lambda}$ and of the effective orbital g-factor $\tilde{g}$ should be taken negative, see Section 3.4; this is the description normally used and which we will use in this book. This leads to the consequence that the order of multiplet levels for this t_{2g}-triplet split by the spin–orbit coupling is reversed in comparison with that of isolated ions with $l = 1$, as discussed in Chapter 2. Thus, for less-than-half-filled t_{2g} *levels* (configurations d^1, d^2, d^3) the multiplet sequence will be inverted and the ground state will be that with the maximum possible total angular momentum $\tilde{J} = \tilde{l} + S$, with $\tilde{l} = 1$ and for more-than-half-filled t_{2g} *subshells* the multiplet sequence would be normal (the lowest state would be $|S - \tilde{l}|$).[4] Below we will discuss the corresponding results in application to particular situations (particular ions, in weak or strong crystal fields). But before doing this, we should discuss which physical factors determine one or the other type of crystal field splitting in different situations.

Consider again the situation of a TM ion in a (regular) octahedral coordination (Fig. 3.1). The wavefunctions (3.1), (3.2) have the form shown in Fig. 3.3, that is the expressions for the corresponding wavefunctions really correspond to the distributions of electron density in the corresponding states. Thus, for example, the $|z^2\rangle$ orbital $\sim (2z^2 - x^2 - y^2)$ has a big lobe along the z-direction and smaller lobes in the xy-plane (Fig. 3.3(a)); the $|x^2 - y^2\rangle$ orbital has lobes along the x- and y-directions, etc. The signs of the corresponding contributions, following from eqs (3.1)–(3.3) and marked in Fig. 3.3, are also important. For a single TM ion (e.g., a TM impurity in a nonmagnetic matrix) these signs can be changed to their opposites; this would simply correspond to a change in phase of the corresponding wavefunctions, which for isolated centers is irrelevant. But for concentrated systems the choice of these signs, that is of phases, becomes important, and for example the respective hopping matrix elements can have different signs, which strongly influences the corresponding band structure.

Comparing the electron distribution of e_g and t_{2g} electrons (Fig. 3.3(a, b) vs Fig. 3.3(c–e)), one can easily understand the origin of crystal field splitting shown in Fig. 3.2. Indeed, for e_g states the lobes of the corresponding wavefunctions, that is (negative) electron charge density, are directed toward the negatively charged anions surrounding the TM ion (these anions are called *ligands*, and consequently the crystal field

---

[4] Strictly speaking, for ions in crystals the standard classification of electronic terms by the orbital moment L and total angular momentum J is not applicable: in the absence of full rotational symmetry L and J are not good quantum numbers. Rather one should classify different terms by the corresponding representations of the point group of the crystal. But in cases when the LS coupling scheme is applicable one can still use these notions – at least they give the correct sequence and multiplicity of corresponding terms. This is usually the case for $3d$ and also for $4d$ elements. The situation with $5d$ ions is more complicated – they may lie between the applicability of LS (Russell–Saunders) and jj coupling schemes. But for them, too, the use of effective values L_{eff} and J_{eff} usually gives a correct qualitative (though not quantitative) description.

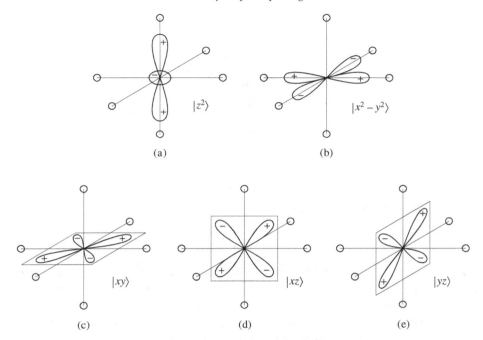

Figure 3.3 Typical shapes (electron densities) of orbitals: (a, b) e_g orbitals; (c, d, e) t_{2g} orbitals.

splitting is sometimes called *ligand field splitting*). Correspondingly, the electron density of e_g-electrons experiences strong Coulomb repulsion from the negatively charged ligands (e.g., O^{2-} ions). This increases the energy of e_g levels. In contrast, the t_{2g} wavefunctions have electron density directed not toward ligands but between them (along the diagonals in the xy-, xz-, and yz-planes for $|xy\rangle$, $|xz\rangle$, and $|yz\rangle$ orbitals, respectively), see Fig. 3.3(c–e). Consequently, the electrons in these orbitals have weaker Coulomb repulsion from ligands and the energy of t_{2g} levels is lower than that of e_g ones (Fig. 3.2). This contribution to the crystal field splitting is known as a point-charge contribution. (As noted in the footnote on p. 37, in crystals one also has to take into account the interaction with further neighbors, and sometimes the latter can change the details of the crystal field splitting; usually, however, the approximation obtained by taking into account mainly the contribution from nearest neighbors gives at least a qualitatively correct picture.)

It is also interesting to look at what would be the distribution of electron density for complex linear combinations of t_{2g} levels (3.4), corresponding to the eigenstates of the real or effective orbital moments l, $\tilde{l}$. One can easily see that the shape of the electron density for two states $|t_{2g}^1\rangle$ and $|t_{2g}^{-1}\rangle$ in (3.4) can be obtained from that of the orbitals xz and yz (Fig. 3.3(d, e)), by rotating them around the z-axis. That is, they have the shape of hollow cones, extended in the z-direction, see Fig. 3.4. The direction of rotation (clockwise or counterclockwise) corresponds to the value of orbital moment $l^z = \pm 1$. We will use this picture later in Section 3.4.

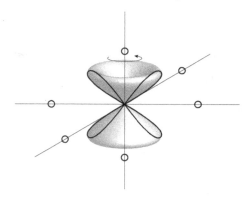

Figure 3.4 The shape (electron density) of states of the type $|xy\rangle \pm i|yz\rangle$ with nonzero orbital moment.

The point-charge contribution is not the only one determining the type and magnitude of crystal field splitting. Another, no less important factor is the covalency between d-electrons and surrounding ligands (for concreteness we will speak below about oxygens, but the results are qualitatively the same also e.g. for halogen ions such as F, Cl or for S, Se, etc.).[5] The d orbitals of TM ions have nonzero overlap with the $2p$ orbitals of surrounding oxygens (and also with the outer $2s$ orbitals, but the covalency with $2p$ oxygen states is more important). Consider first the case of e_g electrons, for example the $|x^2 - y^2\rangle$ orbital (Fig. 3.5). We see that this $|x^2 - y^2\rangle$ d orbital has strong overlap with p_x orbitals of oxygens O1 and O3 and with p_y orbitals of O2 and O4, and correspondingly there will exist a rather large hopping matrix element between this d orbital and the respective p orbitals of oxygens, which can be written in the form

$$\mathcal{H}_{pd} = t_{pd\sigma} \left[d^{\dagger}_{x^2-y^2}(p_{1x} - p_{2y} - p_{3x} + p_{4y}) + \text{h.c.} \right], \tag{3.5}$$

where $d^{\dagger}$, d, $p^{\dagger}$, p are the creation and annihilation operators for the corresponding electrons. The signs here reflect the relative signs of the $|x^2 - y^2\rangle$ orbital and those of the corresponding p orbitals, see Fig. 3.5, so that the product of these signs is the same for each TM–O pair; this guarantees the constructive interference of corresponding hoppings. Correspondingly, for the $|z^2\rangle$ orbital in (3.1), (3.3) we have to choose a combination of respective p orbitals of all six surrounding oxygens, with proper signs and weights, so that the overlap of the $|z^2\rangle$ orbital with each p orbital is the same, both in sign and magnitude. The hybridization in the case when the lobes of d orbitals and those of oxygen p orbitals are directed toward one another, which gives strong overlap, is called σ hybridization, which we have denoted as $t_{pd\sigma}$ in (3.5).

---

[5] In older and in chemical literature the terminology "ligand field splitting" is often used in connection with this covalency contribution to the splitting of d levels, and the term "crystal field splitting" is reserved for the point charge (Coulomb) contribution. In the present physical literature these terms are usually treated as synonyms.

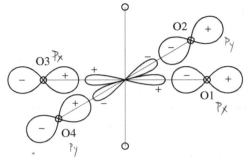

Figure 3.5 Covalency contribution to crystal field splitting for e_g electrons (σ-bonding) due to p–d hybridization.

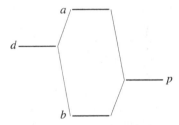

Figure 3.6 *p*–*d* splitting due to *p*–*d* hybridization. Here *b* and *a* denote bonding and antibonding orbitals.

As a result of the *p*–*d* hybridization (3.5) the *d* and *p* levels are mixed and repelled, see the schematic picture in Fig. 3.6 (cf. Fig. 1.10). In normal cases the oxygen *p* levels ε_p lie (much) lower than the *d* levels ε_d, and the bonding and antibonding levels *b* and *a* for e_g orbitals have the energies

$$\varepsilon_b = \varepsilon_p - \frac{t_{pd\sigma}^2}{\varepsilon_d - \varepsilon_p} \,,$$

$$\varepsilon_a = \varepsilon_d + \frac{t_{pd\sigma}^2}{\varepsilon_d - \varepsilon_p} \,,$$

$$(3.6)$$

with the corresponding wavefunctions, which can be written as

$$|\psi_b\rangle = \alpha|p\rangle + \beta|d\rangle \,,$$
$$|\psi_a\rangle = \beta|p\rangle - \alpha|d\rangle,$$

$$(3.7)$$

with $\alpha \gg \beta$.

These results are obtained by diagonalizing the matrix

$$\begin{pmatrix} \varepsilon_d & t_{pd\sigma} \\ t_{pd\sigma} & \varepsilon_p \end{pmatrix}$$

$$(3.8)$$

which gives the levels

$$\varepsilon_\pm = \frac{\varepsilon_d + \varepsilon_p}{2} \pm \sqrt{\left(\frac{\varepsilon_d - \varepsilon_p}{2}\right)^2 + t_{pd\sigma}^2} \, , \tag{3.9}$$

from which for $t_{pd} \ll \varepsilon_d - \varepsilon_p$ we obtain the results (3.6). One can show that for e_g-electrons, taking account of hybridization with all surrounding oxygens, the shift of both $|x^2 - y^2\rangle$ and $|z^2\rangle$ orbitals is the same (this follows, in fact, from symmetry considerations, according to which these two states belong to the two-dimensional E_g-representation, and these levels have to be degenerate as long as the symmetry remains the same, i.e. in the case when the O_6 octahedron around the TM ion is not distorted).

Similar considerations for t_{2g}-electrons show that there will be a similar t_{2g}–$2p$ hybridization, but due to the different form of the t_{2g} orbitals the corresponding d–p overlap will be weaker, see Fig. 3.7: thus, for example, the $|xy\rangle$ d orbital is orthogonal to the p orbitals shown in Fig. 3.5, but there will be nonzero overlap and hybridization with the p orbitals shown in Fig. 3.7. These p orbitals, directed perpendicular to the TM–O bond, are called π orbitals, and the corresponding hopping matrix element is denoted $t_{pd\pi}$. As $t_{pd\pi}$ is smaller than $t_{pd\sigma}$ (typically $t_{pd\pi} \sim \frac{1}{2}t_{pd\sigma}$), the bonding–antibonding splitting such as that shown in Fig. 3.6 will be smaller for t_{2g} orbitals than for e_g orbitals. The resulting picture will then be that shown schematically in Fig. 3.8.[6] We see that the antibonding states having predominantly d-character (e_g and t_{2g} levels in Fig. 3.8) will be split so that the e_g orbitals will lie above the t_{2g} ones. (The bonding orbitals having mainly oxygen $2p$-character will be completely filled.) As a result, the stronger d–p covalency of e_g-electrons leads to a t_{2g}–e_g crystal field splitting of the type shown in Fig. 3.2.[7]

Thus we see that there exist two contributions to the crystal field splitting of d levels, the point-charge contribution (Coulomb repulsion from negatively charged ligands) and that due to d–p covalency or hybridization. Both these contributions lead to the same consequence – that in normal cases the antibonding e_g levels, that is those with predominantly d-character, lie higher than the t_{2g} ones. Numerically the covalency contribution to the crystal field splitting is usually larger than the point-charge contribution. Point-charge effects may still be important for oxides and fluorides, although the covalency contribution definitely dominates for example for ligands such as S, Se, or Te. Also, as the radius of d orbitals increases in going from $3d$ to $4d$ to $5d$ elements (and of p orbitals in the series O, S, Se, Te), the p–d hybridization and with it the value of Δ_{CF} increases when we go down a column in the Periodic Table (see the end of the book), due to the enhanced covalency contribution.

---

[6] As explained above, see for example eq. (3.5), the p orbitals of the six oxygens of the O_6 octahedron should form linear combinations with the same symmetry as the corresponding d orbitals with which they hybridize. This is why we have doubly and triply degenerate orbitals for both bonding and antibonding states of Fig. 3.8. The remaining p orbitals (there are altogether $3 \times 6 = 18$ p orbitals on six oxygens, five of which are used for p–d hybridization) will be nonbonding and remain at the original position of p levels, ε_p.

[7] The situation is reversed and becomes more complicated if the ligand p levels lie *above* the d levels – this is the situation with a *negative charge transfer gap*, see Section 4.3.

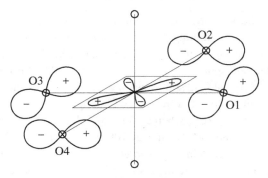

Figure 3.7 Covalency contribution to crystal field splitting for t_{2g} electrons (π-bonding) due to p–d hybridization.

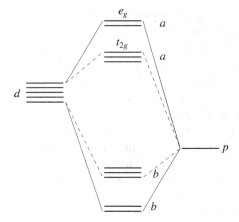

Figure 3.8 The origin of the crystal field splitting due to p–d covalency. b and a stand for bonding and antibonding levels.

Qualitatively, in considering crystal field splitting, it is still usually simpler to deal with the point-charge contribution: this allows us to build a simple physical picture of crystal field splitting in more complicated cases, and the resulting qualitative picture is almost always correct. We will often use this picture later in this book, though one has to be aware that this is not the only, and in most cases not the dominant, contribution.

The covalency contribution to crystal field splitting also largely determines what is known as the *spectrochemical series* (see e.g. Jorgensen, 1971 or Cotton *et al.*, 1995) – the relative strength of different ligands which determines the value of the crystal field splitting Δ_{CF}. This series, with increasing ligand strength from left to right, is

$$I^- < Br^- < S^{2-} < Cl^- < NO_3^- < F^- < OH^- < O^{2-}$$
$$< H_2O < NH_3 < NO_2^- < CN^- < CO \, . \qquad (3.10)$$

The further to the right, the larger is the typical value of Δ_{CF}. This regularity is very important for example in determining the spin state of TM ions, see Sections 3.3 and 5.9. Note that some of the strongest ligands such as NH_3 and CO are neutral, their strong contribution to crystal field splitting is predominantly caused by the covalency (hybridization). The difference in crystal field splittings for different ligands can be quite large. Thus, for example, for Cr^{3+} the value of Δ_{CF} for the weak ligand Cl^- is $13\,800\,cm^{-1}$, for F^- it is $15\,000\,cm^{-1}$, that is 1.7–1.9 eV, whereas for strong ligands it is almost twice as large: for NH_3 it is $21\,600\,cm^{-1}$ and for CN^- it is $26\,700\,cm^{-1}$, that is 2.7–3.3 eV. As mentioned above, crystal field splitting also increases in going down a column in the Periodic Table, from 3d to 4d to 5d elements. There are also semiempirical rules determining the change of Δ_{CF} in the row of d elements. Thus for fluorides the value of Δ_{CF} increases in a series

$$Mn^{2+} < Ni^{2+} < Co^{2+} < Fe^{2+} < V^{2+} < Fe^{3+} < Cr^{3+} < V^{3+}. \qquad (3.11)$$

All these regularities can be useful in analyzing the properties of a series of compounds and in predicting the behavior of new materials.

Using the simple qualitative picture for crystal field splitting described above, one can easily understand the possible effects of symmetry reduction due to distortions of the ligand octahedron. Consider for example the deformation of the O_6 octahedron shown in Fig. 3.9(a), that is a tetragonal elongation in the z-direction, in which the coordinates of the corresponding oxygens change as $z \to z + 2\delta$, $x \to x - \delta$, $y \to y - \delta$ (so that the corresponding volume $V \sim xyz$ in a first approximation remains unchanged). By direct inspection of Fig. 3.3(a, b) one sees that after this distortion the e_g levels should split as shown in Fig. 3.9(b), with the $|z^2\rangle$ level being lower than the $|x^2 - y^2\rangle$ one: the $|z^2\rangle$ wavefunction has bigger lobes in the z-direction, and the increased TM–oxygen distance in this direction decreases the energy of this level. In contrast, the shortening of the TM–O distance in the xy-plane would increase the energy of the $|x^2 - y^2\rangle$ level. Consideration of the d–p covalency would lead to the same conclusion: the stronger d–p hybridization with the in-plane ligands would push up the $|x^2 - y^2\rangle$ level stronger than the $|z^2\rangle$ level, cf. Fig. 3.8. Similar arguments show that the corresponding splitting of t_{2g} levels would be that shown in Fig. 3.9(b): for tetragonal elongation the energy of the $|xy\rangle$ orbital lying predominantly in the xy-plane would increase, and that of the $|xz\rangle$ and $|yz\rangle$ orbitals would go down, cf. Fig. 3.3(c–e). (A simple rule here is that the splitting of t_{2g} levels "follows" that of e_g ones: if the $|x^2 - y^2\rangle$ level goes up, so does $|xy\rangle$, while $|xz\rangle$ and $|yz\rangle$ are doing the same as $|z^2\rangle$.) One can easily understand that in the case not of tetragonal elongation as in Fig. 3.9(a) but of tetragonal *compression* ($\delta < 0$) (Fig. 3.9(c)), the splitting of e_g and t_{2g} levels would be reversed and the resulting sequence of d levels would look as shown in Fig. 3.9(d). Note that, to preserve the center of gravity, the shift of the singlet $|xy\rangle$ will be twice as large as that of the doublet ($|xz\rangle$, $|yz\rangle$). For strong ligands and for very strong deviation from the cubic crystal field, for example for very strong elongation of the ligand octahedra (Fig. 3.9(a)), the extra splitting of t_{2g} and especially e_g levels can be so strong that the z^2 level in Fig. 3.9(b) lies below the xy level, and sometimes even below the (xz, yz) doublet (Huang et al., 2011); such a situation was observed experimentally in some cuprates.

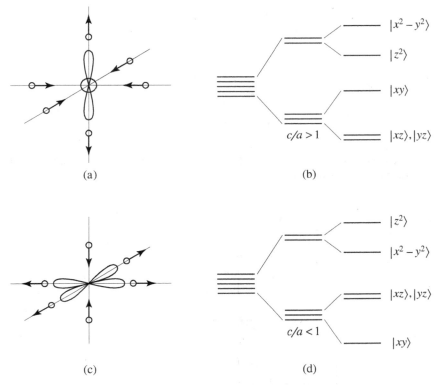

Figure 3.9 Splitting of d levels caused by tetragonal distortion of MO_6 octahedra (Q_3 distortion mode) for tetragonal elongation (a, b) and compression (c, d).

Other types of distortion can lead to extra crystal field splitting of d levels. This is the case for the orthorhombic distortion shown in Fig. 3.10, with $x \to x + \delta$, $y \to y - \delta$, $z \to z$: this distortion also splits the e_g levels, the resulting eigenstates being not the originally chosen $|z^2\rangle$ and $|x^2 - y^2\rangle$ orbitals, but their linear superpositions

$$\frac{1}{\sqrt{2}}\left(|z^2\rangle \pm |x^2 - y^2\rangle\right). \tag{3.12}$$

Similarly, the t_{2g} levels are also split by orthorhombic distortion, not into a singlet and a doublet, but into three singlets.

The t_{2g} levels are also split by *trigonal distortions* – elongation or contraction of MO_6 octahedra along one of the [111] axes (Fig. 3.11(a)). This splitting is shown in Fig. 3.11(b). (The e_g levels are not split by trigonal distortions.) The t_{2g} levels are here split into a singlet a_{1g} and a doublet e_g^π. The corresponding wavefunctions can be written in the form

$$\begin{aligned}
|a_{1g}\rangle &= \frac{1}{\sqrt{3}}\left(|xy\rangle + |xz\rangle + |yz\rangle\right), \\
|e_{g\pm}^\pi\rangle &= \pm\frac{1}{\sqrt{3}}\left(|xy\rangle + e^{\pm 2\pi i/3}|xz\rangle + e^{\mp 2\pi i/3}|yz\rangle\right).
\end{aligned} \tag{3.13}$$

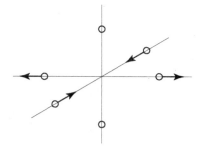

Figure 3.10 The Q_2 distortion mode, which also leads to a splitting of both t_{2g} and e_g levels and gives orthorhombic local symmetry.

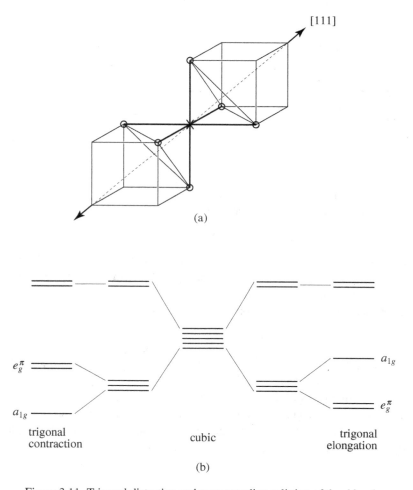

Figure 3.11 Trigonal distortion and corresponding splitting of the d levels.

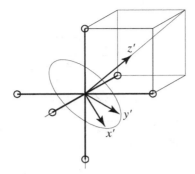

Figure 3.12 Local coordinate axes (x', y', z') for trigonal distortion of $M O_6$ octahedra.

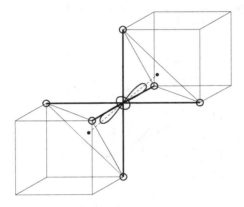

Figure 3.13 The shape of a singlet a_{1g} orbital in a trigonal field.

These orbitals have a rather interesting form. The first one, $|a_{1g}\rangle$, has a very simple shape in local coordinates with the local z'-axis directed along the axis of the trigonal distortion, and with x'- and y'-axes in the perpendicular plane, see Fig. 3.12. In these coordinates the $|a_{1g}\rangle$ orbital is simply

$$|a_{1g}\rangle \sim 3z'^2 - r^2 = 2z'^2 - x'^2 - y'^2 \,, \qquad (3.14)$$

that is it is analogous to the $|z^2\rangle$ e_g orbital (3.1), but with the electron density elongated in the z'-direction, that is pointing from the center of one O_3-triangle in Fig. 3.11(a) to the center of the opposite O_3-triangle, see Fig. 3.13.

The other two t_{2g} orbitals, denoted as $|e_{g\pm}^\pi\rangle$ in eq. (3.13), are complex orbitals. They are analogous to the two lowest orbitals (3.4), and they are actually the states with the orbital moment $|l^{z'} = \pm 1\rangle$, where the quantization axis for the orbital moment is not the tetragonal axis [001], as in (3.4), but the trigonal axis [111]. (The state $|a_{1g}\rangle$ of (3.13), (3.14) is actually the state $|l^{z'} = 0\rangle$ for that axis.) The shape of these e_g^π orbitals is that of a torus with the axis [111] (Fig. 3.14). Similar to the tetragonal case, see (3.2), one can also

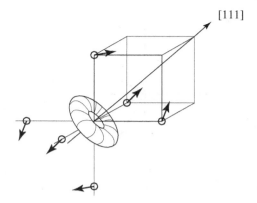

Figure 3.14 Typical shape of the e_g^π state originating from t_{2g} levels. Complex combination with nonzero orbital moment is shown (cf. Fig. 3.4).

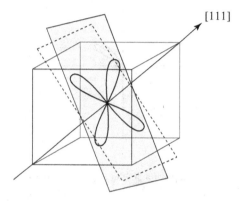

Figure 3.15 Schematic shape of the real e_g^π orbitals (3.15). (The second such orbital lies in a plane tilted around the perpendicular axis, and rotated by 45° around the [111] axis.)

take linear combinations of the e_g^π states of (3.22). They have a somewhat awkward form (Ballhausen, 1962):

$$|e_{g1}^\pi\rangle = \frac{1}{\sqrt{3}}\left(\sqrt{2}(x'^2 - y'^2) - x'z'\right),$$
$$|e_{g2}^\pi\rangle = \frac{1}{\sqrt{3}}\left(\sqrt{2}\,x'y' + y'z'\right) \tag{3.15}$$

(in local coordinates with the z'-axis along the [111] direction). They have the shape of "crosses" (similar to xy and $x^2 - y^2$), lying in planes more or less perpendicular to the z'-axis, but somewhat tilted in two directions, and these orbitals are rotated by 45° with respect to each other, as shown schematically in Fig. 3.15.

From this picture one can easily understand the character of splitting of t_{2g} levels in the trigonal distortion. For local elongation in the z'-direction all three O^{2-} ions in this

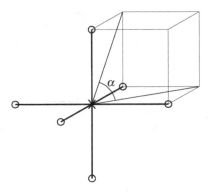

Figure 3.16 The angle α characterizing trigonal distortions.

direction come closer together, and consequently the $|a_{1g}\rangle = |z'^2\rangle$ orbital with the electron density predominantly in this direction would experience stronger Coulomb repulsion from these three oxygens on each side, and its energy would go up; this would give a crystal field splitting as shown on the right of Fig. 3.11(b). In contrast, trigonal contraction would move these three oxygens away from this direction, so that in this case the $|a_{1g}\rangle = |z'^2\rangle$ orbital would point to the "empty space" – to the center of a wide O_3 triangle – and its energy would go down, as shown on the left of Fig. 3.11(b).[8]

Often one describes trigonal distortions by the change of the angle α between diagonals in the xy-, xz-, and yz-planes, see Fig. 3.16. For a regular octahedron this angle is $\alpha = 60°$. Trigonal elongation leads to $\alpha < 60°$, and trigonal compression to $\alpha > 60°$ (see, e.g., Goodenough, 1963).

Above, we have mostly considered crystal field splitting for TM ions in octahedral surrounding. However, this is not the only possible coordination of TM ions. They may be located for example in the centers of ligand (e.g., oxygen) tetrahedra, or they may be located in other polyhedra. The crystal field splitting of d levels depends on this coordination. Thus for tetrahedral coordination (TM ion at the center of a regular tetrahedron) the d levels will also be split into an e_g doublet and a t_{2g} triplet, but the ordering of these levels is inverted compared with the case of octahedral coordination of Fig. 3.2, with the t_{2g}–e_g splitting Δ_{CF}(tetr.) smaller than that in octahedra. In the simplest cases, Δ_{CF}(tetr.) $\simeq \frac{4}{9}\Delta_{CF}$(octahedr.). The resulting crystal field-level scheme looks as shown in Fig. 3.17. The corresponding distortion of a regular ligand tetrahedron would also lead to extra splitting of d levels, similar to that shown in Figs 3.9 and 3.11.

Other coordinations of TM ions lead to yet other crystal field-level schemes. One typical case is fivefold coordination in the form of a pyramid (Fig. 3.18(a)), which is met for example in many vanadates such as CaV_2O_5 and in high-T_c cuprates such as $YBa_2Cu_3O_7$. It can be visualized as an octahedron of Fig. 3.1 with one of the apical oxygens removed. This

---

[8] Note that in a trigonal field the e_g^π levels have the same symmetry as the "real" e_g levels; therefore these levels can easily be mixed by perturbations, etc.

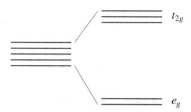

Figure 3.17 Crystal field splitting of d levels for a transition metal ion in a tetrahedral coordination.

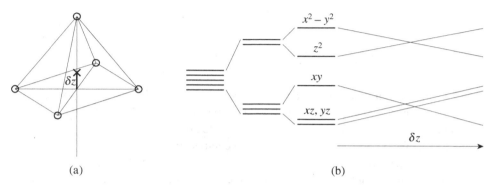

(a)   (b)

Figure 3.18 d levels for a 5-coordinated transition metal ion in a square pyramid as a function of "pulling" the TM ion inside such pyramid.

resembles the case of tetragonal elongation, with the elongation so strong that one oxygen is "moved to infinity." Consequently, one may expect crystal field splitting similar to that of Fig. 3.9. However, often in such cases the TM ion does not remain in the basal plane, but is pulled inside the pyramid, up by δz along the z-direction, as shown in Fig. 3.18(a). If this shift is strong enough, the sequence of levels may be inverted, see Fig. 3.18(b). This is again easy to understand using our qualitative picture described above: if for the TM in the basal plane the absence of the lower apical oxygen decreases the energy of the $|z^2\rangle$ (and $|xz\rangle$, $|yz\rangle$) states, the shift of the TM upward toward the upper apical oxygen gradually increases the Coulomb repulsion of, for example, the $|z^2\rangle$ orbital of Fig. 3.3(a) with this oxygen, thus increasing the energy of this orbital. If strong enough, such a shift can even lead to a crossing of $|z^2\rangle$ and $|x^2 - y^2\rangle$ levels (and of $|xz\rangle$ and $|yz\rangle$ levels with the $|xy\rangle$ level), as shown in Fig. 3.18(b).

Two further interesting coordinations met in TM compounds are the trigonal bipyramid of Fig. 3.19(a), which can be visualized as two tetrahedra "glued together" across a common face (this coordination is met e.g. in hexagonal $YMnO_3$, or in (possibly multiferroic) $LuFe_2O_4$) and the trigonal prism of Fig. 3.19(c) (met e.g. in $Ca_3Co_2O_6$). The crystal field splitting in the first case is shown schematically in Fig. 3.19(b), and in the case of a trigonal prism in Fig. 3.19(d) (the location and distance between the levels in Fig. 3.19(d) depend on the exact shape of this trigonal prism).

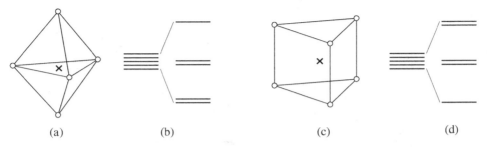

Figure 3.19 (a, b) Crystal field splitting in a trigonal bipyramid. (c, d) Crystal field splitting in a trigonal prism.

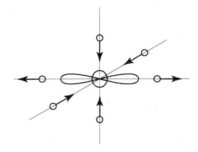

Figure 3.20 Stabilization of z^2-type orbital oriented in a different (here x) direction.

There is one more important factor in this story. Above, when treating the tetragonal elongation of ligand octahedra, we have considered this distortion in the z-direction. But we could equally well consider, for example, the tetragonal elongation or contraction in x- or y-directions. The resulting crystal field splitting would be the same, but the corresponding d orbitals would be different: for example, instead of the orbital $|z^2\rangle \sim 2z^2 - x^2 - y^2$ stabilized by the tetragonal elongation in the z-direction, for similar elongation in the x-direction the corresponding orbital would be $|x^2\rangle \sim 2x^2 - y^2 - z^2$ (Fig. 3.20) and similarly for elongation in the y-direction. For e_g-electrons we can consider any linear combination of the basis orbitals $|z^2\rangle$, $|x^2 - y^2\rangle$ of the form

$$|\theta\rangle = \cos\tfrac{\theta}{2} |z^2\rangle + \sin\tfrac{\theta}{2} |x^2 - y^2\rangle . \tag{3.16}$$

Here the angle θ marks the corresponding states, and the coefficients in (3.16) are chosen so that the state $|\theta\rangle$ is properly normalized.

One can then plot the states $|\theta\rangle$ on a circle, see Fig. 3.21. Each state (3.16) corresponds to a point on this circle, with the azimuthal coordinate θ. One can easily see using for example the expression (3.1) or (3.3) and (3.16) that whereas the state $|\theta = 0\rangle$ corresponds to the orbital $|z^2\rangle$ and $|\theta = \pi\rangle$ to $|x^2 - y^2\rangle$, the orbital $|x^2\rangle \sim |2x^2 - y^2 - z^2\rangle$ would correspond to $|\theta = -\tfrac{2}{3}\pi\rangle$ and $|y^2\rangle$ to $|\theta = \tfrac{2}{3}\pi\rangle$. Similarly, the states $|\pm\tfrac{1}{3}\pi\rangle$ correspond respectively to

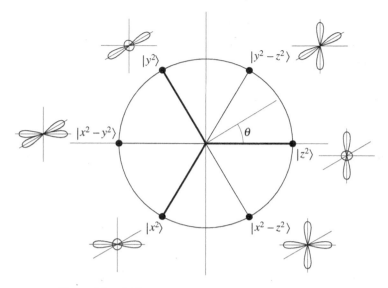

Figure 3.21 Different combinations of e_g orbitals (3.16).

the orbitals $|y^2-z^2\rangle$ and $|x^2-z^2\rangle$. Thus the original cubic symmetry (equivalence of x-, y-, and z-directions) is reflected in the threefold symmetry in the θ plane of Fig. 3.21. One can show that the orthorhombic distortion shown in Fig. 3.10, which stabilizes states of the type (3.12), corresponds in this plot to the states $|\theta = \pm\frac{1}{2}\pi\rangle$ (or states similarly rotated by $\pm\frac{2}{3}\pi$ for orthorhombic distortion with oxygen shifts in (xz) or (yz) planes, instead of (xy) shifts shown in Fig. 3.10).

One can show, in fact, that the equivalence of all states on the circle of Fig. 3.21 is a consequence of the symmetry and of the presence of both tetragonal and orthorhombic distortions (but ignoring lattice anharmonicity and higher-order orbital–lattice coupling, see below). In cubic symmetry not only are the electronic e_g states doubly degenerate, but there also exist *doubly degenerate vibration modes*. These are usually denoted E_g, and in fact these are the tetragonal and orthorhombic distortions described above. The tetragonal deformation is denoted Q_3, and the orthorhombic mode is denoted Q_2 (alternatively, one uses the notation Q_θ and Q_ε). These modes, which should preserve the total volume, can be written, using normalization similar to (3.3), as

$$Q_3 = \frac{1}{\sqrt{6}}(2z - x - y)\,,$$
$$Q_2 = \frac{1}{\sqrt{2}}(x - y)\,, \tag{3.17}$$

that is $Q_3 > 0$ corresponds to tetragonal elongation in the z-direction, $Q_3 < 0$ corresponds to analogous tetragonal compression, and Q_2 corresponds to orthorhombic distortion, see Figs 3.9 and 3.10 (note that sometimes one writes Q_2 with an opposite sign; here the sign of Q_2 is chosen in such a way that the deformation for an arbitrary superposition of distortions, see eq. (3.18) below, corresponds to the wavefunction in the form (3.16)). Thus the doubly degenerate e_g-electrons interact, and can be split, both by Q_3 and Q_2 distortions, and the coupling to these modes is the same, which in fact guarantees that the energies of the states on the whole circle in the θ plane of Fig. 3.21 are the same.

Usually for a single TM ion in a crystal one has one-to-one correspondence between the electronic states (3.16) and the distortions of nearest-neighbor ligands, which can also be represented as linear combinations of the basis distortions Q_3, Q_2:

$$|\tilde{\theta}\rangle = \cos\tilde{\theta}\,|Q_3\rangle + \sin\tilde{\theta}\,|Q_2\rangle\,. \tag{3.18}$$

(Distortions are classical variables, therefore in eq. (3.18) we have $\cos\tilde{\theta}$ and $\sin\tilde{\theta}$ as coefficients, whereas the doubly degenerate electronic states behave as spinors; that is why we have the factors $\cos\frac{\theta}{2}$, $\sin\frac{\theta}{2}$ in the expression for the electron wavefunction (3.16).) Thus, in full analogy with Fig. 3.21, we can represent the local deformations caused by E_g distortions in the (Q_3, Q_2) plane, see Fig. 3.22. From eq. (3.18) with the form of Q_2, Q_3 given by (3.17), we can easily see that, in agreement with Fig. 3.21, the angles $\tilde{\theta} = 0$ and $\tilde{\theta} = \pi$ correspond respectively to elongation and contraction along the z-direction, and, similarly, the angles $\tilde{\theta} = \frac{2}{3}\pi$, $-\frac{1}{3}\pi$ correspond to elongation and contraction along the y-direction, and $\tilde{\theta} = -\frac{2}{3}\pi$, $\frac{1}{3}\pi$ to the same distortions along the x-direction. For $\tilde{\theta} = \theta$, as is almost always the case, these distortions stabilize the orbitals shown in Fig. 3.21. As we have argued above, for an isolated TM ion the mixing angle in (Q_3, Q_2) space is indeed equal to the corresponding angle in electronic space, $\tilde{\theta} = \theta$; that is, there is one-to-one correspondence between local distortions (the distortions of nearest-neighbor oxygen

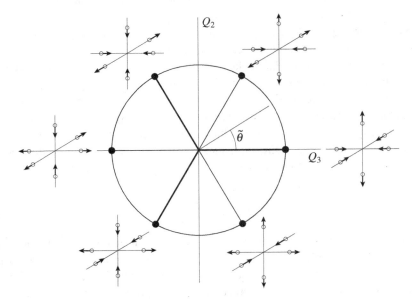

Figure 3.22 Different combinations of Q_3–Q_2 distortions (3.17).

octahedra) and the respective types of orbital (3.16) (and one can easily check that for $\theta = \tilde{\theta}$ with the choice of modes Q_3, Q_2 as in (3.17) the electronic wavefunctions (3.16) would correspond to distortions (3.18)). In most cases the same is true for concentrated systems as well, although in general this need not be the case; and in situations where the crystal field splitting of d levels is determined not only by nearest-neighbor ligands, but where there exists a strong contribution from further neighbors, or when other factors contribute to the electron energy, this rule can in principle break down, and the type of occupied orbital can differ from that which would follow just from the distortion of local ligand octahedra. That is, in those cases $\tilde{\theta} \neq \theta$. These are very rare cases, but such situations have indeed been discovered recently (Wu, 2011). Further on we will not consider such exceptional cases, but assume that the standard correlation holds and that the orbital mixing angle (3.16) is equal to the angle (3.18) determining local distortions, $\tilde{\theta} = \theta$.

The normally valid correspondence between the lattice distortion (3.18) and orbital occupation (3.16) is very often used to determine the latter from the structural data. From eq. (3.18) with (3.17) we find that, in this case,

$$\tan \theta = \frac{Q_2}{Q_3} = \frac{\frac{1}{\sqrt{2}}(y - x)}{\frac{1}{\sqrt{6}}(2z - x - y)} = \frac{\sqrt{3}(y - x)}{2z - x - y} , \tag{3.19}$$

see Fig. 3.22. Here x, y, z can be taken as the corresponding distances from the transition metal to the respective oxygens. Thus, from these distances, which one can usually find from good structural data, one can obtain the mixing angle θ in the (Q_3, Q_2) plane and the type of orbitals occupied using (3.18). This method is very often used to determine orbital occupation, especially in systems with the so-called Jahn–Teller ions with orbital degeneracy, see the next section.

Often one writes eq. (3.19) in a slightly different form. Thus, for example, in manganites such as $LaMnO_3$ the typical orbital occupation is close to that shown in Fig. 3.23, that is different orbitals alternate in the basal plane. However, the exact type of orbitals occupied is somewhat different from that shown in Fig. 3.23 (which would correspond to an alternation of $|x^2\rangle$ and $|y^2\rangle$ orbitals, i.e., in Fig. 3.21, to two sublattices with $\theta = \pm\frac{2}{3}\pi$); the real angles of these sublattices are smaller, $\sim\pm108°$, see for example Rodriguez-Carvajal *et al.* (1998). For such more accurate states we would have, for example, for the site A of Fig. 3.23 a long Mn–O distance l in the y-direction, a short Mn–O distance s in the x-direction, and an intermediate or middle Mn–O distance m in the z-direction. Then (3.19) can be written in the form

$$\tan \theta = \pm\frac{\sqrt{3}(l - s)}{2m - l - s} , \tag{3.20}$$

see for example Goodenough (1963). One has to be somewhat careful in using such expressions: the detailed identification of different Mn–O distances with l, m, s bonds depends on a particular situation, and on the quadrant in the θ (or $\tilde{\theta}$) plane that we are dealing with.

Similarly to e_g-electrons which interact and can be split by the doubly degenerate E_g distortion, the t_{2g}-electrons can be split by both E_g and trigonal distortions, which turn

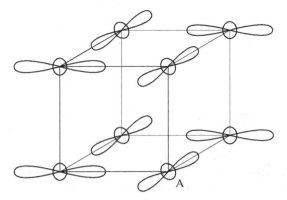

Figure 3.23 Schematic orbital ordering in LaMnO₃.

out to be triply degenerate (denoted T_{2g}). These interactions and their related effects constitute the large and important field of the Jahn–Teller effect, see the next section and Chapter 6.

3.2 Jahn–Teller effect for isolated transition metal ions

When we have a TM ion with several d-electrons, these d-electrons will occupy the available d levels so as to minimize the total energy. If the crystal field splitting is not too large, the filling of d levels follows the Hund's rule discussed in Chapter 2: the electrons occupy the states so as to have the maximum possible total spin. That is, we have to fill the crystal field-split d levels one after another by electrons with parallel spins until all five levels are filled, after which, for $n_d > 5$, we have to start occupying the states with opposite spins. Such atomic configurations are known as *high-spin states*. The effects connected with eventual competition between such high-spin states and an alternative occupations of d levels, leading for example to low-spin states, will be discussed in the next section; here we concentrate on other effects, connected with possible orbital degeneracy.

Consider the crystal field-level scheme for cubic symmetry (regular octahedral coordination of a TM ion) in Fig. 3.2. When we increase the number of d-electrons, we should fill the levels from below one after another by electrons with parallel spins (Hund's rule). Suppose we have four d-electrons, as for example in Mn^{3+}. Three of them will be on t_{2g} levels, so that the t_{2g} subshell will be half-filled. The fourth electron with the same spin, say ↑, should be on e_g levels, see Fig. 3.24(a). But these states are degenerate, and this electron can be on any of these two orbitals (or their linear superposition). The situation will be the same for the d^9 configuration, Fig. 3.24(b), this time however not for an e_g-electron but for an e_g-hole (typical situation for $Cu^{2+}(d^9)$). There are other ionic states with the same feature, for example the low-spin state of Ni^{3+}, in which six d-electrons occupy t_{2g} levels and the last, the seventh, would be formally on a doubly degenerate e_g level.

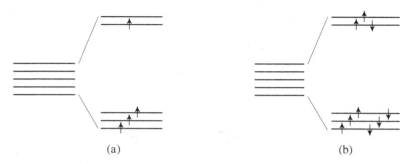

Figure 3.24 Typical electronic configuration leading to double orbital degeneracy and a strong Jahn–Teller effect. (a) d^4 configuration, for example Mn^{3+}; (b) d^9 configuration, for example Cu^{2+}.

Considering only this one e_g-electron or e_g-hole, we see that in a cubic crystal field (regular MO_6 octahedron), besides the usual double-spin degeneracy (the spin of the electron or hole can be ↑ or ↓), there exists an extra double degeneracy – an *orbital degeneracy*.

There is however a well-known statement, known as the Jahn–Teller theorem (Jahn and Teller, 1937), which says that the only degeneracy allowed in the ground state is the spin (Kramers) degeneracy, connected to invariance with respect to time reversal (it can be broken only by an magnetic ordering or by external magnetic field). All other degeneracies are not allowed: any such degenerate state would be unstable and would correspond not to a minimum but to a maximum of the total energy (or to a saddle point), and small distortions reducing the symmetry and lifting the degeneracy would always occur. In our case the distortions lifting the orbital degeneracy would be distortions of the regular O_6 octahedron, for example tetragonal or orthorhombic distortions, in which the e_g levels split as shown in Fig. 3.9. Very generally, one can show that such perturbations (distortions) split the degenerate levels linearly – similar to Zeeman splitting of magnetic levels. Thus for a small distortion u the e_g energy levels will be split as $\pm gu$, where g is the electron–phonon coupling constant.

Of course any distortion will cost elastic energy $\frac{1}{2}Bu^2$, where B is the bulk modulus. Thus the total energy of the ground state as a function of the distortion will be

$$E = \pm gu + \tfrac{1}{2}Bu^2 \tag{3.21}$$

(we assume the coupling constant g to be positive). This energy for two levels, which are degenerate at $u = 0$, is shown in Fig. 3.25. We see from this figure and from eq. (3.21) that in this case the energy minimum would correspond not to a symmetric, undistorted situation $u = 0$, but to a certain finite distortion u_0, which can be found by minimizing the energy (3.21):

$$\frac{\partial E}{\partial u} = 0 \quad \Longrightarrow \quad u_0 = \pm \frac{g}{B} . \tag{3.22}$$

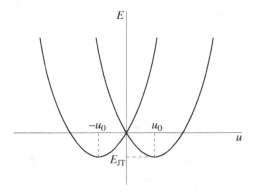

Figure 3.25 Dependence of the energy of doubly degenerate levels on the distortion u lifting the degeneracy, see eq. (3.21).

Putting this value back into the energy (3.21), we obtain that such distortion leads to a decrease of the energy,

$$E_0 = E_{JT} = -\frac{g^2}{2B} , \qquad (3.23)$$

and the elastic energy loss is overcompensated by the electronic energy gain linear in distortion. This is the essence of the Jahn–Teller effect: the symmetric situation leading to (extra) degeneracy is always unstable with respect to distortions lifting the degeneracy and reducing the energy. Thus we see that classically such a system would always distort, so that the electron would occupy a particular level – the one going down in energy (in case of eqs (3.21)–(3.23) this will be one orbital for $u > 0$ and another, orthogonal orbital for $u < 0$). In our concrete case of e_g orbitals we saw above that, for example, for tetragonal elongation the occupied orbital would be $|z^2\rangle$, and for tetragonal contraction it would be $|x^2 - y^2\rangle$, see Fig. 3.9. In effect we would have, for elongated octahedra (the typical situation, see below), not the degenerate situation shown in Fig. 3.24, but the orbital occupation shown in Fig. 3.26. We see that after the Jahn–Teller distortion the electrons now occupy particular orbitals. This is why we speak here both about the Jahn–Teller effect and about orbital ordering. These phenomena always occur simultaneously, although their microscopic mechanisms may be either predominantly due to electron–lattice (in this case Jahn–Teller) interactions, see for example Gehring and Gehring (1975), or largely of electronic (exchange) character (Kugel and Khomskii, 1982), see Chapter 6.

   Classically the system should choose a particular distortion with the corresponding orbital occupation; this is usually the case in concentrated systems with these Jahn–Teller (JT) ions, such as the prototype materials for high-T_c superconductors La_2CuO_4 ($Cu^{2+}(t_{2g}^6 e_g^3)$) and colossal magnetoresistance manganites $LaMnO_3$ ($Mn^{3+}(t_{2g}^3 e_g^1)$). In concentrated systems this phenomenon of cooperative distortions with corresponding orbital occupation is called the cooperative Jahn–Teller effect, or orbital ordering. It will be described in detail in Chapter 6. Here, however, when we consider the case of an isolated

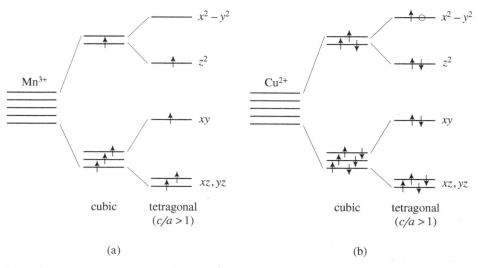

Figure 3.26 The splitting of d levels in cubic and tetragonal crystal fields (elongated octahedra): (a) filling of d levels for d^4 ions (e.g., Mn^{3+}); (b) Filling of d levels for d^9 ions (e.g., Cu^{2+}).

JT ion in a crystal, other phenomena may take place. Inspection of Fig. 3.25 shows immediately that already in this case, when doubly degenerate orbitals interact only with one type of distortion, for example tetragonal along the z-direction characterized by one parameter (distortion u), we have two equivalent minima: one with $u > 0$ (e.g., tetragonal elongation with occupation of the $|z^2\rangle$ orbital) and one with exactly the same energy but corresponding to tetragonal contraction $u < 0$, with occupied orbital $|x^2 - y^2\rangle$. In quantum mechanics in such cases we can have a tunneling of the system from one minimum to the other, so that in effect the system would be in a quantum superposition of these two states, bonding or antibonding superpositions of the type

$$|\Psi_\pm\rangle = \tfrac{1}{\sqrt{2}}\big(|\Psi_1\rangle \pm |\Psi_2\rangle\big), \qquad (3.24)$$

where for example $|\Psi_1\rangle$ is the state with lattice distortion $|u > 0\rangle = |\phi_1\rangle$ and electronic occupation $|z^2\rangle = |\psi_1\rangle$, and $|\Psi_2\rangle$ is the state with lattice distortion $|u < 0\rangle = |\phi_2\rangle$ and electronic occupation $|x^2 - y^2\rangle = |\psi_2\rangle$.

In effect, in contrast to the conventional adiabatic approximation almost always used for describing electrons in solids, in which for the same lattice configuration $|\phi\rangle$ there could exist different electronic states $|\psi_i\rangle$, so that the total wavefunction would be

$$|\Psi_i\rangle_{ad} = |\psi_i\rangle|\phi\rangle. \qquad (3.25)$$

Here, different atomic (lattice) configurations correspond to different electronic states, that is the two basis states are

$$\begin{aligned}|\Psi_1\rangle &= |\psi_1\rangle|\phi_1\rangle, \\ |\Psi_2\rangle &= |\psi_2\rangle|\phi_2\rangle.\end{aligned} \qquad (3.26)$$

Such states are called *vibronic* states. We see that for isolated JT centers (or, similarly, for JT molecules), in contrast to the naive interpretation, the JT effect does not actually lead to a reduction of symmetry: due to quantum-mechanical tunneling the symmetry is restored, and the wavefunctions are symmetric (or antisymmetric) combinations of type (3.24), though of mixed electron–lattice (vibronic) character (3.26). As a result, the average lattice distortion is zero, $\langle u \rangle = \langle \Psi_+ | u | \Psi_+ \rangle = \langle \Psi_- | u | \Psi_- \rangle = 0$, that is it looks as if the JT effect does not lead to measurable consequences (finite distortions). However, the fact that different electronic states $|\psi_i\rangle$ now always go each with their own lattice configurations $|\phi_i\rangle$ strongly influences many physical properties of such systems. Thus it can strongly suppress the values of nondiagonal matrix elements of electronic operators, so that for example for some operator $\hat{A}$, instead of the matrix element

$$A_{12} = {}_{\text{ad}}\langle \Psi_1 | \hat{A} | \Psi_2 \rangle_{\text{ad}} = \langle \psi_1 | \hat{A} | \psi_2 \rangle \tag{3.27}$$

which would exist in the adiabatic case (3.25), we now have, using (3.26),

$$A_{12} = \langle \Psi_1 | \hat{A} | \Psi_2 \rangle = \langle \psi_1 | \hat{A} | \psi_2 \rangle \langle \phi_1 | \phi_2 \rangle , \tag{3.28}$$

that is such nondiagonal matrix elements are reduced by the overlap of lattice wavefunctions describing different distortions, $\langle \phi_1 | \phi_2 \rangle < 1$. This reduction factor is known in the Jahn–Teller field as the *Ham reduction factor* (Sturge, 1967; Ham, 1972); it is analogous to the well-known polaron band narrowing for the motion of polarons (electrons strongly dressed by lattice distortions).

In the real case of e_g-electrons in crystals the situation is even more complicated and much more interesting. As discussed at the end of Section 3.1, doubly degenerate e_g-electrons interact with doubly degenerate E_g vibration modes Q_2, Q_3 and by symmetry these interactions are the same. Thus we can write the corresponding interaction Hamiltonian in this case as

$$\mathcal{H}_{\text{JT}}^{(e_g)} = -\tfrac{1}{2} g \left\{ \left(c_1^\dagger c_1 - c_2^\dagger c_2 \right) Q_3 + \left(c_1^\dagger c_2 Q_2 + \text{h.c.} \right) \right\}, \tag{3.29}$$

where c_1, c_2 are electronic operators at the JT site for orbitals 1, 2, for example $|z^2\rangle$ and $|x^2 - y^2\rangle$ orbitals in the e_g case (here we ignore electron spins); the reason for the factor $\tfrac{1}{2}$ in front will become clear in Chapter 6. In effect, both Q_3 and Q_2 distortions can lift the degeneracy, and instead of the energy of the form of Fig. 3.25 depending only on one lattice coordinate u, we now have an energy surface shown in Fig. 3.27, obtained from Fig. 3.25 by rotation around the vertical axis. (In this approximation the resulting energy would depend only on $Q_2^2 + Q_3^2$, i.e. it would not depend on the azimuthal angle θ in the (Q_3, Q_2) plane; remember here that we always assume that the distortion mixing angle $\tilde{\theta}$ (3.18) is equal to the orbital mixing angle θ (3.16).) This potential surface is known in this field as the *"Mexican hat"* potential. In effect there exist in this case not just *two* equivalent minima $\pm u_0$ (3.22) (see Fig. 3.25) but a whole *continuum* of equivalent states, the trough in Fig. 3.27 (this was already anticipated above, see Figs 3.21, 3.22 and the discussions thereof).

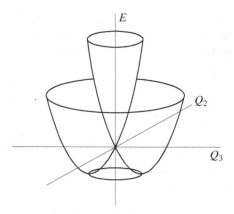

Figure 3.27 Energy surface ("Mexican hat") for doubly degenerate e_g-electrons interacting with doubly degenerate E_g distortions (Q_3, Q_2).

Thus the quantum-mechanical treatment in this case does not just give a tunneling between two minima, as in Fig. 3.25 and eqs (3.23)–(3.28); rather, the state of the system can freely "rotate" in the potential minimum of Fig. 3.27. Correspondingly, the total state of our system would not be simply a vibronic state (3.26), but a whole manifold of such states, and the rotational motion in this configuration space will be, according to the standard quantum-mechanical rules, quantized just as the standard rotational motion of molecules, with corresponding rotational quantum numbers, etc.[9]

There are also very interesting effects connected with the presence in the "Mexican hat" of Fig. 3.27 of the conical point, or *conical intersection*, which is actually a singularity in the $E(Q_2, Q_3)$ surface. Very rich physics is connected with the existence in this coordinate space of trajectories going around this conical intersection, etc., see for example Yarkony (1996) and Bersuker (2006).

Now, all the states at the minimum of the potential surface of Fig. 3.27 are equivalent in the approximation in which we treated the lattice as harmonic, see (3.21), and included only the lowest-order interaction of electronic and lattice degrees of freedom, cf. (3.21), (3.29). If we go beyond these approximations we should include, for example, the lattice anharmonicity (add terms $\sim u^3$ to the energy (3.21), or terms $\sim Q_3^3$, $Q_3^2 Q_2$, etc.) and we could in principle have higher-order electron–lattice interactions. These contributions lead to the dependence of the total energy on the azimuthal angle θ (see (3.16), (3.18) and Figs 3.21, 3.22), which has the form (Kanamori, 1960; Goodenough, 1963)

$$E_2 \sim \gamma \cos 3\theta \ . \tag{3.30}$$

---

[9] Interestingly enough, due to the presence of the electronic wavefunction $|\psi\rangle$ in the total wavefunction $|\Psi\rangle$ (3.26), when we go around the circle in Fig. 3.21 or the bottom of the potential surface of Fig. 3.27 once, that is when we change the phase by 2π, the total wavefunction changes sign. One can easily see this from eq. (3.16): when we change the angle θ by 2π, the wavefunction (3.16) changes sign. As a result, this rotational quantization gives half-integer quantum numbers, that is it will resemble the case of spin-$\frac{1}{2}$ systems. Similar to those cases there appear here effects connected with geometrical phases, which were actually first discovered (Longuet-Higgins *et al.*, 1958) in the physics of JT systems long before the corresponding effects, known now as *Berry phase* effects (Berry, 1984), were discovered in other fields of physics.

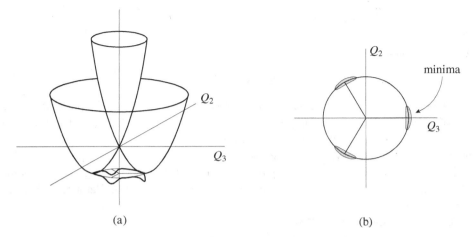

(a)   (b)

Figure 3.28 Warping of the bottom of the "Mexican hat" and the formation of three local minima in the classical approximation, caused by lattice anharmonicity and higher-order Jahn–Teller coupling, eq. (3.30).

In effect, depending on the sign of the coefficient γ in (3.30), there will appear a "warping" of the bottom of the potential trough in Fig. 3.27 (see Fig. 3.28) so that there will appear three minima, for example for $\gamma < 0$ at $\theta = 0, \pm\frac{2}{3}\pi$, with the maxima in between. These minima would correspond to local elongation of MO_6 octahedra along the z-, y-, and x-axes, with corresponding occupation of orbitals $|z^2\rangle$, $|y^2\rangle$, or $|x^2\rangle$ (see Figs 3.21, 3.22), and instead of free rotation the system may be "stuck" in one of these minima. Still, there will exist a tunneling between these minima, similar to that discussed above, and the question of what would be the resulting state was clarified only relatively recently (Koizumi and Bersuker, 1999). The three states corresponding to the three minima, for example $\theta = 0, \pm\frac{2}{3}\pi$, can be shown to split into a singlet state A and a doublet E. There has been controversy in the literature over which state, the nondegenerate singlet or the degenerate doublet, lies lower. Only recently was it shown that both situations are possible: for small anharmonicity, that is low barriers between the three minima, the doublet E is the ground state but for very strong anharmonicity the singlet A will be lower, as originally proposed by Bersuker (1962). It turned out that the conical intersection mentioned above plays a crucial role in determining the type of the resulting state.

For the opposite sign of the constant γ in (3.30) the preferred states would be those with locally compressed octahedra. However, one can give arguments that in typical cases the sign should be such as to give the minima at $\theta = 0, \pm\frac{2}{3}\pi$ (Kanamori, 1960; Khomskii and van den Brink 2000), that is for local elongation. Indeed, the cubic anharmonicity has the form $\zeta u^3/3!$ with $\zeta < 0$, and in effect if we look at the simple case (3.21) with only tetragonal distortions, for the distortion $z \to z + 2u$, $(x, y) \to (x - u, y - u)$ (preserving to volume $\sim xyz$ in the lowest order) the elastic energy would be

$$\sim \frac{B}{2}\left[(\delta x)^2 + (\delta y)^2 + (\delta z)^2\right] - \frac{|\zeta|}{6}\left[(\delta x)^3 + (\delta y)^3 + (\delta z)^3\right], \tag{3.31}$$

so that the change in energy after such a distortion would be

$$\delta E = 3Bu^2 - |\zeta|u^3, \tag{3.32}$$

from which we see that in this case the local elongation ($u > 0$) would be preferable. And indeed, we know hundreds of compounds with strong JT ions (with double degeneracy of e_g type), and practically all of them show local distortions of the elongation type. Only one or two of the known systems are claimed to have locally compressed ligand octahedra. One such example, cited in monographs (Ballhausen, 1962; Goodenough, 1963) as the only exception from this rule for Cu^{2+}, the quasi-two-dimensional insulator K_2CuF_4, was later found still to have a local elongation of octahedra in the x- and y-directions (Fig. 3.29) (Khomskii and Kugel, 1973). This elongation, with long axes alternating in the x- and y-directions, leads to an apparent net contraction of the unit cell in the z (or c)-direction, which was originally taken as a signature of local compression in the c-direction. (As for Cu^{2+} we have not an electron but a hole orbital, the locally elongated octahedra correspond to "cross-shaped" hole orbitals – in this case $|x^2 - z^2\rangle$ and $|y^2 - z^2\rangle$ orbitals.)

All the interesting quantum effects mentioned above are extremely important and actually crucial for small JT systems such as molecules or isolated JT impurities in crystals; they constitute a major part of JT phenomena, which are actively studied in this field. For concentrated systems, with JT ions at every site of the lattice, there may occur a cooperative distortion and a cooperative orbital ordering, which corresponds to a spontaneous breaking of the "rotational" symmetry in the "Mexican hat" of Fig. 3.27. In these cases one usually treats the lattice quasiclassically, ignoring real vibronic effects. This approach will be discussed later in Chapter 6. To what extent the vibronic effects, for which it is crucial to treat the lattice vibrations quantum-mechanically, are relevant for concentrated systems is an open question, definitely deserving further study.

Until now we have mostly considered the JT effect for the case of doubly degenerate e_g orbitals. As is clear from Fig. 3.2, orbital degeneracy can also exist for partially filled triply degenerate t_{2g} levels. Owing to the different shape of t_{2g} orbitals, with weaker π hybridization with neighboring ligands (oxygen, fluorine), the JT coupling with

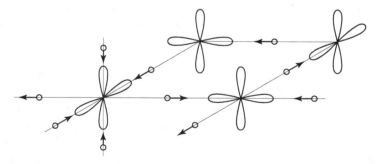

Figure 3.29 Jahn–Teller distortions and hole orbitals for K_2CuF_4, after Khomskii and Kugel (1973).

corresponding lattice distortions for these states is usually (much) weaker than that for e_g-electrons. Still, the JT effect and the corresponding orbital ordering is observed in many compounds with partially filled t_{2g} levels. Specific for these cases is, first, that as discussed in the previous section, the t_{2g} degeneracy is lifted not only by tetragonal and orthorhombic E_g distortions but also by trigonal T_{2g} distortions. Another difference from the e_g case is that for partially filled t_{2g} levels, depending on the number of t_{2g}-electrons, the JT distortion could lead both to local elongation and local compression of MO_6 (MF_6) octahedra. This is already clear from Fig. 3.9: for two or seven d-electrons it is favorable to have a tetragonal elongation $c/a > 1$, with the lowest doublet occupied by two or four electrons, whereas for one or six d-electrons one should rather occupy the singlet state with one electron (for the d^1 configuration) or two electrons (for the d^6 configuration), and the local octahedron would be compressed, $c/a < 1$. (We consider here the TM ions in the high-spin state. We also remember that to preserve the center of gravity of t_{2g} levels the singlet state shifts by E_{JT}, whereas the doublet state shifts by $-\frac{1}{2}E_{JT}$, the sign of E_{JT} being opposite for elongation and contraction, see Fig. 3.9.)

One more complication in the t_{2g} case is that, whereas (as discussed in Section 3.1) for e_g-electrons the orbital moment is quenched, for t_{2g}-electrons it is nonzero and there exists a first-order spin–orbit interaction. As we will see below, this interaction can also lead to occupation of a particular type of orbitals and to corresponding lattice distortions, which turn out to be opposite to those due to the Jahn–Teller effect. These questions will be treated in detail in Sections 3.4 and 6.5.

3.3 High-spin vs low-spin states

As we saw above, the crystal field leads to a splitting of the originally fivefold degenerate d levels, for example into a triplet t_{2g} and a doublet e_g (Fig. 3.2). Correspondingly, the filling of these levels depends on the value of this crystal field splitting $\Delta_{CF} = 10Dq$.

In atoms, according to Hund's rule, the electrons occupy different levels so as to have the largest possible total spin. Consequently, in our case (TM ions in octahedra) the first three electrons will occupy the lowest t_{2g} levels with parallel spins. The problems start with the fourth electron. There is a choice: it can occupy one of the e_g levels, again with the spin parallel to the spins of t_{2g}-electrons (Fig. 3.30(a)). We first make the corresponding estimates using a simple model with Hubbard's repulsion U the same for all orbitals: certain ramifications of this point will be discussed later in this section. The state of Fig. 3.30(a) with the maximum spin (called the high-spin state) is favorable from the point of view of gaining Hund's rule energy: according to our rule formulated above, we gain by this the energy $E_{Hund} = -3J_H$ (the fourth electron with the parallel spin interacts by Hund's rule and gains Hund's energy with the three t_{2g} electrons). However, since this electron now occupies a higher-lying crystal field level e_g, this costs energy Δ_{CF}. There exists another option: the fourth electron can be on one of the t_{2g} levels, but then necessarily with the opposite spin (Fig. 3.30(b)). By that we gain the crystal field energy Δ_{CF}, but lose the Hund's rule energy $-3J_H$. If the crystal field splitting is not too large,

$$\Delta_{CF} < 3J_H \, , \tag{3.33}$$

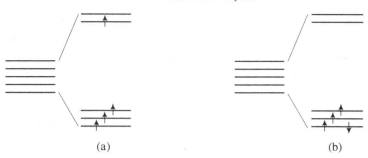

Figure 3.30 (a) High-spin and (b) low-spin states for transition metals with d^4 electron occupation.

then the first state, the one with maximum spin, would be favorable. This is the typical situation for most (but not all) $3d$-ions in TM oxides; such states are called *high-spin (HS) states*. However, if the crystal field splitting in this case is large, $\Delta_{CF} > 3J_H$, it is better to put this fourth electron onto the t_{2g} levels (Fig. 3.30(b)). The resulting spin will be smaller; instead of $S = 2$ in the HS state of Fig. 3.30(a) it will be $S = 1$. Such states are called *low-spin (LS) states*. The crossover from HS to LS states, which would occur with the increase of Δ_{CF}, is actually a change in the multiplet state of the corresponding ion.

LS states are sometimes realized for $3d$ oxides, for example LaCoO$_3$, but they are more typical for $4d$ and $5d$ compounds: owing to the larger extent of $4d$ and $5d$ orbitals, their p–d hybridization and the corresponding crystal field splitting is larger than for $3d$ elements, while the Hund's rule coupling J_H is smaller, see Chapter 2. Whereas for $3d$ elements typically $J_H \sim 0.8$–0.9 eV, for $4d$ it is ~ 0.6–0.7 eV and for $5d$ it is ~ 0.5 eV. Thus, low-spin states are more typical for $4d$ and $5d$ elements. A specific example illustrates this: whereas the $3d$ ion Mn^{3+}(d^4) is usually in the HS state $t_{2g}^3 e_g^1$, $S = 2$ (Fig. 3.30(a)), its $4d$ counterpart Ru^{4+}(d^4) is typically in the LS state $t_{2g}^4 e_g^0$, $S = 1$ (Fig. 3.30(b)).

As mentioned above, low-spin states are also sometimes met in $3d$ compounds, but usually in those containing strong ligands such as NO$_2^-$ or CN$^-$, which give a much larger crystal field splitting, see the spectrochemical series (3.10). This often happens in metallo-organic compounds or in metal clusters, see for example König (1991), Güttlich and Goodwin (2004), or Halcrow (2013). A beautiful example is given by the compound known as "Prussian blue," Fe$_7$(CN)$_{18}$·nH$_2$O, with $n \sim 14$. (The formula for this compound is frequently written Fe$_4^{3+}$[Fe^{2+}(CN)$_6$]$_3$·nH$_2$O.) It was first synthesized in Germany in the 18th century and was one of the first synthetic dyes, with a bright blue color. It has a rather simple crystal structure, shown in Fig. 3.31, with alternating Fe^{2+} and Fe^{3+} ions[10] (and there exist many Prussian blue analogs, e.g. containing Co and Mn, or other combinations

[10] The structure of Prussian blue, Fe$_7$(CN)$_{18}$ · nH$_2$O (Fig. 3.31), resembles that of perovskites ABO_3, with $B =$ Fe, with the A site vacant (or partially occupied by water molecules), and with (CN) groups instead of oxygens at the Fe–Fe bonds. However, there is usually some nonstoichiometry in this system, both in Fe and in CN positions; at certain bonds there are H$_2$O molecules instead of CN.

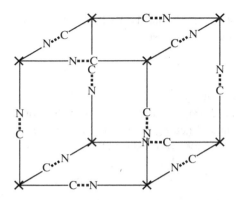

Figure 3.31 Schematic structure of "Prussian blue" containing strong ligands (CN groups). Crosses are Fe ions.

of TM ions). The very strong ligands CN make the state of the corresponding $3d$ ions Fe^{2+}, and in some systems also Fe^{3+}, low-spin.

The relative stability of HS or LS states depends not only on the relative strength of crystal field vs Hund's rule coupling, but also on the number of d-electrons. The criterion (3.33) was written for the particular case of four d-electrons. One can easily study the relative stability of different spin states for other situations, using the same approach. Thus, for example, for ions with configuration d^5 one should compare the energies of the HS configuration $t_{2g}^3 e_g^2$ ($S = \frac{5}{2}$) and the LS one t_{2g}^5 ($S = \frac{1}{2}$). Using the rules formulated above, and taking the energy of t_{2g} levels as zero, we find the energy of the HS state

$$E_{HS}(d^5) = 2\Delta_{CF} - 10J_H \qquad (3.34)$$

(according to our convention, see Section 2.2, the Hund's rule energy is given by the number of pairs of parallel spins, in this case 10). Similarly, the LS state would have the energy

$$E_{LS}(d^5) = -4J_H \qquad (3.35)$$

(three J_H from three up-spins, and one J_H from two down-spins). Comparing eqs (3.34) and (3.35), we see that the HS state is more favorable if

$$2\Delta_{CF} < 6J_H , \qquad \text{i.e.} \qquad \Delta_{CF} < 3J_H , \qquad (3.36)$$

which is the same condition as for four d-electrons (3.33). In the opposite case the LS state would be more favorable.

However, this condition is different for example for six d-electrons (as in Co^{3+} or Fe^{2+}). The HS state in this case would be $t_{2g}^4 e_g^2$ ($S = 2$), and the LS state would be t_{2g}^6 ($S = 0$). Treatment similar to that carried out above gives the energies

$$\begin{aligned} E_{HS}(d^6) &= 2\Delta_{CF} - 10J_H , \\ E_{LS}(d^6) &= -6J_H \end{aligned} \qquad (3.37)$$

(the number of pairs of parallel spins is 10 in the HS state and 6 in the LS state). Correspondingly, the HS state will be stable if

$$2\Delta < 4J_{\mathrm{H}}\,,\qquad \text{i.e.}\qquad \Delta < 2J_{\mathrm{H}}\,,\tag{3.38}$$

and if $\Delta > 2J_{\mathrm{H}}$, the LS state would be preferable. As we see, the condition for having the LS state for d^6 ions, (3.38), is less stringent than that for d^4 or d^5 ((3.33), (3.36)), that is there is more chance of getting the LS states for materials containing for example Co^{3+} or Fe^{2+}: the typical values of crystal field splitting $\Delta_{\mathrm{CF}} \sim 1.5\text{–}2\,\mathrm{eV}$ are already comparable with $2J_{\mathrm{H}} \sim 1.6\text{–}1.8\,\mathrm{eV}$. Indeed, such LS states are realized in the ground state of the perovskite $LaCoO_3$: in this material Co^{3+} is in the low-spin state t_{2g}^6 and is in fact nonmagnetic, $S = 0$. However the magnetic states, for example the HS state $t_{2g}^4 e_g^2$ with $S = 2$, lie close to the nonmagnetic LS state, and they start to become populated with increasing temperature, which leads to a rapid increase in magnetic susceptibility, see the qualitative picture of Fig. 3.32.

The case of six d-electrons illustrates yet another possibility. Besides the HS and LS states, here an intermediate electron occupation is possible: that with $t_{2g}^5 e_g^1$ ($S = 1$). This so-called *intermediate spin (IS) state* was proposed for $LaCoO_3$ by Abbate *et al.* (1963) and by Korotin *et al.* (1996), and some experiments were interpreted as confirming that it is indeed the IS state that becomes thermally populated in $LaCoO_3$ with increasing temperature. The question of which state is actually occupied in $LaCoO_3$ at $T \neq 0$ is still a matter of debate. On the level of an isolated ion, using the estimate of the energy of different spin states carried above (depending only on the ratio of Δ_{CF} and J_{H}), one can easily show that the IS state is never realized: either the HS (for $\Delta_{\mathrm{CF}} < 2J_{\mathrm{H}}$) or the LS (for $\Delta_{\mathrm{CF}} > 2J_{\mathrm{H}}$) state always has the lowest energy. But one could hope that the IS state would become

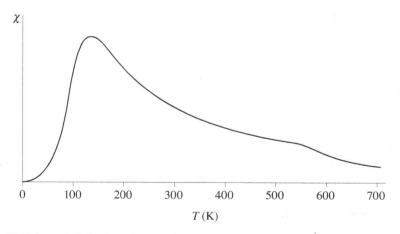

Figure 3.32 Schematic behavior of magnetic susceptibility of $LaCoO_3$ with nonmagnetic ground state (low-spin Co^{3+}), with magnetic states of Co^{3+} lying close in energy.

lower in concentrated systems, when other factors start to contribute, such as hybridization to neighboring sites and band formation. Still, even with these effects included, it seems that it is very difficult to stabilize the IS state of Co^{3+} in $LaCoO_3$, but this state could be realized in doped systems such as $La_{1-x}Sr_xCoO_3$.

In all numerical estimates above in this section we used a simplifying assumption of the constant Hubbard's repulsion U, the same for all orbitals, and calculated the contribution of the Hund's exchange J_H using our recipe (counting the number of pairs of parallel spins). This treatment, though giving qualitatively correct conclusions, may not be quite accurate numerically, in particular as to the coefficients with which the Hund's exchange J_H enters different expressions, see the remark after eqs (2.6), (2.7). If we were to use, instead, the relation (2.7) for U_{mn}, then for example the criteria (3.33) and (3.36) would change to $\Delta_{CF} < 5J_H$, and (3.38) to $\Delta_{CF} < 4J_H$. In a more sophisticated treatment, which should take into account not only U and J_H but also all Slater–Koster or Racah parameters, see for example Ballhausen (1962), the results would lie somewhere in between. The numerical calculations indeed give for the LS–HS transition for the configuration d^4 the criterion $\Delta_{CF} \leqslant 3.69J_H$, and for the configuration d^6 the criterion $\Delta_{CF} \leqslant 2.57J_H$ (R. Green, unpublished). These values lie closer to our original estimates (3.33), (3.38) than those which would follow from the "atomic" relation (2.7), $U_{mn} = U_{nn} - 2J_H$ (known as the Kanamori approximation). Thus our simplified treatment, which assumed one average value of U for all orbitals, turns out to be closer to reality than the seemingly more sophisticated and more accurate Kanamori approximation. We will often use this approximation below. One should, however, always be aware of its limitations, especially as to the exact numerical coefficients with which J_H enters different expressions.

Which particular state of an isolated TM ion in a crystal is stable under what conditions is usually determined from the Tanabe–Sugano diagrams, widely used in spectroscopy (Tanabe and Sugano, 1954, 1956; see also Sugano *et al.*, 1970). An example of such diagrams is shown in Fig. 3.33 for the configuration d^5. On the horizontal axis Δ_{CF} is plotted, normalized to the Racah parameter B (Griffith, 1971). On the vertical axis the energy of different terms, also normalized to B, is shown, relative to the ground state (which is always taken as zero energy). We see that for a certain critical value of Δ_{CF} (in our simplified model it is given by (3.36)) the HS state ($t_{2g}^3 e_g^2$, $S = \frac{5}{2}$, term $^6A_1$ in Fig. 3.33) changes to the LS state ($t_{2g}^5 e_g^0$, $S = \frac{1}{2}$, term $^2T_2$). The IS state ($t_{2g}^4 e_g^1$, $S = \frac{3}{2}$), which can in principle also exist in this case, corresponds to the term $^4T_1$ in Fig. 3.33 and in this approximation always lies higher than the HS and LS states. Similarly, for the configuration d^6 (Co^{3+}) (Fig. 3.34) the HS ground state ($t_{2g}^4 e_g^2$, $S = 2$, term $^5T_2$ in Fig. 3.34) changes for larger Δ_{CF} to the LS state (t_{2g}^6, $S = 0$, term $^1A_1$), the IS state ($t_{2g}^5 e_g^1$, $S = 1$, term $^3T_1$) lying higher.[11]

There is yet another important ingredient in the high-spin/low-spin story. It turns out that the ionic radii (i.e., the sizes) of HS and LS states of the same ion, for example Co^{3+},

---

[11] Here in Figs 3.33, 3.34 we present only a small part of the general Tanabe–Sugano diagrams, to illustrate how they work. The complete set of these diagrams, for different ionic configurations, can be found in the original publications (Tanabe and Sugano, 1954, 1956) and in monographs (Sugano *et al.*, 1970; Griffith, 1971).

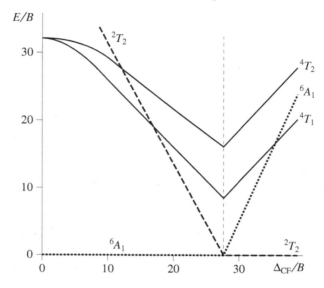

Figure 3.33 Part of a Tanabe–Sugano diagram for the configuration d^5.

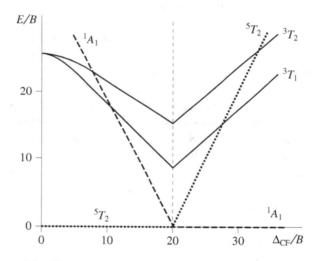

Figure 3.34 Part of a Tanabe–Sugano diagram for the configuration d^6.

are strongly different. The LS ions are always smaller than their HS counterparts, and this difference can reach up to $\sim 15\%$ (according to the tables of ionic radii (Shannon, 1976), the ionic radius of the 6-coordinated HS Co^{3+} is 0.61 Å, and that of its LS counterpart is 0.545 Å). Thus when we transform, for example, HS Co^{3+} into an LS form, we could expect the lattice to shrink significantly. In contrast, if we take a material with HS Co^{2+}

and apply pressure to it, we could shift the equilibrium in the direction of LS. Such effects are indeed observed experimentally, see for example Lengsdorf *et al.* (2004).

For isolated TM ions the competition between HS and LS states is determined by parameters such as Δ_{CF} and J_H, and there may exist at most a thermal occupation of the higher-lying state. In concentrated systems, however, there would exist certain interactions between different ions, with their spin states, and because of that there may occur cooperative phenomena, such as phase transitions with a change of spin states. Such spin-state transitions are indeed observed in a number of materials containing Co^{3+} and Fe^{2+}; they will be discussed in more detail in Sections 5.9 and 9.5.2.

3.4 Role of spin–orbit coupling

As we saw in Chapter 2, spin–orbit coupling $\lambda \boldsymbol{L} \cdot \boldsymbol{S}$ plays a very important role in atoms, largely determining the multiplet structure. When TM ions are in a crystal, the crystal field splitting breaks rotational invariance and, strictly speaking, $\boldsymbol{L}$ and $\boldsymbol{J}$ are no longer good quantum numbers. Nevertheless, spin–orbit coupling can still play a very important role in solids. Its manifestation depends on the ratio between this coupling (and the level splitting produced by it) and the crystal field splitting. If the crystal field splitting is smaller than the spin–orbit coupling, one can still use the atomic nomenclature, including crystal field as a perturbation. However in most cases, at least in the $3d$ series and in most $4d$ compounds, the situation is the opposite and the main crystal field splitting into t_{2g} and e_g levels, Δ_{CF}, is (much) bigger than the spin–orbit coupling. Typically, for $3d$ elements the value of Δ_{CF} in oxides or fluorides is ~ 1.5–$2\,eV$, whereas the spin–orbit coupling is much smaller: thus for Ti $\lambda \sim 20\,meV$, and for heavier Co it is $\sim 70\,meV$. In this case we first have to look at the crystal field levels, and then consider the action of spin–orbit coupling on them. The situation may be different for heavier elements though, especially $5d$ ones.

One has to be somewhat more specific in this general classification. Whereas the cubic crystal field splitting Δ_{CF} is indeed almost always much bigger than λ, the possible further splitting of d levels due to eventual deviations from the cubic symmetry, such as the tetragonal or trigonal splitting shown for example in Figs 3.9, 3.11 (which in particular can be caused by the Jahn–Teller effect), can already be comparable with the splitting due to spin–orbit coupling, even in 3d systems. Usually the JT distortion and that caused by spin–orbit interaction are of opposite sign, and the system can evolve according to either the JT or the spin–orbit "scenario." We will see examples of this later.

Let us now consider the effects of spin–orbit coupling in more detail. As we saw in Section 3.1, fivefold degenerate d levels ($l^z = \pm 2, \pm 1, 0$) are split in the crystal field, typically into a triplet t_{2g} and a doublet e_g, see Fig. 3.2. We saw in that section that the e_g levels are composed of the states $|l^z = 2\rangle$, $|l^z = -2\rangle$, and $|l^z = 0\rangle$, so that $\langle e_g|l|e_g\rangle = 0$, that is the orbital moment is frozen, or quenched in these states. Correspondingly, spin–orbit coupling $\lambda \boldsymbol{l} \cdot \boldsymbol{S}$ does not act on these states in the first order; only at higher order may there appear some influence of spin–orbit coupling on these states. These higher-order effects may still be rather important for some phenomena: for example, they determine the single-site magnetic anisotropy of respective ions, or the anisotropy of g-factors in ESR. Thus, the well-known Jahn–Teller ion Cu^{2+} with configuration d^9 (one hole on the e_g level) usually sits in a strongly distorted ligand octahedron, or even in a fivefold (pyramid) or fourfold (square) coordination, and typically has a strongly anisotropic ESR signal, with $g_\parallel \sim 2.2$ and $g_\perp \sim 2.08$. These deviations from the pure electron value $g = 2$ are due to the contribution of spin–orbit interaction in the second order (Abragam and Bleaney, 1970), see also Section 5.3 below.

The role of spin–orbit interaction is much more important for partially filled t_{2g} levels. As one sees from (3.4), the orbital moment is nonzero for the t_{2g} states and these linear combinations of t_{2g} states correspond to eigenstates of the effective orbital moment $\tilde{l} = 1$. Correspondingly, spin–orbit coupling acts on these states in the first order and can lead to a splitting of these levels.

It is convenient to express spin–orbit interaction on t_{2g} levels through this effective moment $\tilde{l} = 1$, so that it takes the form

$$H_{SO}(t_{2g}) = \tilde{\lambda}\tilde{l} \cdot S, \tag{3.39}$$

where $\tilde{l}$ is the effective orbital moment for the t_{2g} triplet, $\tilde{l} = 1$. One can show (Abragam and Bleaney, 1970) that the coupling constant $\tilde{\lambda}$ is proportional in magnitude to the original spin–orbit coupling constant λ, but has the opposite sign (e.g., $\tilde{\lambda} = -\lambda$ for Fe^{2+} and $\tilde{\lambda} = -\frac{2}{3}\lambda$ for Co^{2+}). Correspondingly, as already mentioned in Section 3.1, the second Hund's rule *for the t_{2g}-subshell* changes to the opposite: for a less-than-half-filled t_{2g}-shell the state with maximum possible degeneracy, or maximum total moment $\tilde{J}$, has the lowest energy[12] (inverted multiplets), whereas for the more-than-half-filled case the states with smaller $\tilde{J}$ lie lower. This has important implications, some of which are rather nontrivial and at first glance unexpected.

A few examples: for one t_{2g}-electron $\tilde{l} = 1$, $S = \frac{1}{2}$, that is the total degeneracy is $(2\tilde{l} + 1)(2S + 1) = 3 \times 2 = 6$. These six levels are split by spin–orbit coupling into a quartet $\tilde{J} = \tilde{l} + S = \frac{3}{2}$ and a doublet $\tilde{J} = \tilde{l} - S = \frac{1}{2}$, the quartet $\tilde{J} = \frac{3}{2}$ lying lower. Similarly, for two t_{2g}-electrons $\tilde{l} = 1$, $S = 1$, $(2\tilde{l} + 1)(2S + 1) = 9$, and the ninefold degenerate states are split into a quintet $\tilde{J} = 2$, a triplet $\tilde{J} = 1$, and a singlet $\tilde{J} = 0$, and the quintet is the ground state. However, if we consider the case of five t_{2g}-electrons, that is one t_{2g}-hole (more-than-half-filled t_{2g}-shell), we would again have, as for the configuration t_{2g}^1, a quartet $\tilde{J} = \frac{3}{2}$ and a doublet $\tilde{J} = \frac{1}{2}$, but the doublet will lie lower. This is apparently what happens in compounds containing $Ir^{4+}(t_{2g}^5)$, for example Sr_2IrO_4 (Kim *et al.*, 2008; Jackeli and Khaliullin, 2009), see below.

This is especially important for the high-spin state of Co^{3+}, see Section 3.3. It has configuration $t_{2g}^4 e_g^2$, that is the t_{2g} levels are more-than-half-filled: it has two t_{2g}-holes instead

---

[12] One often still speaks about the effective total moment $\tilde{J}$, and we will also use this terminology, though one has to be aware that strictly speaking the classification in terms of $\tilde{J}$ is not valid, since for ions in a crystal field $\tilde{J}$ is not a good quantum number. However, the multiplicities of the resulting levels are the same as those obtained using $\tilde{J}$.

of two t_{2g}-electrons as in the example above. The effective orbital moment is $\tilde{l} = 1$, but now the spin is 2 (the two e_g-electrons also contribute to the total S). In effect the total moment is $\tilde{J} = 3; 2; 1$, and according to the "reversed" second Hund's rule, the triplet $\tilde{J} = 1$ would be the lowest-lying state for this more-than-half-filled t_{2g}-shell. Thus if we forget about spin–orbit coupling, we could think that the ground state of HS Co^{3+}, with $S = 2$, would be a quintet; however in fact it is a triplet. This is very important for the interpretation of spin-state transition in LaCoO$_3$ and in similar compounds: from the fitting of magnetic and thermodynamic data one concludes (Zobel *et al.*, 2002) that (with the low-spin singlet ground state) the first excited state is a triplet, from which one often deduces that this excited state is an intermediate-spin Co^{3+}($t_{2g}^5 e_g^1$) with $S = 1$. But we see that in fact the triplet state is the lowest state of the high-spin, not the intermediate-spin Co^{3+}! We will return to this point later, in Sections 5.9 and 9.5.2.

Similarly, for Co^{2+}($t_{2g}^5 e_g^2$) (more-than-half-filled t_{2g} levels) we have $\tilde{l} = 1$, $S = \frac{3}{2}$; the moment $\tilde{J}$ can be $\frac{5}{2}$, $\frac{3}{2}$, and $\frac{1}{2}$, and the doublet $\tilde{J} = \frac{1}{2}$ is the ground state. Thus these ions behave as those with effective spin $S_{\text{eff}} = \frac{1}{2}$ (doublet ground state), but of course with many features, for example anisotropy, different from the case of real spins $\frac{1}{2}$. Such description, in terms of an effective spin S_{eff}, is often used, especially in the treatment of resonance phenomena of TM ions, such as ESR, and it is known as the description in terms of *spin Hamiltonian* (Abragam and Bleaney, 1970). The reduction to the effective spin model (e.g., with effective spin $\frac{1}{2}$) with anisotropic interactions becomes especially important for heavier TM ions, such as low-spin Ir^{4+}($t_{2g}^5 e_g^0$) mentioned above.

As already discussed, when one describes TM ions with partial filling of the t_{2g}-shell, with unquenched orbital moment, by the effective orbital moment $\tilde{l} = 1$, the sign of spin–orbit coupling, written through $\tilde{l}$, is opposite to the normal case and the g-factor for the effective orbital moment, $g_{\tilde{l}}$, is $g_{\tilde{l}} < 0$ instead of the usual value $g_l = 1$ for the real orbital moment. Because of that, the total magnetic moment $M = \mu_B g_{\tilde{J}} \tilde{J}$, which determines the interaction with external magnetic field H, and the corresponding Zeeman splitting

$$\mathcal{H}_Z = -MH = -\mu_B g_{\tilde{J}} \tilde{J} H, \tag{3.40}$$

may be different from what one could naively expect. (Remember that the effective moment μ_{eff} entering the Curie–Weiss law for magnetic susceptibility is $\mu_{\text{eff}} = g_{\tilde{J}} \mu_B \sqrt{\tilde{J}(\tilde{J}+1)}$.) The expression for the effective g-factor $g_{\tilde{J}}$ is (see, e.g., Abragam and Bleaney, 1970)

$$g_{\tilde{J}} = \frac{1}{2}(g_{\tilde{l}} + g_S) + \frac{\tilde{l}(\tilde{l}+1) - S(S+1)}{2\tilde{J}(\tilde{J}+1)}(g_{\tilde{l}} - g_S) \tag{3.41}$$

(here $\tilde{l} = 1$ and $g_S = 2$), see (2.13). For example, for Co^{2+} we have $g_{\tilde{l}} = -\frac{2}{3}$, and in the ground state doublet with $\tilde{J} = \frac{1}{2}$ the g-factor is $g_{\tilde{J}=1/2}(\text{Co}^{2+}) = -\frac{3}{2}g_{\tilde{l}} + \frac{5}{3}g_S = 4.33$, so that the effective moment of Co^{2+}, which determines the Zeeman interaction (3.40),

what about f electrons?

would be $M = g_{\tilde{j}}\mu_B\tilde{J} \sim 2.17\mu_B$, that is it would be reduced compared with the pure-spin moment value $g_S\mu_B S = 2\mu_B \cdot \frac{3}{2} = 3\mu_B$.

An unusual situation arises if we consider the seemingly simpler case of a TM ion with only one d-electron on t_{2g} levels (Ti^{3+}, V^{4+}) in an ideal octahedral coordination; the result turns out to be rather unexpected. This single t_{2g}-electron with spin $\frac{1}{2}$ can again be described by the effective orbital moment $\tilde{l} = 1$ with negative spin–orbit coupling $\tilde{\lambda} = -1$, and with the effective orbital g-factor $g_{\tilde{l}} = -1$. According to the rule formulated above, this ion with less-than-half-filled t_{2g}-shell will have inverted multiplet sequence, that is the ground state would be the quartet $\tilde{J} = \frac{3}{2}$. But then, according to (3.41), the effective $g_{\tilde{j}}$-factor of this apparently magnetic ion ($\tilde{l} = 1$, $S = \frac{1}{2}$) turns out to be zero, $g_{\tilde{j}} = 0$! (One can qualitatively explain this by saying that in this case the orbital moment $\tilde{l} = 1$ points opposite to the spin $S = \frac{1}{2}$ but, as the orbital g-factor is $|g_l| = 1$ and the spin g-factor is $g_S = 2$, the total magnetic moment is zero, $g_s S + g_{\tilde{l}}\tilde{l} = 2S - \tilde{l} = 0$. This, however, is only a crude qualitative picture; to actually prove this result one has to go to the full description such as given by eq. (3.41).)

This exact cancelation of spin and orbital magnetic moments is valid only for the ideal cubic symmetry; even small distortions, tetragonal or trigonal, would destroy the cancelation. Depending on distortion, the effective g-factor here can take values between $+2$ and -4 (Abragam and Bleaney, 1970); the g-factor also becomes anisotropic. It is not actually clear whether these features of d^1 ions such as Ti^{3+} or V^{4+}, especially the possibility of having zero g-factor, can play some role in bulk solids containing such ions; usually the local symmetry is lower than cubic.

A rather nontrivial situation can exist in $4d$ or $5d$ systems, for which the spin–orbit coupling is strong (it can be ~ 0.5 eV, comparable with electron hopping or crystal field splitting due to reduced cubic symmetry). Thus, for example, for low-spin $Ir^{4+}(t_{2g}^5)$ we will have $S = \frac{1}{2}$, $\tilde{l} = 1$ and, according to the general rules, for this more-than-half-filled t_{2g}-subshell the ground state will be the doublet $\tilde{J} = \frac{1}{2}$. Thus the corresponding materials, with many-electron Ir^{4+}-ions, can be mapped onto a nondegenerate Hubbard-like model with effective spin $\frac{1}{2}$ (but with quite nontrivial hopping matrix elements – see, e.g., Shitade *et al.*, 2009).

We meet an even more "exotic" situation in the case of low-spin $Ru^{4+}(d^4)$. This ion has configuration t_{2g}^4, that is a more-than-half-filled t_{2g}-shell, with $\tilde{l} = 1$ and $S = 1$. Thus for this more-than-half-filled t_{2g}-shell from the possible multiplets with $\tilde{J} = 2, 1, 0$ the lowest state would be the nonmagnetic singlet $\tilde{J} = 0$! Thus in this case the apparently magnetic state of Ru^{4+} with $\tilde{l} = 1$ and $S = 1$ can effectively become nonmagnetic – this time not because of zero g-factor, as in the case of d^1-ions considered above, but because the total moment itself is zero in the ground state, $\tilde{J} = 0$. This effect could be even more important for $5d$ ions with the configuration d^4, such as Ir^{5+}, for which the spin–orbit coupling and the corresponding stabilization of the $J = 0$ ground state are much stronger.

Experimentally, however, most materials containing Ru^{4+} (and even some with Ir^{5+}) are magnetic, or are metals with (exchange-enhanced) Pauli paramagnetism. Why does

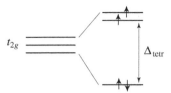

Figure 3.35 The Jahn–Teller splitting of t_{2g} levels for the configuration d^4.

this happen? First of all, in many such materials the exact crystal field symmetry at Ru sites is lower than cubic. In this case there may exist a further splitting of the t_{2g} levels of Ru, which may at least partially quench the orbital moment. Thus if we have a tetragonal distortion, for example contraction, the t_{2g} levels would be split as shown in Fig. 3.35, and we see that the lowest state d_{xy} (or $|\tilde{l}^z = 0\rangle$) would be doubly occupied and the higher-lying states d_{xz}, d_{yz} (or $|\tilde{l}^z = \pm 1\rangle = \frac{1}{\sqrt{2}}(|xz\rangle \pm i|yz\rangle)$) would both be singly occupied, so that the total orbital moment would be zero. Correspondingly, this state would have spin $S = 1$ and would be magnetic (if the extra tetragonal splitting Δ_{tetr} is larger than the spin–orbit coupling). Similarly, if in a concentrated system there exists exchange interaction with neighboring Ru ions and if it is stronger than the spin–orbit coupling, it will also stabilize the magnetic (and magnetically ordered) state. Nevertheless, even in these cases the nonmagnetic Ru^{4+} state with $\tilde{J} = 0$ can lie relatively low in energy, and this state should be taken into account in interpreting the results of different experiments on respective compounds. The systems in which there is magnetic ordering, despite the nominally nonmagnetic ground state of constituent ions, are called systems with *singlet magnetism*, see Section 5.5 below.

In application to the bulk TM compounds, one often uses a mean-field approximation for the treatment of spin–orbit coupling, replacing the interaction $\lambda \boldsymbol{l} \cdot \boldsymbol{S}$ by $\lambda l^z S^z$. In this case, in a magnetically ordered state the orbital moment orients parallel (or antiparallel) to the spin, which leads to the occupation of a particular d orbital and to a corresponding lattice distortion, which appears below the critical temperature of magnetic ordering, T_c or T_N, and which is in fact a form of magnetostriction. Thus, for example, for one electron on the t_{2g} level (Fig. 3.36) with the interaction $\tilde{\lambda} \tilde{\boldsymbol{l}} \cdot \boldsymbol{S}$ ($\tilde{\lambda} < 0$) for spin $S^z = +\frac{1}{2}$ the orbital moment will be $l^z = -1$. The corresponding orbital is

$$|l^z = -1\rangle = \frac{1}{\sqrt{2}}\left(|xz\rangle - i|yz\rangle\right), \tag{3.42}$$

and it has the shape shown in Fig. 3.4, that is it looks like the $|xz\rangle$ orbital rotated around the z-axis: its electron density has the form of a hollow cone (the states with $l^z = +1$ and $l^z = -1$ differ in the sense of rotation, clockwise or anticlockwise, and have the same distribution of electron density). This electronic cloud would experience strong Coulomb repulsion from apical oxygens along the vertical axis, and this would lead to an elongation of the oxygen octahedra. The resulting level structure would look as shown in Fig. 3.36(a). One can see that this distortion and splitting are just opposite to what one would expect

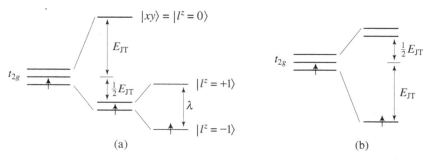

Figure 3.36 Splitting of t_{2g} levels: (a) due to tetragonal elongation and spin–orbit interaction (treated in the mean-field approximation) and (b) due to tetragonal contraction (the orbital moment is quenched in this case).

from the Jahn–Teller effect, Fig. 3.36(b). Indeed, for one t_{2g}-electron the Jahn–Teller effect would lead not to a tetragonal elongation but to a compression: the occupied level in Fig. 3.36(b) would decrease by $-E_{JT}$, whereas in Fig. 3.36(a) this shift is only $-\frac{1}{2}E_{JT}$. However in the case of Fig. 3.36(a) there is an additional splitting due to spin–orbit coupling $\lambda l^z S^z$. In effect the energy of the state of Fig. 3.36(b), favored by the Jahn–Teller interaction, would be

$$E_{(JT)} = -E_{JT} \tag{3.43}$$

and the energy of the state of Fig. 3.36(a), stabilized by the spin–orbit interaction, is

$$E_{(SO)} = -\tfrac{1}{2}E_{JT} - \tfrac{1}{2}\lambda . \tag{3.44}$$

We see from these expressions that when $E_{JT} > \lambda$, the system would evolve according to the "JT scenario," the singlet level $|xy\rangle = |l^z = 0\rangle$ would lie lower, the orbital moment in this state would be quenched, and correspondingly the local distortion of the MO_6 octahedron would be a tetragonal compression. If, however, the spin–orbit interaction is strong enough, $\lambda > E_{JT}$, the spin–orbit coupling would be more important, and the distortion would be the opposite: it would be a tetragonal elongation which would move down the states $|xz\rangle$, $|yz\rangle$, or $|l^z = \pm 1\rangle = \frac{1}{\sqrt{2}}(|xz\rangle \pm i|yz\rangle)$, as in Fig. 3.36(a). We gain less JT energy in this process, but in the resulting state the orbital moment is nonzero, and the spin–orbit coupling would split this doublet still further, decreasing the energy.

Experimentally, both situations are encountered in practice. Thus at the beginning of the 3d series (compounds of Ti, V) the spin–orbit coupling is weak, and the JT effect usually dominates. After JT splitting the orbital moment and spin–orbit coupling effectively disappear (or at least they do not work in the first order). However, for heavier 3d metals the spin–orbit coupling strongly increases (remember that the spin–orbit coupling constant λ is proportional to Z^4 (or Z^2, see footnote 5 on p. 34), where Z is the atomic number of the element), and for example for Co $\lambda \sim 70\,\text{meV}$, whereas it is only $\sim 20\,\text{meV}$ for Ti. Correspondingly, for heavier 3d elements usually the spin–orbit coupling dominates, and

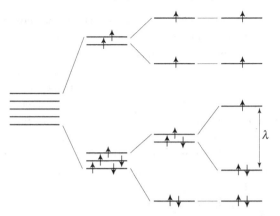

Figure 3.37 Typical crystal field and spin–orbit splitting (treated in a mean-field way) of d levels for ions with configuration d^7 (e.g., Co^{2+}), for tetragonally compressed MO_6 octahedra.

in case of partially filled t_{2g} levels the material distorts according to the "spin–orbit scenario." As we have discussed above, this ordering of orbital moments with corresponding occupation of orbitals of the type of Fig. 3.4, and with corresponding lattice distortion, occurs in magnetically ordered states below T_c or T_N, that is it takes the form of a "giant magnetostriction." Thus, for example, many oxides or fluorides containing Co^{2+} with configuration d^7, that is with electron occupation $t_{2g}^5 e_g^2$, experience a tetragonal compression below T_N, after which the levels split as shown in Fig. 3.37 (i.e., we have three electrons or one hole on the doubly degenerate orbital $|l^z = \pm 1\rangle$). These two orbitals are split further below T_N, which in fact stabilizes this distortion. In effect, for example in CoO below T_N there appears a strong tetragonal compression – so strong that the transition becomes weak I order, with a jump in lattice constants. Accordingly, the spins are oriented predominantly in the z-direction (with some weak canting away from this direction), that is we have a strong easy-axis anisotropy with the easy axis along the [001] direction. This is, in fact, a rather general feature: materials which in a symmetric situation have partially filled t_{2g} orbitals with strong enough spin–orbit coupling, have very strong magnetoelastic coupling and large magnetostriction, and also strong uniaxial anisotropy. These factors are intrinsically connected to the unquenched orbital moment for partially filled t_{2g} levels. If however these levels are half-filled as in $Cr^{3+} (t_{2g}^3)$ or $Mn^{2+} (t_{2g}^3 e_g^2)$ and $Fe^{3+} (t_{2g}^3 e_g^2)$, or completely occupied as in $Ni^{2+} (t_{2g}^6 e_g^2)$ or $Cu^{2+} (t_{2g}^6 e_g^3)$, the corresponding effects are much weaker.

One should mention yet another fact. Until now we mostly spoke about the competition between Jahn–Teller and spin–orbit interactions, and about the interplay between the spin–orbit coupling and tetragonal distortion. However, as mentioned in Section 3.1, t_{2g} orbitals can also be split by trigonal distortions, see Fig. 3.11. Accordingly, all effects discussed above can also occur for this distortion. In effect, for partially filled t_{2g} levels one can have a JT effect with trigonal distortion or, in case of strong spin–orbit coupling, there may exist a trigonal distortion of the opposite sign, accompanied by spin and orbital ordering. But in this case the orbital quantization axes should be chosen differently – along the trigonal axes, or one of the [111] axes in a cubic setting (Fig. 3.12). As shown in Fig. 3.11(b), the corresponding crystal field splitting is in principle similar to that of tetragonal distortion, that is the t_{2g} levels are split into a singlet and a doublet (remember that such distortions do not split the e_g levels). As mentioned in Section 3.1, the singlet a_{1g} has a quenched orbital moment $|l^{z'} = 0\rangle$, where the quantization axis z' is the [111] axis in Fig. 3.11(a). The remaining two states are the eigenstates of this moment with $|l^{z'} = \pm 1\rangle$, their electron density has the form of a "torus" as shown in Fig. 3.14. Thus for trigonal splitting the orbital moment $l^{z'} = \pm 1$ can be oriented in the direction [111] (in a cubic setting) and, accordingly, the spins should also be oriented in this direction. This is the case for many compounds containing Fe^{2+} ($3d^6$, high spin), with configuration $t_{2g}^4 e_g^2$: for example FeO and $KFeF_3$ both have easy axis in the [111] direction, and both have trigonal distortion below T_N. Why Co^{2+} compounds (with one hole on t_{2g} levels) usually have tetragonal distortions, whereas similar compounds with Fe^{2+} (with two t_{2g}-holes or one t_{2g}-electron on top of the half-filled shell t_{2g}^3) typically have trigonal distortions is not actually clear (this is not always the case, but apparently typical for these compounds).

3.5 Some general principles of the formation of typical crystal structures of transition metal compounds

In what follows we will mostly deal with bulk TM compounds, for example TM oxides. These may have quite diverse crystal structures, which strongly influence all their properties. Therefore it makes sense to describe some of the general principles guiding the formation of one or the other type of crystal structure.

This field is quite well developed, and there are many books and reviews devoted to this subject, for example Galasso (1970), Hyde and Andersson (1989), Rao and Raveau (1998). We will not be able to discuss this topic in detail, but will summarize briefly the most important general principles governing the build-up of TM compounds.

When discussing the formation of solids from ions, one usually speaks of four main types of bonding: ionic, covalent, metallic, and van der Waals bonding. Real metallic bonding is rarely met in TM compounds such as oxides or halogenides, which are the main subject of this book. Van der Waals bonding is sometimes encountered, in particular in some layered compounds; it is quite weak and does not strongly influence the electronic structure or magnetic properties of such compounds.

The most typical for our systems is ionic bonding, or some combination of ionic and covalent bonding; the covalent contribution becomes stronger for example in sulfides or selenides, compared with oxides. In the first approximation, it is predominantly ionic bonding that, in combination with the sizes of corresponding ions, determines the build-up of one or another type of crystal for different compounds.

Considering the size of the respective ions is extremely important; often size seems to play a main role. Very often one can get a pretty good description of the structural features of particular compounds on the basis of the sizes of constituent ions. These are usually characterized by the ionic radii, R, which are of course different for individual elements, as well as for different ionization levels of a given element. But they can also depend on the spin state of the corresponding ion, as was the case for Co^{3+}, see Section 3.3. They are also different for different coordinations. Thus, for example, the ionic radius of Fe^{3+} in oxygen octahedra is 0.645 Å, while the ionic radius of the same ion in tetrahedral coordination is 0.40 Å. Ionic radii for some ions, in different coordinations, are presented in the famous tables of Shannon (1976); these values are widely used nowadays.

We want to stress one important point right away. There is a rather widespread misunderstanding, especially among physicists. When we consider TM oxides, we usually pay most attention to TM ions: in the end it is mainly their electronic structure which determines the main properties of these compounds. Therefore, when one presents the crystal structures of these systems, one almost always draws these structures keeping only the TM ions and often showing them as large spheres. The oxygen ions are either omitted entirely, or marked by drawing the corresponding oxygen polyhedra, with the oxygen ions themselves shown at best as small circles. For many purposes this is enough, and we will also often do the same in this book. But when discussing the actual crystal structures one has to realize that in most cases the situation is just the opposite: the oxygen ions O^{2-} are *much bigger* than most TM ions. The ionic radius of O^{2-} is, depending on coordination, ~ 1.35–1.42 Å, whereas the typical values for TM ions are less than half that size, often ~ 0.5–0.7 Å for $3d$ elements. Some of the typical values of ionic radii (in Å) of several important ions forming materials which we will discuss in this book are presented in Table 3.1; the data is taken from Shannon (1976), where one can also find data for other elements. In this table we present simplified information, mainly to give the reader a general feeling for the situation. (The roman numbers in brackets in this table denote the coordination number.) For TM ions we mostly show the radii for octahedral coordination (denoted VI) and sometimes tetrahedral (IV); for some other metal ions such as Ba^{2+} or La^{3+} we also show the radii for coordination XII, typical for these elements in perovskites such as $BaTiO_3$ or $LaMnO_3$.

We see from Table 3.1 that, first of all, indeed $3d$ ions are much smaller than typical anions such as O^{2-} or F^-, Cl^-. Then, the size of the respective ions (with the same valence) decreases from left to right in rows of the Periodic Table (see the Periodic Table at the end of the book), and increases from top to bottom in columns. As mentioned above, for rare earth elements (La, ..., Lu) this is known as lanthanoid contraction.

The "construction" of many crystal structures of TM compounds such as oxides or fluorides can be understood, in a first approximation, as a close packing of large ions, with small ones occupying the interstices, the "cavities" between them. Consider for example binary compounds such as $(TM)_m O_n$. First we form a close packing of large O^{2-} "spheres." In a plane this is just hexagonal packing, see Fig. 3.38. In the ab-plane these ions form a triangular lattice, Fig. 3.38(b) (in order to make this figure not too crowded, we have shown the ions of this (first) layer by small spheres, not by large ones touching one another as they almost do in reality).

Table 3.1 *Ionic radii of some important ions in Å, from Shannon (1976)*

Element	Ionic radius	Element	Ionic radius (coordination VI)
O^{2-}	1.35–1.42	Ti^{3+}	0.67
F^-	1.28–1.33	Ti^{4+}	0.60
Cl^-	1.81	V^{3+}	0.64
Br^-	1.96	V^{4+}	0.58
I^-	2.20	Cr^{3+}	0.615
Mg^{2+}	0.57 (IV); 0.72 (VI)	Mn^{2+}	0.83
Zn^{2+}	0.60 (IV); 0.74 (VI)	Mn^{3+}	0.645
Cd^{2+}	0.78 (IV); 1.31 (XII)	Mn^{4+}	0.53
Ca^{2+}	1.34 (XII)	Fe^{2+}: LS / HS	0.61 / 0.78
Sr^{2+}	1.44 (XII)	Fe^{3+} HS	0.645
Ba^{2+}	1.61 (XII)	Co^{2+}	0.745
La^{3+}	1.36 (XII)	Co^{3+}: LS / HS	0.545 / 0.61
Gd^{3+}	~ 0.95–1.1	Co^{4+}	0.53
Lu^{3+}	~ 0.9–1.0	Cu^{2+}	0.57–0.73 (different coordinations)

The next such close-packed layer, situated just above the given one, can either lie on top of the centers of the "up-triangles" of Fig. 3.38(b) (marked by squares there) or on top of the "down-triangles" (marked by small triangles). If we denote the first layer as A, then the second can be located either in B positions (squares in Fig. 3.38(b)) or in C positions (empty triangles).

Consider for example two anion layers A and B (large "spheres" at the position of circles and squares). The interstices between the first and second layers A and B are of two types: tetrahedral positions below each square or above each circle (ions put into these interstices would have four equivalent O^{2-} ions as nearest neighbors), or octahedral positions (with TM ions between layers A and B occupying the positions marked in Fig. 3.38(b) by small triangles – the positions sandwiched between the lower close-packed layer A and the second layer B, see Fig. 3.39). Thus we see that there are typically two types of coordination of TM ions in a close-packed anion lattice: tetrahedral or octahedral positions. The rest depends on the packing of such anion layers and on which interstices are occupied, in which order, and by which TM ions.

Close-packed anion layers can have different stacking. Thus there may be a sequence $ABAB\ldots$, or $ABCABC\ldots$, or $ABACABAC\ldots$, etc. Different stackings of these layers give different packing of TM–O polyhedra. Thus the stacking $ABABAB\ldots$ would give a structure repeating that of Fig. 3.38(b), with "empty" triangles on top of each other, that is

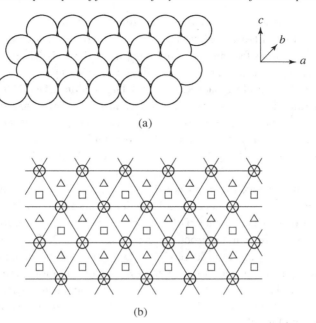

(a)

(b)

Figure 3.38 (a) Close packing of a layer of, usually, large anions; (b) possible locations of ions in the next close-packed layer of such ions (squares or small triangles).

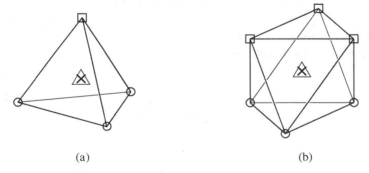

(a)   (b)

Figure 3.39 Typical location of small transition metal ions (crosses) in (a) tetragonal and (b) octahedral interstices formed by large anions. Here, following the notation of Fig. 3.38, we denote the anions in the first layer by circles, and in the second (upper) layer by squares.

forming columns in the c-direction. When we compare this with Fig. 3.39(b), we see that if we occupy these positions with TM ions, we get vertical columns of face-sharing MO_6 octahedra "lying on a face." This is a typical pattern in compounds such as $CsNiCl_3$ or $BaCoO_3$ (large ions Cs^+ and Ba^{2+} occupy positions between these columns, see below). Such materials will thus have a quasi-one-dimensional structure. The crystal structures for the $ABABAB\ldots$ packing of anions are usually hexagonal.

Similarly, if we occupy *all* octahedral positions with anions having close packing *ABAB*..., we will get a hexagonal structure of NiAs-type. In this case the MO_6 octahedra will have a common face in the c-direction, as in the $CsNiCl_3$ structure mentioned above, and in the ab-plane these octahedra will be edge-sharing.

If we have hexagonal close-packed layers of large anions, *ABABAB*..., and if metal ions M occupy the octahedral holes in every second layer, $(AMB)(AMB)$..., we have the CdI_2 structure. The metal ions here form two-dimensional triangular layers of edge-sharing MX_2 octahedra. Many interesting layered TM compounds belong to this class, for example many TM dichalcogenides TaS_2, $NbSe_2$, $NiTe_2$ (see, e.g., Wilson, 1972); we will mention some of these below.

Another typical sequence of anions is *ABCABC*...; this often gives cubic structures. If in such packing we occupy all octahedral interstices with metal ions, we will get a NaCl structure, which is realized in many TM monoxides such as NiO, see Fig. 3.40. (In fact, the "construction" of the standard rock-salt structure of Fig. 3.40 may be understood as a close packing of [111] layers of large Cl^- or O^{2+} ions with the stacking *ABC*..., with Na^+ or Ni^{2+} ions in the octahedral interstices.) We see that TM ions (crosses in Fig. 3.40) also form triangular [111] layers. But one can speak of four such layers, perpendicular to all four [111]-type cube diagonals in Fig. 3.40, so that the resulting structure of the thus formed MnO or NiO is cubic. MO_6 octahedra for nearest-neighboring TM ions have here common edges with 90° M–O–M bonds, and next-nearest neighbors have one common oxygen, that is they have corner-sharing octahedra with 180° M–O–M bonds.

An interesting "derivative" of this structure appears if we replace half of the transition metal ions in, for example, NiO or CoO by nonmagnetic ions such as Li^+ or Na^+, with Ni^{3+} and Li^+ ordered in consecutive [111] layers, see Fig. 3.41 (nonmagnetic ions are denoted in this figure by triangles, and magnetic Ni or Co by crosses; oxygen, as always,

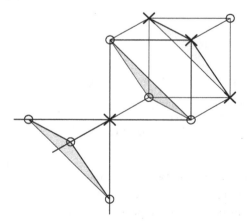

Figure 3.40   The formation of rock-salt (NaCl) structure, for example in NiO. The oxygens (circles) form an *ABCABC* sequence in the [111] direction in a cubic setting, and transition metals such as Ni (crosses) occupy octahedral interstices created by this close packing of oxygens.

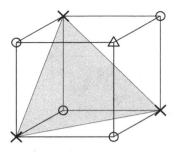

Figure 3.41 Formation of structures in systems such as $LiCoO_2$ or $NaNiO_2$. Here circles are anions, crosses are TM ions (Co, Ni), and the triangle is a nonmagnetic ion (Li, Na).

by circles). The sequence of layers in this case (looking along the [111] axis) would be A(Ni)B(Li)C(Ni)A(Li)B(Ni)C(Li)A ..., with $ABCABC$ close-packed anion (here oxygen) layers. In effect one would form systems with two-dimensional layers made of edge-sharing MO_6 octahedra, with a triangular lattice in these layers. The full symmetry of this structure would be rhombohedral. Such structures exist in systems such as $LiNiO_2$, $LiVO_2$, or Li_xCoO_2 (a very important material, the basis of rechargeable lithium batteries).

If in such a structure the nonmagnetic ions such as Li^+ were absent, we would get the structure known as a $CdCl_2$-type structure ($AMB\,CMA\,BMC$ ...); the bonding between $\{AMB\}$ blocks in this case is usually of van der Waals type, that is these are really layered materials. This, for example, is the structure of $MnBr_2$, $FeCl_2$, $NiCl_2$, etc. If, however, there are some other ions such as Cu^+ or Ag^+ instead of alkali ions (Li^+, Na^+) between these triangular magnetic layers, the magnetic layers would remain the same, but the nonmagnetic ions Cu^+, Ag^+, instead of staying inside the O_6 octahedra (as does Na in Na_xCoO_2), will be in a dumbbell coordination between two oxygens. In effect the sequence of *oxygen* layers will be AB(Cu)BA ... The resulting structure is called a delafossite structure, which is present for example in $Cu^{1+}Fe^{3+}O_2$, $Cu^{1+}CrO_2$, $Ag^{1+}CrO_2$, etc. These materials are very interesting as possible multiferroics, see Chapter 8.

The anion layers A, B in delafossites are not close packed: the layer B will be on top of B. This is caused by the special tendency of some ions with configuration d^{10}, such as Cu^{1+} or Ag^{1+}, to be in a linear coordination (O–Cu–O dumbbell). Just such ions stabilize the delafossite structure. The tendency of Cu^{1+} to be linearly coordinated is also important for high-T_c cuprates.

If, besides small TM ions, there are other, large metal ions in the system, as happens in ternary oxides such as $BaTiO_3$ or $LaMnO_3$, the ions Ba^{2+} and La^{3+} are relatively large – they have sizes comparable with (or even bigger than) the size of O^{2-}, see Table 3.1. Such large ions then occupy the positions of some O^{2-} ions in the close-packed structure of the latter. This is what happens, for example, in the very important class of *perovskites*,[13]

---

[13] The mineral perovskite has formula $CaTiO_3$.

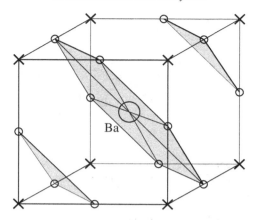

Figure 3.42 Formation of the perovskite structure ABO_3, for example $BaTiO_3$. Large A ions (here Ba, large circle) replace, in a regular way, some oxygen ions (small circles).

Fig. 3.42; this is how one can understand their formation. The perovskite structure, for example that of $BaTiO_3$, is usually treated as a cubic lattice of TM ions (e.g., Ti in $BaTiO_3$, crosses in Fig. 3.42) with oxygens (small circles) between them in the middle of the TM–TM edges, with Ba occupying the center of the cube. But as we have mentioned above, this picture, which is usually quite adequate for the purpose of discussing many properties of such compounds, is in fact somewhat misleading: from the point of view of crystallo-chemistry one should treat the compound as a close packing of *large* ions O^{2-} and Ba^{2+} ([111] layers in a cubic setting, shown by shading in Fig. 3.42) with TM (here small Ti^{4+}) occupying the octahedral interstices between them. We see that some of the positions in the hexagonal close-packed layers of O^{2-} are here occupied by large Ba^{2+} (or La^{3+} in $LaMnO_3$, etc.) ions, which thus will be 12-coordinated by oxygens.

Using similar considerations, one can understand the "construction" of most other oxides: MO_2 (TiO_2, CrO_2, ...) with the rutile structure, spinels AB_2O_4, corundum M_2O_3, etc.

The rutile structure (Fig. 3.43), can be visualized as chains of MO_6 octahedra in the c-direction with common edges, with neighboring chains rotated by $90°$ and having common corners with the first chain. But one can also look at this structure as that of a row of MO_6 octahedra with common edges, lying in close-packed MO_2 layers similar to those shown in Fig. 3.38(b), with the next column of Fig. 3.43 in the next such layer.

Similarly, the corundum structure M_2O_3 (e.g., Cr_2O_3, Fe_2O_3, V_2O_3) can be viewed as analogous to that of quasi-one-dimensional hexagonal $CsNiCl_3$ described above, with face-sharing MO_6 octahedra in the c-direction, however not forming infinite vertical columns as in $CsNiCl_3$, but with only a pair of such octahedra occupied by two TM ions, the next pair being shifted, see Fig. 3.44 (where we again denote TM ions by crosses, and oxygens by circles). In effect, in the $A(M)B$ layers in the corundum structure the TM ions in the ab-plane occupy not all octahedral holes, but only 2/3 of them, so that in this plane they form

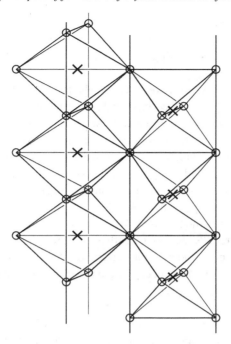

Figure 3.43 The structure of rutile, for example TiO_2, VO_2. As always, crosses are metallic ions and circles are anions, such as oxygens.

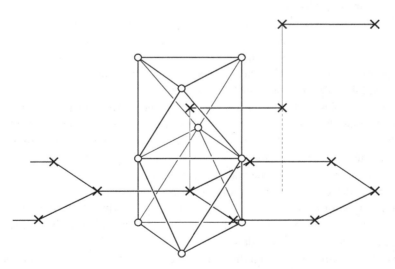

Figure 3.44 Main elements of the corundum structure (e.g., Al_2O_3, Cr_2O_3, V_2O_3). Crosses are metal ions, circles are oxygens.

a honeycomb lattice. The ilmenite structure (e.g., $FeTiO_3$ – the mineral ilmenite) is related to this structure; it can be visualized as the corundum structure of for example Fe_2O_3 in which every second Fe in vertical pairs is replaced by Ti (with Fe^{2+} and Ti^{4+} forming alternating honeycomb layers).

In all these oxides the TM ions are in O_6-octahedra, all the varieties being, first, due to different stacking of close-packed oxygen layers (*ABAB*, or *ABCABC*, etc.), possible replacement of some oxygens by large cations such as K^+, Ba^{2+}, or La^{3+}, and different occupation of octahedral interstices. However, as mentioned at the beginning of this section, in close-packed oxygen sublattices there also exist tetrahedral interstices, Fig. 3.39(a). These may also be occupied, either together with some octahedral ones, or only tetrahedral ones. If we take the *ABCABC* packing of large anions and occupy all tetrahedral interstices, we would get the structure type of fluoride, CaF_2. Spinels[14] AB_2O_4 may be visualized as close packing of oxygens of the type *ABCABC*..., with a particular occupation of tetrahedral *A* ions and octahedral *B* sites. The recently discovered class of compounds of the type $YBaCo_4O_7$ (structure type of the mineral swedenborgite, $SbNaBe_4O_7$) may be viewed as oxygen stacking *ABAC*..., with Co occupying only tetrahedral intersites, with large Ba^{2+} again replacing some of the oxygens.

Until now we have been discussing TM compounds such as oxides or halcogenides, with predominantly ionic bonding and with small TM ions. If we go, for example, to sulfides or selenides, covalency becomes more and more important, and the simple (at least conceptually) scheme described above – close packing of large anions and large nonmagnetic cations, with small TM ions in the interstices – may break down, and other, more complicated structures may appear. Good examples are pyrite FeS_2 ("fool's gold") and NiS_2. In these systems the sulfur ions form dimers, and the formula of these compounds can be represented as $Fe^{2+}(S_2)^{2-}$, $Ni^{2+}(S_2)^{2-}$. The crystal structure is then of NaCl-type, with Fe and Ni occupying the place of Na, and S_2 molecules – that of Cl. In many other sulfides, and especially in selenides and tellurides, the covalent bonds between Se and Te play a very important role. Thus the whole variety of TM compounds is not exhausted by the examples and the rules discussed above. However for a large number of these compounds the general principles of their "construction" described above are valid, and their general understanding can help to devise new materials, understand which possible substitutions work, etc. And even more subtle structural features, such as possible deviations from the ideal perovskite structure, can often be explained using the same notions of close packing of corresponding ions. We will see many such examples in what follows.

As we have discussed above, the ionic radii play a very important role in stabilizing one or other type of crystal structure, for example perovskite ABO_3 such as $SrTiO_3$ or $LaMnO_3$ vs hexagonal materials such as $CsNiCl_3$ or $BaCoO_3$. But within the same class, for example in perovskites ABO_3, depending on the relative radii of *A* and *B* cations, there may occur further distortions from the ideal cubic structure, that is to an orthorhombic or rhombohedral one.

---

[14] The mineral spinel has formula $MgAl_2O_4$.

These transitions are also connected with the tendency toward closer packing, and they depend on the ratio of the corresponding ionic radii R. For ideal close packing in a perovskite ABO_3 one should have $2(R_B + R_O) = a$, $2(R_A + R_O) = a\sqrt{2}$ where a is the lattice parameter of an ideal cubic perovskite lattice. One often characterizes the situation in perovskites by the *tolerance factor t*, defined as

$$t = \frac{R_A + R_O}{\sqrt{2}(R_B + R_O)} . \tag{3.45}$$

In the ideal case, $t = 1$. Empirically one finds that perovskite structures are formed for $0.7 \lesssim t \lesssim 1$. If t is close to 1, perovskites remain cubic. However for smaller t (or, in other words, if A ions are small enough) cubic perovskites transform into orthorhombic (or rhombohedral) structures by tilting and rotation of the BO_6 octahedra, shown for example in Fig. 3.45, in which the typical rotation of MO_6 octahedra around the [110] axis (in a cubic setting) is shown; such rotations lead to the orthorhombic (*Pbnm*) structure.[15] (Typically, there also exist simultaneous rotations around the [001] axis.) The axes a, b, c in a *Pbnm* setting are shown in Fig. 3.45; we see that, compared with the prototype unit cell with the lattice parameter taken as 1, the unit cell in the orthorhombic *Pbnm* structure is $\sim (\sqrt{2}, \sqrt{2}, 2)$. These distortions, which are often called $GdFeO_3$-type distortions, lead to a decrease of the B–O–B angle, which in turn leads to a decrease of the effective d–d hopping and of the corresponding bandwidth.

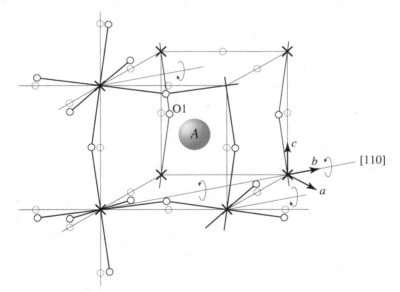

Figure 3.45 Distortions ($GdFeO_3$-type distortions) in a perovskite structure, leading to the formation of an orthorhombic structure. The A atom is shown as a large sphere, B ions are shown by crosses, and oxygens by small circles.

---

[15] Sometimes one describes this structure by the group *Pnma*, which differs from *Pbnm* by the choice of axes: $(a, b, c)_{Pbnm} \rightarrow (c, a, b)_{Pnma}$.

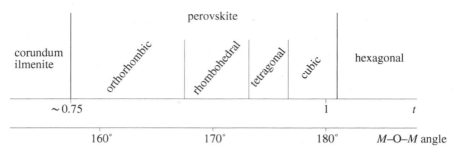

Figure 3.46 Schematic location of different structures of distorted perovskites depending on the tolerance factor t (3.45), and typical values of the corresponding $M–O–M$ angles. The numerical values in this figure are not absolute boundaries; rather they point to a general tendency and serve for general orientation. The detailed boundaries may depend not only on the tolerance factor t, but on various other factors as well.

The typical sequence of different phases in perovskites for different tolerance factors is shown in Fig. 3.46, in which we also show typical values of TM–O–TM angles for different phases, see Goodenough (2004). (The numerical values of tolerance factors for the stability of one or the other structure do not have absolute meaning; as in most such empirical regularities they show the general trends, and serve for general orientation.) For very small tolerance factors, $t \lesssim 0.7$–0.75, the perovskite structure is not stable; thus for manganites $RMnO_3$, where R is a small rare earth (Tm, Eu) or Y, the hexagonal structure of $YMnO_3$-type is stable, see also Chapter 8 – although by special "tricks" one can also stabilize these systems in the perovskite structure. The other common structures for small tolerance factors, with TM still in octahedra, are ilmenites of the type of $FeTiO_3$ and corundum structures. The perovskite structure is also unstable for $t \gtrsim 1.02$. In this region hexagonal structures such as those of $CsNiCl_3$ or $BaCoO_3$, with $NiCl_6$ or CoO_6 octahedra sharing common faces and forming chains in the c-direction, are formed. Thus perovskites typically exist for $0.75 \lesssim t \lesssim 1$, however in most cases they are not cubic, but more often orthorhombic or rhombohedral.

The decrease of the TM–O–TM angle and the concomitant reduction in d–d hopping and the corresponding bandwidth can influence the physical properties of the corresponding materials rather strongly. Thus, for a given class of compounds, for example with the same TM ion B (Ni, or Mn, or Fe, etc.) the materials with smaller A ions, that is with smaller tolerance factor t and correspondingly stronger tilting and smaller bandwidth, have more chance of being insulating. For example the A ions can be different rare earths, the size of which decreases with increasing atomic number, that is going from La to Lu (lanthanoid contraction). Consequently we should expect smaller d–d hopping and more insulating character for smaller rare earth A ions. This is indeed what happens experimentally. Thus, doped $LaMnO_3$ (e.g., $La_{1-x}Ca_xMnO_3$) becomes metallic at a certain doping range ($0.3 \lesssim x \lesssim 0.5$), but analogous composites $Pr_{1-x}Ca_xMnO_3$ with smaller Pr instead of La remain insulating. Or, for example, among nickelates of the type $RNiO_3$ the compound $LaNiO_3$ is metallic, $PrNiO_3$ and $NdNiO_3$ experience metal–insulator transitions at relatively low

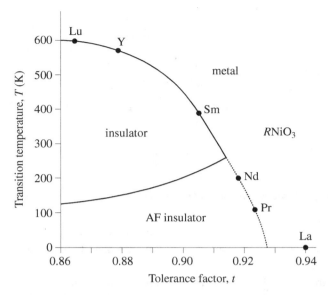

Figure 3.47 Qualitative phase diagram of perovskite nickelates $RNiO_3$ for different rare earths R. The dotted line (Nd, Pr) shows simultaneous I order metal–insulator and paramagnetic–antiferromagnetic transition.

temperatures $T_c \sim 100$–150 K, but for smaller rare earths the region of existence of the insulating phase is broader, and the corresponding critical temperatures strongly increase, see Fig. 3.47.

A very powerful and convenient classification of distortions in perovskites is due to Glazer (1972). In a first approximation all such distortions can be obtained from rotations of rigid BO_6 octahedra. Such rotations can be represented as consecutive rotations around [100], [010], and [001] or a-, b-, and c-axes. If one octahedron rotates for example around the c-axis clockwise, then it is clear that from the condition of "connectivity" of the lattice the neighboring octahedra in a- and b-directions should rotate around the same c-axis in the opposite direction, anticlockwise, see Fig. 3.48. But octahedra in the next, upper plane can rotate both in phase with the first plane, clockwise, or out of phase, anticlockwise. Glazer denoted the first case as c^+ (rotation around the c-axis in phase with neighbouring c-planes) and the second one as c^-. Similar rotations can also occur around other cubic axes, a and b. Thus one can classify all possible distortions of this type by combinations such as for example $(a^- b^+ c^-)$ or $(a^- b^- c^-)$ (not all combinations are allowed; possible combinations are presented in Glazer, 1972). The combinations mentioned above are the most common ones in perovskites, though they are not the only possible ones. Distortions $(a^- b^+ a^-)$ (GdFeO$_3$-type distortions) give the orthorhombic structure $Pbnm$. This structure is met, for example, in LaMnO$_3$. Notation such as $(a^- b^+ a^-)$ indicates that the rotations around the a-axis [100] and the c-axis [001] are equal – therefore one also writes a^- in the third place here. Another very common distortion in perovskites is $(a^- a^- a^-)$;

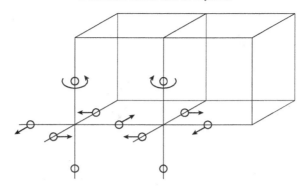

Figure 3.48 A particular contribution of the distortion in perovskites.

this gives a rhombohedral structure $R\bar{3}c$ and is met for example in LaCoO$_3$ or in the high-temperature phase of BiFeO$_3$ – both are important materials which will be discussed in detail below.

Note that, strictly speaking, during these rotations and tiltings one also has to distort somewhat the octahedra themselves, to preserve the "continuity" of the lattice, see for example Zhow *et al.* (2005a). This can sometimes lead to a rather counterintuitive result: that in the new structure the b-axis in a *Pbnm* setting, which is the axis of the main rotation of the initial cubic lattice ([110], see Fig. 3.45) becomes shorter than the perpendicular axis in the basal plane, $b < a$, rather than $b > a$ as should occur for rotations of rigid octahedra. This usually happens for very small A ions in perovskites ABO_3, that is for small tolerance factor (3.45).

Of course the consideration of close packing of large ionic "spheres," with small TM ions occupying the interstices between them, does not take into account all important factors. Thus the Coulomb interaction of the corresponding ions (Madelung energy) also gives an important contribution. Further, as mentioned above, the covalency effects are important in many cases, see for example Garcia-Fernandez *et al.* (2010). In fact, if we go to the real microscopic description, the very sizes of ions, determining the distances between different ions, are determined by the same chemical considerations as in molecules, that is "deep inside" they have the electronic nature. The "mechanical" description, which uses the notions of rigid spheres and close packing, is of course a simplification. Nevertheless, the "geometric" scheme described above often provides a very good general framework for the description of the structure of many TM compounds.

S.3 Summary of Chapter 3

In this chapter we have discussed the main modifications which occur when we place TM ions in a crystal. A typical situation in TM oxides is that with a TM ion surrounded by six oxygens forming an O_6 octahedron, although other coordinations (a TM ion in a tetrahedron, etc.) are also possible. In these cases the spherical symmetry of an isolated atom or ion is lost, and the local symmetry is determined by the symmetry of the crystal lattice.

The interaction of d-electrons with the surrounding anions leads to a splitting of the five-fold degenerate d levels, known as crystal field (CF) or ligand field splitting. There are two physical contributions to the CF splitting: point-charge contribution (the Coulomb interaction of d-electrons with charges of surrounding ions, mostly nearest-neighbor anions) and the hybridization (covalency) with the p states of these anions. In the usual cases these two contributions lead qualitatively to the same order of d levels. Numerically, in most cases the covalency contribution is larger than the point-charge contribution, but for qualitative purposes it is usually more convenient to deal with the point-charge picture.

For the cubic crystal field, for example for regular octahedral coordination such as that in undistorted MO_6 octahedron, five d levels are split into a lower triplet t_{2g} and a higher-lying doublet e_g. For tetrahedral coordination this splitting is inverted, and the doublet e_g lies lower (and the $t_{2g}-e_g$ splitting is approximately twice smaller). The wavefunctions of e_g-electrons, $\sim (2z^2 - x^2 - y^2)$ and $\sim (x^2 - y^2)$, have lobes directed toward the anions (e.g., oxygens), and the strong repulsion of corresponding electronic clouds from negatively charged oxygens O^{2-} (together with stronger $d-p$ hybridization with these oxygens) raises these e_g levels. The wavefunctions of t_{2g}-electrons, $\sim xy$, xz, and yz, are directed between the oxygens, therefore their repulsion from O^{2-} and consequently their energy is lower. The $t_{2g}-e_g$ splitting is denoted as Δ_{CF} (or sometimes as $10Dq$); for $3d$ ions in oxides it is typically $\sim 1.5-2\,\text{eV}$, and it increases for $4d$ and $5d$ elements.

There exist certain empirical rules determining the relative strengths of different ligands, that is the relative magnitude of CF splitting caused by them. Thus, for example, Cl^- are weaker ligands than O^{2-}, the cyan group $(CN)^-$ is one of the strongest ligands (much stronger than oxygen), etc. The ordering of different ligands according to their strength gives the so-called spectrochemical series, presented in eq. (3.10).

Further decrease of symmetry, for example due to distortions of the O_6 octahedra, leads to an extra splitting of the d levels. Tetragonal and orthorhombic distortions split both e_g and t_{2g} levels, and trigonal distortions split only t_{2g} levels.

In most cases the existing d-electrons fill the d levels according to Hund's rule, that is with the maximal possible spin; these are the high-spin (HS) states. However if the CF splitting Δ_{CF} is very large, which may happen for strong ligands such as CN groups, it may be favorable not to occupy the higher-lying e_g levels with electrons having spins parallel to those of t_{2g}-electrons, but rather to put the extra electrons (for $n_d > 3$) with opposite spins on the lower t_{2g} levels. Such states are known as low-spin (LS) states. Examples are some materials containing Co^{3+} or Fe^{2+} ions with six d-electrons: they may exist in a HS state $t_{2g}^4 e_g^2$ with spin $S = 2$, or in a LS state $t_{2g}^6 e_g^0$ with $S = 0$ (and also the intermediate-spin states, e.g. $Co^{3+}(t_{2g}^5 e_g^1)$ with $S = 1$, are in principle possible). There may even occur (phase) transitions between these states. Low-spin states are not very typical for $3d$ systems, but they are much more common for $4d$ and $5d$ systems, for which the crystal field splitting is larger, but the Hund's rule energy J_H is smaller. However, some $3d$ ions can also exist in LS states, though typically for stronger ligands than oxygen.

When we fill the d levels with n d-electrons, there may appear situations in which one electron (or one d-hole) occupies levels that for a symmetric configuration, for example for regular MO_6 octahedra, are degenerate. This may be a double degeneracy (one electron on e_g levels, as e.g. in $Mn^{3+}(t_{2g}^3 e_g^1)$, or one hole on e_g levels, as in $Cu^{2+}(t_{2g}^6 e_g^3)$). Or it may be a triple degeneracy, for partially filled t_{2g} levels. In such cases the system usually turns out to be unstable: it can distort, which would split the degenerate levels, so that one of them, containing an electron, would go down in energy, and another, empty level would go up. The decrease in energy of the occupied level is linear in distortion (as in Zeeman splitting), while the energy loss for lattice distortion is only quadratic, so that such a process would always occur spontaneously. This is the essence of the Jahn–Teller (JT) theorem, or Jahn–Teller effect, which states that symmetric configurations with orbital degeneracy are always unstable, and there should occur in such cases lattice distortions which would reduce the symmetry and lift orbital degeneracy. This effect is especially important for concentrated systems with orbitally degenerate (so-called JT) ions: it would lead in this case to a cooperative structural phase transition – the cooperative Jahn–Teller effect, with corresponding orbital ordering. There are many such materials among TM compounds, especially those containing strong JT ions – Mn^{3+}, forming colossal magnetoresistant and multiferroic manganites; Cu^{2+}, the main ingredient of high-T_c superconducting cuprates; etc. These effects will be discussed in detail in Chapter 6.

The Jahn–Teller effect for isolated TM impurities with orbital degeneracy, or for small molecules, leads to nontrivial quantum effects, such as quantum tunneling of a system between different distorted states. In effect the average symmetry can be restored (or not broken), but the wavefunctions of the system change drastically: each electronic (orbital) state will be associated with the corresponding distortion (vibronic wavefunctions). How important such effects are in concentrated systems is not really clear; in most cases they are ignored in such systems.

As we have discussed in Chapter 2, d-electrons have not only spin, but also an orbital moment, and the corresponding spin–orbit interaction $\lambda l \cdot S$ gives different multiplets for isolated atoms or ions. When TM ions are in a crystal, they do not have spherical symmetry any more, and the role of orbital moment changes. In the e_g doublet the orbital moment is frozen, or quenched: the corresponding wavefunctions are $|3z^2 - r^2\rangle \sim |l^z = 0\rangle$ and $|x^2 - y^2\rangle \sim (|l^z = 2\rangle + |l^z = -2\rangle)$; on this manifold the spin–orbit coupling $\lambda l \cdot S$ is zero, $\langle e_g | \lambda l \cdot S | e_g \rangle = 0$. Spin–orbit interaction acts on these levels only in the second order in perturbation theory. These perturbative contributions are still very important, for example they can cause strong modification of the values of the g-factor, and also give rise to magnetic anisotropy (discussed in more detail in Section 5.6). But the effect of spin–orbit interaction is even stronger for ions with partially filled t_{2g} levels: in this case the orbital moment is not quenched, and the spin–orbit coupling acts already in the first order, and it can influence strongly the properties of corresponding ions with partially filled t_{2g}-shells. The role of spin–orbit interaction is especially strong for heavier elements: for the end members of the $3d$ series (Fe, Co) and especially for $4d$ and $5d$ elements. It can determine, in particular, the multiplicity of corresponding terms, and can in principle even lead to compensation of spin and orbital moments.

A very convenient method of treating spin–orbit coupling in systems with partially filled t_{2g} levels is to describe the t_{2g} triplet by the effective orbital moment $\tilde{l} = 1$. In this way one can rewrite the spin–orbit interaction and describe the resulting energy levels. It turns out that the effective spin–orbit coupling $\tilde{\lambda} \tilde{l} \cdot S$ is negative. As a result, the rule of multiplet order (the second Hund's rule) for the corresponding effective total moment $\tilde{J} = \tilde{l} + S$ becomes opposite to the normal one. We have reversed multiplet order: the configurations with larger value of $\tilde{J}$ have lower energy for less-than-half-filled t_{2g}-*shell* (number of t_{2g}-electrons less than 3), and we have direct multiplet order (states with smaller $\tilde{J}$ have lower energy) for more-than-half-filled t_{2g}-shells (number of t_{2g}-electrons greater than 3).

At the end of this chapter we discussed some general principles governing the formation of different crystal structures of TM compounds such as TM oxides. These are usually considered as predominantly ionic materials, but still with a very important role played by covalency. Here the considerations of ionic sizes, or ionic radii, are important. Typically, TM ions are much smaller than for example O^{2-} or Cl^- ions, so that one can often visualize and understand the structure of TM oxides as a close packing of large anions (O^{2-}, Cl^-) with small TM ions occupying octahedral or tetrahedral interstitial positions between them. For ternary or more complex compounds, such as perovskites ABO_3 (e.g., $BaTiO_3$, $LaMnO_3$, etc.) the large A ions (Ba, La) replace some of the large anions (O^{2-}), the small TM B ions still being in "cavities" between them. In this way one can rationalize the structure of quite a lot of TM oxides and halogenides. However for similar systems with for example S, Se, or Te the situation may be more complicated. Also the Coulomb forces (Madelung energy) can contribute significantly to the stability of one or the other crystal structure. Nevertheless, the considerations of close packing of large ions, with small transition metals in between, provides a very useful general framework for understanding the crystal structures of quite a lot of TM compounds, including also deviations from ideal structures, such as for example $GdFeO_3$-type distortions of perovskites ABO_3, which depend on the ratio of the ionic radii of A and B ions, characterized by the tolerance factor t, and which often transform the structure of perovskites from cubic to orthorhombic or rhombohedral. The decrease in TM–O–TM (B–O–B) angle accompanying such distortions reduces the d–d hopping, and can strongly modify the properties of the corresponding materials.

4

Mott–Hubbard vs charge-transfer insulators

4.1 Charge-transfer insulators

Until now, when considering systems with strongly correlated electrons, we mostly discussed the properties of d-electrons themselves. However most often we are dealing not with systems with only TM elements (pure TM metals), but with different compounds containing, besides TM ions with their d-electrons, also other ions and electrons. These may be itinerant or band electrons, for example in many intermetallic compounds; some of these will be considered below, in Chapter 11. But more often we are dealing with compounds such as TM oxides, fluorides, etc., which are insulators. Still, even in this case we have in principle to include in our discussion not only the correlated d-electrons of transition metals, but also the valence s- and p-electrons of say O or F. This we have already done to some extent when we were considering the crystal field splitting of d levels in Chapter 3, in particular the p–d hybridization contribution to it, see Section 3.1 and Figs 3.5–3.8.

In some cases we can project out these other electrons and reduce the description to that containing only d-electrons, but with effective parameters determined by their interplay with say p-electrons of oxygens.[1] In other cases, however, we have to include these electrons explicitly. This, in particular, is the case when the energy of oxygen $2p$ levels is close to that of d levels. A typical situation, which we meet in systems such as NiO, or in perovskites such as colossal magnetoresistance manganites on the basis of $LaMnO_3$, or in the parent material for high-T_c cuprates La_2CuO_4 ("two-dimensional perovskite"), is one in which the TM ions in the lattice are separated by oxygens (see Fig. 4.1), where, as usual, we denote the TM ions by crosses and oxygen ions by circles. In ideal cases these materials have cubic lattices, with TM ions in the centers of O_6 octahedra, and with oxygens sitting directly between the TM ions, so that the TM–O–TM angle is 180°. In reality even for perovskites there often appear distortions of this ideal structure, in the form of tilting and rotation of rigid MO_6 octahedra, leading to orthorhombic or rhombohedral structures, see for example Section 3.5; in the presence of TM ions with orbital degeneracy these MO_6 octahedra themselves may be somewhat distorted because of Jahn–Teller effects, see

---

[1] We again will mostly speak below of TM oxides, although in principle the same ideas are also applicable, with some modifications, to sulfides, selenides, fluorides, etc.

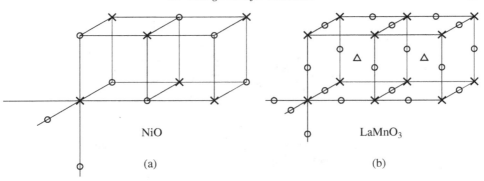

Figure 4.1 Typical crystal structures of transition metal compounds with (a) NaCl structure and (b) perovskite structure, showing the location of anions (e.g., oxygens, circles) between transition metal ions (crosses).

Section 3.2. But the overall features such as the location of oxygens between the TM ions, with the TM–O–TM angle not far from 180°, still remain the same.

The important electrons to be included in this case are the d-electrons of transition metals and the p-electrons of ligands: these have the strongest overlap with the d-electrons, and consequently they influence the properties of such systems the strongest (in principle one also has to include the s-electrons, but they rarely change the qualitative picture). Thus the simplest theoretical model, generalizing the Hubbard model (1.6), would be the so-called d–p model (sometimes, especially in relation to high-T_c superconducting cuprates, also called the three-band model):

$$\mathcal{H} = \varepsilon_d \sum d_{i\sigma}^{\dagger} d_{i\sigma} + \varepsilon_p \sum p_{j\sigma}^{\dagger} p_{j\sigma}$$
$$+ \sum t_{pd,ij} \left(d_{i\sigma}^{\dagger} p_{j\sigma} + \text{h.c.} \right) + U_{dd} \sum n_{di\uparrow} n_{di\downarrow}$$
$$\left(+ U_{pp} \sum n_{pj\uparrow} n_{pj\downarrow} + U_{pd} \sum n_{di\sigma} n_{pj\sigma'} \right). \tag{4.1}$$

Here we consider the simplest case of, say, nondegenerate d levels hybridizing with the corresponding p levels with the strongest p–d overlap and hopping $t_{pd,ij}$, that is p states with the lobes directed toward the d ions. This is for example the case shown in Fig. 4.2 (cf. Fig. 3.5), where we show the situation typical for high-T_c cuprates, with a two-dimensional square lattice of Cu^{2+} ions surrounded by four oxygens (which can be visualized as a "2d perovskite" – the basal, xy-plane of Fig. 4.1(b)). Here we include only one e_g orbital of Cu^{2+} – the $d_{x^2-y^2}$ orbital, which, according to the crystal field splitting for strongly elongated CuO_6 octahedra (typical for the strong JT ion $Cu^{2+}(d^9)$) contains one d-electron, the remaining d levels being doubly occupied, see Fig. 3.26. (In fact one should rather speak here about one (x^2-y^2)-hole, with the other "hole" states empty. However for simplicity we will first speak about one e_g-electron; this situation is realized e.g. in the low-spin state of $Ni^{3+}(t_{2g}^6 e_g^1)$.) We will use this simplified model mostly to illustrate the possible effects connected with explicit inclusion of oxygen p states in our treatment,

Figure 4.2 Typical overlap of e_g orbitals (here $x^2 - y^2$-type orbitals) with the p orbitals of surrounding ligands (so-called $pd\sigma$ overlap).

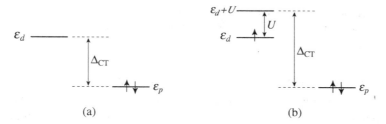

(a)   (b)

Figure 4.3 Typical position of p and d levels in transition metal oxides, showing the charge-transfer gap: (a) for transition metal ions with an empty d-shell (configuration d^0) and (b) for transition metal ions with configuration d^1.

in particular to explain the notion of charge-transfer (CT) insulators, as contrasted with the better-known Mott, or Mott–Hubbard insulators. For real systems, of course, we have to include all relevant d orbitals, and also different p orbitals of oxygen (remember that for p orbitals with $l = 1$ there are three of them for each oxygen, compared with five d orbitals per TM ion).

In typical cases the relative position of d levels of transition metals and p levels of oxygens is as shown in Fig. 4.3, that is the p levels lie below the d levels. The oxygens in compounds are nominally O^{2-}, that is all p levels are completely filled, which we mark on Fig. 4.3. The d levels may be either empty, or contain a certain number of d-electrons, configuration d^n. We see that in contrast to the standard Hubbard model, where the electrons localized at particular sites could be excited by hopping to other, already occupied sites, which would cost energy U,

$$d^n d^n \longrightarrow d^{n-1} d^{n+1} , \qquad \Delta E = U = U_{dd} , \qquad (4.2)$$

here, besides this process, there exists another possibility and another type of excitation: an electron from the occupied $2p$ level of oxygen can be transferred to the TM ion, which costs a certain energy Δ_{CT}, called the *charge-transfer energy*:

$$d^n p^6 \longrightarrow d^{n+1} p^5 , \qquad \Delta E = \Delta_{CT} . \qquad (4.3)$$

In this process there appears an extra d-electron and, most importantly, also an oxygen hole – one missing electron in the filled p-shell p^6. In chemical language this corresponds to going from the normal state O^{2-} to the ionic state O^-. Often the resulting state is denoted $d^{n+1}\underline{L}$, where $\underline{L}$ denotes a ligand (in this case oxygen) hole.

In the simplest case of empty d-shells the charge-transfer energy (4.3) is simply equal to

$$\Delta_{CT}(d^0) = \varepsilon_d - \varepsilon_p , \qquad (4.4)$$

see Fig. 4.4(a). If however there are already electrons on the d-shell, then in the charge-transfer process (4.3) we not only promote an electron from the p level ε_p to the d level ε_d, but this extra electron at the d level will repel from the existing d-electron, that is we have

$$\Delta_{CT}(d^1) = (\varepsilon_d + U_{dd}) - \varepsilon_p . \qquad (4.5)$$

This modification is due to the fact that for correlated electrons, which in this case are d-electrons, a simple single-particle description is inapplicable, and excitation energies should also contain the corresponding change in interactions. The real definition of the charge-transfer energy (4.3) operates with many-electron states, which in general, for specific ionic states, would depend also on particular configurations of the involved states d^n and d^{n+1}, that is on the Hund's energy J_H, etc. Thus the definition of the charge-transfer energy (4.3), which includes all many-particle effects, is the correct one and the only one needed when discussing different processes in these systems. Specifications such as (4.4) and (4.5), which operate with bare energies ε_p, ε_d and which depend on the actual electronic configuration, are rarely needed; they might be used, for example, in *ab initio* band structure calculations.

Sometimes, especially for almost-filled d-shells such as those in cuprates based on $Cu^{2+}(d^9)$ ions, it is convenient to rewrite the whole treatment not in terms of electrons, but in terms of holes. Then the state Cu^{2+} would contain one d-hole, and the original state of oxygen $O^{2-}(2p^6)$ with the filled $2p$-shell would have no holes. The charge-transfer process (4.3) in this case would look like $d^9 p^6 \rightarrow d^{10} p^5$, that is it would correspond simply to a transfer of the original d-hole to the oxygen. In terms of hole levels an energy diagram similar to Fig. 4.3 would look as shown in Fig. 4.4 (in which the arrow $\uparrow$ now denotes a hole); that is the charge-transfer energy would be simply

$$\Delta_{CT} = \tilde{\varepsilon}_p - \tilde{\varepsilon}_d , \qquad (4.6)$$

Figure 4.4 Typical position of p and d levels in the hole picture (convenient e.g. for $Cu^{2+}(d^9)$).

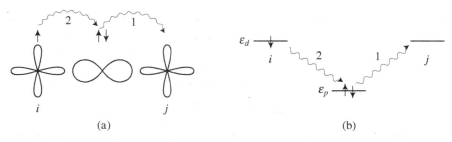

Figure 4.5 The process of hopping of d-electrons from one transition metal site to another via the intermediate oxygen.

where $\tilde{\varepsilon}_p$ and $\tilde{\varepsilon}_d$ are the energies of states with respectively one p-hole and one d-hole. Still, if we want to create for example *two* holes on the same oxygen (e.g., transferring two p-electrons from one oxygen ion to two Cu ions) we should also in principle take into account the Coulomb repulsion of two holes on oxygen, U_{pp}, see eq. (4.1). This contribution is often ignored, but in fact it is not small at all and can be comparable with U_{dd} on early $3d$ metals: $U_{pp} \sim 3\,\text{eV}$. Also, the Hund's rule for oxygen should in principle be included, as it is also quite large, $J_H^{(pp)} \simeq 1.2\,\text{eV}$.

In the simplest cases such as that of perovskites, with oxygens sitting between the TM ions, direct d–d hopping can usually be ignored but there is an overlap and corresponding intersite hopping between the d states of transition metals and the p states of oxygens. This is the term with the coefficient t_{pd} in the model Hamiltonian (4.1). In effect the d-electrons can hop from one TM site to another via oxygens. This process is illustrated in Fig. 4.5: in the first step (1) an electron, for example with spin ↑, hops from the oxygen $2p$ orbital to a TM (j) say to the right, and in the second step (2) a d-electron with the same spin hops from TM ion i to the now vacant p state of oxygen. In effect the d-electron from site i is transferred to the neighboring site j – not directly, but via the intermediate oxygen. The corresponding hopping would occur as a two-step process, as $d_i^n p^6 d_j^n \longrightarrow d_i^n p^5 d_j^{n+1} \longrightarrow d_i^{n-1} p^6 d_j^{n+1}$, through an intermediate state with an oxygen hole. This state, $p^5 d_j^{n+1}$, costs excitation energy Δ_{CT}, cf. (4.3). Thus the resulting transfer element for the effective d–d hopping would be given by

$$t_{dd}^{\text{eff}} = \frac{t_{pd}^2}{\Delta_{\text{CT}}} . \tag{4.7}$$

In effect we can often reduce the more complete d–p model (4.1) to the Hubbard model (1.6), with the d–d hopping $t = t_{dd}^{\text{eff}}$. However, as we will see, this is not always possible and not for all processes.

Before discussing this, though, we should also mention that for a realistic description we should take into account the fact that there is not one d orbital at a TM site, but in general five possible orbitals, and also there are three p orbitals for each oxygen. One can easily see that for the 180° TM–O–TM bond as shown for example in Fig. 4.5, the e_g orbitals have

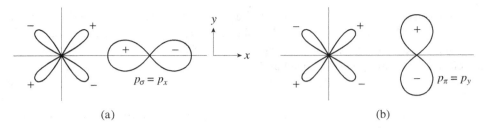

Figure 4.6 Possible d–p overlap for t_{2g}-electrons. (a) The xy t_{2g} orbital is orthogonal to the p_x orbital of oxygen. (b) There exists, however, nonzero hopping to the p_y orbital, called $pd\pi$ hopping.

a strong overlap with the $2p$ orbitals of oxygens directed toward them, that is along the TM–O bonds. These orbitals are called σ orbitals, and the corresponding hopping matrix element is denoted $t_{pd\sigma}$. However the t_{2g} orbitals of transition metals have zero overlap with these σ orbitals, see Fig. 4.6(a): we have to remember that the wavefunction of for example the d_{xy} orbital has an xy form, that is its lobes have different signs, as shown in Fig. 4.6(a), and its overlap with the σ orbital p_x on Fig. 4.6(a) is exactly zero just by symmetry.

In contrast, this d_{xy} orbital has a nonzero overlap with the other orbital of this oxygen, the p_y orbital (Fig. 4.6(b)), see also Section 3.1. This p orbital, perpendicular to the TM–O bond, is called a π orbital, and the corresponding p–d hopping integral is denoted $t_{pd\pi}$. As is clear from Figs 4.5 and 4.6, this hopping is smaller than $t_{pd\sigma}$: $t_{pd\sigma} \sim 1.7 t_{pd\pi}$ to $2 t_{pd\pi}$ (Harrison, 1989). Correspondingly, the effective d–d hopping via oxygens (4.7), and the respective bandwidths $\sim t_{dd}^{\text{eff}}$, would also be bigger for e_g-electrons than for t_{2g}-electrons. This often has important consequences. Thus for example in doped manganites (so-called manganites with colossal magnetoresistance, CMR) of the type La$_{1-x}$A$_x$MnO$_3$ (A = Ca, Sr) with average valence Mn^{3+x}, that is with the average configuration $t_{2g}^3 e_g^{1-x}$ (or containing formally $(1-x)$ ions Mn$^{3+}(t_{2g}^3 e_g^1)$ and x ions Mn$^{4+}(t_{2g}^3 e_g^0)$), we can usually ignore possible hopping of t_{2g}-electrons and consider them as localized, giving localized spins $S = \frac{3}{2}$ per Mn. But the remaining $(1-x)$ e_g-electrons are partially filling a relatively wide e_g band, and these electrons can lead to metallic conductivity of CMR manganites for a certain doping range. These mobile e_g-electrons also provide a special mechanism for the exchange interaction between localized spins, called *double exchange*, see Section 5.2; this double exchange leads to a ferromagnetic ordering in the resulting metallic state of manganites.

We now consider in which situations we can exclude the oxygen states and go over to the effective one-band description of the type of Hubbard model with effective hopping t_{dd}^{eff} (4.7), and when this is not possible; and what will then be the state of the system. At first glance it may seem that we could go from the general d–p model (4.1) to the effective single-band Hubbard model if the p–d hopping t_{pd} is much smaller than both U and Δ_{CT}. Indeed, we see from eqs (4.2), (4.3) that if we start from the "standard" state with d-electrons localized at respective TM ions and with a filled p-shell p^6, the possible excited states of type $d^{n-1}d^{n+1}$ and $d^{n+1}p^5$ cost excitation energies U_{dd} and Δ_{CT}. Similar to the normal Mott insulating state described for example by the Hubbard model (1.6), we can show that if the electron hopping t_{pd} is (much) smaller than both U_{dd} and Δ_{CT}, the electrons would remain localized at their sites, and the material would be an insulator. Thus the resulting ground state would be the same as that of Mott, or Mott–Hubbard insulators.

However depending on the ratio of U_{dd} and Δ_{CT}, the *lowest charge-carrying excited states* could be different. If $\Delta_{CT} > U_{dd}$, the lowest excitations would be $d^n d^n \rightarrow d^{n+1}d^{n-1}$, that is they would be the same as in Mott–Hubbard insulators. In this case we can indeed exclude the oxygen states as is done in eq. (4.7), and go over from the full $d-p$ model (4.1) to the Hubbard model (1.6), with $t = t_{dd}^{\text{eff}}$ (4.7). All the properties of this state would be the same as in the Mott–Hubbard case. This state would be the standard Mott–Hubbard insulator.

If however we have the opposite relation, $\Delta_{CT} < U_{dd}$, the ground state of the system would be the same (provided only that $t_{pd} \ll \Delta_{CT}$), but the lowest charge excitations would be different: these would not be the $d-d$ transitions, but the $p-d$ transitions between the oxygen $2p$-shell and the d states of the transition metal. That is, according to eq. (4.5), the electron excitations in this case would be extra electrons on d levels, but the hole excitations would be the oxygen p-holes. Correspondingly, if we dope such a system by holes, as we do for example in CMR manganites $\text{La}_{1-x}(\text{Ca}, \text{Sr})_x\text{MnO}_3$ or in high-T_c cuprates $\text{La}_{2-x}\text{Sr}_x\text{CuO}_4$, the holes thus created would be predominantly oxygen p-holes.

Of course there would always be some $p-d$ hybridization present (terms with t_{pd} in the Hamiltonian (4.1)), thus there would be no pure d states or pure oxygen p states – such states are always hybridized, see eq. (3.7) and Fig. 3.8, so that the resulting states would be superpositions of the type $\alpha|d\rangle + \beta|p\rangle$, $\alpha^2 + \beta^2 = 1$. If we are considering d- and p-hole states, then in the first case, that is Mott–Hubbard insulators with $\Delta_{CT} \gg U_{dd}$, the holes would be predominantly on d sites, so $\alpha \gg \beta$. In the opposite case of $\Delta_{CT} \ll U_{dd}$ the holes would be mostly on oxygens, so $\beta \gg \alpha$. This second state is called a *charge-transfer (CT) insulator*. This concept was first proposed by Zaanen, Sawatzky, and Allen and is often called the ZSA scheme (Zaanen *et al.*, 1985).

Once again, in both cases we have an insulating ground state with localized d-electrons, that is with corresponding localized magnetic moments and corresponding magnetic ordering, provided $t_{pd} < \{U_{dd}, \Delta_{CT}\}$. But this insulating state can be of two types: a Mott–Hubbard insulator if $\Delta_{CT} > U_{dd}$ and a charge-transfer insulator in the opposite case $\Delta_{CT} < U_{dd}$. The resulting phase diagram thus has the schematic form shown in Fig. 4.7. We have an insulating state in the unhatched region of Fig. 4.7. But it is important, once again, that this insulating region is divided into two subregions, the Mott–Hubbard and charge-transfer insulators. Correspondingly, one can classify all insulators with strongly correlated electrons into two groups, Mott–Hubbard insulators and charge-transfer insulators. And, interestingly enough, it turns out that, for example, most $3d$ oxides belong not to the Mott–Hubbard but rather to the charge-transfer type; only oxides of the earlier TM elements (Ti, V) are of the Mott–Hubbard type, while most oxides of the heavier $3d$

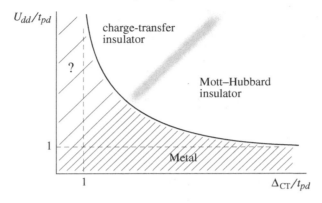

Figure 4.7 Schematic ZSA diagram, after Zaanen *et al.* (1985).

elements (Fe, Co, Ni, Cu) are charge-transfer insulators. This property also depends on the valence of the corresponding ions; these questions will be discussed below.

In the hatched regions of Fig. 4.7, that is when the p–d hopping becomes large, something else should happen. Specifically, if we reduce U_{dd}/t_{pd} in the Mott–Hubbard regime, we would sooner or later end up in a metallic state, going through a Mott transition.

The question of what happens to charge-transfer insulators when we reduce Δ_{CT}/t_{pd} and move to the left in the phase diagram of Fig. 4.7 is much less clear, see for example Khomskii (1997). One can also end up in a metallic state, but still with strongly correlated d-electrons (U_{dd}/t_{pd} still $\gg 1$). This state would then resemble that of heavy fermions, Chapter 11: a metallic state in which itinerant electrons (e.g., weakly correlated oxygen p-electrons) coexist with strongly correlated d-electrons. Alternatively, the resulting state may be an insulator of a special kind (similar to what are sometimes called Kondo insulators). These questions will be discussed in more detail in Sections 4.3 and 10.5.

At the end of this section we will discuss in which cases we typically have which regime. It turns out that the charge-transfer energy Δ_{CT} decreases in going from light to heavy TM ions, and it also decreases for higher-valence states of TM ions. The values of Δ_{CT} are obtained from spectroscopic data (Bocquet *et al.*, 1992). The corresponding results are shown schematically in Fig. 4.8 (this picture was compiled by T. Mizokawa). We see that for early $3d$ metals Δ_{CT} is quite large, so that the corresponding materials, Ti or V oxides, lie in the Mott–Hubbard part of the ZSA phase diagram of Fig. 4.7. In contrast, most heavy $3d$ elements have relatively small $\Delta_{CT} < U_{dd}$, that is they give charge-transfer insulators. Thus most TM compounds with strong electron correlations and with localized electrons that we consider in this book belong strictly speaking not to Mott–Hubbard insulators, but rather to charge-transfer insulators. And for heavy $3d$ ions in high formal oxidation states, such as Cu^{3+} or Fe^{4+}, the charge-transfer energy is very small or even negative. This means that for the corresponding compounds most holes are

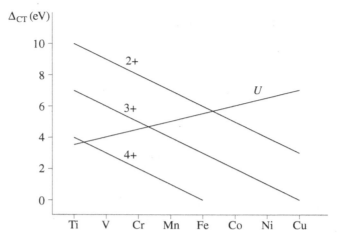

Figure 4.8 Characteristic values of the charge-transfer energy and Hubbard's U for the $3d$ series (courtesy of T. Mizokawa).

actually on oxygens, and one has to represent for example Cu^{3+} rather as $Cu^{2+}\underline{L}$, where $\underline{L}$ stands for a ligand hole (e.g., hole on oxygen) and similarly Fe^{4+} as $Fe^{3+}\underline{L}$, and Co^{4+} as $Co^{3+}\underline{L}$ or even $Co^{2+}\underline{L}^2$ (the actual state is a superposition of these states; we are speaking here rather about the weight with which the corresponding configuration enters the total wavefunction).

Qualitatively one can explain the trends in the behavior of the charge-transfer gap Δ_{CT} shown in Fig. 4.8 as follows. When we go, in a TM ion, from valence 2+ to 3+ and then to 4+, we reduce the filling of the d levels, $d^n \rightarrow d^{n-1} \rightarrow d^{n-2}$. The energy of the transfer of an electron from the p-shell of a ligand to the d levels of a TM ion includes the Hubbard repulsion of the transferred electron with the d-electrons already present on the TM ion, see Fig. 4.3 and eqs (4.4), (4.5). This repulsion is reduced with a decrease in the number of d-electrons, that is crudely one could expect $\Delta_{CT}(M^{3+}) \simeq \Delta_{CT}(M^{2+}) - U$. There are of course different processes of screening, etc. which would "spoil" the numerical agreement, but the general trend is caught correctly: Δ_{CT} decreases with increasing valence (oxidation state) of the TM ion, and more or less by the same amount in going from M^{2+} to M^{3+} as from M^{3+} to M^{4+}, cf. Fig. 4.8.

Another trend seen in Fig. 4.8 is the decrease of Δ_{CT} in going from left to right in the Periodic Table. This can be explained by the increase in nuclear charge Z in this series: for larger Z the d levels move down (i.e., the binding energy of d-electrons, with their small radius and incomplete screening, increases). At the same time the p-electrons of anions are far from TM nuclei, and their energies do not change much. In effect the energy of p–d transfer decreases with increasing Z, that is in going from Ti to Cu for the $3d$ series shown in Fig. 4.8.

4.2 Exchange interaction in charge-transfer insulators

Going back to the ground state of charge-transfer insulators, we have to discuss possible modifications of the exchange interaction compared with the standard superexchange typical for Mott insulators, see Chapter 1. Similarly to the treatment of superexchange in Section 1.3, in the model (4.1) we have to consider possible hoppings of electrons, which would lift spin degeneracy.

One contribution to superexchange is exactly equivalent to that of the Hubbard model (1.6): according to eq. (4.7), there will exist a possibility of hopping of a d-electron from one TM site to another and back, see the detailed scheme of Fig. 4.9(a). Here we mark the sequence of electron hoppings by the numbers 1, ..., 4. One can write this process as a "chemical reaction"

$$d_i^n p^6 d_j^n \xrightarrow{1} d_i^n p^5 d_j^{n+1} \xrightarrow{2} d_i^{n-1} p^6 d_j^{n+1} \xrightarrow{3} d_i^n p^5 d_j^{n+1} \xrightarrow{4} d_i^n p^6 d_j^n . \qquad (4.8)$$

The resulting exchange for nondegenerate d levels would be antiferromagnetic,

$$\mathcal{H} = J \sum_{\langle ij \rangle} S_i \cdot S_j , \qquad (4.9)$$

with the exchange constant $J = J_{dd}$, where

$$J_{dd} = \frac{2t_{dd}^2}{U_{dd}} = \frac{2t_{pd}^4}{\Delta_{CT}^2 U_{dd}} , \qquad (4.10)$$

cf. (1.12), with t_{dd} given by (4.7) (the factor 2 comes from two possible sequences of electron hoppings – either first from site i to site j and then back, as shown in Fig. 4.9(a), or in the opposite order, from j to i and back). This, in fact, is the main mechanism of superexchange in almost all transition metal compounds such as oxides, not necessarily charge-transfer insulators: in the standard Mott–Hubbard insulators the d–d hopping also occurs via the intermediate ligands (Anderson, 1959), see Chapter 5.

However when we include the oxygen p states explicitly, there appears yet another possibility, shown in Fig. 4.9(b): first one electron from the intermediate oxygen hops to, say, the TM ion j on the right, then the other p-electron hops to the other TM ion i, and then these electrons hop back. The "reaction" similar to (4.8), corresponding to Fig. 4.9(b), would look as follows:

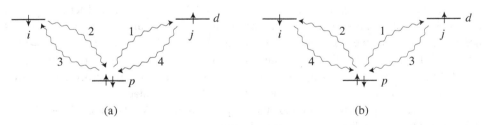

(a)   (b)

Figure 4.9 The mechanism of exchange interaction involving anion p orbitals in an apparent way.

$$d_i^n p^6 d_j^n \xrightarrow{1} d_i^n p^5 d_j^{n+1} \xrightarrow{2} d_i^{n+1} p^4 d_j^{n+1} \xrightarrow{3} d_i^{n+1} p^5 d_j^n \xrightarrow{4} d_i^n p^6 d_j^n . \qquad (4.11)$$

As we see, for nondegenerate d levels and for a 180° TM–O–TM bond this process is possible if the spins of TM ions i and j are antiparallel; thus this process also gives anti-ferromagnetic coupling, adding to the interaction (4.9). The difference from the process of Fig. 4.9(a) and eq. (4.8) is that, in contrast to that process, here the intermediate state after two hoppings is not that with one d-electron transferred to the neighboring TM ion, $d_i^{n-1} p^6 d_j^{n+1}$ with excitation energy U_{dd}, but rather the state $d_i^{n+1} p^4 d_j^{n+1}$, with two holes on oxygen (configuration p^4 rather than p^6). As defined in eq. (4.3), the transfer of one p-electron to a d site costs energy Δ_{CT}. Consequently, the creation of two such holes would cost $2\Delta_{CT}$, but two holes on the same oxygen also repel one another with repulsion U_{pp}, that is the energy of this intermediate state would be $2\Delta_{CT} + U_{pp}$. This quantity would then stand instead of U_{dd} in the expression for the exchange constant J in (4.9).

We can also notice that for the process of Fig. 4.9(b) and eq. (4.11) there exists an extra possibility: after two first hops (1) and (2) the return of electrons back to the oxygen may go in the opposite order, first from site i and then from site j, that is we can interchange the sequence of reverse hops (3) and (4). In effect there will be twice as many different "routes" for the process of Fig. 4.9(b) as for Fig. 4.9(a). Thus this second method of exchange interaction would finally give an extra antiferromagnetic exchange (4.9) with the exchange integral

$$J_{pd} = \frac{4 t_{pd}^4}{\Delta_{CT}^2 (2\Delta_{CT} + U_{pp})} = \frac{2 t_{pd}^4}{\Delta_{CT}^2 (\Delta_{CT} + \frac{1}{2} U_{pp})} . \qquad (4.12)$$

The total antiferromagnetic exchange constant in this case will be the sum of the contributions (4.10) and (4.12), that is

$$J_{total} = \frac{2 t_{pd}^4}{\Delta_{CT}^2} \left(\frac{1}{U_{dd}} + \frac{1}{\Delta_{CT} + \frac{1}{2} U_{pp}} \right) . \qquad (4.13)$$

Depending on the ratio of the Hubbard repulsion of d-electrons $U = U_{dd}$ and the charge-transfer energy Δ_{CT} (or rather $\Delta_{CT} + \frac{1}{2} U_{pp}$), either one or the other term in eq. (4.13) will dominate. We see here again the difference between Mott–Hubbard and charge-transfer insulators, illustrated in Fig. 4.7: in the Mott–Hubbard regime $\Delta_{CT} \gg U_{dd}$; the second term in (4.13) becomes irrelevant, and we return to the standard result for the superexchange in the Hubbard model (1.12) or (4.9), (4.10). In the charge-transfer regime, however, which is realized for $U_{dd} \gg \Delta_{CT}$, we should keep the second term in (4.13), that is the resulting ground state in this case will also be insulating and antiferromag-netic (remember that we consider here the situation with nondegenerate d levels with 180° TM–O–TM bonds), but with the exchange interaction dominated by the process (4.11) of Fig. 4.9(b) and with the exchange constant (4.12). Sometimes the first process is referred to as superexchange, and the second one as semicovalent exchange.

We thus see that the type of ground state is quite similar in Mott–Hubbard and charge-transfer insulators; the difference is mainly in the character of the lowest charge excitations and in the behavior of these two types of system with doping – in Mott–Hubbard insulators the lowest charge excitations are those between the d-shells, whereas in charge-transfer insulators these are transfers of electrons from oxygens to TM ions, with the creation of oxygen (or, more generally, ligand) holes. Although the detailed mechanism of the exchange process may be different, in most cases the resulting type of magnetic ordering in these two regimes is the same.

An interesting question is whether in the insulating state, for $t_{pd} \ll \{U_{dd}, \Delta_{CT}\}$, the details of the ground state, for example the type of magnetic or orbital ordering, are always the same in the Mott–Hubbard and in the charge-transfer regimes. In the simplest case of nondegenerate d levels in simple lattices such as perovskites, considered above, this is indeed the case: the ground state is an antiferromagnetic Mott insulator, the only difference being in the detailed expression for the exchange interaction, cf. eqs (4.10) and (4.12) (and of course the "lowest" high-energy charge excitations are different). However this is not always the case. Thus, in the orbitally degenerate situation even the ground state itself and the type of orbital and magnetic ordering can be different for Mott–Hubbard and charge-transfer insulators (Mostovoy and Khomskii, 2004).

4.3 Systems with small or negative charge-transfer gap

At the end of Section 4.1 we mentioned that for the late $3d$ elements and for high-valence or high-oxidation states of TM ions, the charge-transfer gap may be quite small and may even become negative, see for example Fig. 4.8 and Bocquet *et al.* (1992), Khomskii (1997). (This is even more probable for $4d$ and $5d$ elements, e.g. going in the columns of the Periodic Table from Ni to Pd to Pt, or in the series Cu–Ag–Au; see the Periodic Table at the end of the book.) The properties of the corresponding compounds may be rather nontrivial.

The most important feature in this case is the very important role played by (oxygen) p-bands, which would typically acquire a certain number of holes. This is seen from the very definition of charge-transfer energy (4.3): if $\Delta_{CT} < 0$, the "reaction" $d^n p^6 \longrightarrow d^{n+1} p^5$ would occur spontaneously, since in this process we would not spend but rather gain energy. That is, there should appear oxygen holes (configurations p^5 instead of the "normal" configurations p^6 of O^{2-}). This is also clear from the crystal field splitting scheme (Fig. 4.10), which generalizes to this case the standard level scheme of Fig. 3.8. (As always, one has to be careful in using this picture; the levels shown here are not just single-particle levels of noninteracting electrons, but include the effect of Coulomb (Hubbard) interaction, so that the energy of the p–d transition $d^n p^6 \longrightarrow d^{n+1} p^5$, i.e. $\Delta_{CT} = \varepsilon_d - \varepsilon_p$ (here negative), is $\Delta_{CT} = \varepsilon_d^0 - \varepsilon_p^0 + nU$.)

One sees from the (simplified) picture of Fig. 4.10 that if "originally" there were six p-electrons on p levels, for $\Delta_{CT} < 0$ it would be favorable to transfer (at least some of) them to the d levels, thus creating p-holes. In this situation this would happen even for stoichiometric systems and even in the ground state at $T = 0$. As a result the initial valence state of such a TM ion, for example $Fe^{4+}(d^4)$, would look rather as

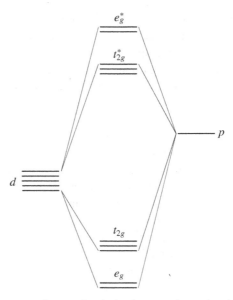

Figure 4.10 Schematic structure of energy levels for the case of negative charge transfer gap (p levels of the anion lie above the d levels of the transition metal).

$Fe^{3+}(d^5)O^-(2p^5) \equiv Fe^{3+}\underline{L}$, where by $\underline{L}$, as always, we denote a ligand (here oxygen) hole. One can call this effect *self-doping* (Korotin *et al.*, 1997). Large participation of oxygen holes in situations with small or negative charge-transfer gaps can modify very strongly the properties of such compounds.

At first glance, one unexpected consequence concerns crystal field splitting in the presence of negative charge-transfer gap. As we see from Fig. 4.10, the order of d orbitals is the standard one (t_{2g} levels lie below e_g) for *antibonding* orbitals (denoted here as e_g^* and t_{2g}^*), but for bonding orbitals the splitting due to covalency, or p–d hybridization, is the opposite. In the usual case with p states lying deeper than d ones (positive charge-transfer gap, see Fig. 3.8), these bonding orbitals have predominantly p character, for example these are p orbitals of oxygens, and we usually do not worry about them – they lie deep below the Fermi level. All the interesting phenomena for "normal" systems with $\Delta_{CT} > 0$ happen on the antibonding orbitals, which are actually the d levels we usually deal with. However the situation changes in the case of negative charge-transfer gap: here it is the bonding orbitals that have predominantly d character, and it may look as though the crystal field splitting of d levels is inverted (Ushakov *et al.*, 2011): the e_g (bonding) levels lie below the t_{2g} ones, contrary to the usual situation. In this rather rare case the Coulomb and covalency contributions to the crystal field splitting would work in the opposite directions. The p–d hybridization gives the order of levels shown in Fig. 4.10, that is the bonding e_g levels are below t_{2g}. But the Coulomb, or point-charge contribution, would tend to produce the opposite effect: the e_g-electrons would still repel more strongly from the negatively charged ligands than the t_{2g}-electrons; this could push the e_g levels (the bonding ones!) *above* t_{2g}. Apparently, the covalency contribution usually dominates (Ushakov *et al.*, 2011).

Returning to the specific features of materials with negative charge-transfer gap and with oxygen holes: a large contribution of these holes can strongly change many properties of these systems. Such *p*-holes may still be localized, or they may behave as itinerant ones – in which case the resulting state can become metallic, with a large contribution of oxygen states at the Fermi level. This is indeed what often happens in TM oxides with, formally, high valence of the heavier $3d$ metals. Thus, $SrFeO_3$ and $SrCoO_3$ are metals, $SrCoO_3$ is ferromagnetic, and $SrFeO_3$ has a spiral magnetic structure. Another similar example is CrO_2 (Korotin *et al.*, 1997) – also a ferromagnetic metal. Apparently the origin of ferromagnetism in these crystals is connected with their metallic conductivity (see Sections 5.2 and 9.4) which, in turn, is largely due to the significant contribution of oxygen holes.

The situation with small or negative charge- transfer gap is especially important for some doped TM compounds. It can happen that the original undoped material has TM ions with positive charge-transfer gap, but doping by holes would create ionic states with higher valence, for which this gap may become negative. Examples are hole-doping of $LaCoO_3$ in systems such as $La_{1-x}Sr_xCoO_3$, or hole-doping of the prototype material for the high-T_c superconductor La_2CuO_4, $La_{2-x}Sr_xCuO_4$. The original ionic states (Co^{3+}, Cu^{2+}) have, according to Fig. 4.8, relatively small yet still positive values of Δ_{CT}. But the formal valence state created by Sr doping, Co^{4+} and Cu^{3+}, would already have $\Delta_{CT} \lesssim 0$. This means that in this case the states should rather be represented as $Co^{3+}\underline{L}$ and $Cu^{2+}\underline{L}$, that is the doped holes would go not so much to the d-shells but rather to the oxygens, creating oxygen holes (though the *quantum numbers* of the respective states would be the same as those of Co^{4+} and Cu^{3+}). The interpretation of the properties of such materials should be done taking into account a large contribution of *p*-holes; this can strongly modify the properties of these compounds.

Yet another interesting phenomenon, for which a small or negative charge transfer gap and a large contribution of oxygen holes seem to be very important, is spontaneous charge disproportionation, sometimes observed in such systems. This will be discussed in more detail in Section 7.5; here we only stress that this phenomenon is typically observed in systems with large participation of oxygen holes. Thus this charge disproportionation is observed in $CaFeO_3$ (Takano *et al.*, 1981), where states which are formally $Fe^{4+}(d^4)$ "decompose" into $Fe^{3+}(d^5)$ and $Fe^{5+}(d^3)$,

$$2Fe^{4+} \longrightarrow Fe^{3+} + Fe^{5+} . \qquad (4.14)$$

The real electronic configurations of the corresponding ions, however, are definitely not Fe^{4+} or Fe^{5+}, but rather $Fe^{3+}\underline{L}$ and $Fe^{3+}\underline{L}^2$, that is the actual holes are on oxygens – although, again, the quantum numbers of the corresponding states coincide with those of Fe^{3+} ($S = \frac{5}{2}$) and Fe^{5+} ($S = \frac{3}{2}$). Thus, one should represent the reaction (4.14) rather as

$$2Fe^{3+}\underline{L} \longrightarrow Fe^{3+} + Fe^{3+}\underline{L}^2 , \qquad (4.15)$$

that is, this reaction involves predominantly a transfer of p-holes. Charge disproportionation (4.14), (4.15) becomes possible precisely because of the large contribution of oxygen
holes: if this process involved a real transfer of d-electrons, $2d^4 \rightarrow d^5 + d^3$, it would cost
very large Hubbard energy U_{dd}. But if the holes are largely on oxygens, their redistribution
costs only the energy of repulsion of two p-holes on the extended oxygen p-shell (made
up of p orbitals of six oxygens) around a TM center – which is of course much smaller.

There are other examples of this phenomenon, for example in perovskite nickelates
$RNiO_3$, especially with small rare earths R, or in the material $Cs_2Au_2Cl_6$, which can be
understood as a perovskite $CsAuCl_3$ in which two "Au^{2+}" disproportionate into $Au^{1+}(d^{10})$
and $Au^{3+}(d^8) = Au^{1+}\underline{L}^2$.

A special and very important group of questions concerns the behavior of systems when
Δ_{CT} changes gradually from positive to negative values, that is when we move "from the
right to the left" in the ZSA phase diagram (Fig. 4.7) while still keeping U_{dd}/t_{pd} large
(i.e. into the region marked in Fig. 4.7 by the question mark). As already mentioned in
Section 4.1, in this case the d-electrons are still strongly correlated ($U/t \gg 1$). But such a
narrow d level would overlap with the p levels which can form relatively broad and much
less correlated p bands. The situation would then resemble the case of mixed-valence or
heavy-fermion rare-earth compounds, discussed in Chapter 11, with the energy spectrum
and density of states appearing for example similar to Fig. 11.7 there, and with all the possible complications we encounter in such materials. The outcome and type of the resulting
state are *a priori* unclear, and may depend on the details of the crystal and electronic
structure of the particular system. Possible resulting states can be metallic, with different
magnetic properties, such as $SrFeO_3$ mentioned above; or they may be insulating, magnetically ordered or diamagnetic, possibly with spontaneous charge disproportionation, etc.
These questions will be discussed in more detail in Section 10.5.

Last, but not least, the large contribution of oxygen holes can be instrumental in the
phenomenon of high-T_c superconductivity in cuprates, for example $La_{2-x}Sr_xCuO_4$. These
materials are obtained mostly by hole-doping of CuO_2-planes, that is by formally creating
states "Cu^{3+}" from the states $Cu^{2+}(d^9)$, which are in fact predominantly states with holes
on oxygens, so $Cu^{2+}\underline{L}$. To what extent this factor is indeed crucial for high-T_c superconductivity is actually an open question, but it seems really important. This will be discussed
in more detail in Section 9.7.

4.4 Zhang–Rice singlets

In the previous section we considered the situation with small or negative charge-transfer
gap, and argued that in that case, when we dope systems by holes, the holes go in the
first approximation not to the d-shells of TM ions, but rather to the p-shells of ligands,
for example oxygens. In this case we definitely have to include these p states in our
description, and use the model of eq. (4.1) type.

However there is still $d–p$ hybridization, which will mix d and p states. In some cases
this can lead to a situation in which the behavior of the system would resemble that of the

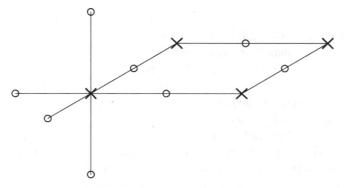

Figure 4.11 Main elements of the crystal structure of CuO_2-planes in high-T_c cuprates such as $La_{2-x}Sr_xCuO_4$. Crosses are Cu ions, circles are oxygens.

simple one-band Hubbard model (1.6). This, in particular, can be the case in hole-doped cuprates – the basis of the high-T_c superconductors. The objects which appear in this case are known as Zhang–Rice singlets (Zhang and Rice, 1988).

4.4.1 d–p bound states and reduction of the d–p model to a single-band model

Suppose we start from the system La_2CuO_4, built out of two-dimensional CuO_2 planes, with the Cu sublattice forming a square lattice, and with oxygens located between the copper ions (see Fig. 4.11). Owing to strong local distortion (elongation in the c-direction) of the CuO_6 octahedra the crystal field splitting is such that the orbital degeneracy of Cu^{2+} ions is lifted, and we have one "active" e_g orbital $x^2 - y^2$, Fig. 3.26(b) (in this case it is more convenient to speak of hole orbitals). Thus in a first approximation we can ignore orbital degeneracy and consider only the system comprising this d orbital and hybridized p orbitals of four neighboring in-plane oxygens; the apical oxygens play no role in this limit.

When we dope such system by holes, we formally create states $Cu^{3+}(d^8)$ from the initial states $Cu^{2+}(d^9)$. There are, in principle, two ways of doing this. We can remove the electron from the higher-lying $x^2 - y^2$ orbital of Fig. 3.26(b), in which case we will be left with the nonmagnetic (low-spin) state of $Cu^{3+}(t_{2g}^6(z^2)^2)$, $S = 0$. Alternatively, we can remove the electron from the z^2 orbital; in this case we would get the high-spin state of $Cu^{3+}(t_{2g}^6 z^2\uparrow(x^2 - y^2)\uparrow)$ with $S = 1$ (and the Jahn–Teller elongation of CuO_6-octahedra should disappear, or at least be strongly suppressed). As the splitting of e_g orbitals in Cu^{2+} (and Cu^{3+}) is usually quite large, the first situation is preferred: it is more favorable to leave two electrons on the z^2 orbital. In effect, formal $Cu^{3+}(d^8)$ – if it existed – would practically always be in the nonmagnetic $S = 0$ state.

However, in fact the situation is different. As discussed above, Cu^{3+} is a state with negative charge-transfer gap (Fig. 4.8). This means that the holes go predominantly to the oxygen p orbitals, that is we obtain the state $Cu^{2+}(d^9)\underline{L}$, which would hybridize with the state $Cu^{3+}(d^8)$.

The state $Cu^{2+}\underline{L}$ would have a Cu^{2+} ion with $S = \frac{1}{2}$ (as mentioned above, it is more convenient in this case to speak about one hole on the $x^2 - y^2$ orbital), and the hole on the p orbitals of surrounding oxygens would also have spin $\frac{1}{2}$. The hybridization of these states would lead to antiferromagnetic exchange between them, and produce a singlet state of the type $(d\uparrow p\downarrow - d\downarrow p\uparrow)$, that is the d-hole on Cu^{2+} and the p-hole on the surrounding oxygens would form a singlet bound state; this is the *Zhang–Rice singlet*. This singlet state would mix with the state Cu^{3+} with $S = 0$, described above.

It is important that the d-hole in the state $x^2 - y^2$ hybridizes with the p states of *four* oxygens. This coherent hybridization strongly increases the binding energy and stabilizes the Zhang–Rice singlets. This can be shown in the following way: as is clear from Fig. 3.5, the $x^2 - y^2$ d state hybridizes with a combination of p states of oxygens with the same symmetry, cf. eq. (3.5):

$$|p\rangle_{x^2-y^2} = |p_{\text{coh.}}\rangle = \tfrac{1}{\sqrt{4}}(+p_1 - p_2 - p_3 + p_4) \tag{4.16}$$

(the signs are chosen in such a way that the s–p overlaps have the same sign). If the p–d hopping matrix element is $t_{pd} = \langle d_{x^2-y^2}|\hat{t}|p_1\rangle$, then hybridization with the coherent superposition (4.16) will be given by

$$t_{pd,\text{coh.}} = \langle d_{x^2-y^2}|\hat{t}|p_{\text{coh.}}\rangle = 4\cdot\tfrac{1}{\sqrt{4}}t_{pd} = 2t_{pd}\,, \tag{4.17}$$

that is the p–d hopping t_{pd} will be doubled. Similarly, if one d state coherently hybridizes with N other states, the effective hopping t (of whatever nature) will be modified to $\sqrt{N}\,t$. This fact will also be important for example for Mott transitions in systems with degenerate bands, see Section 10.4.2.

In effect the d–p hybridization would lead to the formation of a bound state whose energy in this simple approximation would be given by the solution of the two coupled equations for the d state and the coherent p state (4.16), with the energies given, as usual, by the eigenvalues of the matrix

$$\begin{pmatrix} \varepsilon_d & t_{pd,\text{coh.}} \\ t_{pd,\text{coh.}} & \varepsilon_p \end{pmatrix}. \tag{4.18}$$

That is, taking (4.17) into account,

$$\omega_\pm = \tfrac{1}{2}(\varepsilon_p + \varepsilon_d) \pm \sqrt{\tfrac{1}{4}(\varepsilon_p - \varepsilon_d)^2 + t_{pd,\text{coh.}}^2}$$
$$= \tfrac{1}{2}(\varepsilon_p + \varepsilon_d) \pm \sqrt{\tfrac{1}{4}(\varepsilon_p - \varepsilon_d)^2 + 4t_{pd}^2}, \tag{4.19}$$

which for $\varepsilon_p - \varepsilon_d \gg t_{pd}$ gives for the ground state (the Zhang–Rice singlet)

$$\omega_- = \varepsilon_d - \frac{4t_{pd}^2}{\varepsilon_p - \varepsilon_d}. \tag{4.20}$$

We see that, first, we indeed obtain a singlet bound state, in this case of d- and p-holes; and, second, due to coherent hybridization with a proper combination with (here) four p states of oxygens the one-bond hopping matrix element t_{pd}, determining the binding energy, is enhanced, $t_{pd} \longrightarrow t_{pd,\text{coh.}} = 2t_{pd}$, which in the case (4.20) increases the binding energy by a factor of four. (If the d and p states are nearly degenerate, $\varepsilon_p - \varepsilon_d < t_{pd,\text{coh.}}$, we would get from eq. (4.19) the binding energy $\sim 2t_{pd,\text{coh.}}$, i.e. it would be increased by a factor of two.[2])

Let us sum up what we have obtained. In effect, when we start from an undoped cuprate with $\text{Cu}^2(d^9, S = \frac{1}{2})$, for example La_2CuO_4, and hole-dope it, as in $\text{La}_{2-x}\text{Sr}_x\text{CuO}_4$, each hole first goes to the p states of the oxygens, and then forms a singlet bound state with Cu^{2+}. That is, the resulting effect is exactly the same as if we dealt with a system with nondegenerate d states of the transition metal, described by the simple Hubbard model (1.6), and hole-doped it: each hole is a state without charge carrier (an electron, or a hole) and without spin. The Zhang–Rice singlet is just such a state. This was the conclusion of Zhang and Rice (1988): in typical cases for high-T_c cuprates one can reduce the d–p model (4.1) to the nondegenerate Hubbard model, at least for the ground and lowest excited states, with the hole state replaced by the Zhang–Rice singlet state.

One should note that in the case of hole-doped cuprates, in general one also has to take into account the Coulomb interaction of two holes, at least when they are on the same site, for example in the state $\text{Cu}^{3+}(d^8)$. That is, strictly speaking, we cannot describe this situation using the single-particle picture, as we did above. The inclusion of this interaction would change some expressions and numerical estimates (this is analogous to going from the description of molecular bonding in the molecular orbital (MO LCAO) picture to the Heitler–London description). However, qualitative conclusions remain the same.

The mathematical description of this situation follows exactly that of eqs (4.18)–(4.20). But when we include the interactions, we have to deal with many-electron configurations such as $d^n p^6$ and $d^{n+1} p^5 = d^{n+1} \underline{L}$. The secular equation for energy levels would then look exactly as eq. (4.18), but with full energies E_{d^n} and $E_{d^{n+1}\underline{L}}$ in the diagonal positions. Correspondingly, the solution would also be the same as (4.19) and (4.20), with ε_d replaced by E_{d^n} and ε_p replaced by $E_{d^{n+1}\underline{L}}$. That is, in the denominator of eq. (4.20) we would have $E_{d^n} - E_{d^{n+1}\underline{L}}$, which, according to eq. (4.3), is nothing else but the charge-transfer gap $|\Delta_{\text{CT}}|$.

The total density of states in a concentrated system, in the case of a charge-transfer insulator with small but still positive gap (e.g., for undoped La_2CuO_4), would look as shown in Fig. 4.12. The d and p levels would be broadened to bands, and without d–p hybridization the occupied d band would lie below the p band. But because of hybridization a split-off band would be formed, which is nothing else but the Zhang–Rice band. For the situation corresponding to cuprates it is not a bound, but rather an antibound state in the electronic picture (but it would be the bound state in the hole representation, often used for cuprates).

---

[2] One can also obtain these results directly, by writing the secular equation describing the hybridization of p states of four oxygens separately, without going to the coherent superposition (4.16), that is by writing the secular equation in the form of a 5×5 matrix, instead of (4.18). One can then see that the energies of the lowest and highest bonding and antibonding states will be given by the same eqs (4.19), (4.20), the remaining three roots $\omega = \varepsilon_p$ describing the nonbonding oxygen states.

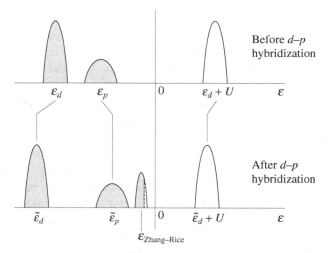

Figure 4.12 Schematic form of the density of states before and after d–p hybridization, illustrating the formation of Zhang–Rice singlets. The dashed line in the lower diagram represents the position of the Fermi level for hole-doped systems, in which holes go to the Zhang–Rice singlet states.

This Zhang–Rice band may be separated from the p band, as shown in Fig. 4.12, or may merge with it. One can say that this band is formed due to a repulsion between occupied d and p bands.

In the undoped cuprates with $Cu^{2+}(d^9)$ all states below the Fermi level (zero in Fig. 4.12) are filled. But when we dope such a system by holes, the first holes go to these Zhang–Rice states (the new Fermi level for the doped system is marked in Fig. 4.12 by a dashed line). And it is the states in this band which can in principle be described by the one-component Hubbard model.

The conclusion of Zhang and Rice (1988) that one can reduce the description of the doped cuprates from the full d–p model (4.1) to the single-band Hubbard model (1.6) (or to the so-called t–J model, see Section 9.7 below) is almost universally accepted now in the description of high-T_c cuprates. However, there are also conflicting claims. Thus, Emery and Reiter (1988) argued that the p-hole introduced by doping is attached not to one Cu^{2+}, forming a singlet state (the Zhang–Rice singlet, Fig. 4.13(a)), but to *two* neighboring Cu ions, "to the left and to the right" of a given oxygen ion with the hole. It would then have strong antiferromagnetic exchange with these two Cu ions, see Fig. 4.13(b). In the Zhang–Rice model the antiferromagnetic d–p coupling leads to the formation of a single-site singlet bound state. But in the Emery–Reiter picture these three sites (three spins) of Fig. 4.13(b) would form a state with total spin $S = \frac{1}{2}$, which can be viewed as a state with Cu spins parallel, and the spin on the oxygen opposite to them; that is, the oxygen hole provides a mechanism of ferromagnetic coupling between the spins of Cu^{2+}, which, according to the considerations of Chapter 3 and Section 4.2, would be $\sim t_{pd}^2/\Delta_{CT}$ – much stronger than the initial antiferromagnetic Cu–Cu coupling $\sim t_{pd}^4/\Delta_{CT}^3$. Thus, in this picture hole doping leads to a very strong frustration of the original antiferromagnetic

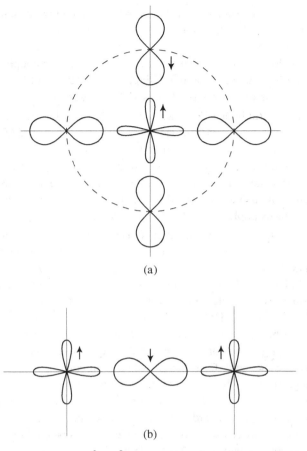

(a)

(b)

Figure 4.13 (a) Hybridization of an $x^2 - y^2$ (hole) orbital of Cu with four p orbitals of surrounding oxygens, showing the formation of Zhang–Rice singlets. (b) An alternative picture (V. Emery) in which the oxygen hole couples to two neighboring Cu ions.

ordering in undoped cuprates, which can explain the rapid suppression of antiferromagnetism with doping. (Note that in the Zhang–Rice picture we can also come to such a conclusion – it would be similar to the suppression of antiferromagnetism by doping in the simple Hubbard model, see Chapter 1.)

Although the controversy between the two pictures described above is still not completely resolved, it seems that the Zhang–Rice picture is closer to reality, and it is widely used nowadays. Similar pictures work not only for cuprates; one can use similar arguments for other systems with small or negative charge-transfer gaps and with p-holes. For other TM ions one can also have a bound state of such a p-hole and a d site, but the total spin of the resulting state would not necessarily be zero. Thus for example CrO_2 with, formally, $Cr^{4+}(d^2, S = 1)$ has in fact a lot of oxygen holes, that is it can be represented as (or at least strongly admixed with) the state $Cr^{3+}(d^3, S = \frac{3}{2})$ forming a bound state with an oxygen p-hole with spin $S = -\frac{1}{2}$, so that the total spin of the resulting state is again $S = 1$. In other words, we can also have here a state resembling the Zhang–Rice state (bound state of a p-hole and a TM ion), in this case also with antiferromagnetic coupling between them. The resulting state here, however, is not a singlet, but a state with $S = 1$. Similarly, in $La_{1-x}Sr_xCoO_3$ we formally have x Co^{4+} ions with configuration t_{2g}^5, $S = \frac{1}{2}$ (the low-spin state, see Section 3.3). But in fact Co^{4+}, according to Fig. 4.8, has negative charge-transfer gap, that is the state would rather be the (low-spin) $Co^{3+}(t_{2g}^6, S = 0)\underline{L}(S = \frac{1}{2})$.

One notices here an important point. If in the case of $\Delta_{CT} < 0$ we initially put n d-electrons and no p-holes on a TM ion, configuration $d^n p^6$, we can usually understand, using arguments presented in Chapters 2 and 3 (especially Section 3.3), what the electronic configuration of such a d ion would be and what the corresponding term (the total spin) would be – for example $Cu^{3+}(d^8, S = 0)$, $Cr^{4+}(d^2, S = 1)$, $Co^{4+}(t_{2g}^5, S = \frac{1}{2})$. If in real cases some holes go to oxygens, that is instead of the configuration $d^n p^6$ we have $d^{n+1} p^5 = d^{n+1}\underline{L}$, the p-hole binds to the corresponding d ion in such a way that the *quantum numbers*, for example the total spin, remain the same as those of the "original" configuration d^n. Thus we have a Zhang–Rice singlet ($S = 0$) for $Cu^{3+} \rightarrow Cu^{2+}\underline{L}$, or a "Zhang–Rice-like" bound state with $S = 1$ for $Cr^{4+} \rightarrow Cr^{3+}\underline{L}$, or a similar state with $S = \frac{1}{2}$ for $Co^{4+} \rightarrow Co^{3+}\underline{L}$. However the *spatial distribution of spin density* would be different; instead of being concentrated on the d shells of for example Cr^{4+} or Co^{4+}, the spin density would largely be concentrated on the oxygens surrounding a given TM ion. Correspondingly, the magnetic form factor, which is important for example in magnetic neutron scattering, would be different from that of the d shell. Such form factors are used for example to extract the values of the magnetic moment from the magnetic neutron scattering. If one were to use the standard tabulated form factors for the d-electrons for systems with negative change-transfer gaps and with oxygen holes, one could get wrong values of the effective magnetic moment μ_{eff}. This has to be taken into account when treating such experimental data.

4.4.2 "Real" p-holes, exchange interaction, and magnetic states in systems with ligand holes

In Section 4.2 we considered modifications of the exchange interaction in charge-transfer insulators. But there we mostly treated the situation with $\Delta_{CT} > 0$, and qualitatively the exchange was the same as in the simple Hubbard model – maybe with orbital degrees of freedom taken into account, and with the replacement of t_{dd}^2/U by $t_{dd}^2/(\Delta_{CT} + \frac{1}{2}U_{pp})$, see eqs (4.10) and (4.12), (4.13). However, in the case of negative charge-transfer gap the situation may be different. Here we would have *real p*-holes on some oxygens. This can lead to strong modification of the magnetic state.

The simplest picture has already been shown above, Fig. 4.13(b), from which it is clear that the real p-holes would make the Cu–Cu exchange ferromagnetic, instead of antiferromagnetic as in undoped cuprates such as La_2CuO_4. And this ferromagnetic exchange is

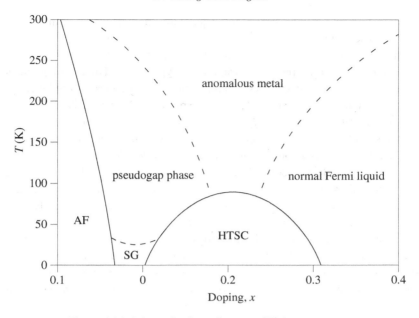

Figure 4.14 Schematic phase diagram of high-T_c cuprates.

stronger than the original antiferromagnetic one: according to (4.13) the antiferromagnetic exchange in the bond Cu^{2+}–Cu^{2+} is $J_{dd} \sim t_{pd}^4/\Delta^2(\Delta_{CT} + \frac{1}{2}U_{pp})$, whereas the exchange between the (supposedly localized) p-hole and the spin of Cu^{2+} is $J_{pd} \sim t_{pd}^2/\Delta_{CT}$ – much larger than J_{dd}. (It is precisely this strong exchange which leads to the formation of the Zhang–Rice singlet.) That is, every such hole appearing at hole doping would lead to *strong frustration* of the original antiferromagnetic state, which can lead to rapid suppression of antiferromagnetism in $La_{2-x}Sr_xCoO_4$ even by small hole concentration. This agrees with experimental observations, see the schematic phase diagram of high-T_c cuprates (Fig. 4.14), shown here on the example of $La_{2-x}Sr_xCuO_4$.

This picture is of course oversimplified; we have treated the p-holes as localized and ignored the possibility of hopping and delocalization of these holes, which occurs due to the same p–d hybridization; such p_1–d–p_2 hopping would be of the same order, $\sim t_{pd}^2/\Delta_{CT}$, as the p–d exchange (and there is an extra channel of p-hole delocalization, due to direct overlap of p states on neighboring oxygens, which would make the p bands even broader). Similarly to the arguments presented in Chapter 1, one can show that such p-hole motion would also be suppressed by the background antiferromagnetic ordering, and the motion of these holes would also destroy antiferromagnetism (this factor is included in the effective one-band model, justified by the formation of Zhang–Rice singlets, see the preceding section). In any case, both these factors, frustrating exchange J_{pd} and kinetic (band) energy of p-holes, would lead to the same effect – a rapid suppression of the underlying antiferromagnetic ordering with doping. The simple picture of Fig. 4.13(b), though missing some details, is usually sufficient for qualitative understanding of the possible role of doping in these systems.

One can go one step further and try to anticipate what the situation would be at high doping, that is in this case, for large concentration of oxygen holes. Different outcomes are in principle possible. One can end up with a simple nonmagnetic Fermi-liquid metallic state, cf. Fig. 1.19. On the contrary, especially when the relevant TM state $d^n \underline{L}$ is itself magnetic, for example $Co^{4+}(d^5) = Co^{3+}(d^6)\underline{L}$ (or even $Co^{2+}(d^7)\underline{L}^2$), the coupling of these localized moments via the itinerant p-electrons (or p-holes) could make the resulting metallic state ferromagnetic. This can be visualized as a form of double exchange, to be discussed in Section 5.2, although with some modifications (in the standard double-exchange picture the mobile electrons are predominantly also d-electrons, whereas here they are rather the p-electrons of ligands). Such a ferromagnetic metallic state is apparently realized in $La_{1-x}Sr_xCoO_3$ for $x \gtrsim 0.2$, see Section 9.2.2, and also in some other systems. We will meet this situation in several places later in this book.

As we saw above, in systems with small or negative charge-transfer gap there are a lot of oxygen holes. An interesting question arises over the spin polarization what would be on oxygens. This question also exists for undoped TM compounds in the charge-transfer regime.

If we were dealing with antiferromagnetic ordering, and if oxygens were located between the TM sites with opposite spins, there would be zero molecular field on oxygens, and we would have the average spin $\langle S_O \rangle = 0$. (There may still be a lot of p-holes on oxygens, but with equal number of up and down spins.) However, the situation may be different for example for ferromagnetic ordering, or for in-plane oxygens in the A-type magnetic structure (ferromagnetic layers stacked antiferromagnetically, where the in-plane oxygens are "sandwiched" between Mn ions with parallel spins, see Fig. 5.24 below), which is the typical situation in undoped manganites of the type $LaMnO_4$. The question is whether the spins on oxygens would be parallel or antiparallel to the spins of neighboring TM ions.

It turns out that the results may be different depending on the specific situation. Consider for example the pair $Fe^{3+}(t_{2g}^3 e_g^2)$–O^{2-}, or $Mn^{2+}(t_{2g}^3 e_g^2)$–O^{2-}, with TM ions with spin ↑. As is clear from Fig. 4.15(a), in this case only the p-electrons with spin ↓ could hop virtually to such TM ions, and the remaining polarization on the oxygen would be ↑, that is *parallel* to the net magnetization. However if we have for example the pair $Cr^{3+}(t_{2g}^3 e_g^0)$–O^{2-}, Fig. 4.15(b), the situation is not so clear. On the one hand, p-electrons with spin ↓ can hop to the t_{2g} states of Cr^{3+} (dashed hoppings in Fig. 4.15(b)), which would also make the oxygen spin polarization parallel to that of Cr. But this hopping to t_{2g} states, $t_{pd\pi}$, is rather small; the hybridization with (here empty) e_g states, $t_{pd\sigma}$, is much more efficient. In this case, however, it is better to transfer virtually the p-electrons with spin up (parallel to the spin of Cr) to these empty e_g states, because then we would have an intermediate state satisfying Hund's rule, which is energetically more favorable. But then the spin remaining on the oxygen would be *opposite* to that of Cr. Thus these processes (the hoppings p–t_{2g} and p–e_g) compete, and it is not *a priori* clear which would dominate. Thus in such cases

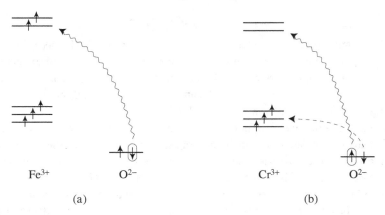

Figure 4.15 Possible spin polarization of oxygen ions for different electron configuration of neighboring transition metal ions: (a) $Fe^{3+}(d^5)$; (b) $Cr^{3+}(d^3)$.

the magnetic polarization on oxygens may be opposite to that of TM ions. A similar situation can exist in many other compounds, for example containing $Ru^{4+}(t_{2g}^4 e_g^0)$, such as $(Ca, Sr)_2RuO_4$ or $(Ca, Sr)_3Ru_2O_7$, or in other systems. And in this case one $p–t_{2g}$ hopping channel is blocked, whereas two channels are available to the hopping into empty e_g orbitals. On the other hand, as mentioned in Chapter 2, for $4d$ and $5d$ elements the Hund's rule coupling J_H, which favors the $p–e_g$ transfer of the parallel ($\uparrow$) spin, becomes smaller; thus here the question of spin polarization on oxygens is also open.

S.4 Summary of Chapter 4

In typical strongly correlated systems such as TM oxides the TM ions in a crystal are usually separated by anions, for example O^{2-} (we are mostly speaking of oxygens, though the same arguments apply also to other ligands such as S, Se, Cl, etc.). In such cases the direct overlap and hopping between the d states may be very small, but the hybridization between the d states of transition metals and the p states of ligands (e.g., oxygens) will be important. The effective d–d hopping will in such cases occur via the intermediate oxygens. In such cases one has to include in our description, besides the d-electrons, also the p-electrons of these oxygens. The resulting model (4.1) would generalize the standard Hubbard model (1.6) and give a more complete description of TM oxides.

When one includes the oxygen p states explicitly, there appear novel charge excitations in the system: besides the d–d transitions $d^n d^n \rightarrow d^{n+1} d^{n-1}$ treated in the Hubbard model, there may appear excitations with the transfer of an electron from the filled $2p$ shell $2p^6$ of O^{2-} to a TM: $d^n p^6 \rightarrow d^{n+1} p^5$. The d–d transitions "cost" energy U_{dd} – the energy of Hubbard repulsion between two d-electrons on a TM ion. In contrast, the excitation energy of p–d transitions is different, and can be both larger or smaller than that of d–d transitions. It is called the *charge-transfer energy* and is denoted Δ_{CT}.

For deep oxygen p levels $\Delta_{CT} \gg U_{dd}$; then we can exclude the p states and reduce our description to the standard Hubbard model (1.6), with an effective d–d hopping occurring via the oxygens, $t_{dd} = t_{pd}^2 / \Delta_{CT}$, where t_{pd} is the p–d hopping. When the charge-transfer energy Δ_{CT} is smaller, $\Delta_{CT} < U_{dd}$, but still with $t_{pd} \ll \{\Delta_{CT}, U_{dd}\}$, the ground state would be the same as in Mott insulators, that is the d-electrons would be localized, with localized magnetic moments and with (antiferro)magnetic ordering. But the lowest charge-carrying excitations would be different: these would not be d–d excitations, but p–d excitations, that is the creation of an extra electron on the d level, with the hole excitation being the oxygen p-hole. Such systems with $\Delta_{CT} < U_{dd}$, which for integer number of d-electrons are still insulators, are called *charge-transfer insulators*. The resulting schematic phase diagram – the Zaanen–Sawatzky–Allen (ZSA) diagram – is shown in Fig. 4.7.

The magnetic exchange in such states is again similar to that of Mott, or Mott–Hubbard insulators, and for nondegenerate d-electrons and for simple geometry with the M–O–M bond angle $\sim 180°$ it would also be antiferromagnetic, but the virtual excited states in the exchange process would be those with two holes on oxygen, of the type $d^{n+1} p^4 d^{n+1}$, instead of the excited state $d^{n-1} d^{n+1}$ in the Hubbard model. Correspondingly, the charge-transfer energy Δ_{CT} (or $\Delta_{CT} + \frac{1}{2} U_{pp}$) will appear in the denominator in the expression for the exchange constant, instead of U_{dd} in the Hubbard case, cf. (4.12) and (4.10).

Charge-transfer energy changes in a regular manner across the $3d$ series (it decreases from Ti to Cu), and it also decreases with increasing valence of the transition metal ions, in going from TM^{2+} to TM^{3+} to TM^{4+}. The corresponding data are presented in Fig. 4.8.

Especially interesting may be the situation for small or negative charge-transfer energy, which may happen for the late $3d$ elements with high valence, such as, formally, Fe^{4+} or Cu^{3+}. In these cases the real electronic configuration would rather be that with oxygen holes, of the type $Fe^{3+}\underline{L}$ instead of Fe^{4+}, or $Cu^{2+}\underline{L}$ instead of Cu^{3+}, where $\underline{L}$ denotes the ligand (here oxygen) hole. One can call this situation *self-doping*: oxygen holes will appear spontaneously even for nominally undoped, stoichiometric compounds. Examples of such systems are $CaFeO_3$ or CrO_2. Similarly, when we hole dope such systems, as for example in high-T_c cuprates such as $La_{2-x}Sr_xCuO_4$, the doped holes would predominantly go to oxygens, that is in this example we would create not so much $Cu^{3+}(d^8)$ but rather $Cu^{2+}(d^9)O^-(p^5) = Cu^{2+}\underline{L}$.

In some cases, in particular in high-T_c cuprates, one can reduce the general description in terms of the $d-p$ model (4.1) to the standard nondegenerate single-band Hubbard model (1.6). This possibility is based on the formation of bound states (Zhang–Rice singlets) made up of Cu^{2+} ($S = \frac{1}{2}$) and a ligand hole with the opposite spin, delocalized over four oxygens surrounding the Cu ion. Coherence effects in this $d-p$ hybridization strongly enhance the binding energy of these singlets. In effect the state created by doping the magnetic Mott (or rather charge-transfer) insulator such as La_2CuO_4 would correspond to one singlet per doped hole – exactly similar to the states created in the simple Hubbard model when we remove electrons, that is dope it with holes. Although there still remain some questions regarding this picture, it is applied successfully in many cases.

In general the states of the systems with small or negative charge-transfer gaps can be different, especially when U_{dd} remains large. They may be metals (of a heavy-fermion type, because they still contain strongly correlated d-electrons coexisting with less correlated p-electrons of oxygens). Or they may be insulators (again probably resembling Kondo insulators). They may have different magnetic properties, including ferromagnetism (examples are CrO_2 and $(La/Sr)CoO_3$). Spontaneous charge disproportionation may occur, as happens in $CaFeO_3$, see more details in Section 7.5. And the very phenomenon of high-T_c superconductivity in cuprates may be connected with this factor.

5

Exchange interaction and magnetic structures

In considering the exchange interaction and magnetic structures in real materials, one has to take into account several factors. The most important factors are the specific geometry of the material and the presence of different possible exchange routes, and the specific orbital structure of particular ions. By considering special cases in the same spirit as we did in Section 4.1, we can deduce general regularities determining the sign and strength of exchange in various particular situations. These are usually formulated as a set of rules, called Goodenough–Kanamori–Anderson (GKA) rules. To be realistic, one has to proceed from a model such as that of eq. (4.1), taking into account both d-electrons of transition metals and p-electrons of ligands.

5.1 Superexchange in insulators and Goodenough–Kanamori–Anderson rules

Consider first the simplest geometry of Fig. 4.5 with oxygens (or other ligands) sitting directly between TM ions, that is with 180° TM–O–TM bonds. If we have "magnetically active" orbitals having one localized d-electron each, both directed toward the oxygen, as in Figs 4.5 and 5.1(b), or, similarly, if we have half-filled t_{2g} orbitals overlapping with the same p_π orbitals of oxygen, Fig. 4.6(b), the situation will be exactly equivalent to that considered in Section 1.3 or 4.1: we will have a rather strong antiferromagnetic exchange

$$J \sim \frac{t_{pd}^4}{\Delta_{CT}^2 U} = \frac{t_{dd}^2}{U} \tag{5.1}$$

for Mott–Hubbard insulators, or

$$J \sim \frac{t_{pd}^4}{\Delta_{CT}^2 (\Delta_{CT} + \frac{1}{2} U_{pp})} = \frac{t_{dd}^2}{\Delta_{CT} + \frac{1}{2} U_{pd}} \tag{5.2}$$

for charge-transfer insulators, where the effective d–d hopping is $t_{dd} = t_{pd}^2/\Delta_{CT}$, cf. (4.10), (4.12), (4.13). (Below, for simplicity, we will mostly speak about the first case, but one has to remember that in many cases one should put in the denominator not $U = U_{dd}$, but $\Delta_{CT} + \frac{1}{2} U_{pp}$.) This mechanism of exchange (Anderson, 1959) is called *superexchange*,

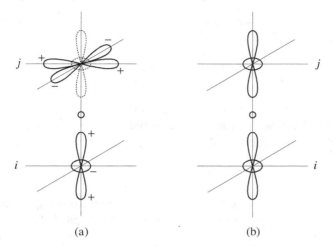

Figure 5.1 Different possible situations for exchange interaction depending on the type of (half-)occupied orbitals (solid lines) and empty orbitals (dashed lines).

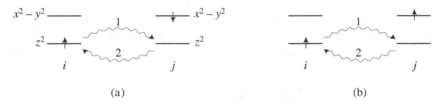

Figure 5.2 The process of hopping between occupied and empty orbitals (the situation of Fig. 5.1(a)) for different spin configurations.

or sometimes *kinetic exchange*; in application to charge-transfer insulators (eq. (5.2)) one sometimes uses the terminology *semicovalent exchange* (Goodenough, 1963).

The situation, however, changes when the orbital occupation is different. Suppose that we have a system with one e_g-electron as in Mn^{3+} or low-spin Ni^{3+}, or with one e_g-hole as in Cu^{2+}. This electron (or hole) can in principle occupy any of the two e_g orbitals d_{z^2}, $d_{x^2-y^2}$, or their linear combination. Suppose that for the TM pair along the z-direction one TM has the electron on the z^2 orbital, and another on the $(x^2 - y^2)$ one, see Fig. 5.1(a). As is clear from this figure, these orbitals are orthogonal (pay attention to the signs of the different lobes of wavefunctions), and there is no hopping between them (one can also draw here the p_z orbital, via which in fact such d–d hopping occurs, but this will not change our conclusion). In this situation the electron from the z^2 orbital at site i can hop to the similar *empty z^2* orbital at site j (dashed orbital in Fig. 5.1(a)) and then back. This virtual hopping would correspond to the process shown in Fig. 5.2. This again will give a negative contribution to the total energy $\sim -t^2/U$, cf. Section 1.3 (here $t = t_{dd}$). But, in contrast to the case of nondegenerate orbitals, or half-filled orbitals directed toward each other as in Fig. 5.1(b), we see that here such hopping is allowed both for antiparallel and parallel spins. The resulting energy gains, however, would not be the same: in the first case (antiparallel spins, Fig. 5.2(a)) the energy for the intermediate state will be U, that is the energy gain in this case will be

$$\Delta E_{\uparrow\downarrow} = -\frac{t^2}{U} \ . \tag{5.3}$$

However in the second case, with parallel spins (Fig. 5.2(b)), in the intermediate state there will be two electrons with parallel spins at site j. According to the considerations of Chapter 2, such electrons gain Hund's rule energy J_H, that is the resulting energy will be[1]

$$\Delta E_{\uparrow\uparrow} = -\frac{t^2}{U - J_H} \ . \tag{5.4}$$

We see that in this case the second possibility, ferromagnetic ordering, is more favorable, the corresponding energy gain compared with the case of antiparallel spins being

$$\Delta E_{\uparrow\uparrow} - \Delta E_{\uparrow\downarrow} = -t^2 \left(\frac{1}{U - J_H} - \frac{1}{U} \right) \simeq -\frac{t^2}{U} \frac{J_H}{U} \ , \tag{5.5}$$

where in the last step we have used the expansion in $J_H/U < 1$ (remember that in typical cases for $3d$ elements $J_H \sim 0.8\text{–}0.9\,\text{eV}$ and $U \sim 3\text{–}5\,\text{eV}$, so that $J_H/U \sim 0.2$). These qualitative considerations can be made rigorous: using perturbation theory in t/U, similar to Section 1.3, one can indeed obtain that in this case the resulting exchange (1.2) would be ferromagnetic,

$$\mathcal{H} = J \sum_{\langle ij \rangle} S_i \cdot S_j \ , \qquad J \sim -\frac{t^2}{U} \frac{J_H}{U} \ . \tag{5.6}$$

Thus we see that if on neighboring sites the "active" half-filled orbitals, having unpaired spins, are directed toward each other, so that there exists effective hopping t between them (Fig. 5.1(b)), the exchange interaction is antiferromagnetic and strong, $J_{af} \sim t^2/U$. This is the essence of the *first GKA rule*. If however the occupied orbital on one site overlaps only with an empty orbital on the other site (Fig. 5.1(a)), the exchange would be ferromagnetic and weaker, $J_f \sim (t^2/U)(J_H/U)$ (with $J_H/U \sim 0.2$) (the *second GKA rule*).[2]

One has to make a few comments regarding this result. As follows from our derivation, the ferromagnetic coupling in the second case can be reduced not only because of the small factor J_H/U, but also because, as is clear from Fig. 5.1(b), the electrons in this case can hop either from site i to site j and back, or vice versa, from j to i and back. This gives a factor of 2 in the expressions for the exchange integrals (1.12), (4.13). However in

---

[1] Here and below for simplicity we take Coulomb (Hubbard) repulsion U the same for all orbitals. In general U_{mn} for orbitals m and n may be different, see the detailed discussion in Section 3.3. The use of different values of U_{mn} would make all the treatment less universal and dependent on the particular orbital occupation. As a result however the expressions obtained below would be valid qualitatively, but the coefficients with which in particular J_H enters the resulting expressions should not be taken as numerically accurate.

[2] To facilitate presentation, I use here the notions of the first, second, and third GKA rules. In general, depending on the detailed lattice geometry, specific electronic configuration and orbital occupation, relative importance of direct d–d exchange or that via ligands, there may be many different situations, therefore exact classification is hardly possible. That is why in the literature one usually speaks of GKA rules in general. Nevertheless, the crude classification used in the text can serve as a general guide and is useful, though not exact.

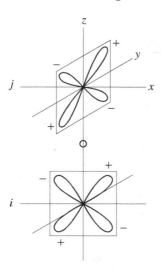

Figure 5.3 Orthogonal t_{2g} orbitals.

the case of Fig. 5.1(a) only one of these processes is allowed: the electron can hop from site i to j and back, but not vice versa. Consequently the ferromagnetic exchange in this case would not have this extra factor 2 and would be weakened further by this effect. This is, however, not necessarily always the case. Thus if we have for example the situation shown in Fig. 5.3, with electrons on t_{2g} orbitals xz and yz on neighboring sites, we see that again only occupied and empty orbitals overlap, so that we have ferromagnetic exchange of type (5.5) (with smaller hopping integrals for t_{2g} orbitals). Here, however, the electrons from the xz orbital at site i can hop to the xz orbital at site j, and the electrons from the yz orbital at site j can hop to the yz orbital at i. Thus the factor 2 in the exchange coupling will appear here; but the small factor $\sim J_H/U$ for the ferromagnetic exchange, compared with the antiferromagnetic one (which would occur if the occupied orbitals on both sites were e.g. xz with direct hopping) would still remain. Nevertheless the situation of Fig. 5.1(a), where the number of channels is reduced in the case of orthogonal orbitals, is more typical.

We should make yet another comment as to how this rule works for charge-transfer insulators. In this case we have to include the oxygen orbitals explicitly: instead of Fig. 5.1(a) we should consider the situation of Fig. 5.4, in which the corresponding hopping processes are marked (i.e. we have to generalize the treatment of Section 4.2 to degenerate orbitals). We see that again in the intermediate state, after transferring two electrons from the oxygen to sites i and j, the energy in the case of parallel spins (Fig. 5.4(a)) is lower, $2\Delta_{CT}+U_{pp}-J_H$, compared with that of antiparallel spins (Fig. 5.4(b)), $2\Delta_{CT}+U_{pp}$, which would favor the ferromagnetic exchange. One has to be a bit more careful here: already at the first step, if we transfer a p-electron with spin $\uparrow$ to the empty orbital at site j (where there is already an electron with spin $\uparrow$), the charge-transfer energy itself would change slightly, $\Delta_{CT} \to \Delta_{CT} - J_H$. Also, in this case several exchange "routes" are possible: the first electron hopping from the oxygen to site j or to i, and the same for the return hops

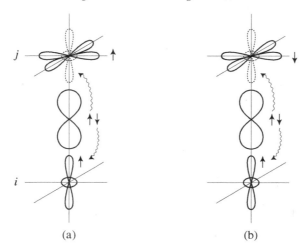

Figure 5.4 Superexchange processes involving oxygen p orbitals in an apparent way, leading to ferromagnetic exchange (p–d hopping between the occupied d orbital of site i and an empty orbital (dashed line) on site j).

(four possibilities altogether). Combining all these factors, we finally get that in the case of charge-transfer insulators, with $t_{pd} \ll \Delta_{CT} < U_{dd}$ for this orbital occupation (the overlap, via oxygen, of occupied and empty orbitals) we have a ferromagnetic exchange, with exchange constant

$$
\begin{aligned}
J &\sim \frac{4t_{pd}^4}{\Delta_{CT}^2(2\Delta_{CT} + U_{pp})} \cdot J_H \left(\frac{1}{\Delta_{CT}} + \frac{1}{2\Delta_{CT} + U_{pp}} \right) \\
&\sim \frac{2t_{dd}^2}{\Delta_{CT} + \frac{1}{2}U_{pp}} \cdot \frac{3}{2} \frac{J_H}{\Delta_{CT}},
\end{aligned}
\tag{5.7}
$$

where, as usual, $t_{dd} = t_{pd}^2/\Delta_{CT}$, and where in the last step we have taken $U_{pp} < \Delta_{CT}$.

We see that in all specific cases, for Mott insulators ($U > \Delta_{CT}$) and for charge-transfer insulators ($U < \Delta_{CT}$), the result is qualitatively the same: for 180° metal–oxygen–metal bonds, in the case of occupation of orthogonal orbitals (when the overlap and electron hopping are only allowed between occupied and empty orbitals), the exchange interaction turns out to be ferromagnetic and weaker by a factor $\sim J_H/U$ or J_H/Δ_{CT} compared with the respective antiferromagnetic exchange for nondegenerate orbitals or occupied orbitals with direct overlap between them.

The third remark is that often we are dealing not with systems with only one "active" localized electron (as in low-spin Ni^{3+}($t_{2g}^6 e_g^1$) where the full t_{2g}^6 shell is magnetically inactive), but with a situation such as that met for example in Mn^{3+}, with configuration $t_{2g}^3 e_g^1$. Here Mn^{3+}, besides one e_g-electron (which e.g. in $LaMnO_3$ has the strongest overlap with the oxygen orbitals and consequently gives the strongest contribution to the exchange) also contains three localized electrons on t_{2g} levels. Besides their possible direct contribution

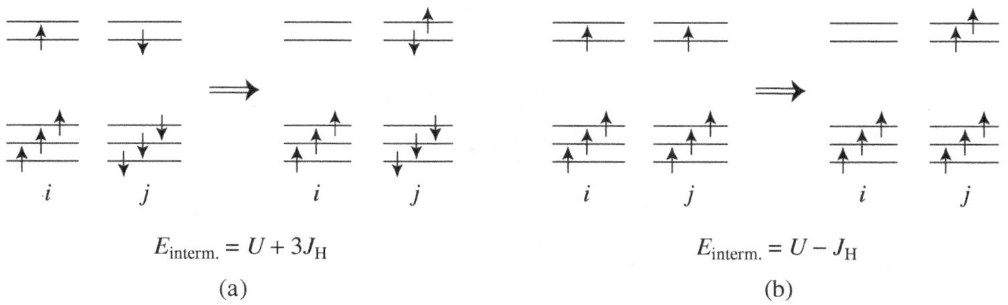

$$E_{\text{interm.}} = U + 3J_{\text{H}}$$

(a)

$$E_{\text{interm.}} = U - J_{\text{H}}$$

(b)

Figure 5.5 Enhancement of ferromagnetic exchange by Hund's coupling to other ("spectator") d-electrons.

to the exchange, which we ignore for a moment, they will also interact with e_g-electrons by Hund's rule interaction. Consequently, if in this case the e_g-electrons at neighboring sites occupy orthogonal orbitals, and we generalize to this case the treatment leading to the ferromagnetic exchange (5.6), (5.7), then for antiparallel spins we would have the energy of the intermediate state (with the e_g-electron transferred to a neighbor) equal to $U + 3J_{\text{H}}$ (by transferring the electron from site i to site j in this case we lose the Hund's energy on site i, coming from the interaction of the e_g-electron with its three t_{2g} "mates," see Fig. 5.5(a)). When however we move this e_g-electron from site i to site j having parallel spins, Fig. 5.5(b), we gain J_{H}, that is the corresponding denominator in the energy (4.2) will be $U - J_{\text{H}}$ (one can easily obtain these results by counting the numbers of pairs of parallel spins, which, according to our convention in Chapter 2, gives the contribution of Hund's energy in different configurations). Correspondingly, instead of eq. (5.5) which gave us the ferromagnetic exchange, we now have

$$\Delta E_{\uparrow\uparrow} - \Delta E_{\uparrow\downarrow} = -t^2 \left(\frac{1}{U - J_{\text{H}}} - \frac{1}{U + 3J_{\text{H}}} \right) \simeq -\frac{t^2}{U} \frac{4J_{\text{H}}}{U} \tag{5.8}$$

(assuming we can still use the perturbation expansion in $J_{\text{H}}/U < 1$, which in this case becomes questionable). We see that the presence of other localized spins can rather strongly enhance the ferromagnetic coupling – in this case by a factor of 4 (see also footnotes on p. 66 and p. 122).

Yet one more comment is in order here. Above we have considered ferromagnetic exchange originating from the hopping of an electron to an *empty* orbital of a neighboring site. One can easily see that the situation would be exactly the same if we have an overlap of a half-filled orbital at one site with a *fully occupied* orbital of a neighbor, for example if we have two neighbors with the configuration $(e_g^1)_i (e_g^3)_j$ with "active" half-filled orbitals orthogonal, see Fig. 5.6 – for example the exchange between Mn^{3+} (e_g^1) and Cu^{2+} (e_g^3), when one e_g-electron of Mn^{3+} (site i) is on the z^2 orbital, and for Cu^{2+} (site j) we assume the usual configuration $(z^2)^2 (x^2 - y^2)^1$. In this case the virtual hopping will consist of the transfer of one down-spin electron from the doubly occupied orbital $(z^2)_j$ to the same orbital on site i, and the Hund's rule interaction of the remaining two spins at site Cu will

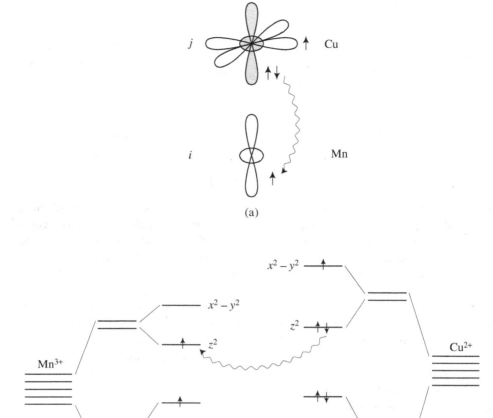

Figure 5.6 A possible exchange process of different ions, with the example of the pair Mn^{3+}–Cu^{2+}. The shaded z^2 orbital on site j (Cu) is doubly occupied by electrons.

make the energy of the ferromagnetic configuration of Fig. 5.6 lower than that of the anti-ferromagnetic configuration of this pair, that is it will stabilize the ferromagnetic exchange $J_{ij} \sim -(t^2/\tilde{U})(J_H/\tilde{U})$ (where $\tilde{U}$ includes the difference in energies of two different ions, here Mn and Cu, i.e. $\tilde{U} = \varepsilon_{Mn} - \varepsilon_{Cu} + U_{Mn}$ is the total energy change in transferring one electron from Cu to Mn for $J_H = 0$) – exactly as in the case of an empty z^2 orbital at site j, considered above (Figs 5.1(a), 5.2).

The situation with TM in oxygen octahedra having a common corner with 180° TM–O–TM bonds (see Figs 5.1, 5.4), though met rather often (e.g. it is typical for

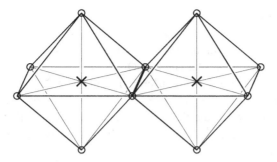

Figure 5.7 A typical situation of two transition metals (crosses) with a common edge and with 90° metal–oxygen–metal exchange paths.

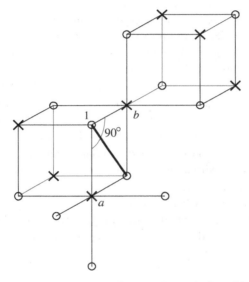

Figure 5.8 Lattice of B sites of a spinel AB_2O_4 (metals are indicated by crosses and anions by circles), illustrating the 90° exchange. The common edge of two BO_6 octahedra is marked by a bold line.

perovskites or layered materials such as La_2CuO_4), is however not the only one. The geometry of the lattice is often much more complicated. The other typical situation is when neighboring MO_6 octahedra have a common edge, with the TM–O–TM angle being 90°, see Fig. 5.7. Such a configuration is met for example in another major class of materials – spinels AB_2O_4, the basic structure elements of which are shown in Fig. 5.8.[3] In Fig. 5.8 we show the sublattice of octahedral B sites and oxygens; the metals at A sites (not shown) have tetrahedral coordination and are located between the "cubes" of Fig. 5.7 (as always

---

[3] There exist also many thiospinels AB_2T_4, containing $T = S$, Se, Te instead of oxygen. For simplicity we will speak below about oxygen, but one has to be aware that the TM covalency with S, Se, Te is (much) stronger than that with oxygen, which can influence the properties of these systems.

we denote transition metal ions by crosses and oxygens by circles). We see that in this structure the neighboring BO_6 octahedra, for example those around sites a and b, have two common oxygens, that is a common edge (thick line in Figs 5.7, 5.8), and in an ideal case the B–O–B angle, for example the angle a–O_1–b, is $90°$. It turns out that the character of superexchange in this case is quite different from that of the $180°$ TM–O–TM bonds. Consider again first the case of e_g-electrons. A typical situation in this case is shown in Fig. 5.9. In Fig. 5.9(a) we show the situation in which the electrons (localized spins) are on the same orbital, for example z^2, and in Fig. 5.9(b) when they are on orthogonal orbitals z^2 and $x^2 - y^2$.

For octahedra with a common edge, as in Figs 5.7, 5.8, the distance between TM ions is shorter (by a factor $\sim\sqrt{2}$) than in the situation with common corner and with $180°$ M–O–M bonds, thus there may appear a direct d–d overlap and direct d–d hopping between the neighboring sites; this can be especially important for t_{2g} orbitals, see below. For e_g orbitals of the type shown in Fig. 5.9, with the lobes of the wavefunctions directed toward the oxygens, the hopping via oxygens is still more important. But, as is clear from Fig. 5.9, the situation in this case is drastically different from that of the $180°$ TM–O–TM bond of Figs 4.5, 5.1, and 5.4. In those cases it was *the same* p orbital which participated in the exchange and from which, or via which, the electrons hopped to neighboring TM ions. In the case of $90°$ bonds as in Fig. 5.9 we see that it is *different* p orbitals which overlap with the two TM ions i and j: the unshaded orbital p_z overlaps with unshaded e_g orbitals at site i, and the shaded orbital p_x overlaps with the corresponding orbitals at site j. Correspondingly, the virtual hopping which can contribute to the exchange should involve these two orthogonal p orbitals. One can easily see that because of this the corresponding exchange in this case will always be ferromagnetic, irrespective of the occupation of particular e_g orbitals. Indeed, in both cases of Fig. 5.9 the virtual hoppings will be from the p orbitals to the corresponding transition metal, and each time only the opposite spin

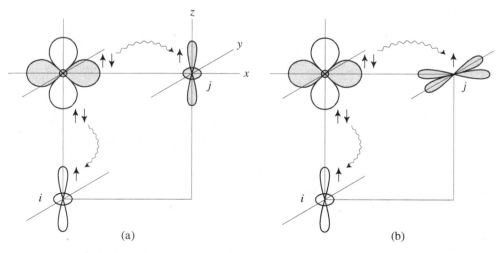

(a)   (b)

Figure 5.9 A possible situation for $90°$ bonds for "active" e_g orbitals of transition metals. Here the shaded d orbitals at the upper right corner are singly occupied.

can hop there because of the Pauli principle. As a result, for example for spins S_1 and S_2 both up, as shown in Fig. 5.9, after the hopping of two spin-down p-electrons *from different p orbitals* to sites i and j, in the intermediate state there will remain two unpaired *parallel* spins on two different oxygen orbitals, which is favorable because of the Hund's rule on oxygen J_H^p. If however the spins of d-electrons on sites i and j were antiparallel, electrons with opposite spins would hop to these sites from the oxygen, and the resulting intermediate state with two oxygen holes with antiparallel spins would be less favorable. Thus the same arguments which we have used for the 180° exchange between orthogonal orbitals would give here the result that the energy of the ferromagnetic state is lower than that of the antiferromagnetic one, the corresponding energy gain being due to *Hund's rule on oxygen*, so that the resulting ferromagnetic exchange constant would be

$$J_{90°} \sim -\frac{t_{pd}^4}{\Delta_{CT}^2(2\Delta_{CT} + U_{pp})} \frac{J_H^p}{(2\Delta_{CT} + U_{pp})} \tag{5.9}$$

(of course the numerical coefficients would depend on the particular type of orbitals occupied, with corresponding hopping integrals). This conclusion is valid both for the occupation of *the same e_g* orbital at sites i and j (Fig. 5.9(a)), and of *different* orbitals (Fig. 5.9(b)).[4] Note that the Hund's rule coupling on oxygen J_H^p is not small at all; actually $J_H^p \sim 1.2\,eV$, even bigger than the Hund's rule constant for transition metal ions (for which $J_H \sim 0.8$–$0.9\,eV$ for $3d$ ions, $\sim 0.7\,eV$ for $4d$ ions, and ~ 0.5–$0.6\,eV$ for $5d$ ions). In any case, the main conclusion remains the same: for 90° TM–O–TM bonds the exchange of e_g-electrons is always ferromagnetic and weak, see (5.9), both for the occupation of the same and of different d orbitals.[5]

The situation may be different if t_{2g} levels participate in the exchange. First of all, as one can see from Fig. 5.10, for the case of edge-sharing octahedra the t_{2g} orbitals on neighboring sites, pointing between the oxygens, are actually directed toward each other, for example the xy orbitals in Fig. 5.10. The resulting d–d hopping turns out to be very important for early $3d$ metals such as Ti, V, Cr. It can give rise to antiferromagnetic exchange if the occupied "active" orbitals are indeed directed toward one another, as in Fig. 5.10. This is the case for example in Cr spinels of the type $CdCr_2O_4$: the Cr^{3+} ions with configuration t_{2g}^3 occupy here the B sites of a spinel structure, shown in Fig. 5.8,

---

[4] This exchange would be antiferromagnetic only in the case when a (half-)filled e_g orbital at one site is exactly orthogonal to the corresponding p_σ orbital of oxygen, for instance if the e_g-electron at site i in Fig. 5.9(b) occupies the $(x^2 - y^2)$ orbital, so that the oxygen p-electron would hop from the p_z orbital to the *empty z^2* orbital at site j. But as one can easily see, this would require the action of Hund's rule *twice*: once on oxygen and once again at the TM site i. Correspondingly, the resulting antiferromagnetic exchange would be of higher order in J_H, $\sim (t_{dd}^2/(2\Delta_{CT} + U_{pp})) \cdot (J_H J_H^p/(2\Delta_{CT} + U_{pp})^2)$, and it can be ignored.

[5] One can notice here yet another difference in the case of 90° exchange for e_g-electrons compared with the 180° one. If in the case of 180° exchange one possible intermediate state was $d^{n-1}p^6d^{n+1}$, which gives the exchange (ferro or antiferro) with $U = U_{dd}$ in the denominator, eq. (5.1), that is, there was a possibility of effective d–d transfer via oxygen – for the 90° case this possibility does not exist anymore, since it is different orthogonal p orbitals which overlap with e_g orbitals of the neighboring TM ions. Correspondingly, in this case the intermediate state will necessarily be the one with two oxygen holes, even in the case of Mott–Hubbard insulators with $\Delta_{CT} \gg U$. This will reduce the corresponding (ferromagnetic) 90° exchange in the Mott–Hubbard case even further.

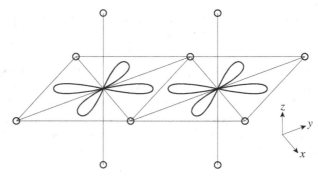

Figure 5.10 Direct d–d overlap and hopping for t_{2g} orbitals in the case of a common edge structure.

and the three t_{2g} orbitals at each site have a direct overlap with the respective orbitals of neighboring sites. But for heavier TM such as Co, Ni, usually hopping via the oxygens again becomes more important.

Note also that, especially in the case of t_{2g} orbitals in the situation of neighbors with a common edge and with $90°$ M–O–M angle, one must keep track of the signs of electron hoppings t_{dd} and t_{pd}. This is not so important if only one particular mechanism of exchange is active, either that due to direct d–d hopping or that due to hopping via oxygens: in both cases the exchange constant is either $\sim t_{dd}^2$ or $\sim t_{pd}^4$, see eq. (1.12) and the expressions earlier in this section. However, this becomes very important if there exist both a direct d–d overlap and that via oxygens, as is the case for t_{2g} orbitals for octahedra with a common edge. In that case the direct hopping t_{dd} and that via the oxygens, with the effective $\tilde{t}_{dd} = t_{pd}^2/\Delta$ (see (5.1), (5.2)), may have either the same or opposite signs (note that direct hopping in the Hubbard model (1.6) is usually taken as negative). In the first case these two processes reinforce each other and increase the total exchange, but in the second case they would (partially) cancel and reduce the corresponding exchange interaction for these particular orbitals.

I will not consider here all types of different situations which could appear in this case; the arguments and examples presented above give enough information, so that using these rules one should be able to deduce what the resulting exchange will be. Thus one can easily see that, depending on the occupation of t_{2g} orbitals, the t_{2g}–t_{2g} exchange via the $90°$ TM–O–TM exchange path may be either relatively strong antiferromagnetic (the case of Fig. 5.11(a), with occupied t_{2g} orbitals at neighboring centers overlapping with *the same* oxygen p orbital), or weak ferromagnetic (if such orbitals overlap with *different* p orbitals, empty with empty and shaded with shaded, Fig. 5.11(b)). But one more important point should be mentioned here: when t_{2g} orbitals are active, for $90°$ exchange there appear important exchange passes involving simultaneously both t_{2g} and e_g orbitals. This is illustrated in Fig. 5.12(a). There we show the situation in which there is one electron occupying the xz orbital at site i, and there is an electron at the $(x^2 - y^2)$ orbital at site j. We see that in this case *the same* p_x orbital overlaps both with the t_{2g} orbital xz at site i (with the hopping integral $t_{pd\pi}$) and with the e_g orbital $(x^2 - y^2)$ at site j (with the hopping integral $t_{pd\sigma}$). Correspondingly, all processes considered for the $180°$ exchange of Figs 1.12, 4.9, and

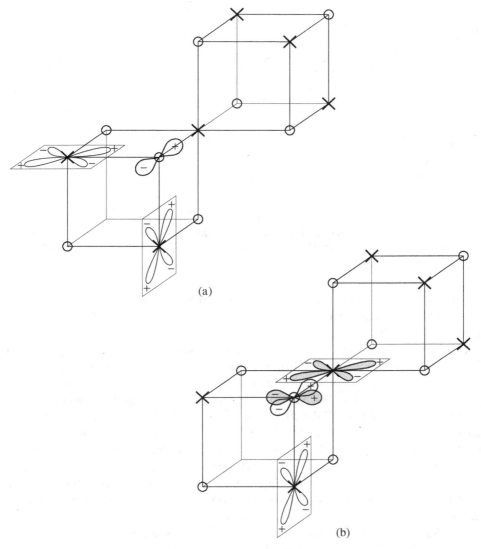

Figure 5.11 (a) Antiferromagnetic coupling of t_{2g} orbitals with 90° metal–oxygen–metal bonds. (b) Ferromagnetic t_{2g}–t_{2g} coupling for 90° bonds; the occupied t_{2g} orbitals overlap with orthogonal p orbitals of the anion.

5.1(a), leading to antiferromagnetic exchange, would also act here, so that in this case the t_{2g}–e_g hopping would lead to a strong antiferromagnetic exchange even in the case of 90° exchange, with the exchange integral

$$J \sim \frac{t_{pd\sigma}^2 t_{pd\pi}^2}{\Delta_{\mathrm{CT}}^2} \left(\frac{1}{U_{dd}} + \frac{1}{\Delta_{\mathrm{CT}} + \frac{1}{2} U_{pp}} \right), \tag{5.10}$$

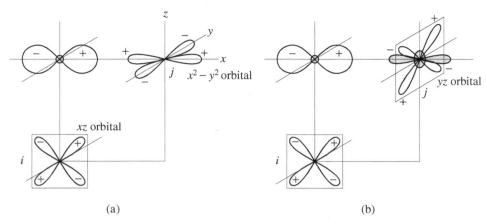

(a)   (b)

Figure 5.12 (a) A possible process of strong antiferromagnetic t_{2g}–e_g exchange via one p orbital of an anion for the 90° geometry. (b) The situation which would lead to a weak ferromagnetic exchange, with the empty e_g orbital at site j shaded (cf. Fig. 5.11(b)).

cf. eq. (4.13). (Remember that the overlap of e_g orbitals with p orbitals of ligands is denoted $t_{pd\sigma}$, and $t_{pd\pi}$ is the smaller overlap of t_{2g} orbitals with p orbitals of ligands; typically $t_{pd\sigma} \sim 2t_{pd\pi}$.)

Similarly, if for example the same xz orbital is occupied at site i, but at site j the electron is on the t_{2g} orbital yz (Fig. 5.12(b)), or on the $|y^2 - z^2\rangle$ orbital, both orthogonal to the O–TM bond, the hopping of an electron to an *empty* e_g orbital at site j (e.g. to the orbital $|x^2\rangle = 3x^2 - r^2$, shaded in Fig. 5.12(b)) would lead to a ferromagnetic exchange.

We cannot discuss here all the different possibilities which could exist in different situations; I hope that the general "recipes" are clear from the examples presented above. Crudely, one can summarize the most typical situation in what we may call the *third GKA rule*. In the case of 90° bonds the e_g–e_g exchange is always ferromagnetic and weak, the direct t_{2g}–t_{2g} interaction could give antiferromagnetic exchange, and depending on the particular orbital occupation, the t_{2g}–t_{2g} exchange via oxygens and the t_{2g}–e_g interaction can be either relatively strong antiferromagnetic or weak ferromagnetic.

The examples considered above do not exhaust all the situations which we meet in TM compounds. Even in the conceptually simplest structures such as perovskites there often exist rotations and tiltings of TM O_6-octahedra, leading for example to transitions from cubic to orthorhombic or rhombohedral structures; as a result the TM–O–TM angle can sometimes deviate strongly from 180° – see Fig. 5.13, where the tiltings of O_6-octahedra around the [110] axis, typical for many perovskites, are shown (to preserve the "integrity" of the lattice the directions of these tiltings should alternate from site to site). We see that as a result of these tiltings (the resulting structure is an orthorhombic one, *Pbnm*) the TM–O–TM angles become smaller than 180° – they can be as small as $\sim$ 150°–160°, see Fig. 3.46. In effect the situation becomes, in a sense, intermediate between that of 180° and 90° exchange. Then, according to the GKA rules, in the case of strongly localized electrons

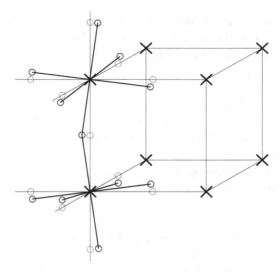

Figure 5.13 Intermediate metal–anion–metal angles in perovskites with orthorhombic or rhombo-hedral structure due to tilting of octahedra ($GdFeO_3$-type distortion).

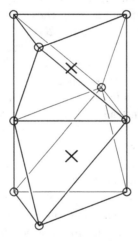

Figure 5.14 Two MO_6 octahedra with a common face (crosses are metals and circles are ligands, e.g. oxygens).

the antiferromagnetic exchange is weakened with reducing TM–O–TM angle, whereas the ferromagnetic interaction is usually strengthened by that. This tendency is important for many phenomena, for example for the magnetostriction mechanism of multiferroic behavior, see Chapter 8.

In other situations the MO_6 octahedra can have neither common corners (with more or less 180° exchange) nor common edges (leading to 90° exchange), but common faces, see Fig. 5.14. Such situations are typical for some hexagonal systems, such as $BaCoO_3$

and $CsCuCl_3$. The recipes formulated above can also be used to understand the details of exchange in these cases, but for complicated geometries and for different orbital occupation the task of finding what will be the superexchange of particular ions in particular configurations becomes more and more cumbersome and difficult to solve "by hand." In such cases one had better use real *ab initio* calculations such as LDA or LDA + U, which usually also allow one to find the exchange constants.

One further general remark is in place here, which concerns the relative importance of the exchange due to direct d–d hopping of Fig. 5.10 and that via ligands (O, S, Se, etc.). The hopping and exchange via ligands (e.g., oxygen), given by expressions of the type (5.1), (5.7), and (5.10), always contain some power of the charge-transfer energy Δ_{CT} in the denominator. Consequently, this mechanism of the exchange is definitely very important, and probably dominates for heavier 3d transition metals such as Mn or Co, with relatively small charge-transfer energy (see, e.g., Fig. 4.8), even though they have partially filled t_{2g} orbitals which may have a direct d–d overlap for 90° bonds. However, as follows from the discussion in Chapter 4, Δ_{CT} becomes quite large for earlier 3d metals such as Ti or V. Consequently, one could think that for these ions, in systems with 90° bonds, for example in Ti and V spinels such as $MgTi_2O_4$ or ZnV_2O_4, the direct d–d hopping t_{dd}^{direct} would be more important than the corresponding effective hopping via the ligands, $t_{dd}^{\text{eff}} = t_{pd}^2/\Delta_{CT}$. One often uses this picture to explain the properties of these systems, see for example Di Matteo *et al.* (2005), Khomskii and Mizokawa (2005).

Along the same lines: as mentioned above, the direct d–d hopping, if present, typically leads to antiferromagnetic coupling, for example for Cr^{3+} with edge-sharing octahedra, as in Cr spinels such as $CdCr_2O_4$ (three occupied t_{2g} orbitals in Cr^{3+} have a direct overlap with the nearest-neighboring Cr in the B sublattice of spinels, see Fig. 5.8). However, in similar spinels with S, Se, Te instead of oxygens, the 90° exchange Cr–O–Cr starts to dominate because of the much larger covalency of S, Se with 3p, 4p orbitals. And, as this 90° exchange is ferromagnetic, such thiospinels are typically ferromagnetic – for example the well-known ferromagnetic semiconductors such as $CdCr_2S_4$ or $HgCr_2Se_4$ (Methfessel and Mattis, 1968); see also Section 5.6.3 below.

To conclude this section, we summarize briefly the main rules determining the type of exchange interaction in different situations (in the expressions below we omit numerical coefficients such as 2, etc.). Different situations are shown schematically in Fig. 5.15, where the active (half-filled) d orbitals are shown as unshaded, or lightly shaded, and empty orbitals are dark shaded.

A. When the main exchange goes via direct d–d hopping t_{dd}:

    1. If there is an overlap and hopping between active (half-filled) orbitals at two centers, the exchange is antiferromagnetic and strong, $J \sim t_{dd}^2/U$ (the first GKA rule). (Here and below, U is always U_{dd}.)

    2. If the occupied orbitals are orthogonal, and the electrons can only hop from occupied to empty orbitals, the exchange is ferromagnetic and weak, $J \sim -(t_{dd}^2/U) \cdot (J_H/U)$ (the second GKA rule).

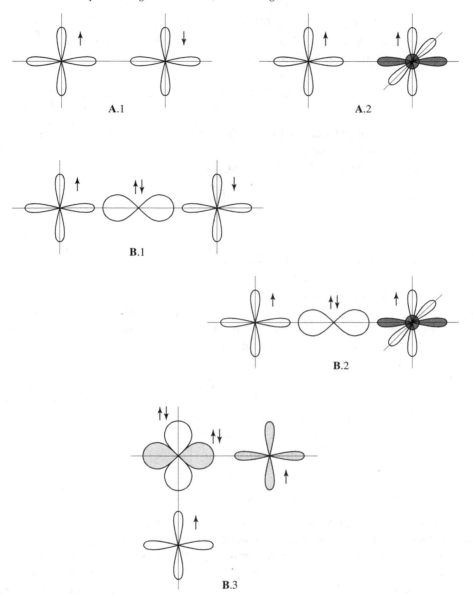

Figure 5.15 Summary of different orbital occupations in different geometries, leading to the GKA rules (see text).

B. When the exchange goes via the p orbitals of ligands (e.g., oxygens):

1. If the occupied orbitals of two sites overlap with the same ligand p orbital, the exchange is strong antiferromagnetic,

$$J \sim \frac{t_{pd}^4}{\Delta_{\mathrm{CT}}} \left(\frac{1}{U} + \frac{1}{\Delta_{\mathrm{CT}} + \frac{1}{2}U_{pp}} \right) = (t_{dd}^{\mathrm{eff}})^2 \left(\frac{1}{U} + \frac{1}{\Delta_{\mathrm{CT}} + \frac{1}{2}U_{pp}} \right),$$

where Δ_{CT} is the charge-transfer gap and $t_{dd}^{\mathrm{eff}} = t_{pd}^2/\Delta_{\mathrm{CT}}$ (the first GKA rule).

2. If the occupied and empty d orbitals on neighboring TM ions overlap with the same p orbital, the exchange is weak ferromagnetic,

$$J \sim -\frac{(t_{dd}^{\mathrm{eff}})^2}{U} \cdot \frac{J_{\mathrm{H}}}{H}$$

for $U < \Delta_{\mathrm{CT}}$ and

$$J \sim -\frac{(t_{dd}^{\mathrm{eff}})^2}{\Delta_{\mathrm{CT}} + \frac{1}{2}U_{pp}} \frac{J_{\mathrm{H}}}{\Delta_{\mathrm{CT}}}$$

for $U > \Delta_{\mathrm{CT}}$ (the second GKA rule).

3. There are situations, occurring especially in the case of 90° TM–O–TM bonds, in which the occupied orbitals overlap with different, orthogonal p orbitals of a ligand. In this case the exchange is also weak ferromagnetic, with the exchange integral

$$J \sim -\frac{(t_{dd}^{\mathrm{eff}})^2}{\Delta_{\mathrm{CT}} + \frac{1}{2}U_{pp}} \frac{J_{\mathrm{H}}^p}{\Delta_{\mathrm{CT}}} ,$$

where J_{H}^p is the Hund's rule interaction on the ligand (oxygen). This expression is valid in this case even for Mott–Hubbard insulators with $U < \Delta_{\mathrm{CT}}$. This rule may be called the third GKA rule.

5.2 Double exchange

The discussion in the previous section was applicable to strong Mott insulators with localized d-electrons and with integer number of d-electrons per site. We saw that antiferromagnetic interactions are most typical for this situation; they are met much more often than ferromagnetic ones. As we saw in Section 5.1, for the latter one needs very special conditions: specific occupation of particular d levels, or very specific geometry. However if we dope Mott insulators, or if there exist in a system, besides localized electrons, also itinerant electrons filling some energy bands, the coupling of localized and itinerant electrons can lead to an effective exchange between localized electrons, which, in particular, can be ferromagnetic.

In the case of itinerant electrons occupying a broad conduction band, filling it up to the Fermi energy ε_{F}, with the Fermi momentum k_{F}, the interaction of localized electrons via such a conduction band is the well-known RKKY (Ruderman–Kittel–Kasuya–Yosida) interaction (see, e.g., Khomskii, 2010), which is long-range and depends in an oscillating manner on the distance r between the localized electrons,

$$J(r) \sim \frac{I^2}{\varepsilon_F} \frac{\cos(2k_F r)}{r^3} . \tag{5.11}$$

This interaction appears to second order in the exchange interaction between localized and itinerant electrons,

$$I_{ik} \, \boldsymbol{S}_i \cdot c_{k\sigma}^\dagger \hat{\boldsymbol{\sigma}} c_{k\sigma} . \tag{5.12}$$

The origin of the "s–d" exchange I in (5.12) may be either s–d hybridization, or a local intra-atomic interaction of the same nature as the Hund's interaction J_H. As the resulting RKKY exchange is $\sim I^2$, the detailed nature, and even the sign of the local s–d exchange I, does not play a crucial role and does not modify significantly the resulting magnetic structure. Rather, what is important is the location of localized electrons, relative to the period of oscillation of the exchange (5.11). The RKKY interaction is the main source of exchange in many rare earth metals and intermetallic compounds (Gschneider and Eyring, 1978), and it is crucial for the formation of the spin glass state in dilute alloys of magnetic (in particular TM) elements in nonmagnetic metals such as Cu, etc. (Mydosh, 1993).

For our purposes, however, it is more important to consider the other situation – not the case of a broad conduction band and large Fermi energy $\varepsilon_F \gg I$, which was assumed in the derivation of the RKKY interaction (5.11), but the opposite case of narrow bands and small Fermi energy, such that the local s–d (or, in our case, $d_{\text{itinerant}}$–d_{local}) exchange I is at least comparable with or larger than ε_F. In this case one cannot use perturbation theory in $I/\varepsilon_F < 1$, as done in the derivation of (5.11), but rather has to consider the opposite limit. This situation leads to a special mechanism of exchange, known as *double exchange* (DE) (Zener, 1951; de Gennes, 1960).

We will consider double exchange having in mind a particular situation of doped Mott insulators such as $La_{1-x}Sr_x MnO_3$. (Actually, the first ideas of double exchange were proposed (Zener, 1951) specifically for this system, although later the same picture was even used to explain ferromagnetism in such metals as Fe and Ni.) Undoped $LaMnO_3$ is a perovskite with localized electrons, and an antiferromagnetic Mott insulator; its Mn^{3+} ions have configuration $t_{2g}^3 e_g^1$. Ignoring for a while the question of orbital ordering of degenerate e_g-electrons (which in fact is very important for determining the detailed crystal and magnetic structure of $LaMnO_3$), we pay most attention here to the effects of doping. When we replace La^{3+} by Sr^{2+} or Ca^{2+}, we remove some e_g-electrons, that is we dope the system with e_g-holes. The e_g band in perovskite systems such as $LaMnO_3$, with large $dp\sigma$ overlap for $\sim 180°$ Mn–O–Mn bonds, gives a relatively wide conduction band, which is nevertheless smaller than (or at least of the order of) the effective Hund's energy of coupling of e_g-electrons with three t_{2g}-electrons, which we can treat as localized.

Experimentally, when we increase the doping this system goes through a number of transitions, see the schematic phase diagram in Fig. 5.16. One sees here insulating phases

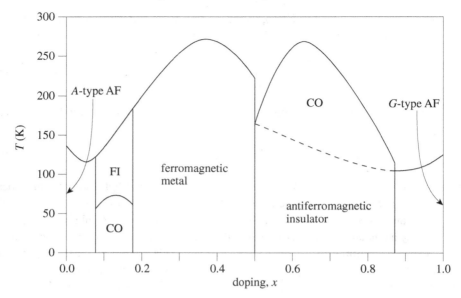

Figure 5.16 Schematic phase diagram of La$_{1-x}$Ca$_x$MnO$_3$ (after S.-W. Cheong). AF, antiferromagnetic state; FI, ferromagnetic insulator; CO, charge ordering.

with different types of magnetic ordering: *A*-type (ferromagnetic layers stacked antiferromagnetically) for undoped LaMnO$_3$ and for doping $x \lesssim 0.1$; Mott insulator with *G*-type magnetic ordering (checkerboard alternation of up and down spins) close to $x = 1$; different charge ordering (CO) phases, etc. For us the most interesting is the phase existing in the range $0.2 \lesssim x \lesssim 0.5$, which is a ferromagnetic metal. It was precisely in this phase that the strong suppression of resistivity by magnetic field was discovered. This colossal (negative) magnetoresistance (CMR) gave a name to these compounds – CMR manganites – and generated significant interest in these systems.

The mechanism of ferromagnetic behavior of this system is evidently connected with doping and with simultaneous appearance of metallic conductivity. This mechanism is the double exchange.

The simplest model to describe this phenomenon is the model dealing with localized electrons (here three localized t_{2g}-electrons with $S = \frac{3}{2}$), with coexisting doped electrons occupying a narrow band and interacting with localized spins by the Hund's rule exchange:

$$\mathcal{H}_{\text{DE}} = -t \sum_{\langle ij \rangle, \sigma} c_{i\sigma}^{\dagger} c_{j\sigma} - J_{\text{H}} \sum_i \mathbf{S}_i \cdot c_{i\sigma}^{\dagger} \hat{\boldsymbol{\sigma}} c_{i\sigma} + J \sum_{\langle ij \rangle} \mathbf{S}_i \cdot \mathbf{S}_j , \qquad (5.13)$$

where $c_{i\sigma}^{\dagger}, c_{i\sigma}$ are operators of conduction electrons with intersite hopping t (these describe e.g. e_g-electrons, or rather e_g-holes in La$_{1-x}$Ca$_x$MnO$_3$), the term with J_{H} is the Hund's rule exchange with localized (t_{2g}) electrons with total spin $\mathbf{S}_i$, and the last term is the antiferromagnetic exchange interaction between localized spins, which, in the absence of doping, would give antiferromagnetic ordering. Here the symbol $\langle ij \rangle$, as always, indicates summation over nearest neighbors.

One can easily see that doping, that is the presence of conduction electrons, would give a tendency to ferromagnetic ordering. This is an effect very similar to the mutual influence of the kinetic energy of electrons and the magnetic structure, discussed for the one-band model (Hubbard model) in Chapter 1. We saw there that background antiferromagnetic ordering suppresses electron hopping and reduces the kinetic, or band, energy. To gain this extra kinetic energy, it was favorable to change the magnetic structure, for example make it ferromagnetic: we would lose the exchange energy of localized electrons, but this energy loss could be overcome by the kinetic energy gain of mobile electrons. This was the reason for the Nagaoka ferromagnetism and for possible formation of ferromagnetic polarons ("ferrons") around charge carriers in the situation considered in Chapter 1.

One can show that the physics here, in the model (5.13), containing not one but two types of electrons – localized electrons with spin S and conduction electrons – is very similar. If the Hund's energy (the second term in the Hamiltonian (5.13)) is large compared with the electron hopping t or bandwidth $W \simeq 2zt$ (where z is the number of nearest neighbors), then at each site the spin of the band electron $\boldsymbol{\sigma}_i$ should be parallel to the spin of the localized electron $\boldsymbol{S}_i$. If the spin on a neighboring site $\boldsymbol{S}_j$ is parallel to $\boldsymbol{S}_i$, the conduction electron can easily hop from i to j, see Fig. 5.17(a). If however the spin of the neighbor is antiparallel, Fig. 5.17(b), this hopping will be forbidden, or at least strongly suppressed: in the final state the two electrons at site j would have opposite spins, which would cost large Hund's energy $\sim J_{\mathrm{H}}$. And if the energy gain $\sim t$ is less than this energy loss, the conduction electron will be "locked" at its site in the antiferromagnet, and we will not be able to gain its kinetic energy. To gain this energy it will again be favorable to change the magnetic structure of localized spins, making it ferromagnetic. This mechanism of stabilization of ferromagnetic ordering is the *double exchange* we are talking about.

Actually, to gain a certain kinetic energy we do not immediately have to make the localized spins ferromagnetic. For low doping it may be sufficient just to cant the antiferromagnetic sublattices somewhat by a certain angle θ, see Fig. 5.17(c). We allow certain hoppings of conduction electrons, but do not lose all the exchange energy right away. Thus at the beginning, at low doping, we may first expect the formation of a *canted* antiferromagnetic state, and only at higher concentration of mobile electrons can the state become really ferromagnetic. This picture was first proposed by de Gennes (1960).

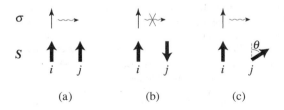

Figure 5.17 Illustration of the mechanism of double exchange.

Mathematically it is easy to consider this problem, treating the localized spins quasi-classically. One can show that in the limit $J_H \gg t$ the hopping of an itinerant electron between two sites with angle θ between their localized spins, as in Fig. 5.17(c), is given by an effective hopping matrix element (Anderson and Hasegawa, 1955; de Gennes, 1960)[6]

$$t_{\text{eff}} = t \cos \frac{\theta}{2} . \tag{5.14}$$

For the ferromagnetic ordering (parallel spins, $\theta = 0$) we have a full, unhindered hopping. For antiparallel spins ($\theta = \pi$) the effective hopping is zero, as follows from our qualitative considerations.[7]

Using the expression (5.14), we can easily consider what should happen with the magnetic structure of our system with doping. Let us assume that we have a homogeneous canted state, with angle θ between the spins of the two sublattices (Fig. 5.18). The average energy of this state, with electron concentration x, is, from (5.13) and (5.14),

$$\frac{E}{N} = J S^2 z \cos \theta - xzt \cos \frac{\theta}{2} . \tag{5.15}$$

Here the first term is the magnetic energy (as always, z is the number of nearest neighbors), and the second is the band energy of electrons with concentration x, moving in a tight-binding band with the hopping matrix element $t \cos \frac{\theta}{2}$ and with the bottom of this band at $-zt \cos \frac{\theta}{2}$ (we put all electrons at the bottom of the band and ignore their Fermi distribution, with the Fermi energy $\sim t_{\text{eff}} x^{5/3}$, which is of higher order in $x < 1$). Minimizing this energy in the angle θ, we find

$$\cos \frac{\theta}{2} = \frac{tx}{4 J S^2} . \tag{5.16}$$

Figure 5.18 A canted magnetic state, which appears in a situation with double exchange.

[6] We ignore here the phase factor in t_{eff} (Anderson and Hasegawa, 1955), which in principle may be important for other effects that we do not consider here.

[7] Strictly speaking, it is not completely correct; it is true only if we treat the localized spins classically. If we take into account the quantum nature of localized spins, the situation changes. Indeed, the Hund's interaction (the second term in (5.13)) does not require that the localized spins and the spin of itinerant electrons should be parallel; it only requires that the total state at a site should be that with maximum total spin $S_{\text{tot}} = S + \sigma$. Thus, for example, for $S = \frac{1}{2}$ this tells us that the resulting state should be a *triplet* with $S_{\text{tot}} = 1$. But such a triplet, besides the components with parallel spins $|S^z \uparrow, \sigma^z \uparrow\rangle$ and $|S^z \downarrow, \sigma^z \downarrow\rangle$, also contains the state with $|S_{\text{tot}}^z = 0\rangle = \frac{1}{\sqrt{2}}(|S^z \uparrow, \sigma^z \downarrow\rangle + |S^z \downarrow, \sigma^z \uparrow\rangle)$. We see that a part of this wavefunction corresponds to a state which one would get from the initial state of Fig. 5.17(b) by transferring the conduction electron to the right. Thus this hopping is not completely forbidden in the antiferromagnetic state, as would follow from the expression (5.14), but is only suppressed by the factor $1/\sqrt{2}$ for $S = \frac{1}{2}$ (or $1/\sqrt{2S+1}$ for higher spin). However, this suppression of hopping is sufficient to cause the same consequences as discussed above, namely that the doping gradually changes the magnetic structure to ferromagnetic, possibly via a canted state; only some numerical factors and the detailed form of the resulting phase diagram would be different.

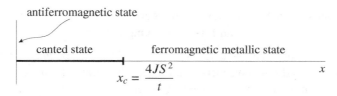

Figure 5.19 A possible region of existence of the canted state depending on the doping x.

Thus indeed the system is antiferromagnetic ($\cos \frac{\theta}{2} = 0$, $\theta = \pi$) for an undoped system, $x = 0$, and spins start to cant as soon as we dope it, $x \neq 0$. The canting angle increases gradually with x, until at

$$x = x_c = \frac{4JS^2}{t} \tag{5.17}$$

$\cos \frac{\theta}{2}$ reaches 1, that is $\theta \to 0$, so the system becomes ferromagnetic. It remains ferromagnetic for larger $x > x_c$. If the hopping energy t is much bigger than the antiferromagnetic coupling of localized electrons J, $t \gg J$, this occurs at yet rather small values of $x \ll 1$, which justifies our approximation of putting all electrons at the bottom of the band and ignoring the Fermi energy $\sim x^{5/3}$. Thus the resulting phase diagram of this system looks schematically as shown in Fig. 5.19.

If we take into account the quantum effects described in footnote 6, one can show that the canted state would not appear for arbitrary small doping x, but there will exist a lower critical concentration $\tilde{x}_c$ below which the material will still remain antiferromagnetic, and the canting will appear only for $x > \tilde{x}_c$ (Nagaev, 1969). There exists also yet another complication: one can show (Kagan *et al.*, 1999) that the state with homogeneous canting is usually unstable with respect to phase separation into an undoped region with pure antiferromagnetic order, and doped metallic and ferromagnetic "droplets." Such an inhomogeneous state seems indeed to be seen in lightly doped CMR manganites; we will discuss these questions in detail later, in Section 9.7.

Nevertheless, despite all these ramifications concerning the intermediate state, the general conclusion remains correct: the presence of itinerant electrons moving in narrow bands, such as those created by doping in CMR manganites, gives a tendency to make the magnetic ordering ferromagnetic, due to a mechanism which in this situation is called double exchange. Thus, summarizing, we can conclude that in systems with strongly correlated electrons the Mott insulating state with localized electrons is typically antiferromagnetic, whereas ferromagnetism is more often observed in doped metallic systems, both in single-band (Chapter 1) and many-band cases. There are also specific situations in which ferromagnetism may appear in insulating cases, but these situations usually require a particular orbital ordering or a very specific geometry, and they are rather the exception than the rule. One can see this even empirically: in a large compilation listing a lot of magnetic oxides, collected in Tsuda *et al.* (1991), only about 10% of materials included are ferromagnetic, 90% being indeed antiferromagnetic.

5.3 Role of spin–orbit interaction: magnetic anisotropy, magnetostriction, and weak ferromagnetism

Until now, when treating the exchange interaction in TM compounds, we have ignored spin–orbit coupling and only considered pure spin interactions. In this case the magnetic exchange turned out to be isotropic, of Heisenberg type, $\sim S_i \cdot S_j$.[8] In this case the spin space has no connection with the real space of the crystal, and the spin quantization axes z, x, y have nothing to do with real directions in the crystal lattice. Such a connection is established only when we take into account spin–orbit coupling $\lambda L \cdot S$: this couples spins to the orbital motion of electrons characterized by the orbital moment L, and the latter already has some particular orientation in the crystal. Thus it is via spin–orbit interaction that spin orientation starts to align with the lattice, and this interaction finally determines such effects as magnetic anisotropy of a system, magnetoelastic coupling, and magnetostriction. In certain cases it can also lead to weak canting of antiferromagnetically ordered spins and can give weak ferromagnetism.

These questions are usually discussed in detail in books devoted to magnetism, see for example Coey (2010), often using a phenomenological approach. We will not dwell too much on these questions here, and discuss them mostly to the extent that they are related to the microscopic electronic structure of the corresponding systems.

In considering the spin–orbit interaction in Section 2.3 we already came across several situations in which the orientation of spins was important. We saw there that we had especially strong effects in the case of unquenched orbital moment, for example for partially filled t_{2g} levels.

Phenomenologically one describes magnetic anisotropy by adding to the Heisenberg exchange (1.12) or (5.6) terms depending on the orientation of spins (or of the average magnetization) with respect to the crystal axes, see Landau and Lifshitz (1960). The form of these terms depends on the lattice symmetry. The magnetocrystalline energy, as all other contributions to the total energy, should be a scalar which does not change under the symmetry operations of the crystal, nor under time inversion. The latter condition tells us that this energy should be an even function of magnetization (or sublattice magnetization). The crystal symmetry determines the specific type of this combination. Thus for example for uniaxial crystals such as those with tetragonal symmetry, the corresponding energy should have the form

$$E_{\text{tetr.}} = -\kappa m_z^2 = -\kappa \cos^2 \theta \quad \text{or} \quad +\kappa (m_x^2 + m_y^2) = \kappa \sin^2 \theta, \tag{5.18}$$

---

[8] Although the main mechanism of exchange in these systems is not the real Heisenberg exchange, which is the exchange part of the Coulomb interaction $\int \Psi_1^*(r)\Psi_2^*(r') \frac{1}{|r-r'|} \Psi_1(r')\Psi_2(r)$. As mentioned above, one can show that this term, though formally present, is always much smaller than the superexchange. In fact, as we have seen in Chapter 1, the real mechanism of spin ordering in our systems is not the gain in the exchange part of the Coulomb energy, but rather the gain in the kinetic energy of electrons, either in the form of virtual hoppings (effects $\sim t^2/U$) in insulators, or real hoppings in double exchange, operating in doped systems.

where m_α are the directional cosines of the magnetization (for simplicity we consider here the case of a ferromagnet), $m_\alpha = M_\alpha / |\mathbf{M}|$, so that $m_x^2 + m_y^2 + m_z^2 = 1$; θ is the angle between magnetization and the z-axis.

The sign of the coefficient κ in eq. (5.18) determines the orientation of magnetization in the system. If $\kappa > 0$, spins are preferentially oriented in the z-direction; this is *easy-axis* anisotropy. If $\kappa < 0$, spins lie in the perpendicular plane; we have the situation of *easy-plane* anisotropy. The coefficient κ may be temperature-dependent, and it may even change sign at a certain temperature (which will cause *spin-reorientation* transitions). In particular, close to T_c $\kappa \sim |M|^2$, that is for II order transitions, in which $|M|^2 \sim (T_c - T)$, the coefficient κ goes to zero. But the corresponding term in the free energy becomes nonzero below T_c, and it dictates that for example for the easy-axis case close to T_c all spins are oriented along the z-axis. In that sense easy-axis anisotropy makes the system Ising-like. At lower temperatures, where nonlinear effects start to play a role, the detailed type of ordering may change. This is, in particular, the case in systems in which the exchange interaction would lead to a spiral magnetic structure, see Chapter 8 below. In the case of easy-axis anisotropy close to T_c the magnetic ordering would be collinear, but possibly with a certain period (which may even be incommensurate with the lattice) – this will be the so-called sinusoidal spin density wave, Fig. 5.20(a). Only at lower temperatures will it go over to a helicoidal structure, for example the spiral shown in Fig. 5.20(b). Such a crossover is seen in many materials with a spiral ground state, for example in several multiferroic systems such as $TbMnO_3$ or $Ni_3V_2O_8$, see Chapter 8.

In principle there also exist terms in the magnetocrystalline anisotropy of higher order in m, for example of order m^4, m^6, etc. The form of these terms is also determined by the lattice symmetry. We will not present here the corresponding expressions, which can be found for example in Landau and Lifshitz (1960) or in specialized monographs on magnetism. We present here only an important case of a cubic system. As all the main directions x, y, and z are equivalent here, there are no second-order terms in the anisotropy energy, and the lowest symmetry-allowed term is

$$E_{\text{cubic}} = \kappa_1(m_x^2 m_y^2 + m_y^2 m_z^2 + m_x^2 m_z^2) \quad \text{or} \quad E_{\text{cubic}} = -\tfrac{1}{2}\kappa_1(m_x^4 + m_y^4 + m_z^4). \quad (5.19)$$

The positive values of the coefficient κ_1 lead to structures in which the magnetization is directed along one of the cubic axes, m_x, m_y, or $m_z \neq 0$. That is, in this case we would have three equivalent easy axes. For $\kappa_i < 0$ the energy (5.19) is minimized when the magnetization is directed along one of the cubic diagonals (one of four equivalent [111]

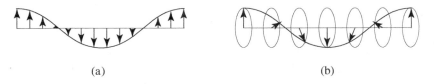

(a)   (b)

Figure 5.20 (a) Sinusoidal and (b) helicoidal spin density waves.

directions); in this case $m_x = m_y = m_z$, that is from the condition $m_x^2 + m_y^2 + m_z^2 = 1$ we get $m_x^2 = m_y^2 = m_z^2 = \frac{1}{3}$, and the anisotropy energy (5.19) of this state would be minimal for $\kappa_1 < 0$. Close to T_c the constant of cubic anisotropy behaves as $\kappa \sim M^4$.

There exist different mechanisms of magnetic anisotropy. The most common, and usually the strongest mechanism, is the *single-site anisotropy*. It is determined by the electronic configuration of the corresponding TM ions, in particular by crystal field splitting. When we take spin–orbit coupling into account, in certain cases there can appear *exchange anisotropy* – the exchange interaction can in principle differ from the simple Heisenberg form $\sim \boldsymbol{S}_i \cdot \boldsymbol{S}_j$. These effects will be discussed in the following sections. In some cases the classical dipole–dipole interaction can also contribute to magnetic anisotropy, though in most concentrated systems based on transition metals this effect is usually weak (although it may be significant in some rare earth compounds).

5.3.1 Orbital singlets: magnetic anisotropy

As discussed above, the microscopic origin of magnetocrystalline anisotropy is the spin–orbit coupling. Its action, and the resulting magnetic anisotropy, depend strongly on the electronic configuration of the corresponding TM ions. One should discriminate between the situation with orbital singlets and that with unquenched orbital moments of TM ions.

In the case of an orbital singlet, such as in the ions with half-filled or fully occupied t_{2g} levels, that is with quenched orbital moments, the spin–orbit coupling acts to second order of perturbation theory in $\lambda/\delta E$, where δE is the excitation energy, corresponding for example to the transfer of t_{2g}-electrons to e_g levels (e.g., the transition from the ground state configuration t_{2g}^3 to $t_{2g}^2 e_g^1$ in $Cr^{3+}(t_{2g}^3)$); in this case $\delta E = \Delta_{CF} = 10Dq$. In other cases, for example for partially occupied t_{2g} levels but with rather strong tetragonal or trigonal distortion, the t_{2g} levels may split into a singlet $|xy\rangle$ (or $|l^z = 0\rangle$) and a doublet ($|xz\rangle$, $|yz\rangle$) (or $|l^z = \pm 1\rangle$), cf. Fig. 3.9 and eq. (3.4). If in this case we have one t_{2g}-electron in a singlet in the ground state (xy or a_{1g}), the excitation could be the transition of this electron to the e_g^π doublet, thus in this case the excitation energy δE would be the distance between xy and (xz, yz), or between a_{1q} and e_g^π levels, that is it would be determined by the strength of tetragonal or trigonal distortion. Still, if this extra splitting δE is larger than the spin–orbit coupling λ, one can apply perturbation theory in $\lambda/\delta E$ and all the results discussed below will be applicable. The situation becomes more complicated, however, if $\delta E \sim \lambda$.

Using perturbation theory in $\lambda/\delta E$, one can calculate different effects. One of them is the single-site anisotropy which, for tetragonal symmetry, can be written as

$$\mathcal{H}_{\text{single-site}} = -K S_z^2 \,. \tag{5.20}$$

This term contributes to the magnetic anisotropy of magnetically ordered states, (5.18), with the coefficient κ given in the mean-field approximation by the relation

$$\kappa = K \langle S_z \rangle^2 = K M_z^2 \,, \tag{5.21}$$

where $M_z = \langle S_z \rangle$ is the average magnetization (or sublattice magnetization). However terms of the type (5.20) contribute to magnetic properties above T_c as well, bringing about for example the anisotropy of the magnetic susceptibility $\chi(T)$. They also give anisotropy of the g-factor probed for example by ESR, so that it deviates from the pure electron value $g = 2$ and becomes anisotropic. The resulting change in g-factor δg is of the order

$$\frac{\delta g}{g} \sim \frac{\lambda}{\delta E}, \tag{5.22}$$

where $\delta g = |g - 2|$.

The strength of the single-site anisotropy (5.20) depends on the details of the electronic occupation of corresponding ions, but it can be estimated crudely as[9]

$$K \sim \frac{\lambda^2}{\delta E} \sim \left(\frac{\lambda}{\delta E}\right)^2 \delta E \sim \left(\frac{\delta g}{g}\right)^2 \delta E. \tag{5.23}$$

One can also notice that the quantity S_z^2 entering the expression (5.20) is nontrivial only if the spin of the ion is $S > \frac{1}{2}$; for $S = \frac{1}{2}$ this is a trivial constant. This also agrees with the expression for the anisotropy constant κ entering (5.18) (its contribution from single-site anisotropy), valid at $T = 0$ (Abragam and Bleaney, 1970):

$$\kappa = K S(S - \tfrac{1}{2}). \tag{5.24}$$

(At finite temperatures in a mean-field approximation the corresponding expression goes over to (5.21).) In the case $S = \frac{1}{2}$ it is the other contributions which give rise to uni-axial anisotropy (5.18), for example anisotropy in the exchange interaction, see below. But for $S > \frac{1}{2}$, when single-site anisotropy (5.20) is present, it usually gives dominating contributions to the magnetic anisotropy.

As mentioned above, the other consequence of the spin–orbit interaction, which makes magnetic systems anisotropic, is the appearance of exchange anisotropy in the magnetic exchange interaction. Without spin–orbit coupling the exchange interaction has the form of the Heisenberg exchange, $J \mathbf{S}_i \cdot \mathbf{S}_j$, that is, it is a scalar product of two spins (theoretically we are speaking of SU(2)-invariance of this interaction – it is invariant with respect to rotations). But in general the exchange interaction is a convolution of two vectors $\mathbf{S}_i$ and $\mathbf{S}_j$ with the tensorial exchange "constant"

$$\mathcal{H}_{ij} = \mathbf{S}_i \cdot \hat{J} \cdot \mathbf{S}_j = \sum_{\alpha\beta} S_{i\alpha} J_{\alpha\beta} S_{j\beta}, \tag{5.25}$$

---

[9] The general expression for the single-site terms in the Hamiltonian can be written (Abragam and Bleaney, 1970) as

$$\mathcal{H}_{\text{anis.}} = \lambda^2 (\Lambda_x S_x^2 + \Lambda_y S_y^2 + \Lambda_z S_z^2),$$

with

$$\Lambda_x = \sum_n \frac{\langle 0|L_z|n\rangle \langle n|L_x|0\rangle}{E_n - E_0},$$

and similarly for Λ_y, Λ_z. Here $|n\rangle$ are the excited levels with energy E_n. For a uniaxial system one can reduce this expression to (5.20), and we get the estimate (5.23) for the anisotropy constant.

where the indices $\alpha, \beta = \{x, y, z\}$ and $J_{\alpha\beta}$ is a 3×3 tensor. If this tensor is diagonal and $J_{\alpha\beta} = J\delta_{\alpha\beta}$, we come back to the Heisenberg exchange interaction (1.12). But in general it may contain nondiagonal terms, both symmetric and antisymmetric.

The symmetric part of the tensor $J_{\alpha\beta}$ can always be diagonalized, and for example in the tetragonal case leads to the interaction

$$\mathcal{H} = \sum_{ij} J_{\parallel} S_i^z S_j^z + J_{\perp}(S_i^x S_j^x + S_i^y S_j^y) \tag{5.26}$$

with $J_{\parallel} \neq J_{\perp}$. If $J_{\parallel} > J_{\perp}$, we have an Ising-like interaction; in the opposite case the interaction is xy-like. If one writes $J_{\parallel} = J$, $J_{\perp} = J + \delta J$, one can get an estimate for δJ:

$$\delta J \sim \left(\frac{\lambda}{\delta E}\right)^2 J \sim \left(\frac{\delta g}{g}\right)^2 J . \tag{5.27}$$

In a mean-field approximation, when we replace $S_i \cdot S_j$ by $\langle S \rangle^2$, the anisotropic exchange (5.26), (5.27) would also contribute to the total phenomenological exchange anisotropy (5.18), giving the contribution $\delta\kappa \sim \delta J$. Comparing the expressions for single-site and exchange anisotropy (5.23), (5.22), we see that typically the single-site contribution (5.23) is much larger than the exchange contribution (5.27) (usually $\delta E \gg J$). But for example for systems with $S = \frac{1}{2}$, for which single-site anisotropy is absent, all anisotropy comes from the exchange terms (5.26), (5.27).[10]

5.3.2 Antisymmetric exchange and weak ferromagnetism

Another very important difference between the general interaction (5.25) and the Heisenberg interaction $\sim S \cdot S$ is a possible existence of *antisymmetric* exchange. As always, one can associate with an antisymmetric 3×3 matrix a dual vector

$$D_{\alpha} = \sum_{\beta\gamma} \varepsilon_{\alpha\beta\gamma} J_{\beta\gamma} , \tag{5.28}$$

where $\varepsilon_{\alpha\beta\gamma}$ is the totally antisymmetric tensor, and then the antisymmetric part of the exchange interaction (5.25) can written as

$$\mathcal{H}_{ij}^{\text{as}} = D_{ij} \cdot (S_i \times S_j) . \tag{5.29}$$

---

[10] There exists in principle yet another, purely classical source of magnetic anisotropy – classical dipole–dipole interaction of magnetic moments

$$\frac{(M_i \cdot M_j) - (M_i \cdot r_{ij})(M_j \cdot r_{ij})/|r_{ij}|^2}{|r_{ij}|^3} .$$

This also depends on the direction of moments and on the relative orientation of respective sites. Usually this part of the anisotropy is smaller than single-site anisotropy, but it may become comparable with exchange anisotropy, especially for systems with weakly coupled spins (small J in (5.27)).

This interaction was first obtained on symmetry grounds by Dzyaloshinskii (1958) for an interaction between two antiferromagnetic sublattices S_1, S_2 and written as

$$\boldsymbol{D}_{\mathrm{D}} \cdot (S_1 \times S_2) \,. \tag{5.30}$$

Later this expression was derived microscopically in Moriya (1960), and it is often called the Dzyaloshinskii–Moriya (DM) interaction. We will call the vector $\boldsymbol{D}_{\mathrm{D}}$ of (5.30) the Dzyaloshinskii vector; the corresponding vector $\boldsymbol{D}_{ij}$ for a given bond (ij) in (5.29) is called the Dzyaloshinskii–Moriya vector, or sometimes simply the Moriya vector.

Thus, one has to discriminate between the Dzyaloshinskii–Moriya vector $\boldsymbol{D}_{ij}$ *for a given bond* and the corresponding Dzyaloshinskii vector $\boldsymbol{D}_{\mathrm{D}}$ *for the whole crystal*. The vector $\boldsymbol{D}_{\mathrm{D}}$ is obtained by the appropriate summation of local vectors $\boldsymbol{D}_{ij}$; in some cases these can (at least partially) cancel. The presence and direction of the full vector $\boldsymbol{D}_{\mathrm{D}}$ and of the corresponding interaction (5.30) can be determined from symmetry considerations, as in fact was done originally by Dzyaloshinskii (1958).

The perturbation theory treatment of Moriya gives for $\boldsymbol{D}$ the estimate

$$D \sim \frac{\lambda}{\delta E} J \sim \frac{\delta g}{g} J \,. \tag{5.31}$$

Thus, when this interaction is present (when it is allowed by symmetry), it is stronger than the symmetric exchange anisotropy (5.27), and can in general be comparable with single-site anisotropy (5.23).

A very important consequence of the DM interaction (5.29), (5.30) is that, whenever present (see below), it leads to a canting of neighboring spins, and eventually to the possible appearance of a weak ferromagnetic moment in an antiferromagnet. Indeed, if a pair of spins S_1, S_2 which otherwise would be collinear (ferro- or antiferromagnetic due to the exchange $J S_1 \cdot S_2$) experience the DM interaction (5.29), for example with the Dzyaloshinskii–Moriya vector as shown in Fig. 5.21(a), then these spins would cant (Fig. 5.21(a)) so as to make the vector product $S_1 \times S_2$ nonzero and parallel (or rather antiparallel) to $\boldsymbol{D}_{12}$. This allows us to gain a certain energy – the energy of antisymmetric exchange (5.29). One can easily see that this energy gain will be linear in the canting angle θ, $\delta E_{\mathrm{DM}} \sim D \sin \theta \sim \theta$, whereas the energy loss in the exchange interaction will be quadratic, $\delta E_{\mathrm{exch}} \sim J \theta^2$. In effect there will be canting of these spins for any $D \neq 0$ ($D \ll J$), with the canting angle $\theta \sim D/J$ ($\sim \lambda/\delta E \sim \delta g/g$, see (5.31)). For parallel spins this would only reduce the total magnetization slightly (and introduce a weak antiferromagnetic component, since the "horizontal" projections of spins S_1 and S_2 in Fig. 5.21(a) are antiparallel). But if the original spins are antiparallel, that is ordered antiferromagnetically, the effect will be much more interesting: as we see from Fig. 5.21(b), the same arguments as above show that the canting of spins S_1, S_2 will lead to the appearance of a net *ferromagnetic* moment of this pair, $\boldsymbol{M} = S_1 + S_2$.

As above, each spin would cant by a small angle $\frac{1}{2}\theta$, with $\theta \sim D/J$, so that the total ferromagnetic moment M would be $\sim S\theta \sim SD/J$ (or, at finite temperature, $\sim \langle S \rangle D/J$, where $\langle S \rangle$ is the average *sublattice* magnetization). If, as sometimes happens, the net

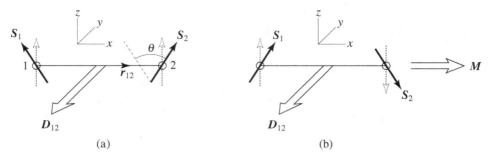

Figure 5.21 Spin canting due to antisymmetric exchange (Dzyaloshinskii–Moriya interaction). (a) Original ferromagnetic ordering. (b) Original antiferromagnetic ordering.

moments of different pairs of spins in a crystal add rather than cancel, then the whole sample will develop, in addition to the antiferromagnetic ordering for example with $\langle S_1 \rangle = -\langle S_2 \rangle$, a ferromagnetic component $M \perp \langle S_1 \rangle$, $\langle S_2 \rangle$, with $|M| \sim |\langle S_1 \rangle| D/J$. As in most cases $D/J \ll 1$ (typically $D/J \sim 10^{-2}$), this moment is much smaller than the sublattice magnetization, therefore we are talking in this case about *weak ferromagnetism*. But, as experimentally the external magnetic field couples much more strongly to the ferromagnetic component than to the antiferromagnetic one, the presence of even a weak ferromagnetic moment leads to very strong effects.

In which particular situations and in which systems do we have this antisymmetric DM exchange (5.29), that in particular can lead to weak ferromagnetism? This is largely determined by symmetry considerations. Moriya (1960) formulated a set of rules determining in which situations the interaction (5.29) is nonzero for a particular bond, and in which directions the vector D_{ij} could point. These rules follow from symmetry considerations. Thus, for example, it is clear that if there exists an inversion center between spins S_1 and S_2, the DM coupling of these two spins should be zero; under inversion the spins S_1 and S_2 change places, and the vector product $S_1 \times S_2$ entering (5.29) would change sign, from which it follows that the corresponding term in the Hamiltonian should be zero. This and similar arguments lead to a set of rules, formulated by Moriya:

1. When there is a center of inversion, located between spins S_1 and S_2, $D_{12} = 0$.
2. When there exists a mirror plane between sites 1 and 2, perpendicular to the line (12) (i.e., to the vector r_{12}), the vector D_{12} is parallel to this mirror plane.
3. When there is a mirror plane including r_{12}, the vector D_{12} is perpendicular to this mirror plane.
4. When there exists a twofold axis perpendicular to r_{12} and going through the middle of r_{12}, the vector D_{12} is perpendicular to such twofold axis.
5. When there exists an n-fold axis ($n > 2$) along r_{12}, the vector D_{12} is parallel to r_{12}.

These rules are usually sufficient to understand the direction in which the Dzyaloshinskii–Moriya vector and the Dzyaloshinskii vector may lie in a particular

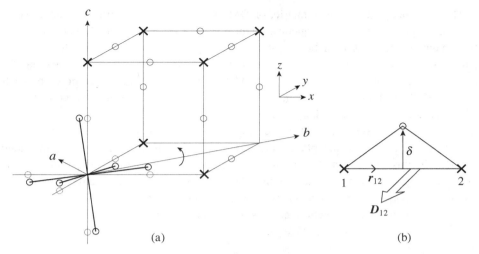

Figure 5.22 The appearance of the Dzyaloshinskii–Moriya interaction in perovskites due to tilting of MO_6 octahedra.

situation, though they do not tell us anything about its magnitude and, most importantly, about its sign. One typical case, very important for example in many multiferroics, is the situation illustrated in Fig. 5.22. There we show the perovskite lattice ABO_3 (as always, we denote TM ions by crosses, and oxygen ions by circles), and we show a typical distortion (GdFeO$_3$-type distortion) very often occurring in many perovskites with small A ions, which consists of tilting the rigid BO_6 octahedra around the basal plane diagonals [110] (in alternating directions from site to site), leading to a transition from cubic to orthorhombic structure $Pbnm$ (as mentioned in Section 3.5, usually this is also accompanied by rotations around the vertical axis [001]). After such a distortion the local situation for two sites 1 and 2 becomes that shown in Fig. 5.22(b). We see that there is now no inversion center, but there is a mirror plane perpendicular to r_{12}, going through the middle oxygen. According to rule 2 above, the Dzyaloshinskii–Moriya vector D_{12} should lie in this mirror plane. Simultaneously there also exists a mirror plane containing the triangle (1–O–2), and rule 3 tells us that D_{12} should be perpendicular to this mirror plane. As a result we see that D_{12} should be perpendicular both to r_{12} and to the vector δ by which the central oxygen in Fig. 5.22(b) shifts from the middle of the line (12). And indeed it turns out that in this case

$$D_{12} \sim r_{12} \times \delta \,, \tag{5.32}$$

and the increase in tilting of the BO_6 octahedra in Fig. 5.22(a) (i.e., the increase in δ) leads to a strengthening of the DM interaction, and the change in sign of δ leads to a change in sign of D_{12} and, correspondingly, to the opposite sign of the resulting canting of spins S_1 and S_2 and to the opposite weak ferromagnetic moment $M = S_1 + S_2$ (see Fig. 5.21(b)).

The existence of the antisymmetric, or DM exchange (5.29) for a given pair of spins is rather typical for many magnetic systems. However it does not always give a net ferromagnetic moment in a bulk system: in many cases such moments for different pairs cancel. This is for example the case in La_2CuO_4 – the prototype material for high-T_c superconducting cuprates. Undoped La_2CuO_4 is a Mott insulator with localized spins $S = \frac{1}{2}$ on each Cu^{2+} ion, forming a "layered perovskite" structure (layers with a structure identical to that of the basal xy-plane in Fig. 5.22(a), with similar distortions). In this system each plane develops a net moment in the c-direction, but the moments in the neighboring planes are opposite, so that there is no net ferromagnetic moment in this system in the absence of magnetic field. But a field parallel to the c-direction can flip such moments, so that $LaCuO_4$ shows a typical metamagnetic behavior, with a sharp increase in magnetization at a certain critical field $H_c \parallel c$. But there are also many antiferromagnetic materials, in which this mechanism leads to a net magnetization in the ground state. One such typical weak ferromagnet is for example the hematite Fe_2O_3.

There could also exist situations in which the DM interaction transforms the material, which would otherwise be a collinear ferromagnet, into an antiferromagnetic (spiral) system. This is illustrated schematically in Fig. 5.23. (This is an extension of the situation shown in Fig. 5.21(a) to a larger system – here a chain of magnetic atoms.) If the exchange interaction is ferromagnetic, and if there exists a DM interaction (5.29) with the vector D that is the same for each pair of spins, the resulting magnetic structure will be a spiral (of cycloidal type), as shown in Fig. 5.23. That is the system, instead of being ferromagnetic, would become a large-period spiral antiferromagnet (with a small pitch angle $\theta \sim D/J$ between neighboring spins and, correspondingly, with a large period). This could be the reason for the appearance of such a magnetic structure for example in MnSi – a very interesting material with a long-period spiral structure, with a large non-Fermi-liquid phase under pressure, in which novel magnetic textures – (skyrmions) – were found, see Section 5.8. A similar effect explains the appearance of the long-period spiral state instead of simple collinear antiferromagnetism in the classical multiferroic material $BiFeO_3$: the ferroelectric transition, taking place in $BiFeO_3$ below $T_{FE} = 1100\,K$, leads to a breaking of the inversion symmetry and to the appearance of the DM interaction (5.29) with the DM vectors pointing as in Fig. 5.23, which, in turn, modifies the simple two-sublattice antiferromagnetic ordering that would have existed in $BiFeO_3$ into a spiral with a long period $\sim 700\,Å$.

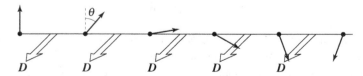

Figure 5.23 The formation of a cycloidal spiral in the case of the same Dzyaloshinskii vector for every spin pair.

As mentioned above, in many magnetic systems, such as perovskites, the Dzyaloshin-skii–Moriya vectors, even when present, can be different for different pairs of spins, which is clear from Fig. 5.22. Whether a net ferromagnetic moment due to spin canting will appear depends both on the type of magnetic structure and on the direction of spins, that is on spin anisotropy. It is clear for example from Fig. 5.22 that the direction of the vector $\boldsymbol{D}_{ij}$ depends on the local crystal structure, but if the spin orientations and magnetic anisotropy are such that spins are not allowed to bend in the direction of $\boldsymbol{D}$, weak ferromagnetism would not appear. Again, symmetry considerations tell us in this case which types of magnetic ordering and which spin orientations can coexist, in particular which ones can give (weak) ferromagnetism.

For perovskites which have the orthorhombic *Pbnm* structure due to tilting of the MO_6 octahedra, see Fig. 5.22, such rules have been established by Bertaut (1965). (Note that in the *Pbnm* structure, obtained e.g. by tilting and rotation of BO_6 octahedra in Fig. 5.22, the new orthorhombic unit cell will be $\sqrt{2} \times \sqrt{2} \times 2$ with respect to the original one, with new axes a, b, c marked in that figure. In what follows, after Bertaut (1965), we denote the orthorhombic axes a, b, c as x', y', z', which are *not* the original cubic axes x, y, z of Fig. 5.22(a).)

There exist four basic types of magnetic ordering in perovskites, denoted F-, A-, C-, and G-types, see Fig. 5.24. Here F-type ordering is simple ferromagnetism, A-type ordering consists of ferromagnetic layers stacked antiferromagnetically, C-type ordering contains ferromagnetic chains antiparallel to one another, and G-type ordering is simple two-sublattice (checkerboard) antiferromagnetism with all nearest neighbors of a given spin antiparallel to it.[11] Depending on a particular situation (spin anisotropy of respective ions), spins can be ordered along different axes, which is denoted for example $F_{x'}$ (ferro-magnetic ordering with spins parallel to the x'-axis) or $G_{z'}$ (ordering of G-type with spins parallel to the z'-axis). Symmetry relations show which particular types of ordering can coexist. Bertaut presents the corresponding results in the form of a table:

Γ_1	$A_{x'}$	$G_{y'}$	$C_{z'}$
Γ_2	$F_{x'}$	$C_{y'}$	$G_{z'}$
Γ_3	$C_{x'}$	$F_{y'}$	$A_{z'}$
Γ_4	$G_{x'}$	$A_{y'}$	$F_{z'}$

Here each row corresponds to a particular representation, denoted Γ_1, Γ_2, etc. using the Bethe notation. Thus for example if we have two-sublattice antiferromagnetism with spins along the z'-axis, $G_{z'}$ (included in the second row of the table), there may (and will) appear ferromagnetism $F_{x'}$ with the ferromagnetic moment $\boldsymbol{M} \parallel x'$ (and simultaneously

---

[11] In principle there may exist more complicated types of magnetic ordering in perovskites, such as E-type (spins $\uparrow\uparrow\downarrow\downarrow$ in x- and y-directions) or CE-type ordering typical for half-doped manganites with charge ordering; these cases will be considered later in corresponding chapters. There may also occur different types of spiral structures, etc. But these four structures – F, A, C, and G – are the main basic types of ordering in perovskites.

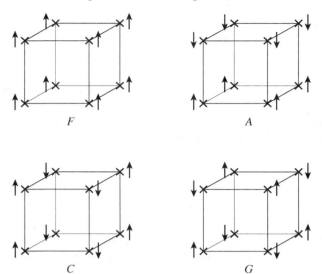

Figure 5.24 Four basic types of magnetic ordering in a cubic lattice, such as the magnetic lattices of B ions in ABO_3 perovskites.

there may appear an admixture of C-type ordering $C_{y'}$). But for example the same $G_{z'}$ antiferromagnetism cannot give weak ferromagnetism in the y'-direction, etc.

Note also that, as always in the physics of solids, if some terms or some types of ordering are allowed by symmetry, they *do* always appear in practice, although the amplitude or intensity of these extra types of ordering may be very small and difficult to detect; there will always be some (maybe very weak) interactions which would lead to the appearance of such extra orderings. Experimentally this situation is usually addressed by neutron scattering: magnetic neutron scattering first allows us to determine the dominant magnetic ordering, in our example $G_{z'}$, and then, guided by the table above, one can look specifically for much weaker signals for instance of $F_{x'}$ or $C_{y'}$ ordering.

The antisymmetric DM exchange is not the only possible source of weak ferromagnetism in magnetic insulators. Another cause could be the (single-site) magnetic anisotropy discussed above. The easiest qualitative way to understand this could be again the situation shown in Fig. 5.22(a): as we see, due to the tilting of TM–O_6 octahedra the local axes of neighboring octahedra are not parallel anymore. Suppose that the local single-site anisotropy is such that the easy axis is along the direction from the TM to an apical oxygen (we do not discuss here how realistic this assumption is, the example serves only as an illustration). As a result of the GdFeO$_3$-type distortion, the two sites along the c-axis have easy axes tilted in opposite directions. Then, with for instance antiferromagnetic interaction between these spins, we would have the situation shown in Fig. 5.25 – the spins on these sites would be canted, and there would appear a net ferromagnetic moment of the pair, M.

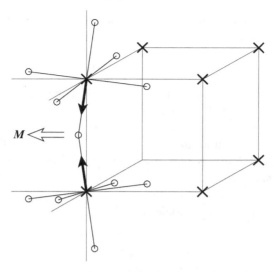

Figure 5.25 Possible appearance of weak ferromagnetism due to single-site anisotropy in orthorhombic perovskites.

Again, in a bulk system such moments may cancel, etc. But there are real materials for which this mechanism is responsible for weak ferromagnetism – for example rutile NiF_2 (Moriya, 1960). Sometimes these two mechanisms of spin canting, local anisotropy and DM antisymmetric exchange can coexist and even compete in the same material, as for example in YVO_3 (Ren *et al.*, 2000).

Two remarks should be made in connection with the anisotropic terms in the exchange Hamiltonian (5.25), (5.29). First, as already mentioned above, one has to distinguish between the Dzyaloshinskii–Moriya vector D_{ij} entering the antisymmetric interaction for a given pair of spins, $D_{ij} \cdot (S_i \times S_j)$, and the Dzyaloshinskii vector for the whole crystal, D_D; cf. eqs (5.29) and (5.30). The vectors D_{ij} may be different for different bonds, they can have different directions and magnitudes. It is for these local vectors that the Moriya rules have been formulated. However, the Dzyaloshinskii vector D_D is a property of the whole crystal, and its existence is determined by the symmetry of the crystal as a whole. The local vectors D_{ij} can sum up to zero, and though they may lead to a local canting of spins, this does not necessarily produce a net ferromagnetic moment. As mentioned above, this is the case for example in the layered material La_2CuO_4, in which such ferromagnetic moments are say up in one layer and down in the neighboring layer, so that there is no net moment. Thus the nonzero value of the Dzyaloshinskii–Moriya vector on particular bond(s) does not yet guarantee that the material will have nonzero net magnetization. Rather, the possibility of having net weak ferromagnetism is determined by the vector D_D, which has to be nonzero. It is this vector D_D which would enter for example into the Landau expansion of the free energy F for antiferromagnets with weak ferromagnetism, and which would couple to the antiferromagnetic vector $L = S_1 - S_2$ (where S_1

and S_2 are the average magnetizations of the two antiferromagnetic sublattices) and to the net magnetization $M = S_1 + S_2$:

$$F \sim D_\mathrm{D} \cdot (L \times M) \tag{5.33}$$

(note that this expression is equivalent to (5.30)). If this term is nonzero, there will always appear weak ferromagnetism ($M \neq 0$) in a system.

The second, very important fact was noticed by Shekhtman *et al.* (1992). They have shown that if one proceeds from the nondegenerate Hubbard model with the inclusion of spin–orbit interaction, that is considering systems without orbital degeneracy and with spin $\frac{1}{2}$, then *for a given bond* the anisotropy caused by the DM antisymmetric exchange $D_{ij} \cdot (S_i \times S_j)$ and that due to anisotropic symmetric exchange (5.25), (5.26) in fact cancel. In this case the anisotropic symmetric terms for a given bond (ij) typically have the form

$$\frac{1}{2J}(D_{ij} \cdot S_i)(D_{ij} \cdot S_j) \,. \tag{5.34}$$

The DM interaction seems to be much stronger: as mentioned above, see (5.31), it is $\sim (\lambda/\delta E)J \sim (\delta g/g)J$, whereas symmetric anisotropy is weaker, $\sim (\lambda/\delta E)^2 J \sim (\delta g/g)^2 J$, see (5.27). But in fact the DM interaction contributes to the total energy only in the second order, $\sim D^2/J \sim (\delta g/g)^2 J$. Indeed, the energy of this interaction for $D/J \ll 1$ is $\sim D \sin\theta \sim D\theta$, where θ is the canting angle. But this canting angle itself is determined by the competition between the Heisenberg exchange $J S_i \cdot S_j$ (which requires $\theta = 0$) and the antiferromagnetic exchange $D_{ij} \cdot (S_i \times S_j)$. The angle θ should be found by minimizing the total energy, consisting of both these contributions, and it turns out that $\langle S_i \times S_j \rangle \sim \theta \sim D/J$, which gives the energy $\sim D^2/J \sim (\delta g/g)^2 J$ – that is, of the same order as symmetric anisotropy. For a given bond these two contributions cancel exactly. A total weak ferromagnetism in the whole sample may still exist because in many cases the contributions of different bonds cannot be simultaneously reduced to zero; only in these cases can one get a net ferromagnetic moment.

An important manifestation of anisotropic terms in magnetic systems exists in the spectrum of magnetic excitations, which contribute to their thermodynamic properties, and which can be measured for example by neutron scattering and by resonance experiments. For an isotropic case with Heisenberg interaction of type $J S \cdot S$ the magnetic excitations – magnons, or spin waves – have a gapless spectrum of the type $\omega \sim Jk^2$ (for small momenta k) for the ferromagnetic case and $\omega \sim Jk$ for the antiferromagnetic one. This is connected with the isotropy of spin space: for example for a Heisenberg ferromagnet the orientation of magnetization in the ordered state is arbitrary, that is ferromagnetic states with different orientation of M are all equivalent and have the same energy. The excitations with $k \to 0$ correspond to rotations of the magnetization of the sample as a whole. Indeed, such excitations transform one of the degenerate ground states into another, which should cost no energy. That is why spin waves in isotropic magnets are gapless, with the spectrum $\omega(k)$ starting from zero. Theoreticians call this result the Goldstone theorem: if an ordered state breaks a *continuous* symmetry, then in systems with short-range interactions there should exist in such an ordered state a gapless collective excitation, called the Goldstone mode (see many examples in Khomskii, 2010).

However the situation changes if we have a certain anisotropy in the system, for example an easy-axis anisotropy $-\kappa m_z^2$ (5.18) with $\kappa > 0$. In this case there exists in the system a *spin gap* – the spin-wave spectrum starts from a finite value ω_0. For an easy-axis ferromagnet $\omega_0 \sim H_A \sim \kappa S$, where H_A is the so-called anisotropy field. For an easy-axis antiferromagnet $\omega_0 \sim \sqrt{2 H_A H_{\text{exch.}}} \sim \sqrt{2\kappa J}$, where $H_{\text{exch.}}$ is the exchange or molecular field. The antisymmetric DM interaction and the symmetric anisotropic exchange in general also contribute to the spin gap. As mentioned above, the standard ways to measure these spin gaps are neutron scattering and ferromagnetic or antiferromagnetic resonance; but they can also be obtained from the bulk magnetic measurements.

One should also note that the same antisymmetric (DM) interaction plays a very important role in the mechanism of the magnetoelectric effect and in the multiferroic behavior of some magnetic systems, see Chapter 8. A detailed treatment, such as that carried out in Shekhtman *et al.* (1992), which would include both the DM exchange and the anisotropic symmetric exchange, has not been carried out yet in application to the magnetoelectric and multiferroic effects.

5.3.3 Magnetoelastic coupling and magnetostriction

As we have seen above, spin–orbit interaction couples spins with orbitals, and through these with the crystal lattice. Consequently, it gives rise to an effective interaction between spin and lattice subsystems, or spin–phonon interaction, and the changes in one subsystem can cause modifications in the other. This is one of the causes of *magnetostriction* – lattice distortion caused by a change in the spin subsystem.

Spin–orbit coupling and the resulting magnetic anisotropy and weak ferromagnetism are not the only mechanisms of magnetostriction. A simpler mechanism comes for example from the distance dependence of the exchange interaction $J_{ij} S_i \cdot S_j$, with $J_{ij} = J(r_{ij})$ being a function of both distance and direction between sites i and j. Correspondingly, when we change the lattice (the distance r_{ij}), there will appear an effective coupling of spins and lattice displacements, of the type $g u_{ij} S_i \cdot S_j$, with the coupling constant $g \sim \partial J/\partial r$ (more accurately, one has to speak here of the dependence of J on different strain components e_{ij} and on $\partial J/\partial e_{ij}$). Moreover, as we saw in Section 5.1, in superexchange occurring through intermediate ligands the shift of these ligands, for example oxygens, sitting more or less between two spins would also modify the exchange. This follows, for example, from the GKA rules: if the angle M_1–O–M_2 decreases, that is the oxygen moves away from the line connecting the magnetic ions M_1 and M_2, the antiferromagnetic exchange will typically decrease (and vice versa, the effective intersite hopping t_{eff} and therefore the double exchange would increase when this angle approaches $180°$). Therefore the magnetoelastic coupling due to exchange striction would involve not only the dependence of the exchange interaction J_{ij} on the distance between spins S_i and S_j, but also on "internal coordinates" such as the M_1–O–M_2 angle, that is such magnetoelastic coupling would involve transverse phonon modes.

These mechanisms of spin–lattice coupling can sometimes lead to a change in magnetic transition from II order to I order (Bean–Rodbell mechanism; Bean and Rodbell, 1962): for example the lattice may contract on entering the magnetically ordered phase, this contraction may lead to an increase in exchange interaction and effective magnetization, and these effects may enhance each other to the extent that this process becomes avalanche-like, that is the transition becomes I order.

Another important consequence of such magnetostriction is that it can be one of the mechanisms by which magnetic ordering couples to electric polarization in *multiferroics* – materials which are simultaneously magnetic and ferroelectric. These questions will be discussed in detail in Chapter 8.

Going back to the role of spin–orbit coupling, we have seen above that it produces, in particular, magnetic anisotropy – that is the direction of magnetization (or sublattice magnetization in antiferromagnets) is tied to particular directions in the crystal. Consequently, when we change, for example, the magnetic structure by changing the temperature or applying a magnetic field, these changes also modify the crystal lattice. In other words, the energy of magnetocrystalline anisotropy also depends on the lattice distortion, or on strain e_{ij}, and it is this dependence which contributes to magnetoelastic coupling and to magnetostriction. If the dependence of the isotropic exchange energy on distortions gives the dependence of striction on the value of (sublattice) magnetization, the corresponding dependence of anisotropy terms leads also to a coupling of strain with the *direction* of magnetization.

Whereas the direct dependence of exchange contact on lattice coordinates gives magnetoelastic coupling $\sim g \sim \partial J / \partial r \sim J/a$, where a is the typical distance (lattice constant), the magnetoelastic coupling and magnetostriction caused by the spin–orbit interaction will be (for singlet orbital states with quenched orbital moment $\langle 0|L|0 \rangle = 0$) proportional to $(\lambda/\delta E)^2 \sim (\delta g/g)^2$, as in (5.27).

The distortion of a crystal can change its symmetry. Thus a nonzero value of the strain e_{zz} will change cubic symmetry to tetragonal. Consequently, if we start with a cubic crystal with anisotropy of type (5.19), in a strained crystal the uniaxial anisotropy (5.18) will appear. This will be important especially in systems with degenerate orbital states and unquenched orbital moments, which we discuss now.

5.4 Systems with unquenched orbital moments

The situation with magnetic anisotropy and with magnetostriction is quite different for TM ions with unquenched orbital moments. For TM in octahedra these are the ions with partially filled t_{2g} levels, if the octahedra are regular or only slightly distorted. This situation was discussed in some detail in Section 3.4; here we repeat the main conclusions and compare the behavior of such systems with those having quenched orbital moments.

As discussed in Section 3.1, the t_{2g} orbitals can be chosen as real, $|xy\rangle$, $|xz\rangle$, $|yz\rangle$; alternatively one can choose complex combinations $|xy\rangle$ and $|\pm\rangle \sim \frac{1}{\sqrt{2}}(|xz\rangle \pm i|yz\rangle)$. The state $|xy\rangle$ is the same as $\frac{1}{\sqrt{2}}(|l^z=2\rangle - |l^z=-2\rangle)$, and the states $|\pm\rangle$ are $|l^z = \pm 1\rangle$, see eqs (3.2), (3.4).

The triplet t_{2g} can be described by the effective orbital moment $\tilde{l} = 1$, so that $|xy\rangle$ corresponds to $|\tilde{l}^z = 0\rangle$ and $|\pm\rangle = \frac{1}{\sqrt{2}}(|xz\rangle \pm i|yz\rangle)$ correspond to $|\tilde{l}^z = \pm 1\rangle$. Correspondingly, the spin–orbit coupling $\lambda \boldsymbol{L} \cdot \boldsymbol{S}$ also goes over to $\tilde{\lambda}\tilde{\boldsymbol{l}} \cdot \boldsymbol{S}$, with $\tilde{\lambda} = \alpha\lambda$, with $\alpha = -1$ for Fe^{2+} and $\alpha = -\frac{2}{3}$ for Co^{2+} (two typical ions to which this treatment is usually applied).

In contrast to the case of orbital singlets, in which we treated the spin–orbit coupling in perturbation theory in $\lambda/\delta E$, here the spin–orbit interaction acts in the first order, leading to the ground state with $\tilde{l}$ parallel (or antiparallel) to $\boldsymbol{S}$. This is especially important in magnetically ordered states in which the average spins take certain values in particular directions. The corresponding orientation of the orbitals, with their specific shape of electron density (see Fig. 3.4), leads, first of all, to strong magnetostriction. Thus, as already mentioned in Section 3.4, in CoO below T_N the spins are oriented predominantly along one of the cubic axes, which causes the corresponding orientation of the electron cloud, with the t_{2g}-*hole* orbital having the shape shown in Fig. 3.4; and because of that the lattice distorts, leading to a tetragonal state with $c/a < 1$. The corresponding level scheme for $Co^{2+}O$ is shown in Fig. 3.37: tetragonal compression splits the t_{2g}-triplet into a lower singlet $|xy\rangle = |\tilde{l}^z = 0\rangle$ and an upper doublet $|\tilde{l}^z = \pm 1\rangle$ which will be occupied by three electrons, that is at this stage the degeneracy still remains. But spin–orbit coupling splits the upper doublet still further, and it is this extra energy gain which in effect stabilizes this distortion (the standard Jahn–Teller effect would have caused the opposite distortion with $c/a > 0$ and with the t_{2g}-doublet $|\tilde{l}^z = \pm 1\rangle$ or $|xz, yz\rangle$ lying lower and being occupied by four electrons). It is precisely because of spin–orbit coupling that most Co^{2+} compounds are distorted with locally compressed octahedra, and not elongated ones, as the JT effect would require. The net energy gain for a JT-type distortion (elongation) would be E_{JT}, while for that observed experimentally (compression) it is $\frac{1}{2}(E_{JT} + |\tilde{\lambda}|)$; and if $|\tilde{\lambda}| > E_{JT}$ (which is the case for CoO or $KCoF_3$), the system evolves "according to the spin–orbit scenario" and not according to the Jahn–Teller one (cf. eqs (3.43), (3.44) above). Spin–orbit coupling seems to dominate for heavier $3d$ elements, such as Co or Fe, but for example in the early $3d$ elements such as V or Ti the JT effect is usually stronger, and the corresponding JT distortion determines the type of orbital ordering (and quenches the orbital moment for these systems); see also the discussion in Section 6.5.

As t_{2g} levels can be split both by tetragonal E_g and trigonal T_{2g} distortions, magnetostriction in such compounds can lead either to tetragonal or trigonal structures. Which one will be realized in a particular compound depends on the detailed electronic and crystal structure, and also on the strength of the coupling of t_{2g}-electrons to tetragonal and trigonal modes $gc^\dagger_{i\alpha\sigma}c_{i\alpha\sigma}u_i$ and on the hardness of the corresponding distortion mode $\frac{1}{2}Bu_i^2$: the energy gain due to distortion is $\sim g^2/B$, and if this energy gain is larger for tetragonal distortion (i.e., the coupling with tetragonal mode g_{tetr} is stronger and/or the "spring constant" of this mode B_{tetr} is smaller), the system would become tetragonal below T_c or T_N. If the corresponding gain is larger for trigonal distortion, the trigonal one would be realized. Experimentally it seems that most of the cubic Co^{2+} compounds (with Co in octahedra) become tetragonal (CoO, $KCoF_3$), whereas for Fe^{2+} usually the trigonal distortion is realized (FeO, $KFeF_3$).

It should be noted that the actual magnetic anisotropy, that is the dependence of magnetic energy on the orientation of spins relative to the crystal lattice, involves not only the spin–orbit coupling $\lambda L \cdot S$ (which gives the dependence of spins on the shape of the electronic cloud of the magnetic ion) but also the interaction of this electronic cloud with the lattice itself. In other words, magnetic anisotropy and magnetostriction, or magnetoelastic coupling, usually go together: either all are relatively weak, or all are strong. And it is for example the strong magnetostriction which makes systems such as CoO or FeO uniaxial below T_N: in tetragonally compressed CoO both spins and orbitals are kept parallel to the z-axis by the spin–orbit coupling to first order, that is the anisotropy constant κ (5.18) here is $\kappa \sim \lambda$, and not $\sim \lambda^2/\delta E$ as for quenched orbital moments. Consequently, the materials which contain ions with partially filled degenerate t_{2g} levels and with unquenched orbital moments have much stronger magnetostriction and magnetic anisotropy. When we recall that for the nondegenerate case the anisotropy constant is $\sim \lambda^2/\delta E$, we can easily understand that typically we would have three situations:

1. Ions with completely filled t_{2g} shells (e.g., Ni^{2+} ($t_{2g}^6 e_g^2$)) or with half-filled such shells (Mn^{2+}; Fe^{3+} ($t_{2g}^3 e_g^3$)) would have as the lowest excited states those with $t_{2g}-e_g$ excitations, so $\delta E \sim 10Dq \sim 2\,eV$; in these systems we expect the weakest anisotropy and magnetostriction.
2. Systems with partially filled t_{2g} levels, but in which these levels are split by local distortions, so that the resulting states are not degenerate. Examples are some Ti or V compounds, in which the crystal field splitting of t_{2g} levels, leading to a nondegenerate state, is typically $\delta E \sim 0.2\,eV$. In this case we can still use perturbation theory in $\lambda/\delta E$, though with some caution. In any case, crude estimates show that the anisotropy constants would be ~ 10 times larger than for Ni^{2+} or Mn^{2+}.
3. And finally, if we have really degenerate partially filled t_{2g} levels, as in CoO or FeO, anisotropy and magnetostriction would be yet an order of magnitude stronger.

Experiment indeed confirms these expectations, see for example Kanamori (1960). Thus experimentally magnetic insulators with Co^{2+} and Fe^{2+} have magnetostriction and magnetic anisotropy $\sim 10^2$ larger than similar compounds with Ni^{2+}, Mn^{2+}, or Fe^{3+}. This feature is used in some practical applications, when Co^{2+}-containing compounds are used for example for magnetoacoustic devices, etc. And microscopically all these strong effects are due to the intrinsic orbital (t_{2g}) degeneracy of the corresponding ions, with unquenched orbital moments, because of which spin–orbit coupling acts already in the lowest order.

5.5 Singlet magnetism

As discussed above, there may occur situations where the ground state of a transition metal (or rare earth) ion is a nonmagnetic singlet. This may happen due to particular crystal field splitting – thus some rare earth ions with an even number of electrons, for example Pr^{3+}, Tb^{3+}, or Tm^{3+}, often have a singlet ground state (e.g., Γ_1 in the Bethe notation which is always used for rare earths), separated by a certain energy gap Δ (typically 10–100 K) from the next excited state, for example a singlet Γ_3 or a triplet Γ_4. Another such situation could exist for example in transition metals with d^2 configuration in a very strong tetragonally compressed octahedron, so that the electron occupation would be $(xy)\uparrow(xy)\downarrow$ (not a very realistic situation for $3d$ elements, but not excluded e.g. for $5d$). This could actually be an example of a low-spin state, see Section 3.3, with the crystal field splitting exceeding the Hund's rule energy which tends to stabilize the high-spin state. As discussed in that chapter, low-spin states indeed are realized for example for Co^{3+}. Yet another possibility, mentioned in Section 4.3, could be the case when, due to spin–orbit coupling, the ground state of an ion is the state with $J = 0$ ($4d$ or $5d$ ions with the low-spin configuration t_{2g}^4, such as Ru^{4+} or Ir^{5+}).

A similar situation can occur even for nominally magnetic ions in the presence of strong positive single-site anisotropy $+K S_z^2$, cf. (5.20), if this anisotropy is stronger than the exchange interaction $I S_i \cdot S_j$. That is, the total Hamiltonian of such a system is

$$\mathcal{H} = I \sum_{\langle ij \rangle} S_i \cdot S_j + K \sum_i (S_z)_i^2 . \tag{5.35}$$

(Here we denote the exchange interaction by I, to distinguish it from the total moment of the ion, J, which we use in some examples.) The ground state of this system would be the state with $\langle S_z \rangle = 0$ at every site; it would have lower energy than the best magnetically ordered state, which in this case would be an xy ordering (spins in the xy-plane). By creating such an order we gain the exchange energy $\sim I$, but lose the anisotropy energy $\sim K$, and for $K > I$ such an ordered state is not realized and the system remain nonmagnetic, with the $|S_z = 0\rangle$ state at each site.

Yet another similar situation is met in materials in which the magnetic ions form dimers, which are often in a singlet state, with the triplet excited state lying higher in energy. If the intradimer coupling I_{intra} is larger than the interdimer exchange I_{inter}, the system would remain nonmagnetic and may be treated as composed of singlet "units."

As already mentioned above, there usually exists intersite coupling of magnetic excited states. And if this becomes strong enough and exceeds the energy of promotion of the ion (or dimer) to an excited magnetic state, there may appear a magnetic ordering in the system, despite the fact that the ground state of an *isolated* ion (or dimer) is a nonmagnetic singlet. This is the situation of *singlet magnetism* (sometimes also called *induced magnetism*), see for example Cooper (1972).

The easiest way to understand this phenomenon is to consider first the behavior of such a singlet system (at first even without the intersite interaction) in an external magnetic field.

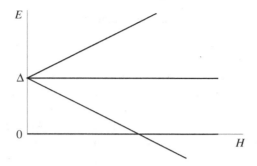

Figure 5.26 Schematic behavior of levels of a singlet–triplet system in a magnetic field.

Consider for example the situation in which the ground state of an ion is a $|J = 0\rangle$ singlet, and the first excited state is a $|J = 1\rangle$ triplet, and apply to such a system a magnetic field in the z-direction. The triplet state would experience the usual Zeeman splitting, see Fig. 5.26, and the $|J_z = \pm 1\rangle$ states would be split as

$$E_{\pm} = \Delta \pm g_J \mu_B H \qquad (5.36)$$

where Δ is the difference between the ground-state singlet $|J = 0\rangle$ and the excited triplet $|J = 1\rangle$. We see that for the field $H > H_c = \Delta / g_J \mu_B$ the magnetic state $|+1\rangle$ would lie below the singlet $|J = 0\rangle$, that is there would occur an abrupt transition to a magnetic (in this case fully polarized) state. The details of such a process depend on the specific situation, in particular on the presence or absence of nondiagonal matrix elements of the magnetic dipole between the ground and exited state(s), see below, but the essence of this possible phenomenon is clear from the simplified treatment illustrated in Fig. 5.26.

It is easy to understand that the same role, played by the external magnetic field in the example above, may also be played by the internal, or exchange field $H_{\text{int}} = I \langle J \rangle$, where $\langle J \rangle$ is the average magnetization of the system (it could be the average spin $\langle S \rangle$ in a spin system with Hamiltonian of type (5.35) or in a system of singlet dimers). It is clear that when the intersite exchange I is stronger than the excitation energy Δ, the corresponding exchange field would be larger than H_c, and the ground state of the system would be magnetically ordered.

There exist different specific situations in which the detailed behavior of such singlet magnets is different. Mostly, this behavior depends on the presence of nondiagonal matrix elements of magnetic dipoles between the nonmagnetic ground state and the excited magnetic state(s). More precisely, we have to discriminate between the matrix elements $\langle 1|J_z|0\rangle$ or $\langle 1|S_z|0\rangle$, where $|0\rangle$ is the singlet ground state and $|1\rangle$ is the typical excited state (it may also be a singlet, see below, or a triplet, etc.), and the similar nondiagonal matrix elements of $J^{\pm}$ or $S^{\pm}$. For spherically symmetric systems we can always choose the magnetic field and moment orientation so that it would be the $\langle 1|J_z|0\rangle$ matrix elements which appear. However for tetragonal systems, for example those described by the Hamiltonian (5.35), the behavior would be different for the magnetic field parallel to the z-axis and for a field perpendicular to it. In the parallel field H_z the combination $H_z S_z$ will appear, which in this model does not have nonzero matrix elements between the state $|0\rangle = |S_z = 0\rangle$ and the excited states $|\pm 1\rangle = |S^z = \pm S\rangle$ at energy K (for example, the states $|\pm 1\rangle$ of Ni in $NiCl_2-4SC(NH_2)_2$ (Zapf *et al.*, 2006)). In this case the parallel magnetic field would lead to a crossing of nonmagnetic and magnetic terms such as that shown in Fig. 5.26.[12] For induced magnetism, which would occur when the exchange interaction of the excited states is strong enough, we would get in this specific case an xy magnetic ordering; though in the case of isotropic systems without such transitions the singlet magnetism could appear in a jump-like fashion, as an I order phase transition.

However in other situations, for example for the two lowest singlet states of Pr^{3+} in a hexagonal crystal field, there will be a nondiagonal matrix element of J_z between these two singlets, $\alpha = \langle 1|J_z|0\rangle$. One can easily understand this possibility for the example not of $4f$ electrons, but of d electrons. Thus, for instance, there exists a nonzero matrix element of the orbital moment l_z between the states $|xz\rangle$ and $|yz\rangle$, that is $\langle xz|l_z|yz\rangle \neq 0$. Indeed, we immediately see this from the fact that the magnetic states are $|l_z = \pm 1\rangle \sim |xz\rangle \pm i|yz\rangle$. The situation here strongly resembles that of Van Vleck paramagnetism. The Van Vleck paramagnetic response of systems with a nonmagnetic ground state appears due to an admixture to the ground state $|0\rangle$ of magnetic excited states $|n\rangle$ in an external magnetic field, which in perturbation theory is

$$|\Psi_0\rangle = |0\rangle + \sum_n |n\rangle \frac{\langle n|\boldsymbol{H} \cdot \boldsymbol{J}|0\rangle}{E_n - E_0} , \qquad (5.37)$$

that is there will be a contribution to the energy $\sim H^2$, or there would appear a moment in the ground state $\sim H$. For the excitation energies $\Delta \sim E_n - E_0 \gg T$ the corresponding paramagnetic contribution to susceptibility will be temperature-independent, but for temperatures of the order of the excitation energies this Van Vleck contribution also becomes strongly temperature-dependent, approaching the Curie behavior $\chi \sim 1/T$. Such a temperature-dependent Van Vleck contribution is for example very typical for Eu^{3+}.

The character of singlet magnetism would also be different in this case; it was studied in detail in connection with some Pr compounds in Andres *et al.* (1972), Birgenau *et al.* (1972), and Buyers *et al.* (1975). The mean-field treatment of this situation for the case of two singlets connected by the matrix element $\alpha = \langle 1|J_z|0\rangle$ leads to a self-consistency equation (Andres *et al.*, 1972; Cooper, 1972), from which one indeed first sees that the magnetic solution exists for strong enough intersite exchange interaction,

$$\frac{4Iz|\alpha|^2}{\Delta} > 1 , \qquad (5.38)$$

---

[12] However in this case, in the perpendicular field $H_\perp$ there would be nondiagonal matrix elements of the type $\langle \pm 1|H_\perp S_{x,y}|0\rangle$ or $\langle \pm 1|H_\perp S^\pm|0\rangle$ and as a result there will be a repulsion of these terms in $H_\perp$, that is a gradual admixture of magnetic excited states to the ground state $|S_z = 0\rangle$.

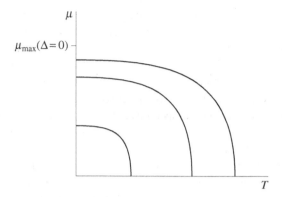

Figure 5.27 Schematic behavior of magnetization in a two-singlet model for different values of I/Δ, after Cooper (1972). The energy gap from the ground to the excited state Δ increases from upper to lower curves.

where z is the number of nearest neighbors. The magnetic transition in this case is continuous, II order, and both critical-temperature and zero-temperature magnetization decrease with increasing Δ, see Fig. 5.27 (Cooper, 1972), until they disappear at $\Delta = 4Iz|\alpha|^2$.

The phenomenon of singlet magnetism was initially studied mostly on the example of rare earth systems, in which indeed the excitation energies are typically relatively small, ~ 10–$100\,\mathrm{K}$. The systems in which it was observed include, for example, Pr_3Tl, Pr_3In, $TbSb$, $TbVO_4$, etc. But apparently some transition metal compounds also belong to this class. As already mentioned above, isolated ions Ru^{4+} and Ir^{5+} have a singlet ground state with total moment $J = 0$. Apparently the magnetic ordering often seen in Ru compounds, and even in some compounds of Ir^{5+} (with its much stronger spin–orbit coupling and correspondingly larger excitation energy Δ) is of this type, although the details of this behavior are not yet well studied (see Khaliullin, 2013). Here an important role can be played by the additional crystal field splitting, for example tetragonal or trigonal, which can at least partially quench the orbital moment and "spoil" the pure $J = 0$ ground state.

Yet another example of this phenomenon in transition metal systems could be the appearance of magnetically ordered states in systems of type $YBaCo_2O_{5.5}$ (Plakhty, 2005), where the low-spin state of Co^{3+} ($S = 0$) is transformed in some phases to a magnetic state, although the detailed situation in these systems is still not completely clarified.

One more related aspect should be mentioned here. As discussed above, when the intersite interaction is not strong enough and the system remains nonmagnetic, one can make it magnetic by applying a magnetic field. As is seen for example schematically in Fig. 5.26, in a critical magnetic field there occurs a crossing of nonmagnetic and magnetic levels. In real situations of course there always exists some intersite interaction, even though it may not be strong enough to make the system magnetic. In any case, it would lead to a certain dispersion of magnetic excited states – the magnetic excitons, similar to the singlet–triplet transitions in Fig. 5.26, or the transition from the state with $S_z = 0$ to the upper magnetic doublet $S_z = \pm 1$ in the model (5.35). When we increase the external magnetic field (or increase the exchange interaction I) the gap in the magnetic excitation spectrum would decrease, and at a critical field $H = H_c$ (or at a critical exchange interaction $I/\Delta = (I/\Delta)_c$) this gap would go to zero at a certain wave vector Q_0, and after that it would become negative. When this happens, the state of the system would change to magnetic (with an ordering with the wave vector Q_0 – e.g. ferromagnetic, antiferromagnetic, or spiral), which can be interpreted as the creation of a macroscopic number of magnons with $q = Q_0$, that is as a Bose-condensation of magnons. This phenomenon was studied in many systems, especially those consisting of singlet dimers ($TlCuCl_3$; $BaCuSi_2O_8$; $SrCu(BO_3)_2$), and the analogy with Bose-condensation was widely used in interpreting the results obtained, see for example Gianmarchi *et al.* (2008) and Zapf *et al.* (2014).[13] In particular, in systems described by the Hamiltonian (5.35), the critical indices at the quantum phase transition at $H = H_c$ coincide with those for Bose-condensation.

The appearance of a soft mode (magnetic excitons with the energy going to zero at the transition) was also studied in systems with real singlet magnetism, such as Pr_3Tl, see for example Birgenau *et al.* (1972) and Buyers *et al.* (1975).

5.6 Magnetic ordering in some typical situations

In this section we discuss briefly, using the basic notions described above, some typical representative cases of magnetic ordering in magnetic insulators. In principle there exist quite a lot of different substances with different types of magnetic sublattices. The detailed form of magnetic ordering depends both on these structural details and on the electronic structure of the corresponding ions. Each of these systems should be considered separately. We will not be able, of course, to discuss all the cases; only some representative cases will be treated. One can find more examples in Goodenough (1963) and many later publications.

Magnetically the simplest case is, of course, ferromagnetic ordering. But as we have already discussed, ferromagnetism usually occurs in metallic systems. Ferromagnetism in insulators is rather rare, and is usually caused by specific types of crystal packing and orbital ordering.

Much richer and much more interesting are situations with different types of antiferromagnetic ordering (among which there can also be ferrimagnets or weak ferromagnets). These systems can show the standard long-range orderings of different types, but antiferromagnetic interactions can also lead to new types of states such as spin glass or spin liquid.

---

[13] In the standard systems with Bose-condensation we consider the systems with a fixed number of particles N, or fixed density $n = N/V$, see for example Landau and Lifshitz (1969); such is the situation in ^{4}He or in cold gases. Here the situation is in some sense different: the number of magnons is in principle not fixed, in which case we usually do not have the conventional Bose-condensation, see for example Sonin (2010). But in this case the magnetic field plays the role of the chemical potential, and for systems with fixed chemical potential some features of Bose-condensation can also be seen.

5.6.1 Antiferromagnets with bipartite lattices; magnetic and structural features

Conceptually the simplest antiferromagnets are those with bipartite lattices, that is lattices in which the magnetic sites can be subdivided into two sublattices so that the nearest neighbors of one belong to the other. Examples are cubic lattices (we mean here lattices formed by the magnetic sites) such as that in perovskites ABO_3, with B being a magnetic ion; or in their layered analogs A_2BO_4 which crudely can be treated as having a two-dimensional square lattice. Other examples are bcc lattices or body-centered tetragonal ones, such as rutile structures (examples are NiF_2 or MnO_2). To bipartite lattices belong also hexagonal systems in which the magnetic ions form a honeycomb lattice in the basal plane, or systems with the corundum structure such as Cr_2O_3 and Fe_2O_3.

Despite the conceptual simplicity of these systems, their magnetic ordering is not always self-evident. Thus, for cubic lattices with predominantly antiferromagnetic interaction the most natural is the antiferromagnetic ordering with alternating spins (checkerboard-like), type G in the terminology presented in Section 5.3.2. However, as discussed there, there exist other types of antiferromagnetic ordering even in this simplest case: type A ordering (ferromagnetic planes stacked antiferromagnetically) and type C ordering (ferromagnetic chains antiparallel to each other), see Fig. 5.24. There may exist in this case also more complicated types of ordering: for example E-type (spins $\uparrow \uparrow \downarrow \downarrow$ in x- and y-directions) or CE-type, etc. All these different types of magnetic ordering are mostly caused by particular orbital ordering, such as alternating $|x^2\rangle$ and $|y^2\rangle$ orbitals in $LaMnO_3$ (giving A-type magnetic ordering), or by a distortion of the crystal lattice (tilting and rotation of the BO_6 octahedra – $GdFeO_3$-type distortion). The situation is similar in bcc or bct lattices: if the exchange interactions are predominantly nearest-neighbor, and the same for all nearest-neighbor pairs, we expect a simple two-sublattice structure with spins at the corners of the cubes and at the centers being antiparallel. This is indeed the case in most rutile-type antiferromagnets such as NiF_2 or MnF_2. But the ordering is different in $CrCl_2$ (the so-called body-centered ordering of the third kind; Goodenough, 1963): $Cr^{2+}(t_{2g}^3 e_g^1)$ is a strong Jahn–Teller ion, and apparently the cooperative Jahn–Teller distortion and the corresponding orbital ordering make the magnetic structure more complicated. Another example of a more complicated magnetic structure in rutile-type crystals is MnO_2, where the magnetic ordering is of spiral type.

Thus even in bipartite lattices the resulting magnetic structures may be quite different. Often this brings about a fundamental difference in the behavior of the corresponding substances. A striking example is provided by two classical antiferromagnets with the corundum structure, Fe_2O_3 and Cr_2O_3. Despite a very similar crystal structure, their magnetic ordering is different, see Fig. 5.28. In this figure we show spin ordering in the main "chain" of Fe or Cr ions in the c-direction; the star denotes the inversion center in the crystalline structure.

We see that the magnetic structure of Fe_2O_3 does not break this inversion (remember that the magnetic moment is a pseudovector, i.e. it does not change sign under inversion – for more details see Chapter 8). In contrast, the magnetic ordering in Cr_2O_3 does break

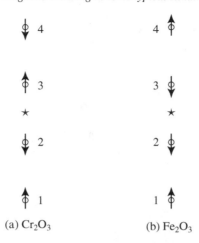

Figure 5.28 Schematic magnetic structure of (a) Cr_2O_3 and (b) Fe_2O_3, demonstrating their different symmetries with respect to inversion (the inversion center is marked by a star).

the inversion symmetry (under inversion up-spins go over to down-spins). This leads to profound differences in the behavior of these materials: Fe_2O_3 turns out to be a weak ferromagnet, due to the Dzyaloshinskii–Moriya antisymmetric exchange, see Section 5.3.2, while Cr_2O_3 does not have such weak ferromagnetism; instead it exhibits a linear magnetoelectric effect, in which an external magnetic field H produces electric polarization $P \sim \alpha H$ and, vice versa, an electric field E produces magnetization $M \sim \alpha E$ (which can in principle be parallel or perpendicular to fields); this will be discussed in detail in Section 8.1. The detailed properties of these (and other) materials depend not only on the type of magnetic ordering, but also on the spin directions, determined by magnetic anisotropy (spins in Fig. 5.28 may lie parallel to the "chain" or perpendicular to it, which would change the behavior of these materials).

When discussing materials with bipartite lattices, one has to specify some details of the crystal structure of the corresponding systems, which may also influence their magnetic properties. In particular, an important class of materials are the perovskites ABO_3, or their layered analogs such as single-layer A_2BO_4, double-layer $A_3B_2O_7$, etc. Often, perovskites show a structural transition from cubic to orthorhombic structure, see for example Section 3.5. As mentioned in that section, for relatively small A ions and larger TM B ions, that is for small tolerance factor

$$t = \frac{R_A + R_O}{\sqrt{2}(R_B + R_O)} , \qquad (5.39)$$

the BO_6 octahedra usually rotate and tilt, so that the B–O–B angle becomes smaller than the ideal angle of $180°$. Consequently, the effective d–d hopping (via oxygens) and the resulting bandwidth become smaller, which stabilizes the insulating state – compare with the case of rare earth nickelates, mentioned in Section 3.5 (Fig. 3.47).

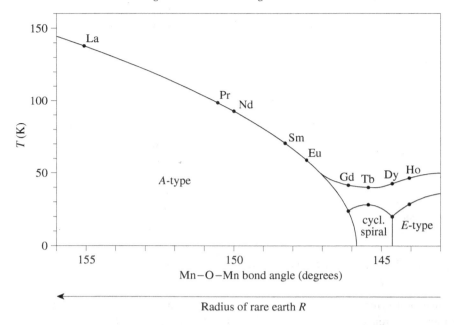

Figure 5.29 Schematic phase diagram of perovskites $RMnO_3$ for different rare earths R.

The same factors – a decrease of the ionic radii of A ions and of the corresponding tolerance factor (5.39), resulting in a stronger tilting and a smaller B–O–B angle – in rare earth manganites $RMnO_3$ lead (together with orbital ordering) to a regular change of the magnetic structure, from the A type (ferromagnetic layers stacked antiferromagnetically) for large R ions (La, Pr) to the E type (spins $\uparrow\uparrow\downarrow\downarrow$ in x- and y-directions) for small ones (Fig. 5.29). There is also a very interesting intermediate region around DyMnO$_3$, TbMnO$_3$, see Fig. 5.29, in which the magnetic structure is spiral-like, which, in particular, makes those materials multiferroic (Kimura *et al.*, 2003a,b), see Chapter 8.[14]

5.6.2 fcc lattices

An important class of magnetic oxides are the mono-oxides such as NiO or MnO. This may not be such a big class as that of perovskites, but many concepts in the quantum theory of magnetism were developed and checked on these materials. They have a rock salt (NaCl) structure, with magnetic ions forming an fcc lattice (Fig. 5.30), with oxygens sitting in the middle of TM–TM bonds in x-, y-, and z-directions (small circles in Fig. 5.30). (This structure can be represented as four simple cubic lattices, starting e.g. from sites a, b, c, d, etc., placed inside each other.) One sees that this lattice is not bipartite: if we include only nearest-neighbor antiferromagnetic exchange, such as between sites a and b

---

[14] The E-type manganites also turn out to be multiferroic, but due to a different mechanism, see Chapter 8.

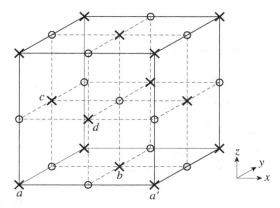

Figure 5.30 Four magnetic sublattices in compounds with rock salt structure, in which the magnetic ions (crosses) form an fcc lattice. Circles denote oxygen ions.

Figure 5.31 The origin of magnetic frustration for an antiferromagnetic interaction.

in Fig. 5.31, we have frustration. Thus, in the triangle (a, b, c) all bonds are equivalent and should be antiferromagnetic, but the very fcc structure does not allow us to form a simple two-sublattice antiferromagnetism: if for example we take spins S_a and S_b to be antiferromagnetic, then whatever we do with spin S_c – whether we orient it up or down – one bond will always be "wrong" with parallel spins, see Fig. 5.31.

This is the essence of the problems in the so-called magnets with *geometric frustration*; we will discuss the situation with frustrated magnets in more detail in Section 5.7. The fcc lattice of TM mono-oxides is, maybe, the first example of such frustrated magnets, but in fact the frustration here is not so severe. This is mainly due to the fact that in reality the exchange interaction in these systems is not just the antiferromagnetic exchange between nearest neighbors; much more important in most of these materials is the interaction between next-nearest neighbors (nnn), for example between ions a and a' in Fig. 5.30, going along a "straight line" through the intermediate oxygen. Indeed, according to the rules discussed in Section 5.1 (Goodenough–Kanamori–Anderson rules), in most cases, at least when the TM ions have partially filled e_g orbitals directed toward the oxygens (with large $t_{pd\sigma}$ overlap with the corresponding p orbitals of oxygens), the 180° TM–O–TM exchange such as that between sites a and a' in Fig. 5.30 is strongest. The nearest-neighbor exchange, for example J_{ab}, when it goes via the common oxygen, is a 90° exchange and according to the GKA rules it is much weaker and may even be ferromagnetic. Only when we are dealing with the early 3d metals such as Ti and V, with only t_{2g} orbitals occupied, can direct overlap of these orbitals (e.g. overlap of xy orbitals on sites a and b in Fig. 5.30) give a strong enough antiferromagnetic exchange competing with that occurring via the oxygens. But the mono-oxides TiO and VO are not at all simple oxides, though they also have a NaCl structure: they always have a lot (~ 10–20%) of vacancies in both metal and oxygen sublattices, they often have varying stoichiometry, and their properties are quite nontrivial. Further, the mono-oxide CrO does not exist in nature.

If indeed the next-nearest-neighbor antiferromagnetic exchange is the strongest one, as is the case in NiO, MnO, CoO, FeO, then what we have to do first is make each of the four cubic sublattices of Fig. 5.30 antiferromagnetic, and then order these sublattices with respect to each other in some way. This intrasublattice ordering is still undetermined: one easily sees that if we make for example the cubic sublattice starting from site a in Fig. 5.30 antiferromagnetic, then its molecular field acting on site b and on the corresponding sublattice will be zero, see Fig. 5.32, that is in the mean-field approximation any orientation of spins in the b sublattice would be possible; they would all be equivalent. But at least this will be only a degeneracy between four sublattices, not an extensive degeneracy proportional to the number of sites (N), as is the case in many other frustrated systems, see below. Correspondingly, this extra degeneracy is usually lifted by magnetic anisotropy, by further-neighbor interactions, etc., so that in effect most of these mono-oxides with this structure have long-range ordering with rather high transition temperatures. Thus the Néel temperature of MnO is $T_N = 122\,\mathrm{K}$, for CoO it is $T_N = 290\,\mathrm{K}$, and for NiO it is $T_N = 520\,\mathrm{K}$. The type of ordering is most often what one calls fcc of second kind, type I (Goodenough, 1963): it can be visualized as antiparallel stacking of ferromagnetic [111] layers, see Fig. 5.33.

Thus, summarizing, we see that although formally for nearest-neighbor antiferromagnetic interaction the fcc lattice is frustrated, due to the specific features of superexchange the second-neighbor exchange typically turns out to be (much) stronger than the nearest-neighbor one, and as a result the degree of frustration is significantly reduced, and fcc antiferromagnets of the type of MnO or NiO behave as more or less decent antiferromagnets.

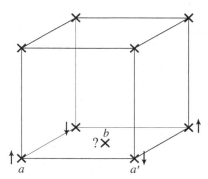

Figure 5.32 Magnetic frustrations in an fcc lattice of magnetic ions in systems with the NaCl structure, such as MnO (cf. Fig. 5.30).

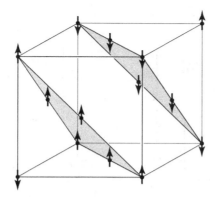

Figure 5.33 Typical magnetic ordering in an fcc lattice.

5.6.3 Spinels

Yet another quite large and representative class of magnetic materials are the spinels AB_2O_4. Magnetic spinels were used widely as the main materials for magnetic memory ("ferrites") during the early stages of the development of the computer industry in the 1950s, and they still present significant interest today. Spinels (as in fact most other oxides) can be understood as a particular close-packing of O^{2-} ions, with metal ions A and B occupying respectively tetrahedral and octahedral interstices, see Section 3.5.

In an ideal case, without extra distortions which sometimes occur, spinels are cubic. One of the easiest ways to depict their crystal structure is shown in Fig. 5.34, where we show the B and O sublattices, and mark the position of A sites; as always we denote metallic B ions by crosses and oxygens by circles. The A ions, sitting inside the oxygen tetrahedra, are shown "in boxes" and their bonds with the surrounding oxygens are marked by wavy lines. We see from this figure that the B–O network resembles somewhat that of rock salt mono-oxides of Fig. 5.30, but the B_4O_4 "cubes" do not fill the whole space, instead some A ions are in their place.[15]

---

[15] Actually the local or point symmetry at each B site is lower than cubic; it is trigonal. This is caused by two factors: first, the oxygen ions in spinels do not necessarily form a regular octahedron, but they can be "pushed" for example inside each B_4O_4 cube in Fig. 5.34, especially if the A ion is large, which leads to a trigonal distortion of the BO_6-octahedra. This shift of oxygens from their ideal positions is usually characterized by a special parameter u. But even if there is no such shift of oxygens, there is a contribution to the crystal field coming from the farther neighbors: in this case one can easily see that for a given B ion, say 1, at the corner of two B_4 tetrahedra, there will be two B triangles on both sides of it along the direction from one tetrahedron to another, the shaded triangles in Fig. 5.35. For a given site this direction, which is actually one of the [111] diagonals of the B_4O_4 cube, is different from the other three such directions. This leads to a trigonal local symmetry of site 1. But for other B ions in Figs 5.34, 5.35 such local trigonal axes will be in other [111] directions, so that in the whole crystal the cubic symmetry is restored, though the point symmetry at each B ion is in principle trigonal.

How important this factor is, and what the corresponding crystal field splitting is, depends on the specific situation. Often in the theoretical treatment it is ignored in a first approximation. It can also be that these two factors leading to such trigonal splitting, local trigonal distortion of BO_6 octahedra, and the role of more distant B ions partially cancel, reducing the total effect.

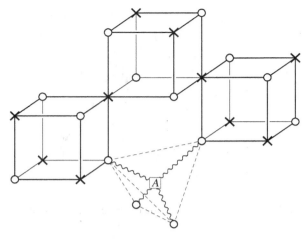

Figure 5.34 Location of octahedral B sites (crosses) and tetrahedral A sites in a spinel AB_2O_4.

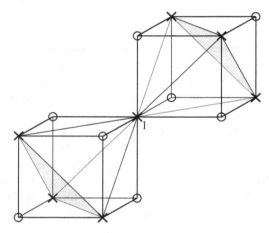

Figure 5.35 A strongly frustrated lattice of corner-sharing tetrahedra of B sites in a spinel AB_2O_4 (this sublattice is often called a pyrochlore sublattice).

A very interesting feature of the spinel structure, seen from Figs 5.34, 5.35, is that the B sites form a network of corner-shared tetrahedra. This is a highly frustrated situation, and it determines the special properties of many spinels, in which only the B sites are magnetic, but the A sites are occupied by nonmagnetic ions. Examples are $CdCr_2O_4$ and similar compounds.

Similar topology of the magnetic sublattice – a network of corner-shared tetrahedra – is also met in another class of compounds: the pyrochlores $A_2B_2O_7$,[16] where both A and

---

[16] The mineral *pyrochlore* is $(Na,Ca)_2Nb_2O_6(OH,F)$.

B sites form such networks. That is why one often calls the *B* site sublattice in spinels a pyrochlore sublattice.

In most spinels studied in detail up to now, both *A* and *B* ions were magnetic. Especially useful are materials containing heavier $3d$ metals such as Mn, Fe, Co. As in this case the main superexchange goes via the oxygens, we see that the *B–O–B* exchange is relatively weak: it is a 90° exchange, which is usually weak (and often ferromagnetic) according to the GKA rules. The *A–O–B* angle is larger, $\sim 125°$, and in most cases, when both *A* and *B* sites are magnetic, it is the stronger interaction. If J_{AB} is antiferromagnetic, which is often the case, then the spins of the *A* sublattice will be opposite to those of the *B* sublattice and, as *A* and *B* spins are different and there are twice as many *B* sites as *A* sites, the resulting magnetic state will be *ferrimagnetic*; this makes these materials especially attractive for applications. This is, for example, the case for magnetite or lodestone $Fe_3O_4 = Fe[Fe_2]O_4$ (for spinels one often denotes the ions at *B* sites using square brackets), where Fe^{3+} ions are on *A* sites and mixed Fe^{2+}/Fe^{3+} ions with 50/50 ratio (or iron ions with average valence $Fe^{2.5+}$) occupy *B* sites. This is a very famous material: it was the first magnetic material known to mankind, which gave us the very word "magnetism," from Mount Magnesia in Asia Minor (modern Turkey) where this stone was first discovered by Europeans (apparently it was known even earlier in ancient China, where it was used for magnetic needles in the first compasses). The magnetism of Fe_3O_4 is in fact not ferro- but ferrimagnetism: due to strong antiferromagnetic *A–B* exchange, Fe_3O_4 orders magnetically at a high temperature $T_c = 858$ K and has a large spontaneous magnetization – theoretically for one formula unit the spin should be $2S_B - S_A$, and with $S_{Fe^{3+}} = \frac{5}{2}$, $S_{Fe^{2+}} = 2$ we would have an uncompensated spin $S = 2$ per formula unit, or, correspondingly, a magnetization $M \sim g\mu_B S = 4\mu_B$, quite a significant value.[17] We meet a similar situation in many other magnetic spinels.

But this is not the only possible type of magnetic ordering in spinels. If the *B–B* exchange is not negligible, there may appear more complicated magnetic structures – for example of the type shown in Fig. 5.36, which is known as a Yafet–Kittel structure: the spins of *A* ions are for example ↓, and the spins of *B* ions are canted so that their net spin is opposite to that of *A* sites but they also have a certain antiferromagnetism between themselves. One can find some extra details for example in Goodenough (1963).

One meets an interesting and nontrivial situation when only the *B* ions are magnetic. The strong frustration in this case makes the magnetic properties very complicated. We will discuss this in the next section.

One should also notice that, similar to other structures, in spinels there may exist TM ions with orbital degeneracy, which can lead to orbital ordering. In some cases it may

---

[17] By the way, the magnetite Fe_3O_4 is famous also because it was apparently the first TM compound in which a temperature-induced insulator–metal transition was observed: the famous Verwey transition at ~ 120 K, discovered by Verwey in 1939 (Verwey, 1939; Verwey and Haaymann, 1941; Verwey *et al.*, 1947). Above 120 K Fe_3O_4 is conducting, although it is not a good metal: the resistivity still continues to decrease with temperature up to ~ 400–500 K. But at $T_V \sim 120$ K the resistivity jumps up by $\sim 10^2$, and below T_V the material is definitely insulating (or a small-gap semiconductor). Even more interestingly, Fe_3O_4 may be the first known multiferroic: apparently it becomes ferroelectric below the Verwey transition temperature. These questions will be addressed in Chapters 7 and 8.

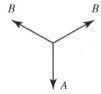

Figure 5.36 A possible ordering of different sublattices in AB_2O_4 spinels with both A and B ions magnetic (Yafet–Kittel structure).

be quite simple: thus in Mn_3O_4 with the JT ions $Mn^{3+}(t_{2g}^3 e_g^1)$ at the B sites we have a ferro. orbital ordering with all occupied e_g orbitals of the $|z^2\rangle$-type and with concomitant (and quite strong) tetragonal elongation, with $c/a \sim 1.14$. But in other cases, again due to frustrations, the orbital ordering can be more complicated, see for example Mostovoy and Khomskii (2003).

One more general remark is in order here. There are many spinels among oxides, but many sulfides or selenides also have the spinel structure, for example $CdCr_2S_4$ or $HgCr_2Se_4$. First, due to the larger radius of $3p$, $4p$, and $5p$ orbitals in chalcogenides (S, Se, Te) compared with $2p$ orbitals of oxygen, there is stronger d–p covalency and correspondingly broader bands in this case. Consequently, there are higher chances of having metallic conductivity in sulfides and selenides compared with oxides. This is in fact a general trend valid for all TM compounds, but it is especially clearly seen in spinels. Second, as already mentioned above, if the magnetic ions are at the B sites, the 90° B–O–B interaction becomes stronger and as it is often ferromagnetic, we have a better chance of the B sites being ferromagnetic in such systems. Typical examples are Cr spinels: whereas for example in $CdCr_2O_4$ the direct t_{2g}–t_{2g} interaction of Cr ions dominates and is antiferromagnetic, and Cr–O spinels have a very complicated magnetic structure due to frustration, similar compounds with S or Se, such as $CdCr_2S_4$, are well-known ferromagnetic semiconductors (Methfessel and Mattis, 1968).

At the end of this section we should make one general remark concerning a rather evident, but not always appreciated, point. In analyzing the magnetic structure of different compounds, one has to be careful in extracting the values of the exchange constants: one has to take into account not only the exchange of a particular pair of spins, but also the interactions with other neighbors. Sometimes one comes across a situation which may be called the situation of "common enemies." Thus, the fact that a certain pair of spins is parallel does not yet imply that the exchange between these two sites is ferromagnetic: it can be that the antiferromagnetic exchange between other sites dominates, forcing the spins of this particular pair to be parallel, see for example the schematic picture in Fig. 5.37. In this figure we show a possible arrangement of spins on two neighboring rows, with the ferromagnetic ordering between rows (parallel spins at sites 2 and 2′) appearing not because

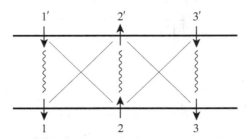

Figure 5.37 Illustration of "forced" parallel spin ordering of spins at sites 2 and $2'$ due to a combination of strong antiferromagnetic exchange in chains $(1, 2, 3)$ and $(1', 2', 3')$ and reasonably strong antiferromagnetic "diagonal" coupling $(1'2)$, $(3'2)$, etc.

the interchain exchange $J_{22'}$ is ferromagnetic, but rather because of stronger antiferromagnetic couplings J_{12}, J_{23} and $J_{21'}$, $J_{23'}$. Assume, first, that the antiferromagnetic exchange in rows (bold lines in Fig. 5.37) is the strongest. It can then happen that, first, the spins on sites $1'$ and $3'$ order antiparallel to the spin of site 2, and then the antiferromagnetic couplings $J_{1'2'}$, $J_{2'3'}$ force $S_{2'}$ to be parallel to S_2, even though the exchange $J_{22'}$ could still be antiferromagnetic (though weaker than $\frac{1}{2}(J_{21'} = J_{23'})$). That is, the spins S_2 and $S_{2'}$ order parallel not because of ferromagnetic exchange between them, but because they have "common enemies" $S_{1'}$ and $S_{3'}$ (and S_1 and S_3). This situation is not at all unrealistic: it is met for example in the brownmillerite $A_2GaMnO_{5+\delta}$, $A = $ Ca, Sr (Pomyakushin et al., 2002), in $BiCoO_3$ (Sudayama et al., 2011), and most probably it is responsible for the parallel orientation spins in the V^{3+} pairs in the insulating and antiferromagnetic phase of V_2O_3 (Moon, 1970; see also Ezhov et al., 1999). One also sees that different interactions here compete with one another: the exchange interaction $J_{22'}$ in Fig. 5.37 wants to make spins 2 and $2'$ antiparallel, whereas the other interactions ($J_{21'}$, $J_{23'}$ and $J_{2'1'}$, $J_{2'3'}$) oppose it. This leads us directly to the topic of the next section – magnetic frustrations.

5.7 Frustrated magnets

A special, major class of magnetic systems are the so-called frustrated magnets. The study of these is a very active field, see for example Diep (2004), Lacroix et al. (2011), and many interesting and important questions do not have answers yet.

When we talk about frustrated systems, we usually have in mind lattices with geometric frustrations, containing as building blocks triangles, or tetrahedra, or maybe pentagons. The arguments illustrated in Fig. 5.31 above show that at least for nearest-neighbor antiferromagnetic interaction and for easy-axis (Ising-like) anisotropy we have a large degeneracy: many spin configurations can have the same energy. But we can have a similar situation also for (at first glance) unfrustrated lattices such as the square lattice of Fig. 5.38, if we add to the nearest-neighbor interaction J also the next-nearest-neighbor interaction J'. Thus for example if $J \gg J'$, we can expect a simple 2-sublattice magnetic

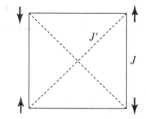

Figure 5.38 Magnetic frustrations in a square lattice with competing nearest-neighbor and next-nearest-neighbor interactions.

structure (similar to the G-type structure shown in Fig. 5.24), but we see that in this case the spins along the diagonals are parallel, which for antiferromagnetic next-nearest-neighbor exchange J' is not favorable. Consequently we can expect that if J' increases, sooner or later the state shown in Fig. 5.38 will become unstable, and some other, more complicated type of magnetic ordering may appear. This is actually also a potentially frustrated situation.

But most often when we talk about frustrated magnets we indeed have in mind materials with real geometric frustration due to a special type of lattice. The most commonly studied are four types of frustrated systems, namely those with triangular, kagome, GGG (or hyperkagome), and pyrochlore lattices. These will be considered below.

The very notion of frustrations is that in such systems it is difficult to form an ordered state such that all interactions, on all bonds, are satisfied. Correspondingly, in some cases the standard magnetic order cannot be established at all. Or, if it is established, the critical temperature of such ordering will be strongly suppressed compared with what one would expect from the strength of the interaction. As is well known, at high temperatures the magnetic susceptibility in systems with localized spins obeys the Curie–Weiss law, $\chi = C/(T - \Theta)$, with the Weiss constant Θ given by the sum of all interactions of a given site with its neighbors, $\Theta \sim \sum_j J_{ij}$. Consequently, for example in systems consisting of triangles with strong antiferromagnetic exchange J, the constant Θ may be quite large, $\sim Jz$ (where z is the number of nearest neighbors). In nonfrustrated systems, for example those with bipartite lattices, one would have an ordering at $T_N \sim \Theta$ (theoretically in the mean-field approximation in this case $T_N = \Theta$; in reality it is usually somewhat different, but not much). In frustrated systems, however, despite potentially strong exchange, an ordered state (if realized at all) would gain only a small fraction of this exchange energy, so that in effect in such cases $T_N \ll \Theta$. The parameter T_N/Θ, or rather Θ/T_N, is often used as an empirical parameter characterizing the degree of frustration: for nonfrustrated systems $\Theta/T_N \sim 1$, for frustrated ones $\Theta/T_N \gg 1$. This quantity is often presented in the discussion of particular frustrated materials, and serves as a criterion for the degree of frustration: strongly frustrated systems have a large value of Θ/T_N.

In the strongly frustrated case the system behaves as a normal paramagnet for temperatures $T > \Theta$, and it is (or may be) ordered below T_N. In the broad region of $T_N < T < \Theta$

the short-range (antiferro)magnetic correlations are already well-developed, but because of the existence of many states with close energy there is still no long-range ordering. This state is usually classified as a *collective paramagnet*.

5.7.1 Systems with triangular lattices

There exist quite a lot of layered magnetic materials with triangular magnetic layers, for example some ordered systems based on the rock salt structure, such as $LiVO_2$, $LiNiO_2$, $NaCoO_2$ (see also Section 3.5). One can visualize these systems, for example $LiNiO_2$, as a rock salt (NiO) with an fcc structure of metallic ions, in which half the magnetic Ni ions are replaced by nonmagnetic Li. If these are ordered, the ordering usually occurs in consecutive [111] planes, similar to the magnetic ordering shown in Fig. 5.33. As a result we have triangular [111] layers of Ni separated by similar layers of nonmagnetic Li, see Fig. 5.39 (crosses are Ni ions, circles are oxygens, and the triangle is Li). Materials such as $CuFeO_2$ or $AgCrO_2$ have similar two-dimensional triangular layers of magnetic ions: they differ from the ordered mono-oxides by different positions of nonmagnetic Cu^{1+} or Ag^{1+} between triangular layers. These systems are known as delafossites. There are also very interesting layered triangular magnets on the basis of organic molecules.

With nearest-neighbor antiferromagnetic interactions the triangular lattice, shown in Fig. 5.40, is indeed frustrated. Thus, if we have an Ising model on such a lattice, one can show exactly (Wannier, 1960) that the ground state has no unique ordering and a large degeneracy, so that it has a finite entropy at $T = 0$. The easiest way to understand this is to split the triangular lattice into a honeycomb lattice with extra spins at the centers of each hexagon (Fig. 5.41). The honeycomb lattice is bipartite – we can order spins on it in an antiferromagnetic fashion so that all neighbors are antiparallel. But then the spins at the centers will lie in a zero molecular field, and they are free to point in either direction, up or down. That is, there will be a finite number of such degenerate states, $2^{N/3}$, as each of these central spins has two possible states ($\uparrow$ and $\downarrow$) and the total number of these sites is $\frac{1}{3}$

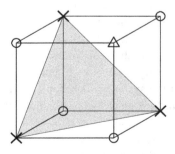

Figure 5.39 The formation of a triangular lattice in ordered magnetic oxides with the NaCl structure (systems of the type $LiNiO_2 = (Li_{1/2}Ni_{1/2})O$). Crosses are magnetic ions, for example Ni, and triangles are nonmagnetic ions, for example Li. Anions are shown by circles.

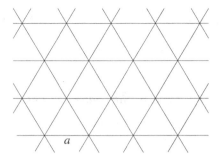

Figure 5.40 A triangular lattice formed by magnetic ions of Fig. 5.39.

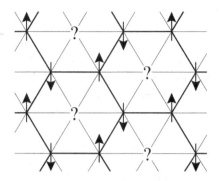

Figure 5.41 Demonstration of the frustration in Ising systems on a triangular lattice.

of the total number of spins, N. Consequently one would have a finite entropy of this state, $S = \frac{1}{3} N \ln 2$. This quantity is not very far from that found exactly by Wannier (1960).

One can illustrate with this example yet another phenomenon, rather typical for frustrated magnets – the appearance of *magnetization plateaux* in the $M(H)$ dependence. Indeed, if we apply an arbitrary weak positive magnetic field to the state of Fig. 5.41, the "undecided" central spins will all order up, and the total magnetization will be $\frac{1}{3} M_{max}$ (where M_{max} is the total magnetization when all spins point in the same direction). Similarly, for a small negative magnetic field all such spins will point down, and the magnetization will be $-\frac{1}{3} M_{max}$. With a further increase of the (say positive) field the magnetization will remain unchanged (at $T = 0$) until at another critical field, $H_c \sim J$, *all* ↓ spins will flip to ↑ and the system will become ferromagnetic. Thus the resulting behavior of magnetization will look as shown in Fig. 5.42, that is it will have a plateau at $M = \frac{1}{3} M_{max}$ between (in this case) $H = 0$ and $H = H_c$ (the solid line shows the behavior at $T = 0$, while the dashed line shows the "smoothed" behavior at $T \neq 0$). Such magnetization plateaux are quite typical for many frustrated systems, and the "superstructures" such as the honeycomb lattice built in the original triangular lattice, see Fig. 5.41, are also rather typical for the states at such plateaux.

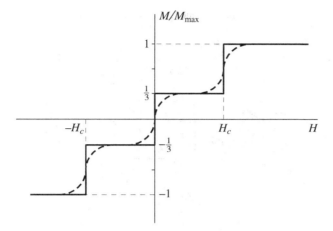

Figure 5.42 Formation of magnetization plateaux in a triangular lattice of Ising spins.

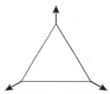

Figure 5.43 The best classical spin ordering in a spin triangle with antiferromagnetic interactions.

All this was valid for Ising spins on a triangular lattice. What about a more realistic situation of real spins S with, say, Heisenberg interaction $J S_i \cdot S_j$? This case is much more complicated, and some questions still remain open. If we consider Heisenberg spins classically, we can easily show that the best classical state at each triangle would be that with spins at $120°$ to each other (Fig. 5.43). This spin configuration gives the lowest classical energy, $3J S^2 \cos \frac{2}{3}\pi = -\frac{3}{2}J S^2$ per triangle.[18] In a concentrated system, in a 2d triangular lattice, this will lead to a long-range order with three sublattices, shown in Fig. 5.44, with the new unit cell shown there by dashed lines, the dimension of the new cell being $\sqrt{3} \times \sqrt{3}$ compared with the original unit cell (solid lines).

Note also that such a triangular lattice of classical spins is rather strongly constrained: as soon as we fix the directions of the spins (at $120°$ to each other) at one triangle, for example the shaded triangle in Fig. 5.44, the direction of the rest of the spins is determined uniquely: we can proceed from one triangle to an adjacent one, and we see immediately that the direction of the remaining free spin at each such step is unique. This will not be the case for other types of lattice, such as kagome or pyrochloric lattices, even classically (see

---

[18] Note that for the Heisenberg interaction $J S \cdot S$ without any anisotropy the spin axes have nothing to do with the axes of the crystal, in our case the triangle: the spins at $120°$ to each other do not have to lie in the plane of the triangle, but may point to any direction in space.

new magnetic $\sqrt{3} \times \sqrt{3}$ unit cell

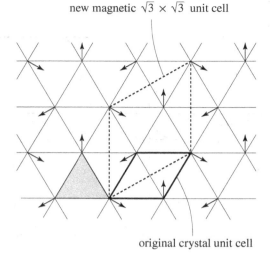

original crystal unit cell

Figure 5.44 120° ordering of classical spins on a triangular lattice.

below). In this sense we may say that the triangular lattice is *overconstrained*, whereas for example the kagome lattice is *underconstrained*.

But this is not at all the end of the story, even for the triangular lattice. We know what it does classically, but the real spins are quantum objects, and quantum effects and fluctuations can change the situation even at $T = 0$. Quantum fluctuations are especially strong for $S = \frac{1}{2}$ (for larger spins the situation gradually approaches the classical one as $S \to \infty$). This can lead to quite different types of state.

5.7.2 Resonating valence bonds

The actual ground state of a two-dimensional triangular lattice of a spin-$\frac{1}{2}$ Heisenberg antiferromagnet is still a matter of some debate. Many numerical calculations seem to show that the same 120° ordering as for classical spins, shown in Fig. 5.44, may be realized in the quantum case as well. However there are also opposite claims, one being that there is no long-range order in the ground state and this system will be some kind of *spin liquid*. By spin liquid in general one refers to a state without the conventional long-range order, such that at each site the average spin is $\langle S_i \rangle = 0$ but in which there are strong short-range antiferromagnetic correlations. Such states may have gapless excitations, or they may have spin gaps. One also has to discriminate between singlet and triplet excitations in such states.

One such state was proposed for the triangular Heisenberg antiferromagnets with $S = \frac{1}{2}$ by Anderson (1973); it is called the *resonating valence bond* (RVB) state. One can explain the main ideas in the following way (see a more detailed exposition in Khomskii, 2010).

Figure 5.45 Comparison of the Néel structure (a) and the valence bond structure (b) in an antiferromagnetic Heisenberg chain.

Let us first start with a one-dimensional chain of $S = \frac{1}{2}$ spins with antiferromagnetic Heisenberg exchange

$$\mathcal{H} = J \sum_i S_i \cdot S_{i+1} \tag{5.40}$$

(note that this definition of the exchange constant J is different from the one we usually employ when we sum independently over sites $\{i, j\}$ in $S_i \cdot S_j$, i.e. in which every spin pair is counted twice, e.g. $S_1 \cdot S_2 + S_2 \cdot S_1$; in contrast, in (5.40) each pair enters only once).

The simplest antiferromagnetic state one can think about – the so-called Néel state – would be that with alternating spins, $\uparrow \downarrow \uparrow \downarrow \uparrow \downarrow \cdots$ (Fig. 5.45(a)). Its energy is

$$E_{\text{Néel}} = -\frac{1}{4} N J . \tag{5.41}$$

But we can form a different state, that of Fig. 5.45(b), by making real singlets of the type $\frac{1}{\sqrt{2}}(1\uparrow 2\downarrow - 1\downarrow 2\uparrow)$ at every other bond. As follows from quantum mechanics, the energy of one such bond is $J\langle S_i \cdot S_j \rangle = -\frac{3}{4}J$. (One can easily obtain this result by recalling that $S_{\text{tot}}^2 = S_{\text{tot}}(S_{\text{tot}} + 1) = (S_1 + S_2)^2 = S_1^2 + S_2^2 + 2(S_1 \cdot S_2) = S_1(S_1 + 1) + S_2(S_2 + 1) + 2(S_1 \cdot S_2)$, from which for $S_1 = S_2 = \frac{1}{2}$ and for the singlet state $S_{\text{tot}} = 0$ we obtain $\langle S_1 \cdot S_2 \rangle = -\frac{3}{4}$.) Qualitatively this is easy to understand: if in the Néel state only the z components of the Heisenberg exchange $J(S_1^z S_2^z + S_1^x S_2^x + S_1^y S_2^y)$ contribute, which gives the factor $\frac{1}{4}$ in (5.41), for a real singlet, which is actually isotropic in spin space, all three components $S_1^z S_2^z$, $S_1^x S_2^x$, and $S_1^y S_2^y$ contribute equally, giving energy $-\frac{3}{4}J$ per bond.

In effect the total energy of such a state, which we can call a *valence bond* (VB) state (the singlets shown in Fig. 5.45(b) are really the same spin singlets as in valence bonds, e.g. in the H_2 molecule) is

$$E_{\text{VB}} = -\frac{3}{4}J \cdot \frac{1}{2}N = -\frac{3}{8}NJ \tag{5.42}$$

(each bond contributes $-\frac{3}{4}J$, but the number of such bonds is $\frac{1}{2}N$, half the number of sites). But, despite the fact that we have lost half of the bonds, we have gained a factor 3 at each singlet bond, compared with the Néel state of Fig. 5.45(a). Thus in this case the VB state has lower energy than the Néel state.[19]

---

[19] Note that the one-dimensional spin chain with antiferromagnetic Heisenberg interaction in a compressible lattice is unstable with respect to lattice dimerization, after which the exchange constants in consecutive bonds alternate, J_+, J_-, J_+, J_-, ..., and the ground state of the system indeed becomes that of Fig. 5.45(b). This phenomenon of spontaneous dimerization is analogous to the well-known Peierls dimerization in one-dimensional metals, see for example Khomskii (2010); for spin chains it is known as *spin-Peierls transition*.

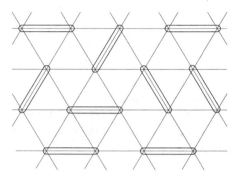

Figure 5.46 The resonating valence bond picture of the benzene molecule C_6H_6.

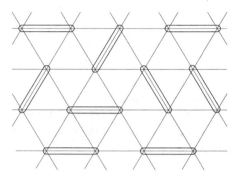

Figure 5.47 A typical configuration of the (short-range) RVB state on a triangular lattice. There are many such equivalent configurations, which could mix by quantum tunneling.

We can decrease the energy of the VB state even further if we allow the so-called *resonance* between different VB configurations: for example instead of connecting by bonds the sites (12)(34)(56)..., we can connect other pairs ...(23)(45)... These different states can mix, which reduces the energy still further. This would be the resonating valence bond (RVB) state. The very concept of such a resonance was suggested in chemistry by L. Pauling, for example for the benzene molecule C_6H_6 (Fig. 5.46), where two resonating configurations of the C_6 hexagon are shown (double lines in Fig. 5.46 correspond to singlet bonds of Fig. 5.45(b)). In application to the triangular lattice the corresponding picture would look as shown in Fig. 5.47, where we have put singlet bonds on the lattice in some particular way. In fact there are many ways to cover a triangular lattice by such bonds (and actually longer, further-neighbor bonds are also allowed, Fig. 5.48). All these different configurations can contribute to the ground state, which will be the final RVB state.

One has to compare the energy of such states with the best state with real long-range magnetic ordering, which in this case is the 120° state of Fig. 5.44. The energy of that state is, classically,

$$E_{120°} = J \cdot \tfrac{1}{4} \cos \tfrac{2}{3}\pi \cdot \tfrac{1}{2} Nz = -\tfrac{3}{8} J N \qquad (5.43)$$

(here $\tfrac{1}{4} \cos \tfrac{2}{3}\pi$ is the value of $\langle \boldsymbol{S} \cdot \boldsymbol{S} \rangle$ for that state, and z is the number of nearest neighbors, in our case $z = 6$). However, the energy of the simplest representative VB state, for example that shown in Fig. 5.47, is

$$E_{\text{VB}} = -\tfrac{3}{4} J \cdot \tfrac{1}{2} N = -\tfrac{3}{8} J N \qquad (5.44)$$

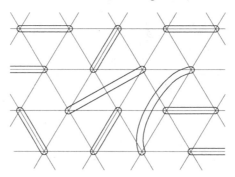

Figure 5.48 A more general RVB state, including singlets on distant bonds.

(each bond gives $-\frac{3}{4}J$, but the number of such bonds is $\frac{1}{2}N$). Thus we see that, whereas for 1d systems the VB state was already lower in energy than the Néel state (even without taking into account the resonance between different VB states which would decrease the energy of such a state even further), here, for the triangular lattice, the energies of the best classical ordered state (5.43) and of one particular VB state (5.44) coincide. When we include the resonance between different possible singlet configurations, the energy would become lower, $E_{RVB} < E_{VB}$. Thus one could think that the RVB state in this case would be a better state than that with long-range order, and that such a triangular system would be in a spin liquid state.

This conclusion, however, is not straightforward. The same quantum fluctuations which decrease the energy of the RVB state in comparison with the VB one would also act on the 120° state of Fig. 5.44: the classical picture would be that shown in the figure, but this is still a Heisenberg system, and there will definitely be quantum fluctuations in this case, which would reduce the average sublattice magnetization $\langle S_i \rangle$ and also decrease the energy compared with the classical value (5.43). Which of these energies – that of the ordered state or that of the RVB state – would as a result become lower in the presence of quantum fluctuations is not clear *a priori*. As noted above, many numerical calculations are in favor of the 120° state, but there are also alternative claims.

In any case, even irrespective of this particular situation (2d triangular lattice), the very notion of an RVB state as one of the possible states of frustrated systems is very important. This notion is widely used to describe the properties of different materials, including even high-T_c superconducting cuprates (Anderson, 1997).

The application of this concept to high-T_c superconductors uses yet another interesting feature of RVB states, resembling the situation mentioned in footnote 3 of Chapter 1. Suppose we start with an RVB state of the type shown in Fig. 5.47, and create an excited state by exiting one of the singlets, say on the bond (*ab*), see Fig. 5.49(a), to a triplet state, that is reversing the direction of one of the spins on the valence bond. We can then "recommute" the spins, to form the singlet on the bond (*bc*) from the spin ↑ of the site *b* and the spin ↓ from the singlet (*cd*) of Fig. 5.49(a). As a result we will have the situation

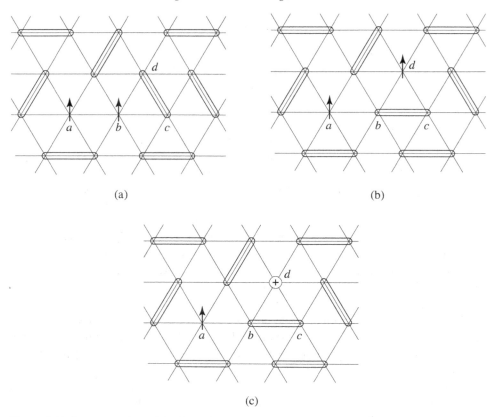

Figure 5.49 Demonstration of the formation of spin excitations with spin $\frac{1}{2}$ appearing by breaking one singlet bond (a) and then "recommuting" some other bonds (b). As a result the isolated spins (neutral objects with spin $\frac{1}{2}$ – *spinons*) move apart. If we remove an electron from such a site (c) there will remain a positive charge without spin – a *holon*.

of Fig. 5.49(b), with two isolated spins $\frac{1}{2}$ separated from each other. Thus we have split the initial excitation – the triplet with $S = 1$ on the bond (*ab*) in Fig. 5.49(a) – into *two* equivalent excitations, each with spin $\frac{1}{2}$. We can continue this process of "recommuting" the spins, as a result of which each such spin-$\frac{1}{2}$ excitation would move through the crystal. Such excitations are called *spinons*. Note that these excitations carry spin ($S = \frac{1}{2}$), but do not carry any charge: at each site we still have a positively charged nucleus and a negatively charged electron.

But now we can do something else. Let us start from the state of Fig. 5.49(b) and *remove* an electron, say from site d. We will end up in the situation of Fig. 5.49(c), in which on site d there is now an uncompensated charge $+e$ – the charge of the ion, but there is no spin at this site. We can also get such a state by removing one electron, for example with spin $\downarrow$, from the singlet bond (*ab*) of the original RVB state of Fig. 5.46, and then moving the resulting "empty state" to site d in Fig. 5.49(c) by recommuting again some singlet bonds. We see that the resulting "empty state" can also move through the crystal, that is it is also an elementary excitation, which carries (positive) charge but no spin. Such an excitation is called a *holon*. We thus see that the elementary excitations in the RVB states are spinons, neutral excitations with spin $\frac{1}{2}$, and holons, charged but spinless excitations.[20] This is different from the normal excitations in the conventional band-like metals or semiconductors, in which the electron or hole excitations have both charge and spin. Thus there occurs in this case *spin–charge separation*. All these features resemble the properties of one-dimensional correlated systems mentioned in footnote 3 of Chapter 1. To what extent these features are relevant for the phenomenon of high-T_c superconductivity is still an open question, but there are serious arguments that they may be really important (Anderson, 1997).

5.7.3 Strongly frustrated lattices

Other typical examples of frustrated systems, in which the role of frustration can be even stronger than in triangular systems, are materials with kagome, pyrochlore, and "GGG" lattices.

(*a*) *Kagome lattice*. The Kagome lattice is shown in Fig. 5.50. (The origin of the name comes from a specific Japanese weaving pattern; one can also observe similar patterns in ancient Roman mosaics and in Islamic architecture.) We see that the Kagome lattice also consist of triangles, but in contrast to the triangular lattice of Fig. 5.40, in which different triangles have common edges, that is two common sites, here they have common corners. This leads immediately to profound differences. If for example we try to form a

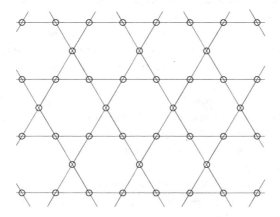

Figure 5.50 Kagome lattice (here circles are the magnetic sites).

---

[20] We can also *add* an electron to this system, say with spin $\downarrow$, putting it on a site with spin $\uparrow$ (labeled *a* in Fig. 5.49). We would again get a spinless object, but with charge $-e$. Typically the notions discussed above are applied to hole-doped cuprates, and thus one usually speaks of holons with charge $+e$, rather than of similar objects with charge $-e$.

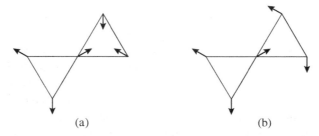

Figure 5.51 Two equivalent magnetic configurations of two corner-sharing triangles (building blocks of the Kagome lattice), both with the optimal classical 120° spin ordering at each triangle.

state similar to the 120°-ordered state of Fig. 5.44, we see that even if we have perfect 120° relative orientation of spins at each triangle, a lot of degeneracy still remains, see Fig. 5.51: if we have for example three spins pointing out of the first triangle, in the second one we have two possibilities shown in Fig. 5.51(a) and (b); the common spin is of course fixed, but the two remaining spins can always be interchanged.

As a result, if we try to form a structure with 120° spin orientation at each triangle, there will be an enormous number of such possible states, all of which would be classically equivalent. Thus, whereas in a triangular lattice the requirement of having 120° orientation practically uniquely determines such a ground state (up to an interchange of three sub-lattices – but this is the usual situation in any antiferromagnet), in a kagome lattice this requirement leaves a lot of degeneracy even at the classical level – the kagome lattice is in this sense *underconstrained*. In effect the situation in kagome lattices becomes much more complicated and much more interesting than in triangular lattices. Most probably there is no long-range ordering in the ground state (although suggestions of rather intricate ordered structures have also been made). The spectrum of elementary excitations is also very unusual: the triplet excitations seem to have a gap, but there are many singlet excitations, and with an increase in size of the system these singlet excitations accumulate at low energies so that finally there may be no gap in these excitations, and their density of states at $\omega \to 0$ may be extremely large.

Among the ordered states discussed in the literature some have VB-like singlets ordered in a particular fashion, or real magnetically ordered states, which may for example be stabilized by a certain anisotropy. Two particular states *with* long-range order, which are often discussed in connection with kagome systems, in particular experimentally, are the homogeneous $q = 0$ state shown in Fig. 5.52(a) and the staggered state ($\sqrt{3} \times \sqrt{3}$ state) of Fig. 5.52(b). In the $q = 0$ state of Fig. 5.52(a) all spins point either out or into a triangle, and the unit cell remains unchanged. In the staggered state of Fig. 5.52(b) the triangles alternate in vector spin *chirality*

$$\chi = S_1 \times S_2 + S_2 \times S_3 + S_3 \times S_1 ; \qquad (5.45)$$

this chirality points either up or down, which we have marked in Fig. 5.52(b) by $+$ and $-$ (in the homogeneous state of Fig. 5.52(a) these chiralities are the same, e.g. $\chi = +$, at each triangle).

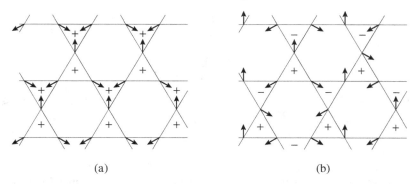

Figure 5.52 Two typical ordered states of the kagome lattice: (a) $q = 0$ structure with constant spin chirality (5.45); (b) $\sqrt{3} \times \sqrt{3}$ structure with staggered vector spin chirality.

There are real materials in which magnetic ions form a kagome lattice. Experimentally in some of them the $q = 0$ state was observed (Grohol *et al.*, 2005), in others a state more resembling a staggered one (Schweika *et al.*, 2007).

(*b*) *GGG lattice.* Another example of strongly frustrated underconstrained systems is "GGG" – gadolinium gallium garnet (the general formula of garnets is $A_3B_5O_{12}$; these are cubic systems, many of which are good ferromagnets). The magnetic Gd ions in GGG form a three-dimensional network of corner-shared triangles, so these are 3d analogs of kagome systems (sometimes one calls this lattice a *hyperkagome*). Consequently, for an antiferromagnetic interaction these would also be very strongly frustrated systems. Closely related to these are some spinels in which one has ordered B' and B'' ions in B sublattices, of the type $A[B'_{1/4}B''_{3/4}]_2O_4$. If the ions B' and B'' order, typically the ordering is such that at each tetrahedron in the B sublattice, see Fig. 5.34, one ion is B' and three are B'', and the triangles formed by the B'' ions share common corners and form a 3d network resembling that of GGG. One such striking system, $Na_4Ir_3O_8$, was shown (Okamoto *et al.*, 2007) to have no long-range order down to very low temperatures, that is it is extremely strongly frustrated (the parameter Θ/T_N, mentioned above, is here very large).

(*c*) *Pyrochlore lattice.* Finally, we will discuss briefly here yet another type of strongly frustrated lattices – the pyrochlore lattice, consisting of corner-sharing tetrahedra. We have already discussed this lattice in Section 5.6.3, when we were talking about the B sublattice of spinels. The real mineral pyrochlores, with general formula $A_2B_2O_7$ (or $A_2B_2O_6X$), have two such sublattices: both A and B sites form pyrochlore lattices. By changing the composition, one can make only one sublattice magnetic, for example in materials such as $R_2Ti_2O_7$ in which the rare earth ions R^{3+} are magnetic, but $Ti^{4+}(d^0)$ are nonmagnetic. Or one can have the B sublattice magnetic, or both A and B sublattices can be magnetic.

Among these materials there are crystals with quite different properties. Some of them are metallic, some even superconducting. There are also many good insulators, and their magnetic properties are often very unusual. This is mostly connected with frustrations in the pyrochlore lattice. Already at the level of one tetrahedron we have a large

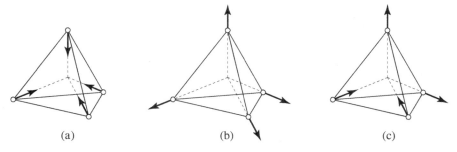

Figure 5.53 Magnetic tetrahedra which are the building blocks of the pyrochlore lattice $A_2B_2O_7$ and of the lattice of B sites in spinels AB_2O_4. Different possible spin configurations for Ising spins with the easy axes pointing toward the centers of the tetrahedra are shown: (a) 4-in state; (b) 4-out state; (c) 2-in, 2-out state, called spin ice.

degeneracy: for Ising spins with antiferromagnetic interaction we have for example the ground state with two spins ↑ and two ↓. But the distribution of these spins among four sites already gives six different possible states. The connection of these tetrahedra through common corners leads to a very large degeneracy of the ground state.

The real situation, even with nominally very anisotropic (Ising) ions, is actually very interesting. As we have discussed in Section 5.6.3, each ion in this case has a natural axis, pointing for example in or out of a tetrahedron. Correspondingly the easy axes, even for very anisotropic ions, would be different for different ions – for example all pointing to the center of a tetrahedron (Fig. 5.53). Then, depending on the sign of the exchange, we can have either the situation of Fig 5.53(a,b) (4-in, 4-out) or that of Fig. 5.53(c) (2-in, 2-out): the first situation occurs for an antiferromagnetic interaction, and the second for a ferromagnetic one.

The most interesting situation, unexpectedly, is that of ferromagnetic interaction: the 2-in, 2-out structure of Fig. 5.53(c) is highly frustrated.[21] This is the famous *spin ice*: the rule "2-in, 2-out" is the same as in real ice, where it applies to hydrogens in the crystallized H_2O molecules. The study of spin-ice systems (a typical example being $Dy_2Ti_2O_7$) has produced a lot of interesting results (see e.g. the article by M. J. P. Gingras in Lacroix *et al.*, 2011), up to a discovery of "magnetic monopoles" (Castelnovo *et al.*, 2008) – specific spin configurations such as spin strings, the endpoints of which behave much like magnetic monopoles in field theory (which in high-energy physics have not been discovered yet, and maybe never will).

Heisenberg systems on pyrochlore lattices have their own rather specific features, which are not really fully clarified yet. An interesting feature in this case is that the ground state for an isolated tetrahedron of spins $\frac{1}{2}$ with antiferromagnetic Heisenberg interaction is a singlet (this problem can be solved exactly), but it is still doubly degenerate. This extra degeneracy is connected with the spin chirality mentioned above, in this case the *scalar spin chirality*

---

[21] Surprisingly, the situation with antiferromagnetic interaction here is simpler: the ground state will have a long-range magnetic ordering with alternating tetrahedra with 4-in, 4-out spins.

$$\kappa_{123} = S_1 \cdot (S_2 \times S_3) . \tag{5.46}$$

This quantity is defined on a triangle (123), but in the exact solution for an isolated tetrahedron one can take any of the four triangles, and the result will be the same. There are some ideas that the ground state of the bulk pyrochlore lattice with Heisenberg spins can be described in terms of these extra degrees of freedom – the chiralities (5.46) – though in many real systems the normal magnetic ordering can be realized, especially if we take into account magnetic anisotropy, which is usually present, and the antisymmetric DM interaction.

(*d*) *"Magnetic Jahn–Teller effect" in frustrated systems.* Yet another possibility in a frustrated system is that there may occur a structural transition which lifts the frustrations and stabilizes a particular ground state. A simple example could be the distortion of a triangular lattice, after which the bonds in one of the three directions become longer, and the corresponding exchange constant smaller, see Fig. 5.54 (if we have a direct d–d interaction without orbital degeneracy, which may be the case e.g. for $Cr^{3+}(t_{2g}^3)$ compounds). In this case the strong bonds with the exchange constant J (solid lines in Fig. 5.54) form effectively a square lattice, and at least if $J \gg J'$ (where J' is the exchange constant on longer, weak dashed bonds of Fig. 5.54), we will have a simple two-sublattice antiferromagnetism on this square lattice (cf. the discussion at the beginning of Section 5.7). A similar phenomenon may also occur in other lattices, for example in pyrochlores or spinels. Thus for example a tetragonal compression of the B-site lattice of spinels with a corresponding increase of the exchange in (xz) and (yz) directions, and a decrease in (xy), could make this lattice bipartite, Fig. 5.55 (again strong bonds are shown by solid lines, and weak bonds by dashed lines), see for example Tchernyshyov *et al.* (2002).

There may also exist other types of distortion, lifting the degeneracy; but in any case the possibility of lifting the frustrations and stabilizing a long-range magnetic ordering is definitely present (and seems to be realized in a number of compounds, e.g. in Cr^{3+} spinels such as $CdCr_2O_4$, although the detailed type of distortion and the resulting magnetic structures in this case seem to be more complicated than the simple examples of Figs 5.54, 5.55).

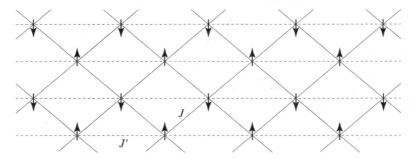

Figure 5.54 Anisotropic triangular lattice.

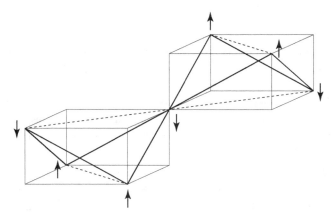

Figure 5.55 Anisotropic pyrochlore lattice. Such deformations could in principle lift magnetic frustrations and lead to an ordered magnetic state.

The general tendency toward such distortions in frustrated magnets has much in common with the Jahn–Teller effect described in Section 3.2: as in that case, the ideal frustrated systems have a large degeneracy in the ground state, with many states having the same energy. It may be favorable to reduce the symmetry of the system by certain distortions, so as to lift, at least partially, this degeneracy. If the energy gain in the magnetic subsystem after such a distortion exceeded the elastic energy loss, such a process should occur spontaneously. In the Jahn–Teller systems of Section 3.2 this was always the case. In frustrated systems the degree of degeneracy is much higher, we should deal not only with discrete but also with continuous energy spectra, and in general we are not guaranteed that for simple distortions the mechanism described above would work. At least I am not aware of a general proof that frustrated systems would always be unstable with respect to some distortion reducing the frustration and (at least partially) lifting the degeneracy; one has to check this for each particular case. But this is a definite possibility, which one has to keep in mind in dealing with frustrated systems.

(*e*) *Spiral structures in systems with competing interactions.* This is a convenient place to discuss yet one more general phenomenon often occurring in frustrated systems with competing interactions, such as that shown in Fig. 5.40 or in anisotropic triangular lattices of Fig. 5.54. Often in such cases the resulting magnetic ordering, if it occurs, is not of a simple two-sublattice type as shown for example in Figs 5.54, 5.55, but is of the spiral type, with magnetic periodicity which can be incommensurate with the lattice period. This can happen especially for larger spins, which behave practically classically. One can explain the origin of such states on a simple one-dimensional model with nearest-neighbor (nn) and next-nearest-neighbor (nnn) antiferromagnetic exchange,

$$\mathcal{H} = J \sum_i \mathbf{S}_i \cdot \mathbf{S}_{i+1} + J' \sum_i \mathbf{S}_i \cdot \mathbf{S}_{i+2} , \qquad (5.47)$$

with J, $J' > 0$. For $J \gg J'$ we expect classically a simple two-sublattice antiferromagnetic ordering as shown in Fig. 5.56(a). But we see that in this structure

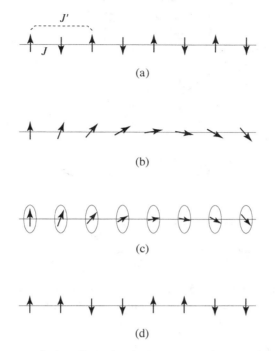

Figure 5.56 Possible states of a one-dimensional spin system with competing nearest-neighbor and next-nearest-neighbor antiferromagnetic exchange.

the second-neighboring spins are parallel, which is unfavorable for the nnn exchange J' in (5.47); this introduces frustration into the system.

In the opposite limit of $J' \gg J$ we first have to make nnn spins antiparallel, as in Fig. 5.56(d), "sacrificing" the nn exchange J. But in principle, between these two limits, we can try to find a magnetic structure optimizing both nn and nnn interactions. The result turns out to be a spiral with a particular period l, or a wave vector $Q = 2\pi/l$, shown for example in Fig. 5.56(b) (here we take the lattice parameter as 1). The plane in which the spins rotate is in this approximation arbitrary: it may be for example a plane containing the chain axis, as drawn in Fig. 5.56(b), or the perpendicular plane (the "proper screw" of Fig. 5.56(c)), or any other plane.

The period of such a spiral can be found in the following way. The exchange Hamiltonian (5.47) can be rewritten in the momentum space as

$$\mathcal{H} = \sum_q J(q)\, S_q \cdot S_{-q} = \sum_q J(q)\, |S_q|^2 \;, \tag{5.48}$$

with $J(q)$ being the Fourier transform of the general exchange interaction J_{ij}. In our case of nn and nnn interactions, the model (5.47),

$$J(q) = 2J \cos q + 2J' \cos 2q \;. \tag{5.49}$$

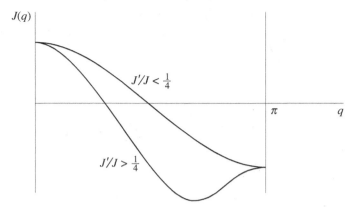

Figure 5.57 Possible forms of the Fourier transform $J(q)$ (5.49) of the exchange interaction in the nearest-neighbor vs next-nearest-neighbor one-dimensional model. The minimum of $J(q)$ gives in the mean-field approximation the wave vector of the resulting magnetic structure.

As we see from the expression (5.48), the minimum energy $E_0 = \langle \mathcal{H} \rangle$ is reached for values of q for which the exchange $J(q)$ is minimal. Thus we have to find the minimum of $J(q)$ (5.49):

$$\frac{\partial J}{\partial q} = -2J \sin q - 4J' \sin 2q = -2J \sin q \left(1 + 4 \frac{J'}{J} \cos q \right) = 0 . \qquad (5.50)$$

The easiest way to analyze this equation is graphically, see Fig. 5.57: we plot $J(q)$ for some representative values of J'/J. We see that there exist solutions of eq. (5.50) with $\sin q = 0$, that is $q = 0$ or $q = \pi$. From the expression (5.49) we see that the solution with $q = 0$ (ferromagnetic ordering) corresponds to a maximum of the energy. The solution $q = \pi$ is the antiferromagnetic state shown in Fig. 5.56(a). It corresponds to a minimum of the energy if $(1 + 4J'/J) > 0$, that is if $J'/J < \frac{1}{4}$. However if $J'/J > \frac{1}{4}$, this solution, as is clear from Fig. 5.57, becomes a maximum, and the minimum of $J(q)$ will be given by the condition

$$1 + 4 \frac{J'}{J} \cos q = 0 , \qquad (5.51)$$

that is

$$\cos q_0 \equiv \cos Q = -\frac{1}{4J'/J} . \qquad (5.52)$$

Thus one sees that as the nnn exchange increases, for $J'/J > \frac{1}{4}$ the simple two-sublattice antiferromagnetism of Fig. 5.56(a) will be changed into a helicoidal solution with the wave vector Q given by the expression (5.52). And for $J'/J \to \infty$ we see that $Q \to \frac{1}{2}\pi$, that is $l = 2\pi/Q = 4$, and the magnetic structure will approach that of Fig. 5.56(d) asymptotically.

In the particular 1d model (5.47) for $S = \frac{1}{2}$ in fact the situation is more complicated and more interesting (which is often the case in 1d systems): one can show (Majumdar and Ghosh, 1969) that in fact starting from $J'/J \simeq 0.241$ the system develops a gap in the spin-wave spectrum, and its ground state resembles very much the state with singlet bonds shown in Fig. 5.45(b); such a state with singlets at every second bond becomes an exact state for $J'/J = \frac{1}{2}$. But in any case, our simple mean-field treatment has predicted correctly that for large enough J'/J the original Néel state of Fig. 5.56(a) (or, in the 1d case, with such spin correlations) would become unstable, and the magnetic state would change; it almost correctly gave the critical value of J'/J for that to happen ($J'/J = \frac{1}{4} = 0.25$ in our classical, mean-field treatment, compared with $J'/J = 0.241$ for the real case of $S = \frac{1}{2}$). The same mean-field-like treatment would work even better for 2d or 3d systems with frustrations and with competing interactions. To find a good and plausible candidate for magnetic ordering we should apply the same procedure – go to momentum representation and find minima of $J(\boldsymbol{q})$ (with $\boldsymbol{q}$ now a 2d or 3d vector). Often this approach describes correctly, or predicts, the types of helical ordering observed in such systems. And the detailed type of this helicoidal structure, in particular the plane of rotation of spins (the proper screw of Fig. 5.56(c), or the cycloidal structure of Fig. 5.56(b)) – a very important question for example for possible multiferroic behavior, see Chapter 8 – will be determined by the magnetic anisotropy, that is by the spin–orbit coupling (Section 5.3).

5.8 Different magnetic textures

As we have seen, different magnetic structures are possible in different situations. Besides the simple ferro- or antiferromagnetic structures, more complicated collinear magnetic structures can appear, but also those with noncollinear spins. Probably the most interesting, but also the most common of those are different types of helicoidal magnetic structures, with spins forming for example spiral structures of the proper screw type (spins rotating in the plane perpendicular to the direction of the spiral) or cycloidal spirals (spins rotating in the plane containing the spiral direction), see Fig. 8.12 in Chapter 8. There are also different types of conical spiral, etc. They are probably more common among rare earth metals and compounds, but are also encountered in many transition metal compounds. Such spiral structures play, in particular, a special role in the multiferroic behavior of some materials, see Chapter 8.

Besides these homogeneous bulk magnetic structures, there exist also different and very rich inhomogeneous magnetic configurations. They are usually considered in specialized books on magnetism; here we will mention some of them only briefly, especially those which play important roles in particular in magnetoelectric phenomena and which promise interesting applications in spintronics, see for example Ziese and Thornton (2001).

The simplest such magnetic inhomogeneities are magnetic domain walls, dividing for example ferromagnetic domains. Such domains can have different, often very complicated

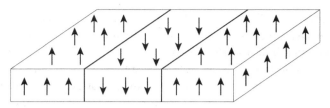

Figure 5.58 Stripe domains in a ferromagnet.

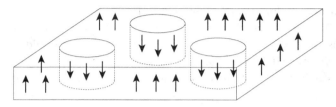

Figure 5.59 Cylindrical magnetic domains in a ferromagnet.

shapes. The simplest ones are stripe domains (Fig. 5.58) and cylindrical magnetic domains (CMD) (Fig. 5.59). The domain walls dividing these domains may have a particular internal structure. Thus typically one discriminates between Bloch and Néel domain walls, see Fig. 8.19 later. In the Bloch domain walls the spins rotate in the plane of the domain wall, that is they resemble a bulk proper screw. The analog of a cycloidal structure is given by the Néel domain walls, in which the spins rotate in the plane containing the vector perpendicular to the domain wall. Besides having different magnetic characteristics, such domain walls may also have interesting magnetoelectric properties, see Chapter 8 for more details.

However, besides two-dimensional defects such as domain walls, there may also exist one-dimensional (line) defects – magnetic vortices. These can be visualized, for example, as cylindrical domains with the radius of the domains tending to zero. Depending on the type of domain wall in such cylindrical domains (Bloch-like or Néel-like) we would have vortices of the type of Fig. 5.60(a) or (b). Usually in the core of such vortices the spins point parallel to the vortex axis, that is perpendicular to the plane in which the spins rotate.[22]

There may also exist point-like magnetic defects. These are known as skyrmions (magnetic "hedgehogs"). A typical skyrmion configuration in a magnetic thin film is shown in Fig. 5.61; it resembles that of the magnetic vortex, Fig. 5.60, but the skyrmions can be real point-like defects in a 3d system. In a skyrmion the spins are pointing for example up far from the center of the skyrmion, down in the center, and gradually rotate

---

[22] Strictly speaking, magnetic vortices, for example similar to vortices in superconductors or in superfluid He, are topological defects, with the circulation around each vortex quantized. That means that for a real vortex we cannot just go to a collinear structure far away from the vortex core. In this sense real vortices are different from the cylindrical magnetic domains with small radius, and also from skyrmions, to be discussed below. Qualitatively, however, this picture can be used.

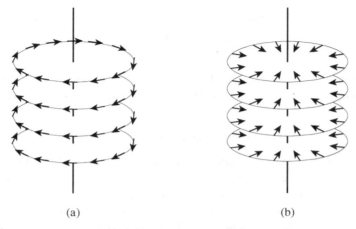

Figure 5.60 Two types of magnetic vortex: (a) Bloch-like structure; (b) Néel-like structure.

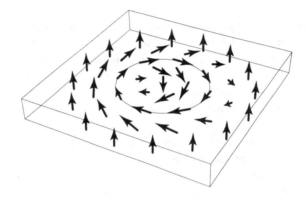

Figure 5.61 Schematic form of distribution of spins in a magnetic skyrmion.

between these two directions in between, as shown in Fig. 5.61. In some cases the spins in the intermediate region lying in the plane can all point for example to the center, or they can rotate "head to tail" as shown in Fig. 5.61; these two cases correspond to different types of magnetic vortex, cf. Fig. 5.60. Such objects can be isolated defects or, at certain conditions, for example for certain values of the temperature and magnetic field, they can form a periodic lattice – a skyrmion crystal. Such crystals can be represented as a combination of three spin helices with wave vectors q_1, q_2, and q_3 at 120° to each other, so that $q_1 + q_2 + q_3 = 0$. This state was first suggested by Bogdanov and Yablonskii (1989) and found experimentally in bulk samples of MnSi by Mülbauer *et al.* (2009) and in Fe monolayers on Ir (111) surface by Heinze *et al.* (2012). As one can deduce from the treatment of Chapter 8, such magnetic whirls could show magnetoelectric activity (Delaney *et al.*, 2009); as is clear from Fig. 5.61, a skyrmion of such type would have a nonzero toroidal moment $T \sim \sum r_i \times S_i$ (see e.g. the inner "belt" in Fig. 5.61), and then it would produce an electric polarization induced by the external magnetic field,

$$P = T \times H , \qquad (5.53)$$

see (8.14), which was indeed observed in the skyrmion crystal phase (so-called *A* phase) of the insulating material Cu_2OSeO_3 by Bos *et al.* (1998) and Seki *et al.* (2012).

5.9 Spin-state transitions

In this section we will discuss certain phenomena observed in some transition metal compounds containing for example ions Co^{3+} or Fe^{2+} (but not only those), namely phenomena connected with the possible existence of states having similar energies, but with different total spin (or different multiplets) – so the high-spin (HS) and low-spin (LS) states, discussed in Section 3.3. The energetic proximity of these states can cause, in concentrated systems, a special type of phase transitions – spin-state transitions. (They can also occur gradually, as a smooth crossover, not necessarily as real phase transitions.) Such transitions lead to strong modifications of magnetic properties of the corresponding systems – actually not only modifying the exchange interaction and the possible type of magnetic ordering, but also changing the magnetic state of transition metal ions themselves, which had not occurred in our previous discussion. In this sense this section stands somewhat apart from the rest of this chapter; we are dealing here not only with purely magnetic phenomena, but other factors and other interactions, in particular the interaction with the lattice, will become important.

Spin-state transitions can occur as a function of the temperature, but they can also be induced for example by pressure, irradiation, doping, etc. The last situation (specific features of spin-state transitions in doped systems) will be discussed in more detail in Section 9.5.2; here we concentrate on the "pure" compounds.

Probably the most famous, and the best studied example of spin-state transition in TM compounds is provided by the perovskite $LaCoO_3$. At low temperatures the Co^{3+} ions in this material are in the LS state ($t_{2g}^6 e_g^0$) and are nonmagnetic, $S = 0$. With increasing temperature some magnetic states become thermally populated, at ~ 60–100 K the magnetic susceptibility strongly increases, and after passing through a maximum it falls again, more or less following the Curie law, see Fig. 3.32. However the fit to the Curie law is actually not very satisfactory.

There occurs in $LaCoO_3$ another transition, or again rather a smooth crossover, to yet another state, at ~ 500 K. One of the early interpretations of this behavior (Racah and Goodenough, 1967) was that at the first smooth transition at ~ 60–100 K the LS Co^{3+} ions are partially transformed to the HS state, and at ~ 500 K this LS–HS transition occurs on all Co ions.

As discussed in Section 3.3, Co^{3+} can exist not only in the LS ($t_{2g}^6 e_g^0$, $S = 0$) and in the HS ($t_{2g}^4 e_g^2$, $S = 2$) states, but also in the intermediate-spin (IS) state, with the configuration $t_{2g}^5 e_g^1$, $S = 1$. *Ab initio* calculations by Korotin *et al.* (1996) suggested that at the first crossover (~ 100 K) the LS Co^{3+} transform partially to this IS state, and only at ~ 500 K do the HS states of Co^{3+} become populated. Much experimental data after the publication of that result was interpreted in this picture, see for example Yamaguchi *et al.* (1997), Ishikawa *et al.* (2004). However more recent experimental and theoretical studies (Noguchi *et al.*, 2002; Ropka and Radwanski, 2003; Haverkort *et al.*, 2006) have cast some doubt on this interpretation. The spectroscopic data (Haverkort *et al.*, 2006) agree better not with the picture of an IS state at $T \gtrsim 100$ K, but rather with a $\sim 50 : 50$ mixture of LS and HS states. Also, theoretically one can show, see Section 3.4 (see also Ropka and Radwanski, 2003; Haverkort *et al.*, 2006), that, taking into account the spin–orbit coupling, the ground state of the HS state of Co^{3+} turns out to be a triplet $\tilde{J} = 1$, whereas that of the IS state would be a quintet (this is, in fact, exactly opposite to what one would naïvely expect from the fact that the HS state has $S = 2$ and the IS state has $S = 1$). Experimental data (Zobel *et al.*, 2002) definitely show that the first excited state in $LaCoO_3$ is a triplet; this agrees with the conclusion that most probably it is a HS state. Still, the exact interpretation of the phenomena occurring in $LaCoO_3$ is under debate.

As we have seen above, in $LaCoO_3$ the spin-state transition is not a real phase transition, but rather a smooth crossover. But there are TM compounds in which the change of spin state occurs as a real phase transition, accompanied by a change of lattice and magnetic properties. This is, for example, the situation in compounds of the type $RBaCo_2O_{5+\delta}$ (where R is usually a small rare earth or Y, and where $0 \lesssim \delta \lesssim 1$), for example in $YBaCo_2O_{5.5}$, which can be considered as a perovskite ABO_3 with ordered Y, Ba on the A sites and with ordered oxygen vacancies. These systems experience a sequence of phase transitions (Fauth *et al.*, 2002; Plakhty, 2005) at which apparently both the structure and magnetic ordering, but also the spin state of Co^{3+}, do change.

In trying to understand the features of spin-state transitions we have to take into account several factors. One is the fact that typically the LS states of ions such as Fe^{2+} or Co^{3+} are much smaller than the corresponding HS states. Thus the ionic radius of the LS state of Co^{3+} in octahedral coordination is 0.545 Å, and that of the HS state is 0.61 Å. Consequently, spin-state transitions are accompanied by a strong change in volume, and this coupling to the lattice is one of the main mechanisms of interaction between different sites. First, this interaction can make the spin-state transition a cooperative phenomenon. Another possible consequence of this interaction could be a stabilization of a "mixed" state, in which the ions with different spin states alternate (Khomskii and Löw, 2004), see Fig. 5.62. From this figure we see immediately that the change in spin state of one site, for example site A in Fig. 5.62, from the (small-size) LS state to the (large-size) HS state would stabilize the small LS state at neighboring sites, so that the resulting state can be an ordered arrangement of HS and LS states. Experimentally this phenomenon seems to be observed in $TlSr_2CoO_5$ (Doumerc *et al.*, 1999, 2001) and possibly in $NdBaCo_2O_{5.5}$ (Fauth *et al.*, 2002).

This factor is also important for understanding the tendency toward spin-state transitions under pressure. Pressure would stabilize the smaller state, that is the LS state, thus under

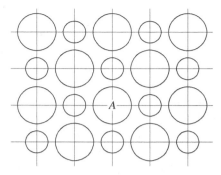

Figure 5.62 Possible origin of the state with alternating spin states of ions such as Co^{3+}, with large circles representing the high-spin states and small circles representing the low-spin states.

pressure there should occur a HS–LS transition. This is observed in many TM materials, for example containing Fe^{2+}, Fe^{3+}, Co^{3+} (see, e.g., Slichter and Drickamer, 1972).

An interesting situation is met in $La_{2-x}Sr_xCoO_3$. For $x \gtrsim 0.2$ this system becomes metallic and ferromagnetic, see Section 9.5.2 for more details. The undoped $LaCoO_3$ with LS Co^{3+} is insulating. The doped material, with some Co ions transferred to a magnetic (IS or HS) state, however, is metallic.

Typically, materials with strongly correlated electrons become more metallic under pressure. Simply speaking, this is connected with the increase of electron hopping and electron bandwidth under pressure. We will discuss many such examples in Chapter 10. In $(La,Sr)CoO_3$, however, the situation is different. Pressure induces here a HS–LS transition, but the LS states are "more insulating" (in the HS or IS states the relatively broad e_g bands are partially filled and participate in metallic conduction, whereas in the LS state with the configuration $t_{2g}^6 e_g^0$ this conduction channel is switched off). Correspondingly, the metallic state of $La_{1-x}Sr_xCoO_3$ with $x \gtrsim 0.2$ transforms under pressure to an insulator (Lengsdorf *et al.*, 2004). The factors described above explain this counterintuitive behavior.

Another factor important for treating systems with spin-state transitions is that the states with nonzero spins have larger entropy than the LS states. As the stability of one or the other state is determined by the condition of the minimum of the free energy $F = E - TS$, the states with higher entropy S, that is the HS or IS states, are usually stabilized at higher temperature.

The mean-field treatment of spin-state transitions which takes into account these factors (interaction between different sites, mainly via the lattice, and spin entropy) leads to the conclusion that sometimes such transitions can be smooth crossovers, but they can also be real phase transitions (Slichter and Drickamer, 1972; König, 1991).

Yet another interesting effect connected with this picture is that in some systems the spin state of some ions can be modified, for example by external illumination. Probably the most spectacular such effect was observed in the Prussian blue analogs, see Fig. 3.31 in Chapter 3. In these systems the spin-state transitions are accompanied by charge transfer, and can be caused both by temperature and illumination. At ordinary conditions in one

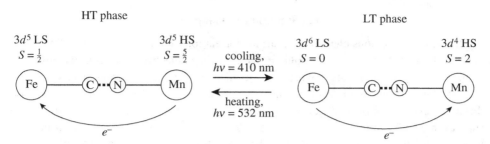

Figure 5.63 Schematic illustration of the process of charge transfer and change of spins with temperature or illumination, occurring in the Mn–Fe Prussian blue analog, after Lummen *et al.* (2008).

such system, $K_{0.2}Co_{1.4}Fe(CN)_6 \cdot nH_2O$, $n \sim 7$, the Co ions are Co^{3+} and the Fe ions are Fe^{2+}, both low-spin and nonmagnetic. At ~ 200 K this system experiences a charge-transfer transition, in which the valences of the TM ions change, Co^{3+} becoming Co^{2+} and Fe^{2+} becoming Fe^{3+}, the resulting states both being magnetic. Even more spectacular is that such a transition due to Fe–Co electron transfer, with the resulting creation of a magnetic state, can also be induced by optical illumination (Sato *et al.*, 1996). Similar effects are observed also for example in the Mn–Fe Prussian blue analog with general formula $Rb_xMn[Fe(CN)_6]_{(2+x)/3} \cdot nH_2O$ (Lummen *et al.*, 2008), in which the electron charge transfer and the change of spins, illustrated in Fig. 5.63, can also be caused both by temperature and illuminations.

S.5 Summary of Chapter 5

As discussed in previous chapters, for strong correlations and integer number of electrons per site the materials are Mott insulators with localized electrons and consequently with localized magnetic moments. Owing to the exchange interaction, predominantly superexchange, the system should develop some type of magnetic ordering. In the simplest case, discussed in Chapter 1 (nondegenerate electrons, one electron per site, simple lattices) this superexchange is antiferromagnetic, $\sim J S_i \cdot S_j$, $J = 2t^2/U$, where t is the d–d hopping matrix element and U is the on-site Hubbard repulsion, and the material would develop antiferromagnetic ordering.

In realistic situations one has to take into account several factors which determine the character, sign and strength of the exchange interaction, and which finally lead to all the variety of different magnetic structures observed experimentally. These factors are mainly the geometry of the lattice and the specific type of orbital occupation of different TM ions. We also have to take into account that, as explained in Chapter 4, in most real cases the exchange of TM ions goes via the intermediate ligands; O^{2-}, F^-, etc. Thus by the lattice geometry we have in mind both the type of lattice formed by the TM ions themselves (which may be simple cubic, as in perovskites, but also quite complicated, as in many frustrated magnets) and the local geometry – the type of TM–O–TM bonds. These may be $180°$ bonds, where the oxygen ions are located on the line connecting two TM ions, which we meet in perovskites ABO_3 or in their layered analogs A_2BO_4, $A_3B_2O_7$, etc. Alternatively they could be $90°$ bonds, which exist when neighboring MO_6 octahedra share a common edge, that is two oxygens. This is for example the situation for B sites in spinels AB_2O_4, or in two-dimensional layered systems with the CdI_2 structure, such as $NaCoO_2$, $LiNiO_2$, etc. Other possibilities also exist.

Several rules, known as the Goodenough–Kanamori–Anderson (GKA) rules, determine in these cases the sign and strength of the exchange interaction between localized electrons. In their simplest form these rules are as follows:

(1) When the orbital occupation and local geometry are such that there is an overlap, or hopping, between occupied (half-filled, having one d-electron or hole) orbitals at neighboring TM sites, or when such occupied orbitals overlap via the same p orbital of the ligand (e.g., oxygen), virtual hopping between these sites is allowed only for antiparallel spins; in this case the exchange is strong and antiferromagnetic. Depending on the specific situation, this antiferromagnetic exchange is $J \sim t_{dd}^2/U$ for direct d–d hopping, $J \sim t_{pd}^4/\Delta_{CT}^2 U$ for hopping via the oxygen in Mott–Hubbard systems (for which the d–d repulsion U is smaller than the charge-transfer energy Δ_{CT}), or $J \sim t_{pd}^4/\Delta_{CT}^2(\Delta_{CT} + \frac{1}{2}U_{pp})$ in charge-transfer insulators ($\Delta_{CT} < U$). (The situation in which the first GKA rule applies is, in fact, exactly equivalent to that of the simple Hubbard model of Chapter 1, with effective values of $\tilde{t}_{dd}$ and U_{eff}.)

(2) If the hopping of electrons between magnetically active (half-filled) orbitals is not possible, the exchange can occur by virtual transfer of electrons from one site to the empty orbital of another, either by direct d–d hopping or via the same ligand orbital. The exchange then requires the action of the Hund's rule coupling at the TM ion, and it is ferromagnetic but weak (weakened by the factor $\sim J_H/U$ compared with the antiferromagnetic exchange). This is the second GKA rule.

(3) In some situations, for example for 90° TM–O–TM bonds, the occupied orbitals of different TM ions overlap with *different* p orbitals of an anion. In this case the exchange occurs by virtual transfer of electrons from different p orbitals each to "its own" TM ion, and the resulting exchange is also ferromagnetic and weak, reduced by the factor $\sim J_H^p/\Delta_{CT}$, where J_H^p is the Hund's rule coupling on the oxygen (or another ligand) which may not be small at all, for example for oxygen $J_H^p \sim 1.2\,\text{eV}$.

There are some further details to these rules. The local geometry can be intermediate between the two limiting cases of 180° and 90° bonds: for various crystal structures the corresponding TM–O–TM angles can be in between, for example 160°, etc. One has to study the detailed character of the exchange interaction separately each time. The GKA rules presented above give only general guidelines, but very often they are sufficient to get a rough understanding of what one should expect in one or the other situation.

When we start to dope the system, introducing mobile charge carriers (electrons or holes) into the system, or when just "by construction" there exist both localized and itinerant electrons in the system, there appears yet another mechanism of exchange interaction – that via conduction electrons. If the bandwidth of the conduction band and the corresponding Fermi energy are large, this mechanism is what is known as the RKKY (Ruderman–Kittel–Kasuya–Yosida) exchange; it is $\sim \cos(2k_F r)/r^3$, where r is the distance between the magnetic ions and k_F is the Fermi momentum. This exchange is oscillating in space and can be both ferromagnetic and antiferromagnetic. In regular systems such as rare earth metals or compounds it can give different types of magnetic ordering, including for example helicoidal. For random systems, such as TM impurities in nonmagnetic metals, it can lead to the formation of a spin-glass phase.

But when the mobile carriers move in a narrow band, the same coupling of localized spins via the conduction electrons turns out to be ferromagnetic. This mechanism of ferromagnetism is called *double exchange*. Its nature is due to the fact, already mentioned in Chapter 1, that to gain the kinetic energy of electrons it is better to make the system ferromagnetic. For strong Hund's coupling J_H the extra doped electrons at each site should have their spins parallel to the spin of the "core" – of the localized electron at a given site. These localized electrons typically have antiferromagnetic exchange between themselves, and if they indeed order antiferromagnetically, the doped electron would not be able to hop from one site to a neighboring one, which would have the localized spin pointing in the opposite direction. Thus to gain the kinetic (band) energy of doped carriers it could be favorable to change the ordering of the localized spins to ferromagnetic (or at least canted): we lose the energy of the exchange interaction of the localized spins J, but gain the kinetic energy of the doped carriers $\sim tx$, where t is their hopping and x their concentration. Thus, crudely speaking, if $x \gtrsim J/t$, it would be favorable for a system to become ferromagnetic – this is the double-exchange mechanism of ferromagnetism.[23] As in most cases $J \ll t$ (typically $J \sim 100\,\text{K} \sim 0.01\,\text{eV}$ and $t \sim 0.5\,\text{eV}$), ferromagnetism can appear in such systems at a

---

[23] Another simple qualitative way to explain this mechanism is to say that the doped electron at a given site, due to the Hund's rule exchange, has spin parallel to the localized spin of the site. Then, when this electron hops

relatively small doping, for example in the CMR manganites such as $La_{1-x}Ca_xMnO_3$ it appears at $x \sim 0.2$.

As to the situation at lower doping, not sufficient to make the whole sample ferromagnetic, there exist different options. One is to create a homogeneous *canted state*, with the coexistence of ferromagnetic and antiferromagnetic ordering (and with the canting angle increasing with doping until we reach a fully ferromagnetic state). Another option is phase separation (see Section 9.7 for more details), with the creation of ferromagnetic metallic droplets in the antiferromagnetic matrix. In any case, we again see in this example the tendency already mentioned above: that ferromagnetism prefers to coexist with metallic conductivity.

Many fine, but very important details of the magnetic behavior of TM compounds are determined by the spin–orbit interaction. Without it the spin system is essentially decoupled from the lattice: the orientation of spins in the ordered states can be arbitrary with respect to the crystal lattice. Thus in this case the magnetic system would be isotropic. Magnetic anisotropy appears when we take into account spin–orbit coupling.

The possible form of magnetic anisotropy in a phenomenological description is determined by the symmetry of the crystal. Thus for tetragonal systems we can have terms $\sim -\kappa m_z^2$, which, depending on the sign of the constant κ, can give either easy-axis or easy-plane magnets. Similarly one can analyze the cases of other symmetries.

Microscopically there are two main sources of magnetic anisotropy: single-site anisotropy, for example terms $\sim -K(S^z)^2$ which are present for ions with $S > \frac{1}{2}$, and exchange anisotropy of the type $J_\parallel S^z S^z + J_\perp (S^x S^x + S^y S^y)$. The g-factor of an ion also changes because of spin–orbit coupling. For TM ions without t_{2g} degeneracy (with configurations t_{2g}^3 or t_{2g}^6, or with t_{2g} levels relatively strongly split by a noncubic crystal field), for which the orbital moment is quenched, we have $\delta g/g \sim \lambda/\Delta_{CF}$, where λ is the constant of the spin–orbit interaction $\lambda L \cdot S$. The single-site and exchange anisotropy in this case are both $\sim (\lambda/\Delta_{CF})^2 \sim (\delta g/g)^2$ (with $K \sim (\delta g/g)^2 \Delta_{CF}$, and $|J_\parallel - J_\perp| \sim (\delta g/g)^2 J$).

However for TM ions with partially filled t_{2g} levels and with unquenched orbital moments the situation is different. In this case the spin–orbit interaction acts in the first order, and especially in magnetically ordered states the spins and orbital moments are parallel (or antiparallel). In effect, below T_c or T_N in such systems the orbital moments also orient in a particular direction, the orbital occupation changes, there appears a nonspherical (quadrupolar) distribution of the electron density, with the corresponding lattice distortion. Thus there appears in such cases a very strong *magnetoelastic coupling* and corresponding *magnetostriction* – much stronger than in TM systems with quenched orbital moments. Simultaneously the magnetic anisotropy becomes quite large, driven to a large extent by this strong magnetoelastic coupling. Thus the very strong magnetic anisotropy, magnetoelastic coupling, and magnetostriction typically observed in such systems, for example those containing $Fe^{2+}(t_{2g}^4 e_g^2)$ or $Co^{2+}(t_{2g}^5 e_g^2)$, are actually

to a neighboring site (of course conserving its spin during the hopping process), it "pulls up" the spin of the neighbor in the same direction, thus finally promoting net ferromagnetism.

caused by the incipient t_{2g} degeneracy and the corresponding large role of the spin–orbit interaction.

The change of orbital occupation and corresponding lattice distortion caused by the spin–orbit coupling, together with the corresponding crystal field splitting of the t_{2g} levels, are reminiscent of the orbital ordering phenomena due to the Jahn–Teller effect, see Chapters 3 and 6. However the type of orbital occupation is different, and the lattice distortions caused by the Jahn–Teller effect and by the spin–orbit interaction turn out to be opposite. Thus by measuring for example the lattice distortion in a particular compound with t_{2g} degeneracy, one can understand which interaction, the Jahn–Teller or the spin–orbit one, dominates in a particular case. As the spin–orbit coupling becomes stronger for heavier atoms (e.g. $\lambda \sim Z^4$, where Z is the atomic number), typically for $3d$ elements the Jahn–Teller effect turns out to be stronger for lighter $3d$ elements (Ti, V, Cr), but spin–orbit coupling dominates for the late $3d$ elements (Fe, Co).

Besides the symmetric but possibly anisotropic exchange, in some systems there may appear an *antisymmetric exchange* $\sim \boldsymbol{D}_{ij} \cdot [\boldsymbol{S}_i \times \boldsymbol{S}_j]$ (the Dzyaloshinskii–Moriya (DM) interaction). It also requires the presence of the spin–orbit interaction, $|\boldsymbol{D}| \sim (\lambda/\Delta_{\mathrm{CF}})J \sim (\delta g/g)J$. Such interaction exists only for particular situations, for example when there is no inversion symmetry center between sites i and j. The rules determining the conditions for the existence of the DM interaction and determining the direction of the Dzyaloshinskii vector $\boldsymbol{D}$, are formulated by Moriya and are presented in Section 5.3.2.

The important consequence of the presence of the DM interaction is that it leads to a canting of spins (for collinear spins we would not gain the energy of the DM interaction). This can lead, in particular, to the appearance of *weak ferromagnetism* in some antiferromagnets. There are quite a few such systems among TM compounds, for example Fe_2O_3, etc. Yet another consequence of the DM interaction can be the appearance of a *linear magnetoelectric effect*, which will be discussed in Chapter 8.

The rules formulated above allow one in principle to understand all the variety of magnetic structures observed in different TM compounds. A special case presents the so-called *frustrated* magnetic materials. Most often these are systems with frustrated lattices, such as triangular, kagome, GGG type (the latter two consist of a network of corner-sharing triangles), or pyrochlore lattice (consisting of corner-sharing tetrahedra). But sometimes frustrations can also exist for example in a square lattice, if we include further-neighbor interactions (e.g. an exchange along square diagonals). In simple terms frustration means that it is impossible to build a magnetic structure which simultaneously satisfies antiferromagnetic interactions on all bonds. In such cases the magnetic properties can become quite complicated. There may still exist long-range ordered states, but possibly rather complicated, and the ordering would occur usually at temperatures much lower than the typical values of the exchange interaction, the measure of which is the Weiss temperature in the Curie–Weiss susceptibility $\chi \sim C/(T - \Theta)$: $\Theta \sim zJS(S + 1)$, where z is the number of nearest neighbors. Whereas in the standard cases we have magnetic ordering temperature $T_{\mathrm{N}} \sim \Theta$, in frustrated systems typically we have $T_{\mathrm{N}} \ll \Theta$. The ratio Θ/T_{N} is taken as an empirical measure of the degree of frustration: the bigger it is, the more frustrated is the system.

In even more frustrated situations there may be no long-range magnetic ordering at all. There may still exist rather well-developed antiferromagnetic correlations, but of a short-range type. Such states are called *spin liquids*. There may be different types of such spin liquid states, with or without gaps in singlet or triplet excitations.

One of the most "popular" types of such states with spin singlets is the *resonating valence bond* (RVB) state. It gives a good (though not exact) description of one-dimensional antiferromagnets, and it can in principle also exist in some higher-dimensional systems, especially those with frustrations and/or with a certain doping. There were suggestions, chiefly by P. W. Anderson, that such states may be relevant for high-T_c super-conductivity in cuprates, see Chapter 9. Such states can have quite nontrivial excitations, for example neutral excitations with spin $\frac{1}{2}$ – *spinons* (instead of the normal spin waves or magnons, which have $S = 1$) or charged but spinless excitations – *holons*. That is, there may occur in such systems *spin–charge separation*: the hole with charge $+e$ and spin $\frac{1}{2}$ is split into a spinon with charge zero and $S = \frac{1}{2}$, and a holon with charge $+e$ but with $S = 0$.

Besides different magnetic states which can exist in the bulk materials as homogeneous states, there may appear in magnets different types of magnetic defects, or complicated magnetic textures. These are for example the well-known domain walls separating magnetic domains with different order parameters, for example different orientation of ferromagnetic magnetization. Domain walls in three-dimensional materials are two-dimensional objects. But there can also appear other types of defect, for example one-dimensional magnetic vortices or point-like defects of the type of "magnetic hedgehogs," called skyrmions. Depending on the particular spin configuration within these defects they may display not only nontrivial magnetic properties, but also interesting electric (e.g., magnetoelectric) properties. Thus, some magnetic vortices and skyrmions can have a toroidal moment. These features are especially important as a source of coupling between electricity and magnetism, to be discussed in more detail in Chapter 8.

A special kind of magnetic phenomenon in TM compounds appears in systems in which the TM ions can exist in different spin states (different multiplets). In some such systems there may appear special types of transitions – spin-state transitions – in which not only the magnetic moments order in some fashion, but the very moment of an ion changes. Such transitions can occur with change of temperature, pressure, composition, or even under illumination. They can be real phase transitions, even I order, but they can also occur as a smooth crossover. The best known example of the latter is presented by the perovskite LaCoO$_3$. In special cases there may appear an ordering, for example of checkerboard type, of different spin states, that is low-spin and high-spin states of Co^{3+}. The main physical mechanism which can lead to cooperative spin-state transitions and to eventual spin-state ordering is usually coupling through the lattice, originating from the fact that different spin states of TM ions typically have very different ionic radii.

A special type of magnetism can exist in systems in which the ground state of a transition metal (or rare earth) ion is a singlet, but in which there are low-lying magnetic excited states (e.g., triplet states). This can happen for example due to the fact that the ground state is that with a low spin, for example the $S = 0$ state of Co^{2+}, or because it is a singlet $J = 0$ state due to the spin–orbit interaction. If in this situation the exchange interaction of TM ions in an excited state is larger than the singlet–triplet excitation energy, the resulting ground state of a concentrated system may still be magnetic, despite the nonmagnetic singlet state of an isolated ion: the effective Zeeman splitting of the excited triplet in the internal molecular field in this case would be larger than the singlet–triplet excitation energy, and the Zeeman-split magnetic level would lie below the singlet level. This situation is known as *singlet magnetism*. It is most often observed in rare earth compounds, for example those with Pr^{3+}, but can also exist in TM compounds. The same physical effect – crossing of a nonmagnetic singlet state and a Zeeman-split excited state – can also occur not in an internal, but in an external magnetic field. After this crossing the state of the system becomes magnetic, which can often be described as a generation of magnons in a coherent state, in which case one sometimes talks about *magnon Bose-condensation*.

6

Cooperative Jahn–Teller effect and orbital ordering

We have already seen in Sections 3.1, 3.2 that in many cases the ground state of transition metal ions in a symmetric (e.g., cubic) crystal field such as that of a regular MO_6 octahedron can have an orbital degeneracy. Such are for example the states with double (e_g) degeneracy in $Mn^{3+}(t_{2g}^4 e_g^1)$, $Cr^{2+}(t_{2g}^4 e_g^1)$, low-spin $Ni^{3+}(t_{2g}^6 e_g^1)$, $Cu^{2+}(t_{2g}^6 e_g^3)$, or with triple degeneracy (partially filled t_{2g} levels). We have seen that in principle all such states are unstable toward lattice distortions lifting this orbital degeneracy – this is the essence of the Jahn–Teller (JT) theorem explained in Section 3.2.

For t_{2g} ions the spin–orbit coupling may also play a significant role, and it can lift this orbital degeneracy. Therefore the manifestations of the JT effect for these systems may be different from those present in the e_g case. Therefore we will first consider the "clean" case of e_g degeneracy, and discuss what would be its consequences in concentrated bulk systems; after that we will turn to the case of t_{2g} systems with their own specific features.

Some aspects of the JT effect for isolated centers, in particular the important role of vibronic effects, have been discussed briefly in Section 3.2. In this chapter we will consider the behavior of concentrated systems with orbital degeneracy. We will see that in such systems, where we have not just isolated JT centers but rather a large amount of them (ideally there are such ions at each unit cell), there will appear an interaction between different JT sites, and due to that the JT effects acquire a cooperative character – we are talking here of the *cooperative Jahn–Teller effect* (CJTE). As a result of such interactions typically there occurs a phase transition with a change of structure, which lifts the orbital degeneracy and simultaneously leads to the occupation of particular orbitals at each site – that is we have *orbital ordering* (O.O.)

In the 1960s and 1970s, when discussing these phenomena, the terminology of the CJTE was used mostly; nowadays we more often talk about orbital ordering. But in fact these are just different terms describing the same physics (while stressing different elements) – Jahn–Teller lattice distortions, with the corresponding interactions, or orbital ordering in the electronic subsystem. One should however realize that one phenomenon cannot occur without the other: even if orbital ordering is mostly driven by electronic effects, see below, the lattice would always follow. Thus in a sense the relation between the CJTE and orbital ordering is a "chicken-and-egg problem"; these phenomena always occur together.

Yet one more general remark. In the structural transition with corresponding orbital ordering the role of order parameter can be played either by the distortion u_{ij} (or the corresponding strain) or by orbital occupation. In the process of orbital ordering the electron density at an ion, which in the high-temperature disordered state is spherical (or has e.g. cubic symmetry, dictated by the crystal field), becomes anisotropic – for example with the electron density elongated in one direction, as in Fig. 3.9(a) or compressed, as in Fig. 3.9(b). Accordingly, we see that in this process there appears no real charge ordering and no electric dipoles, but rather electric quadrupoles are formed at each such site. Thus, as already mentioned in the previous chapter, the real order parameter in the CJTE or in orbital ordering is a quadrupole $Q_{\alpha\beta}$, a second-rank tensor. However in particular cases one may reduce the description to that with a simpler order parameter, for example the effective pseudospin $\tau = \frac{1}{2}$, as we will see below.

The study of orbital ordering is rather well-developed at present. A detailed discussion can be found in several books and reviews: see Englman (1972), Gehring and Gehring (1975), Kugel and Khomskii (1982), Kaplan and Vekhter (1995), Tokura and Nagaosa (2000), Khomskii (2005), Bersuker (2006). Below we will describe the main effects, but of course we cannot present all the details and discuss all the possible cases of orbital ordering.

6.1 Cooperative Jahn–Teller effect and orbital ordering in e_g systems

At the beginning of this section we should recall the main description of orbitals, presented in Sections 3.1 and 3.2. As discussed there, we meet situations with e_g degeneracy for TM ions in octahedral coordinations, typically d^4 (Mn^{3+}, Cr^{2+}) and d^9 (Cu^{2+}). In these cases we have one electron or one hole in the doubly degenerate e_g orbital (Fig. 6.1). Remember that the corresponding e_g orbitals, which are degenerate for the cubic crystal field (regular MO_6 octahedra), can be taken as $|z^2\rangle$ and $|x^2 - y^2\rangle$ orbitals, Fig. 6.1(c, d). But one can take as a basis any linear combination of these orbitals, for example we can use the orbital $|x^2\rangle$, with the same shape as $|z^2\rangle$ in Fig. 6.1(c) but oriented along the x-direction, etc. The general form of any such linear combination can be written as[1]

$$|\theta\rangle = \cos\frac{\theta}{2}\,|z^2\rangle + \sin\frac{\theta}{2}\,|x^2 - y^2\rangle . \tag{6.1}$$

One can represent such states in the θ plane, see Fig. 6.2 (this representation has already been introduced in Section 3.1). Using the form of the e_g wavefunction, which in our normalization can be written as (see Section 3.1)

---

[1] More generally, such linear combinations can also have complex coefficients, $\alpha|z^2\rangle + \beta|x^2 - y^2\rangle$, with $|\alpha|^2 + |\beta|^2 = 1$. In the standard treatment presented here such complex combinations do not appear, but in principle one can think of special situations in which these coefficients indeed are complex (van den Brink and Khomskii, 2001). But the properties of such states are very different from those with real coefficients of the type (6.1). Thus, for example, the combinations $\frac{1}{\sqrt{2}}(|z^2\rangle \pm i|x^2 - y^2\rangle)$ have spherically symmetric electron density, that is they are not associated with any lattice distortion (that is why they do not appear in the standard treatment of the CJTE, which relies on lattice distortions lifting the JT degeneracy). But such complex states, as all complex states in quantum mechanics, change the sign to the opposite under time reversal, that is they break the time reversal invariance and consequently are magnetic. However, they do not have a nonzero magnetic (dipole) moment. One can show that what is nonzero for such states is in fact a *magnetic octupole* (van den Brink and Khomskii, 2001).

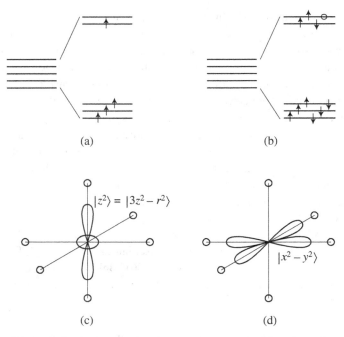

Figure 6.1 Possible orbitally degenerate situations leading to the strong Jahn–Teller effect, and the form of two basic e_g orbitals.

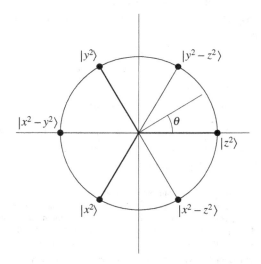

Figure 6.2 θ plane on which one can conveniently locate different possible orbital states (6.1).

$$|z^2\rangle = \tfrac{1}{\sqrt{6}}(2z^2 - x^2 - y^2)\,, \qquad |x^2 - y^2\rangle = \tfrac{1}{\sqrt{2}}(x^2 - y^2)\,, \tag{6.2}$$

we can calculate that different angles in the θ plane correspond to the following states: $|0\rangle \Longleftrightarrow |z^2\rangle$, $|\pi\rangle \Longleftrightarrow |x^2 - y^2\rangle$, $|\tfrac{2}{3}\pi\rangle \Longleftrightarrow |y^2\rangle$, $|\tfrac{4}{3}\pi = -\tfrac{2}{3}\pi\rangle \Longleftrightarrow |x^2\rangle$, $|\tfrac{1}{3}\pi\rangle \Longleftrightarrow |y^2 - z^2\rangle$, $|-\tfrac{1}{3}\pi\rangle \Longleftrightarrow |x^2 - z^2\rangle$, etc., where our short notation is $|z^2\rangle = |3z^2 - r^2\rangle$, $|x^2\rangle = |3x^2 - r^2\rangle$, $|y^2\rangle = |3y^2 - r^2\rangle$.

A convenient way to describe these doubly degenerate states is by using pseudospin-$\tfrac{1}{2}$ operators $\boldsymbol{\tau}$, such that

$$\begin{aligned} |\tau^z = +\tfrac{1}{2}\rangle &\Longleftrightarrow |z^2\rangle\,, \\ |\tau^z = -\tfrac{1}{2}\rangle &\Longleftrightarrow |x^2 - y^2\rangle\,. \end{aligned} \tag{6.3}$$

The θ plane of Fig. 6.2 is then mapped to a (τ^z, τ^x) plane (Fig. 6.3), and an arbitrary state $|\theta\rangle$ (6.1) is represented by a certain orientation of this pseudospin $\boldsymbol{\tau}$ in the (τ^z, τ^x) plane.[2] These pseudospin operators obey the standard commutation relations of spin-$\tfrac{1}{2}$ operators, or of corresponding Pauli matrices. Thus one can deal with these pseudospins in exactly the same way as with ordinary spins $\tfrac{1}{2}$, using the same approximations, but also getting in principle all the same complications such as quantum effects (though due to strong coupling to the lattice the latter can be less severe for orbitals than for real spins, see below). Strictly speaking, however, there is one difference between the pseudospins $\boldsymbol{\tau}$ and the normal spins $\boldsymbol{S}$: the normal spins describe magnetic states, that is formally all components of $\boldsymbol{S}$ are odd under time reversal $\mathcal{T}$. At the same time τ^x, τ^z describe in fact a particular charge distribution, that is these operators are even under time reversal, and only the third component τ^y (which, as we point out in footnote 1, actually almost never enters the description of JT systems) is odd under $\mathcal{T}$.

In Section 3.2 we already discussed the consequences of orbital degeneracy for isolated TM ions and the description of some corresponding phenomena, such as coupling between

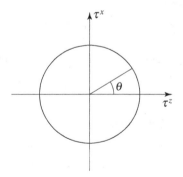

Figure 6.3 The mapping of orbital subspaces of doubly degenerate e_g orbitals to the space of pseudospins $\tau = \tfrac{1}{2}$.

---

[2] The complex states discussed in the previous footnote would correspond to eigenstates of the third Pauli matrix, τ^y.

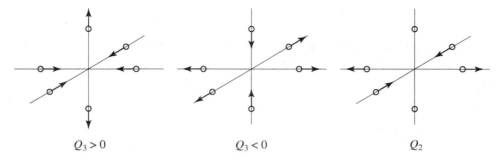

$$Q_3 > 0 \qquad\qquad Q_3 < 0 \qquad\qquad Q_2$$

Figure 6.4 Possible local deformations of MO_6 octahedra lifting the e_g degeneracy.

orbital degrees of freedom and corresponding lattice distortions. The e_g levels can be split by doubly degenerate distortions or phonons Q_3 and Q_2, with Q_3 being a tetragonal elongation ($Q_3 > 0$) or compression ($Q_3 < 0$), and with Q_2 corresponding to orthorhombic distortions, see Fig. 6.4 (to make the discussion in this chapter self-contained, we repeat here some of the material presented in Sections 3.1, 3.2). These local deformations couple to the orbital occupation of the corresponding ion i by the interaction (3.29), which can be rewritten through the pseudospin operators (6.3) using the relations

$$
\begin{aligned}
c_{i1}^{\dagger} c_{i1} &= n_{i1} = \tfrac{1}{2} + \tau_i^z\,, \\
c_{i2}^{\dagger} c_{i2} &= n_{i2} = \tfrac{1}{2} - \tau_i^z\,, \\
c_{i1}^{\dagger} c_{i2} &= \tau_i^+\,, \\
c_{i2}^{\dagger} c_{i1} &= \tau_i^-\,,
\end{aligned}
\tag{6.4}
$$

where the indices 1, 2 stand for $|z^2\rangle$ and $|x^2 - y^2\rangle$ orbitals, respectively. The resulting on-site interaction takes the form

$$
\mathcal{H}_{\mathrm{JT},i} = -g(\tau_i^z Q_{3i} + \tau_i^x Q_{2i})\,,
\tag{6.5}
$$

where, because of symmetry requirements, the coupling constant g of e_g orbitals to Q_3 and Q_2 modes is the same.

6.1.1 Jahn–Teller mechanism of intersite coupling

Now, when we make a concentrated solid out of JT ions, such as $LaMnO_3$ or $KCuF_3$, there will appear an interaction between the orbital states of different ions and between the corresponding distortions. The easiest way to see this is for example by looking at the situation of two neighboring JT ions sharing a common oxygen (or common corner of the respective MO_6 octahedra), Fig. 6.5. It is clear from this picture that for example if we occupy the orbital $|z^2\rangle$ at site A, with corresponding local elongation of the AO_6 octahedron in the z-direction, it will be unfavorable to occupy the same orbital and have the

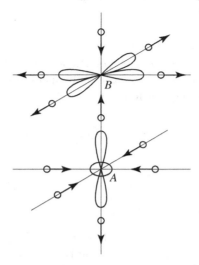

Figure 6.5 Illustration of the coupling of Jahn–Teller distortions to the respective orbital ordering for a pair of neighboring Jahn–Teller ions with common corners of their O_6 octahedra.

same distortion at site B: whereas the AO_6 octahedron is elongated, the octahedron BO_6 would prefer to be compressed, and so one could expect that on site B the other orbital, $|x^2 - y^2\rangle$, would be occupied. In any case, we see that the very fact that in concentrated systems different octahedra share common oxygens immediately leads to a coupling of distortions and of orbital occupations on different centers, as a result of which the process of orbital ordering becomes cooperative and takes the form of a real phase transition.

In fact, for this mechanism to work it is not even required that different JT ions have common oxygens: in a crystal any local distortion leads to a strain field which extends to large distances, and another JT ion, even not necessarily very close to the first one, would feel this strain and adjust its orbital occupation accordingly.

Mathematically we can describe this situation by generalizing the local Jahn–Teller Hamiltonian (6.5) to a large system:

$$
\begin{aligned}
\mathcal{H}_{JT} = &-\sum_{i,q} g_{i,q}\left(\tau_i^z Q_3(q) + \tau_i^x Q_2(q)\right) \\
&+ \sum_q \omega_{3,q} b_{3,q}^\dagger b_{3,q} + \sum_q \omega_{2,q} b_{2,q}^\dagger b_{2,q} \\
= &-\sum_{i,q} \tilde{g}_{i,q}\left(\tau_i^z (b_{3,q}^\dagger + b_{3,-q}) + \tau_i^x (b_{2,q}^\dagger + b_{2,-q})\right) \\
&+ \sum_q \omega_{3,q} b_{3,q}^\dagger b_{3,q} + \sum_q \omega_{2,q} b_{2,q}^\dagger b_{2,q} \,,
\end{aligned}
\tag{6.6}
$$

where $g_{i,q} = g(q)e^{iq \cdot R_i}$ (and similarly for $\tilde{g}$), and where we have introduced phonon operators $b_{\alpha,q}^\dagger, b_{\alpha,q}$ ($\alpha = 3, 2$) so that

$$Q_3(q) \Longrightarrow b^\dagger_{3,q} + b_{3,-q} \, , \qquad Q_2(q) \Longrightarrow b^\dagger_{2,q} + b_{2,-q} \tag{6.7}$$

(we have incorporated numerical factors, containing e.g. the atomic masses, etc., into the effective coupling constants g, $\tilde{g}$); $\omega_{\alpha,q}$ are the frequencies of the corresponding phonons.

One can exclude the phonons from the Hamiltonian (6.6), and we obtain the effective orbital Hamiltonian having the form of an exchange interaction of pseudospins τ. Schematically it has the form

$$\mathcal{H}_{\text{eff}} = \sum_{ij} J_{ij} \, \tau_i \tau_j \, , \tag{6.8}$$

where

$$J_{ij} = -\sum_q e^{iq \cdot (R_i - R_j)} \frac{|\tilde{g}(q)|^2}{\omega(q)} \, . \tag{6.9}$$

Note that to get an interaction between different JT centers, it is necessary to have dispersion (q-dependence) in the interaction matrix element $g(q)$ and/or the phonon frequencies $\omega(q)$: without such dispersion, for constant g or $\tilde{g}$ and ω (i.e., for local phonons), the interaction (6.9) would become $J_{ij} = \delta_{ij} \, \tilde{g}^2/\omega$ and we would have a description of the local JT effect, without any coupling between different sites, that is without any cooperative nature. We also see that the coupling via acoustic phonons leads to a long-range interaction, and that via optical phonons to a short-range interaction (predominantly the coupling of nearest neighbors).

In real cases we have to take into account both modes, Q_2 and Q_3, and the resulting expression depends on the detailed type of the crystal lattice, etc. In effect the orbital interaction may be quite anisotropic in general, containing terms $\tau^z \tau^z$, $\tau^x \tau^x$ and $\tau^z \tau^x$ with different coefficients. One can see this for example by considering a chain of JT ions along the z-direction, the first segment of which is shown in Fig. 6.5. We immediately see that for this chain only the Q_3 modes enter, and as they couple only to τ^z-operators, the resulting effective Hamiltonian (6.8), written through the τ-operators, should contain only τ^z, that is it will be of the Ising type (an Ising "antiferromagnet", as is clear from Fig. 6.5). Only for a very symmetric model may the interaction (6.8) resemble the Heisenberg interaction.

In any case, the result is some form of pseudospin exchange Hamiltonian and, similarly to magnetic systems, one can expect this interaction to lead to certain orbital ordering, which may be "ferromagnetic" (or rather ferro. orbital), or "antiferromagnetic" (i.e., antiferro. orbital), etc. The mechanism of this ordering described above is based on the JT interaction with lattice distortions, or rather through lattice distortions: this is the mechanism which was usually invoked in the treatment of the CJTE. This mechanism is definitely present in all JT systems, and it may be very important.

Interestingly enough, one can give this mechanism a purely classical interpretation. At least the interaction via long-wavelength acoustic phonons can be described using the language of classical elasticity theory, and often one can get a pretty good description of the resulting cooperative effect in the JT distortion and of the corresponding orbital ordering using this language, without any sophisticated calculations (Khomskii and Kugel, 2001,

2003). In elasticity theory, when we remove an atom ("a sphere") from the lattice and put in its place another atom (a sphere of different radius, as e.g. in charge ordering, or an ellipsoid, as in the JT case), this extra atom will produce a strain in the crystal, which depends both on the shape of the "inclusion" and on the elastic constants of the matrix such as c_{11}, c_{12}, c_{44}. One can show that a spherical inclusion of a different radius in a cubic crystal at $r = 0$ ("a sphere in a hole"; Eshelby, 1956) will create a strain field $F(r)$ of the type

$$F(r) \sim d\, Q\, \frac{\Gamma(n)}{|r|^3}, \tag{6.10}$$

where Q is the "strength" of the impurity ($Q \sim v - v_0$, where v is the volume of the inclusion and v_0 is the volume of the part of the matrix it replaces), and where

$$d = c_{11} - c_{12} - 2c_{44}, \tag{6.11}$$

with c_{ij} the elastic moduli of the crystal. The important factor $\Gamma(n)$, where $n = r/|r|$, describes the angular dependence of the strain field – the dependence on the directional cosines n_x, n_y, n_z of the vector r:

$$\Gamma(n) = n_x^4 + n_y^4 + n_z^4 - \tfrac{3}{5} . \tag{6.12}$$

This strain field is illustrated in Fig. 6.6: the strain is positive close to x- and y-axes (shaded regions in Fig. 6.6) and negative in the diagonal directions (or vice versa, depending on the size of the inclusion and on the sign of the combination $c_{11}+c_{12}-2c_{44}$). Accordingly, when we put another such impurity somewhere else in the crystal, it will experience attraction or repulsion from the first one, depending on their relative positions. If the impurities are mobile, as is the case for charge ordering, this can lead to a formation of stripe structures (the charges order along the lines of attraction, e.g. along the x- and y-axes in the "windmill" of Fig. 6.6); we will discuss this in more detail in Chapter 7. Similarly, if we have JT ions which induce a local distortion, most often local elongation of the MO_6 octahedra, the resulting ellipsoids will interact via a similar strain field, and the minimum energy will be reached for a particular packing of such distortions, which minimizes the total strain. Thus

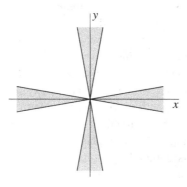

Figure 6.6 "Windmill"-like regions corresponding to attraction or repulsion of impurities in elastic media.

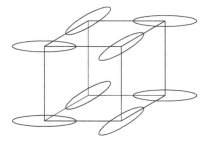

Figure 6.7 Optimal packing of elongated octahedra ("cigars") in a cubic lattice.

one can show (Khomskii and Kugel, 2003) that for systems such as LaMnO$_3$, in which each Mn^{3+}($t_{2g}^3 e_g^1$) induces a local elongation of the MnO$_6$ octahedra, the best packing of these long axes would be that shown in Fig. 6.7 (cf. also Fig. 3.23) – the structure almost coinciding with the one observed experimentally in LaMnO$_3$ below T_{JT} (in this approach two structures, in which different xy-planes are either in phase, as shown in Fig. 6.7, or out of phase have almost the same energy).

Yet one more general feature, discussed in Section 3.2, should be recalled here: as was shown in that section, for e_g-electrons, with almost no exception, the local JT effect leads to *elongation* of the MO_6 (or MF_6, etc.) octahedra. Thus indeed one can treat this situation by considering close packing of long "cigar-shaped" objects, which in particular leads to the structure of Fig. 6.7, with the orbital occupation shown in Fig. 3.23. The standard arguments which give rise to effective orbital interaction similar to (6.8), and which ignore anharmonicity effects, stabilizing elongated octahedra, may give an interaction which would lead to "antiferro. orbital ordering," like the alternation of $|z^2\rangle$ and $|x^2 - y^2\rangle$ orbitals shown in Fig. 6.5. But this would correspond to local elongation (along the z-axis) at A sites and compression at B sites, which, as we argued in Section 3.2, is very unfavorable from the point of view of anharmonicity effects and higher-order Jahn–Teller interactions. As we have already mentioned, experimentally local compression is practically never observed in e_g systems: out of hundreds of known insulating JT systems with e_g degeneracy there may be one or two with local compression, and even these cases are debatable. The outcome is that when we take into account these anharmonic effects, in the situation of Fig. 6.5 the resulting ordering will be not of the $(z^2, x^2 - y^2)$-type shown there, with one MO_6 octahedron elongated and the other compressed, but rather of the (z^2, x^2)- or (z^2, y^2)-type (Fig. 6.8): the shift of the common apical oxygen can stabilize local distortion on the B site which also corresponds to *local elongation* of the BO_6 octahedron, but with the long axis perpendicular to the z-direction; that is again local elongation but along the x- or y-direction. The situation in the basal plane in Figs 3.23, 6.7 is exactly of this type.

In any case, lattice effects definitely play a very important role in the physics of JT systems. They can provide a mechanism of orbital ordering and determine its features. But this is not the only mechanism of orbital ordering. There exists yet another, purely electronic, or superexchange mechanism, also leading to orbital ordering.

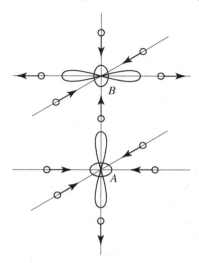

Figure 6.8 Local deformations and orbital occupation of a pair of Jahn–Teller ions when only locally elongated octahedra are allowed, cf. Fig. 6.5.

6.1.2 Superexchange mechanism of orbital ordering ("Kugel–Khomskii model")

In the previous sections we have considered the JT coupling of orbitals with lattice distortions. This coupling dealt only with charge degrees of freedom and did not depend on the spins of the corresponding ions. Of course, as explained in Section 5.1, after a certain orbital ordering is established, it strongly influences the type of exchange interaction and the resulting magnetic properties – this constitutes a large part of the Goodenough–Kanamori–Anderson rules. But in principle orbital ordering in this mechanism occurs independently of spin ordering.

There exists yet another, purely electronic mechanism of orbital ordering, suggested in Kugel and Khomskii (1973), which is actually very close to the superexchange mechanism of magnetic ordering. We can describe e_g-electrons by the Hubbard model (1.6) generalized to the doubly degenerate case

$$\mathcal{H} = -\sum_{\substack{\langle ij \rangle, \\ \alpha\beta,\sigma}} t_{ij}^{\alpha\beta} c_{i\alpha\sigma}^{\dagger} c_{j\beta\sigma} + \sum_{\substack{i, \\ \alpha\sigma \neq \beta\sigma'}} U_{\alpha\beta} n_{i\alpha\sigma} n_{i\beta\sigma'} - J_{\mathrm{H}} \sum_{i} (2\mathbf{S}_{i1} \cdot \mathbf{S}_{i2} + \tfrac{1}{2}), \qquad (6.13)$$

where $\alpha, \beta = 1, 2$ are the orbital indices (e.g., $1 \Longleftrightarrow |z^2\rangle$, $2 \Longleftrightarrow |x^2 - y^2\rangle$). Generally speaking, the hopping integrals $t_{ij}^{\alpha\beta}$ depend on the type of orbitals involved and on the direction between pairs of sites $\{i, j\}$. In principle the Coulomb matrix elements U may also depend on the orbitals: thus the repulsion on the same orbital, $U_{11} = U_{22}$ is larger than on different orbitals, U_{12}. However in the first approximation this difference can often be ignored, see the discussion in Section 3.3. This is sufficient for our qualitative purposes, although some numerical coefficients, in particular those with which the Hund's coupling J_{H} enters the resulting expression, may differ. The Hund's rule coupling in (6.13)

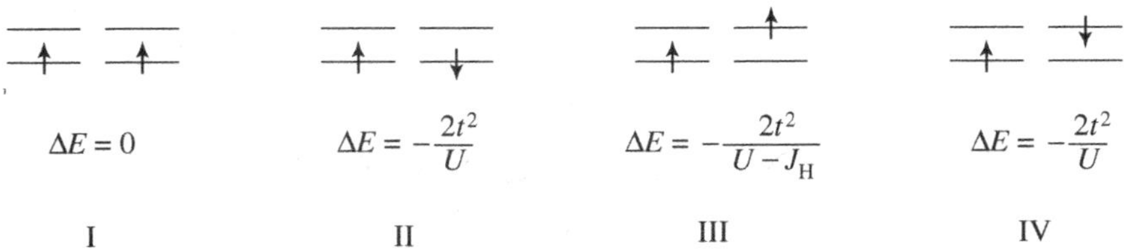

$$\Delta E = 0 \qquad\qquad \Delta E = -\frac{2t^2}{U} \qquad\qquad \Delta E = -\frac{2t^2}{U-J_{\mathrm{H}}} \qquad\qquad \Delta E = -\frac{2t^2}{U}$$

I　　　　　　　　　II　　　　　　　　　III　　　　　　　　　IV

Figure 6.9 Mechanism of simultaneous spin and orbital ordering in the doubly degenerate Hubbard model (the superexchange, or Kugel–Khomskii mechanism).

is written in a way that corresponds to our rule formulated in Section 2.2: it gives energy $-J_{\mathrm{H}}$ for parallel spins, $\boldsymbol{S} \cdot \boldsymbol{S} = +\frac{1}{4}$, and energy zero for antiparallel spins treated in a mean-field approximation, $\boldsymbol{S} \cdot \boldsymbol{S} = -\frac{1}{4}$.

Using this model we can consider the situation with $U \gg t$ and with one electron per site on a doubly degenerate orbital. The localized electrons at each site then have four possible states: $(1\uparrow)$, $(1\downarrow)$, $(2\uparrow)$, and $(2\downarrow)$, which can be described by the values of the pseudospin τ_i and spin S_i. Proceeding in exact analogy with the treatment of superexchange, Section 1.3, we see that in this case the degeneracy will be lifted by virtual hoppings in the second order with perturbation theory in t/U. But, in contrast to the nondegenerate case, illustrated in Fig. 1.12, here for two neighboring sites we have to consider not two, but four different possibilities, as shown in Fig. 6.9.

Consider for simplicity the case with only diagonal hoppings $t^{11} = t^{22} = t$, $t^{12} = 0$. Then one sees that the energy gains in the second order with perturbation theory will be those shown in Fig. 6.9. The first case (same spins, same orbitals) does not give any energy gain, as virtual hoppings in this case are forbidden by the Pauli principle. The other three configurations give a certain energy gain, and we see that the biggest gain will be that of the third configuration: with the same spins but opposite orbitals. This configuration becomes the most favorable one because in the intermediate state with two electrons on the same site those electrons would have parallel spins, which decreases the energy of this state by J_{H} (and also the repulsion of two electrons on different orbitals in the intermediate state in cases III, IV would be reduced, see e.g. eq. (2.7); for simplicity we ignore this effect here, its inclusion would not change the qualitative conclusions). Thus in this simplified model we should expect that the ground state is that with alternating orbital occupations, that is with "antiferro. orbital ordering," but ferromagnetic in spins. This is a rather general tendency, at least in JT systems with 180° TM–O–TM bonds, that is with MO_6 octahedra having common corners (one common oxygen): usually antiferro. orbital ordering gives spin ferromagnetism and, vice versa, ferro. orbital ordering usually gives antiferromagnetism. One can notice here a close relationship to the GKA rules discussed in Section 5.1: indeed, from the GKA rules it follows that the sign of the exchange is determined by orbital occupation, and the overlap of (half-)filled orbitals gives antiferromagnetism, while the overlap of an occupied orbital and an empty one gives ferromagnetism. The first case resembles case II of Fig. 6.9, and the second one case III. The difference is that in the GKA rules we *assume* a certain orbital occupation, but here we *obtain* it together with the corresponding spin ordering.

Again in analogy with the nondegenerate case, where for $n = 1$ and $t/U \ll 1$ we could go from the Hubbard model (1.6) to the effective spin model (1.12), here in this limit we can also go from the degenerate Hubbard model (6.13) to an effective model for localized electrons. But, in contrast to the nondegenerate case, where the state of such localized electrons was characterized by spin S_i, which gave the Heisenberg Hamiltonian (1.12), here the state of each site with one localized electron is characterized by *two* quantum numbers: its spin and its orbital state, that is by two operators S_i and τ_i. Correspondingly, the effective model would have the schematic form

$$\mathcal{H}_{\text{eff}} = J_1 \sum_{\langle ij \rangle} S_i \cdot S_j + J_2 \sum_{\langle ij \rangle} \tau_i \tau_j + 4J_3 \sum_{\langle ij \rangle} (S_i \cdot S_j)(\tau_i \tau_j) . \tag{6.14}$$

The spin part of the exchange has the Heisenberg form $S \cdot S$, but the orbital part $(\tau \tau)$ is in general anisotropic, which in fact is connected with the directional character of orbitals. The coefficients J_1, J_2, J_3 contain terms $\sim t^2/U$ and $(t^2/U)(J_{\text{H}}/U)$. This spin–orbital model is sometimes called the Kugel–Khomskii (KK) model.

In simple symmetric cases, for example the one used in Fig. 6.9 (with $t^{11} = t^{22} = t$, $t^{12} = 0$ independent of the direction of the bond ij and with all Us the same), the effective Hamiltonian (6.14) takes a simpler form (Kugel and Khomskii, 1973, 1982) with the real scalar product $\tau \cdot \tau$,

$$\mathcal{H}_{\text{eff}} = \sum_{\langle ij \rangle} \left[J_1 S_i \cdot S_j + J_2 \tau_i \cdot \tau_j + 4J_3 (S_i \cdot S_j)(\tau_i \cdot \tau_j) \right] \tag{6.15}$$

and with

$$J_1 = \frac{2t^2}{U} \left(1 - \frac{J_{\text{H}}}{U} \right) , \qquad J_2 = J_3 = \frac{2t^2}{U} \left(1 + \frac{J_{\text{H}}}{U} \right) . \tag{6.16}$$

That is, in this case the orbital operators τ_i, similar to spins, also have a Heisenberg interaction. Theoreticians speak in this case about SU(2) × SU(2) symmetry. For $J_{\text{H}} = 0$ the symmetry of this model is even higher: it is SU(4). But for the general case, and in particular for actual e_g orbitals with their particular overlaps and hopping, the resulting Hamiltonian (6.15) is very anisotropic in τ variables; in particular it contains terms $\sim \tau^z \tau^z$, $\tau^z \tau^x$, $\tau^x \tau^x$, but does not contain τ^y at all, see Kugel and Khomskii (1973, 1982).

The situation shown in Fig. 6.9 and described by the Hamiltonian (6.15), (6.16), which leads simultaneously to ferro. orbital and antiferro. spin ordering, is rather typical for materials with e_g degeneracy and with perovskite-type lattices with 180° metal–oxygen–metal bonds. However one has to be careful: this is not a universal tendency. For other geometries the outcome may be different. Thus, as explained in Section 5.1, for e_g systems with edge-sharing octahedra and with 90° TM–O–TM bonds the magnetic ordering would be ferromagnetic independently of the orbital occupation of e_g-electrons. This also agrees with the GKA rules for this geometry, see Section 5.1. Nevertheless, for many cases the trend is indeed such that ferro. orbital ordering gives antiferromagnetic spin exchange, and antiferro. orbitals give ferromagnetic spins. However, one has to be aware that this rule does not apply in all cases.

As just mentioned, in real systems, even in those with simple lattices such as the cubic lattice of perovskites, the effective KK Hamiltonian may look rather complicated, see the examples in Kugel and Khomskii (1982). It is not only highly anisotropic in the τ-operators, but the detailed form of the corresponding terms depends on the direction of the bond between sites i, j, see also Section 6.6 later. Nevertheless, working with the resulting Hamiltonian, one can rather successfully describe spin and orbital structures of a number of materials. As an example we will discuss below the orbital and spin structure of $KCuF_3$, K_2CuF_4, and $LaMnO_3$.

6.1.3 Typical examples of orbital ordering

In $KCuF_3$ the ions $Cu^{2+}(t_{2g}^6 e_g^3)$ contain three e_g-electrons, or one e_g-hole compared with the filled e_g-shell. One usually works with this hole orbital for Cu^{2+}. (Remember that for a hole orbital the sign of local distortion would be opposite to the corresponding distortion for electrons.)

The orbital structure obtained theoretically for $KCuF_3$ (Kugel and Khomskii, 1973) and coinciding with experiment is shown in Fig. 6.10, where the hole orbitals are shown. There exists an equivalent structure in which the upper xy-plane has the same orbitals as the lower one; and experimentally these two polytypes are found with almost equal probability.[3]

An extremely interesting feature of this orbital structure is that, as is clear from the GKA rules, there will be a strong antiferromagnetic exchange in the z-direction (strong overlap of occupied orbitals – the hole orbitals play here the role of active orbitals carrying spin). In the basal plane, however, the occupied hole orbitals are orthogonal to each other; there exists only a hopping between occupied and empty orbitals, and consequently the exchange will be ferromagnetic and weak. The resulting magnetic structure, shown in Fig. 6.10, is

Figure 6.10   Orbital and spin ordering in $KCuF_3$ (the hole orbitals of Cu^{2+} are shown).

---

[3] Note once again that the orbitals plotted in Fig. 6.10 are in this case the *d-hole orbitals*, that is the local distortion of the CuF_6 octahedra here is again local *elongation*, with the long axes perpendicular to the planes of the orbitals shown in Fig. 6.10, which alternate in the x- and y-directions, similar to the distortions shown in Fig. 6.7.

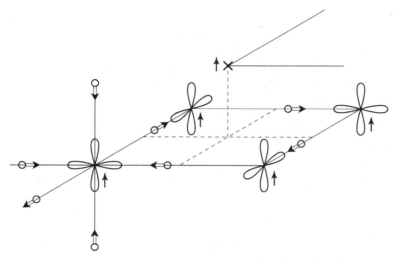

Figure 6.11 Orbital and spin structure of K_2CuF_4. The hole $x^2 - z^2$ and $y^2 - z^2$ orbitals are shown. The shifts of oxygen ions are shown by double arrows.

of A type (ferromagnetic layers stacked antiferromagnetically), but in fact the antiferro-magnetic exchange in the z (or c)-direction is much stronger than the exchange in the xy (or ab)-plane. In effect this material, which is almost cubic, magnetically is one of the best one-dimensional antiferromagnets! And this is completely caused by the corresponding orbital ordering.

Another similar material is the "layered perovskite" K_2CuF_4 (Fig. 6.11). The orbital ordering in the 2d layers here is the same as in the basal plane of $KCuF_3$ shown in Fig. 6.10.[4] Similarly, the magnetic ordering in the planes is ferromagnetic as well. Interest-ingly enough, the coupling between these planes in K_2CuF_4 turns out to be ferromagnetic as well, so that this material is a real ferromagnet. It is also a good insulator, so it presents a rare example of a transparent ferromagnet (unfortunately, magnetically ordered only at low temperatures).

Interestingly enough, this type of orbital ordering in a "layered perovskite" is not the only one possible, even for the same JT ion Cu^{2+}. The most famous case with a different orbital ordering is La_2CuO_4 – the prototype material for high-T_c superconductors. It has basically the same crystal structure as K_2CuF_4, but, in contrast to the latter, it has a ferro. orbital ordering: at every site the occupied *hole* orbitals are $x^2 - y^2$ (Fig. 6.12). The local

---

[4]   Initially one considered the orbital structure of K_2CuF_4 to be quite different, with z^2 orbitals at each site, which, for holes, would imply a compression of CuF_6 octahedra in the c-direction. The material K_2CuF_4 was even cited in some books, for example Goodenough (1963), Ballhausen (1962), as the only material in which Cu^{2+} appears not in an elongated, but in a compressed octahedron; this seems to follow from the crystal structure (the unit cell of K_2CuF_4 is indeed compressed perpendicular to the planes). However this turned out not to be the case, and the real orbital structure of this material is that shown in Fig. 6.11 (Khomskii and Kugel, 1973; Ito and Akimitsu, 1976), which means that each Cu^{2+} is still in a locally elongated F_6 octahedron, but with the long axes alternating in x- and y-directions (perpendicular to the hole orbitals shown in Fig. 6.11). As a result of this there appeared a net compression in the c-direction, which was caused not by the compression of CuO_6 octahedra, but because their long axes lie in the ab (or xy)-plane.

Figure 6.12 Orbital and spin structure of La_2CuO_4 – the prototype material for high-T_c superconductors. The hole $x^2 - y^2$ orbitals are shown.

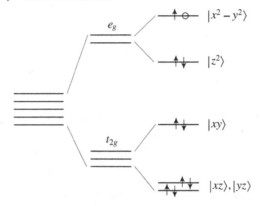

Figure 6.13 Crystal field splitting of d levels of Cu in La_2CuO_4.

distortion of CuO_6 octahedra is again an elongation, but the long axes here are all parallel, oriented in the z (or c)-direction, whereas they alternate in x- and y-directions in K_2CuF_4. Apparently this is connected to the initial distortion and corresponding splitting of e_g levels often present in such "214" structures: their apical oxygens, at least in oxides, are shifted away from the TM ions, so that locally the MO_6 octahedra are already tetragonally elongated with local $\widetilde{c/a} > 1$, even in the absence of the JT effect. Thus there occurs local elongation of NiO_6 octahedra in the c-direction in La_2NiO_4 with a non-JT ion $Ni^{2+}(t_{2g}^6 e_g^2)$. Apparently this "external force" is sufficient to push the JT distortion in La_2CuO_4 in this direction, and the JT effect then elongates the CuO_6 octahedra in the same direction still further, with a strong splitting of e_g levels and with the resulting occupation of $(x^2 - y^2)$ orbitals at each Cu site (Fig. 6.13), thus stabilizing ferro. distortion and ferro. orbital ordering in La_2CuO_4, in contrast to antiferro. ordering in K_2CuF_4.[5] Note that according to the GKA rules such orbital ordering would lead to a rather strong antiferromagnetic exchange in the ab-plane, and indeed the magnetic ordering in La_2CuO_4 is a simple two-sublattice antiferromagnetism, see Fig. 6.12.

[5] The Jahn–Teller splitting of e_g orbitals can be so large that the z^2 and xy orbitals in Fig. 6.13 may cross.

The splitting of z^2 and $(x^2 - y^2)$ levels in La_2CuO_4 and in other high-T_c cuprates is so strong that one often treats the resulting electronic structure keeping only the $(x^2 - y^2)$ band and ignoring the original orbital degeneracy – although in some theories of high-T_c superconductors one tries to use the JT degrees of freedom to explain the origin of super-conductivity in these systems. For one of the discoverers of high-T_c cuprates, K. A. Müller, the notions of JT physics actually served as a guideline for that very discovery.

Another important class of systems for which orbital ordering plays a very important role are the manganites of type $La_{1-x}A_xMnO_3$ (A = Ca, Sr), which in particular have attracted considerable attention due to the discovery that at certain doping levels (typically $0.3 \lesssim x \lesssim 0.5$) these systems exhibit the property of *colossal magnetoresistance* (CMR). We will discuss these doped compounds in more detail in Chapter 9. Here we only mention that in undoped $LaMnO_3$ (but also in some phases with $x \neq 0$) the orbital degrees of freedom, present in $Mn^{3+}(t_{2g}^3 e_g^1)$, largely determine the properties of these systems.

The material $LaMnO_3$, and many other materials of the type $RMnO_3$ (R = Pr, Nd, etc.) have a perovskite structure, and the JT nature of Mn^{3+} ions leads to a structural transition with orbital ordering, at $T_{JT} \sim 800\,K$ for $LaMnO_3$ and at higher temperatures for smaller rare earths. At a lower temperature, $T_N \sim 140\,K$, there appears a magnetic ordering in $LaMnO_3$, an A-type antiferromagnetism (ferromagnetic layers stacked antiferromagnetically). The orbital structure of $LaMnO_3$ in the ground state is usually depicted as in Fig. 6.14(a): the local distortions correspond to an alternating packing of elongated MO_6 octahedra with the long axes in the x- and y-directions (Fig. 6.7). From the point of view of the orbital angle $|\theta\rangle$ (6.1), such states would correspond to two sublattices with $\theta = \pm\frac{2}{3}\pi$.

Theoretical calculations, based both on electron–lattice coupling of the type (6.8), (6.9) and on the KK superexchange (6.14), give somewhat different orbitals: either the angles $\pm\frac{1}{3}\pi$, or $\pm\frac{1}{2}\pi$. When we take into account the local tendency toward elongated octahedra due to anharmonicity, which would rather stabilize the states $\pm\frac{2}{3}\pi$, the sublattices would

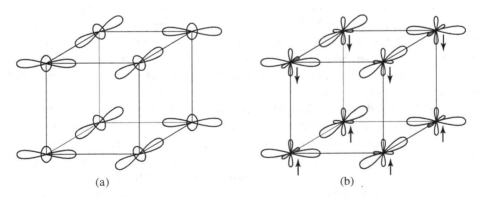

(a)   (b)

Figure 6.14 Orbital occupation in $LaMnO_3$: (a) schematic; (b) more realistic.

tend to "bend" from $\pm\frac{1}{3}\pi$ or $\pm\frac{1}{2}\pi$ in the direction of $\pm\frac{2}{3}\pi$. The resulting state turns out to be something "in between": from the lattice parameters and also from the ESR spectra one finds the orbital angles $\sim\pm105$–$110°$. The orbitals in LaMnO$_3$ are shown conventionally as in Fig. 6.14(a) (cf. also Fig. 3.23). The actual orbitals are indeed predominantly elongated in the x- or y-directions, but they have somewhat larger lobes in the z-direction and, instead, smaller lobes in the third (y or x)-direction, Fig. 6.14(b). If we were to continue this process (i.e., the "unbending", in the θ plane, of the orbital ordering of Fig. 6.14(a), which corresponds to $\theta = \pm120°$), we would end up with the "cross-shaped" orbitals such as those shown in Fig. 6.10 ($\theta = \pm60°$). LaMnO$_3$ is still far from this limit, but this modification away from the simple state of Fig. 6.14(a) helps to make the in-plane interaction more ferromagnetic, and also helps to strengthen the antiferromagnetic coupling in the z (or c)-direction, which finally gives the A-type magnetic ordering shown in Fig. 6.14(b).

Above we have discussed two main mechanisms which could lead to cooperative orbital ordering with the corresponding lattice distortion: the Jahn–Teller interaction via the lattice (Section 6.1.1) and the superexchange (or KK) interaction (Section 6.1.2). The question arises, which of these mechanisms is more important in real systems? This question is not so simple. The point is that in most real systems both these mechanisms usually lead to the same type of orbital ordering. Thus the question becomes rather a quantitative one.

In real materials of course both mechanisms are present; one cannot just turn one of them off. But we can do this in calculations. The corresponding theoretical calculations, carried out using *ab initio* methods, did show that in typical cases both mechanisms are at work and are approximately equally efficient, the electron–lattice (Jahn–Teller) interaction being perhaps somewhat stronger. The most detailed such calculations have demonstrated this for KCuF$_3$ and LaMnO$_3$ (Pavarini *et al.*, 2008; Pavarini and Koch, 2010). Thus in general for a proper description of these effects one should include both mechanisms, but, in contrast, one can get some qualitative understanding keeping only one of them; usually the qualitative features of the solution are reproduced in such a simplified approach reasonably well.

An important point is that due to the fact that orbital ordering can be driven by both superexchange (KK) and the Jahn–Teller interaction (whereas magnetic ordering is only due to the exchange interaction), orbital ordering usually occurs at higher temperatures than magnetic ordering. Sometimes this is taken as proof of the dominant role of Jahn–Teller interactions. However, one should notice that purely electronic interactions, such as those of eqs (6.15), (6.16), can lead to the same consequence (Kugel and Khomskii, 1973). Often different spin and orbital patterns (e.g., ferro. spin / antiferro. orbital and antiferro. spin / ferro. orbital) have, in the main approximation, that is in the absence of Hund's rule coupling J_H in (6.16), the same energy, and this degeneracy is lifted by the terms in the Hamiltonian $\sim J_H/U$. In effect, orbital ordering occurs at $T_{O.O.} \sim t^2/U$, see eqs (6.15), (6.16), but magnetic (spin) ordering occurs at much lower temperatures $T_N \sim (t^2/U)(J_H/U)$ – typically ~ 5 times lower. Still, in real situations of course the electron–lattice (Jahn–Teller) interaction definitely plays a very important role.

6.2 Reduction of dimensionality due to orbital ordering

As we have seen above, a very characteristic feature of orbital ordering is the *directional character* of orbitals; this strongly differentiates orbital ordering, and its mathematical description, from spin ordering. One very specific feature of orbital ordering is that due to this directional character it can effectively reduce the dimensionality of the electronic subsystem, making for example a three-dimensional system electronically two- or even one-dimensional.

The simplest way to see this is for example from Fig. 6.12: if we occupy the $(x^2 - y^2)$ orbitals at each site of the lattice, the electrons can hop in x- and y-directions, but not in the z-direction. Thus if we were to form such orbital ordering in a three-dimensional system, the electronic structure would become essentially two-dimensional. We have already seen a similar example in $KCuF_3$ (Fig. 6.10): as discussed there, due to a particular orbital ordering this practically cubic 3d crystal becomes magnetically very one-dimensional.

There are also other examples of such behavior. An interesting example is provided by spinels with t_{2g}-electrons on B sites, such as $MgTi_2O_4$, see Fig. 6.15. One can visualize this B sublattice as consisting of chains running in (xy)-, (xz)-, and (yz)-directions. Three t_{2g} orbitals have their lobes along these chains: the xy orbital along the xy chain, the xz orbital

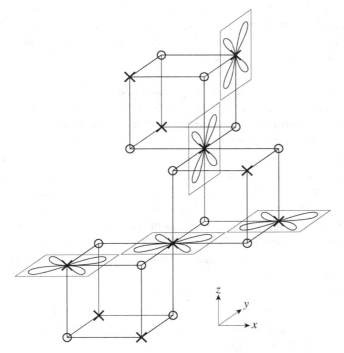

Figure 6.15 Formation of one-dimensional chains for electron hopping in the B sites of a spinel AB_2O_4, for t_{2g} orbitals with direct d–d hopping. Here crosses are transition metal ions and circles are oxygens.

along the xz chain, and the yz orbital along the yz chain, see Fig. 6.15. Consequently, if the main d–d hopping should occur due to direct d–d overlap, which is apparently the case for early 3d metals such as Ti, V, or Cr, then an electron placed for example on the xy orbital will only hop to a similar xy orbital in the corresponding xy chain, that is it will become essentially one-dimensional (and similarly there will appear one-dimensional xz and yz bands).

As is well known, one-dimensional systems are always unstable, for example with respect to Peierls distortions: Peierls dimerization for one electron per site, or tetrameriza-tion for $\frac{1}{2}$ electrons per site, etc., see Khomskii (2010). As a result there appear in such 1d systems singlet dimers, resembling valence bond states discussed in Section 5.7.2. Appar-ently this is what happens in $MgTi_2O_4$: the effective one-dimensionality of the spectrum, caused by the corresponding orbital ordering, leads in effect to a Peierls-like transition with the formation of a spin singlets and with the opening of spin gap. Thus in this case orbital ordering induces Peierls distortion (Khomskii and Mizokawa, 2005).

Orbital ordering can also strongly influence metal–insulator transitions in TM com-pounds. These transitions and their description will be discussed in detail in Chapter 10. Here we only mention the role played by orbitals. The clearest example of the influence of orbital ordering on the metal–insulator transition, also connected with the reduction of dimensionality, is provided by VO_2. In VO_2, which has a rutile structure, there occurs a very strong metal–insulator transition at $T_c = 68°C$, with a jump in resistivity up to 10^4. Structural modifications accompanying this transition consist of dimerization (and also tilt-ing – the so-called "twisting" of dimers) in the V chains running in the c-direction, see the schematic picture in Fig. 6.16.

For a long time there was active discussion in the literature whether this transition is a Mott transition caused by electronic correlations, or whether it is predominantly a lattice-driven Peierls-like transition with dimerization of 1d chains, as shown in Fig. 6.16. Most experiments now demonstrate that Mott physics is indeed relevant for VO_2. But what hap-pens then with the structure, how can we explain the observed changes shown in Fig. 6.16 and why do they occur? The point is that above T_c the electronic properties of VO_2, for example conductivity, are reasonably isotropic, not at all one-dimensional, which is required to explain the pattern shown in Fig. 6.16.

The answer, found in spectroscopic experiments (Haverkort *et al.*, 2005), is that, indeed, above T_c the orbital occupation of $V^{4+}(t_{2g}^1)$ in VO_2 is such that the electronic structure is rather isotropic (all three t_{2g} orbitals are populated more or less equally). However, at the first-order metal–insulator transition at 68°C, simultaneously with the change of structure and conductivity, the orbital occupation also changes drastically: below T_c predominantly the orbitals with lobes in the c-direction (in the direction of V chains in Fig. 6.16, see also Fig. 10.8) become occupied (and the "twisting" of dimers – actually antiferroelectric-like shifts of V ions toward the oxygens – turns out to be very important as well; Goode-nough, 1971). In effect the electronic structure becomes one-dimensional, and because of that Peierls dimerization appears in these 1d chains, which is seen in experiments. (See also the related discussion in Section 10.2.2.)

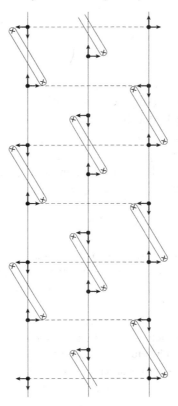

Figure 6.16 Schematic form of distortions in the insulating phase of VO_2, with the formation of V–V singlet dimers and with corresponding tilting of dimers ("twisting").

These phenomena in fact occur not consecutively, but all together, as one single first-order transition, and they enhance one another. In effect the metal–insulator transition in VO_2 has a "combined" nature: it occurs in the presence of strong electron correlations, that is it has many features of Mott transitions, but the lattice distortions also play a very important role. For the latter, the orbital reorientation turns out to be crucial.

Thus, as we have seen above, after a particular orbital ordering is established, due to the directional character of orbitals the electronic structure of the system sometimes becomes essentially one-dimensional, and because of the Peierls distortion there may appear dimers in the system – singlet dimers for one d-electron per site, but sometimes also "ferromagnetic" dimers for ions with larger spins. The examples of the first type, besides the already mentioned VO_2 and $MgTi_2O_4$, include TiOCl (Blanco-Canosa *et al.*, 2009) and some other systems. For some geometries, for example systems with zigzag chains, the orbital ordering by itself can give singlets, even without extra lattice distortions (which, though, also occur but do not play such a crucial role). This is for example the situation in the pyroxene $NaTiSi_2O_6$ (Isobe *et al.*, 2002; Streltsov *et al.*, 2006) or in $La_4Ru_2O_{10}$ (Khalifah *et al.*,

2002; Wu, 2006). But for ionic spins $S > \frac{1}{2}$ the corresponding dimers may turn out to be ferromagnetic due to double exchange; this is presumably what happens in ZnV_2O_4 (Pardo *et al.*, 2008).

Moreover, in some situations orbital ordering can lead to the formation not of singlet *dimers*, but of similar *trimers*, or of more complicated clusters. Thus, in layered materials $LiVO_2$ (Pen *et al.*, 1997), $LiVS_2$ (Katayama *et al.*, 2009), TiI_2 (Meyer *et al.*, 2009) with two-dimensional triangular lattices and with ions with d^2 configurations ($V^{3+}(t_{2g}^2)$, $Ti^{2+}(t_{2g}^2)$) there occurs orbital ordering with three sublattices (Pen *et al.*, 1997), see Fig. 6.17, after which there appear in the system strongly coupled trimers (shaded triangles in Fig. 6.17), in which due to the orbital ordering there is a very strong antiferromagnetic interaction of the three sites in a triangle, the interaction between these triangles being, according to the GKA rules, much weaker (and presumably ferromagnetic). As a result these three sites with $S = 1$ each, with strong antiferromagnetic exchange between them, have a singlet ground state with $S_{tot} = 0$ (so to say "$1 + 1 + 1 = 0$"), and the magnetic susceptibility of these systems shows a dramatic downturn below the respective orbital ordering transitions (in $LiVS_2$ it is simultaneously also a metal–insulator transition), and the systems actually become diamagnetic (maybe still with some Van Vleck paramagnetism).

Larger singlet clusters due to orbital ordering appear for example in CaV_4O_9, where they are singlet squares (Taniguchi *et al.*, 1995; Starykh *et al.*, 1996) or in AlV_2O_4, where they are singlet heptamers (singlet clusters of seven V sites) (Horibe *et al.*, 2006). The same orbitally induced exchange anisotropy apparently makes the pyrochlore $Tl_2Ru_2O_7$ magnetically one-dimensional, as a result of which there appears a Haldane gap[6] in such a 1d chain

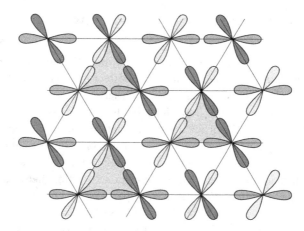

Figure 6.17 Orbital ordering in $LiVO_2$, leading to the formation of tightly bound V_3 triangles (trimers, shaded) with the singlet ground state (after Pen *et al.*, 1997).

---

[6] The Haldane gap is a gap in the spin excitation spectrum, present in one-dimensional magnetic chains with even spins (e.g., $S = 1$) with Heisenberg interaction (Haldane, 1983).

of spin-1 Ru^{4+} ions (Lee *et al.*, 2006). Thus we see that the directional character of orbitals can lead to quite nontrivial magnetic structures, including, in particular, singlet states.

There are other examples of similar physics – of reduced dimensionality, appearance of spin gaps, spin–orbit entanglement, and modification of metal–insulator transitions due to orbital ordering. One can find more examples in the reviews, for example Khomskii (2005) or Oles (2012).

6.3 Orbitals and frustration

Similar to magnetic ordering, the detailed features of orbital ordering depend strongly on the type of underlying lattice. In particular, we have seen in Section 5.7 that very specific phenomena are met in magnetic systems with geometric frustrations. Similarly, one can expect such complications also for orbital ordering in such systems. Sometimes this is indeed the case. But often, due to the specific nature of orbitals, the effective orbital–orbital interaction (orbital "exchange" of type (6.8) or (6.14), though often quite anisotropic) may turn out to be "ferromagnetic," in which case the nominally frustrated nature of the lattice does not play a role. This is for example the case in $NaNiO_2$ with a triangular lattice (low-spin $Ni^{3+}(t_{2g}^6 e_g^1)$ has one electron on a doubly degenerate e_g level) or in the spinel Mn_3O_4, where Mn^{3+} ions with configuration $t_{2g}^3 e_g^1$ occupy the B sublattice of corner-sharing tetrahedra (the pyrochlore-type sublattice). In both these cases the effective orbital interaction is "ferromagnetic" and there is ferro. orbital ordering (the case of $LiNiO_2$, which in principle should behave similarly to $NaNiO_2$, is special and will be discussed later, in Section 6.6.)

In contrast, the very specific features of orbitals, notably their spatial anisotropy, stressed in the previous section, can lead to the appearance of frustrations in apparently simple lattices such as square or cubic ones, see for example Mostovoy and Khomskii (2003). Consider for example doubly degenerate e_g-electrons, and assume that due to strong anharmonicity (Section 3.2), only "elongated" orbitals of the type $|z^2\rangle$, $|x^2\rangle$, or $|y^2\rangle$ are allowed. Thus, consider for example the square lattice of Fig. 6.18 with "ferromagnetic" orbital coupling, so that neighboring orbitals want to point toward one another. Then for a pair of sites along the z-direction (like the pair of sites (a, b) in Fig. 6.18), both sites should have $|z^2\rangle$ orbitals occupied. But for a similar pair along the x-direction the corresponding orbitals should be $|x^2\rangle$ (sites (c, d) in Fig. 6.18). As a result, the situation on the bond (ac) is "wrong": the corresponding orbitals do not point toward each other but are orthogonal. In effect this seemingly simple system turns out to be strongly frustrated, and the resulting type of orbital ordering is not clear at all.

The model describing this situation was introduced by Kugel and Khomskii (1973) and called the "compass model." It describes an Ising-like interaction of (pseudo)spins, but with different spin components for bonds in different directions:

$$\mathcal{H}_{\text{comp.}} = J \left(\sum_{\langle ij \rangle \| z} \tau_i^z \tau_j^z + \sum_{\langle ij \rangle \| x} \tau_i^x \tau_j^x + \sum_{\langle ij \rangle \| y} \tau_i^y \tau_j^y \right). \tag{6.17}$$

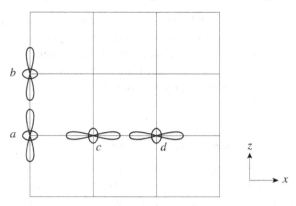

Figure 6.18 Illustration of frustration caused by the directional nature of orbitals even in a simple square or cubic lattice.

It indeed resembles the dipole interaction of needles in magnetic compasses: these want to order "head-to-tail," but the orientation of these needles would depend on the relative direction between the "compasses." The resulting situation would be simple in the one-dimensional case (e.g., the chain of "compasses" along the z-direction would order so that all "needles" (all τ_i in (6.17)) would point say up, along the z-direction). But similar chains in the x-direction would have all the "compasses" pointing, say, to the right. Such compasses forming for example a square or cubic lattice "would not know" which direction to point in. Thus, this seemingly simple model (6.17), which for each pair of sites looks trivial (an Ising interaction) and which naturally emerges in orbital physics, in 2d and 3d cases turns out to be quite nontrivial, even in the classical limit. A very detailed discussion of this and a related model is given in Nussinov (2013).

Recently one version of this model, on a two-dimensional honeycomb lattice, was solved exactly (Kitaev, 2006) and it was shown that the solution has quite nontrivial features, which, in particular, can be used for topologically protected quantum computing. A possible experimental realization of this situation could be provided by the material Na_2IrO_3, in which the spin and orbital degrees of freedom in the presence of strong spin–orbit coupling can be described by a model of type (6.17) or its generalization (Jackeli and Khaliullin, 2009).

A similar situation, in which the effective frustrations are caused by the directional character of orbitals in nominally simple lattices, also occurs in other cases, for example for t_{2g}-electrons. As already mentioned in Section 6.2, for t_{2g}-electrons in perovskite lattices only two of three t_{2g} orbitals interact in each direction, for example xz and yz orbitals for a pair of sites along the z-direction, see Fig. 6.20 below. But for other directions, other pairs of orbitals are active (e.g., xy and xz orbitals for sites in the x-direction). We see that the resulting situation resembles somewhat that of the model (6.17), with different operators entering the expression for the orbital interaction in different directions. One can show that there exist a lot of degeneracies in this situation (Harris *et al.*, 2003), and again the type of ground state is far from trivial (see also Sections 6.5, 6.6).

The directional character of orbitals and the resulting frustrating interactions can also strongly influence the properties of dilute Jahn–Teller systems – materials in which orbitally degenerate JT ions are not at each site of a regular lattice, but are present as random impurities in a non-JT matrix. As discussed in Section 6.1.1, the interaction of such centers via the lattice (via strain) is long-range, but with the sign depending on the direction between the centers and on the orientation of respective orbitals. In effect one can get a situation resembling that of spin glasses – an orbital glass state (Ivanov *et al.*, 1983). The physical properties of such states can also be quite special.

6.4 Orbital excitations

As always when we are dealing with certain degrees of freedom, there may exist collective excitations related to them. For magnetic ordering caused by the exchange interaction of type (1.12) these are spin waves, or magnons. In orbital systems the ordering is characterized by pseudospins τ (6.3), (6.4) instead of spins S, with nonzero averages $\langle \tau \rangle \neq 0$ in orbitally ordered states. Consequently, we can also think of special excitations – orbital waves, or orbitons.

Formally, it may look as if the situation with orbitals is exactly equivalent to that with spins: for the doubly degenerate case they are described by (pseudo)spins $\frac{1}{2}$, with effective exchange interaction of type (6.8). This interaction may be anisotropic, and orbital excitations may be coupled to spin excitations, cf. the interaction (6.14); but for magnetic systems there may also exist different types of anisotropy, as discussed in Section 5.3. Thus in general one may expect the existence of orbital excitations.

However, there are extra factors which make the situation with orbital excitations different from that with spin waves. One is the intrinsically strong coupling of orbitals with the lattice. In fact, orbital excitations may be visualized as transitions between "formerly degenerate" *d* levels, which are split below the Jahn–Teller (or orbital) ordering temperature by the orbital "molecular field"; a large contribution to this splitting is provided by the corresponding lattice distortion, accompanying (or causing) the cooperative JT transition – for example strong tetragonal elongation in the case of ferro. distortion, as in Mn_3O_4. The lattice is "very inert," the electronic transitions such as orbital excitations occur predominantly in a fixed lattice (this is analogous to the Frank–Condon optical transitions). In fact, orbital excitations look like crystal field excitations – transitions between different crystal field levels,[7] in this case those split by the cooperative JT effect. In this sense the situation with orbitons may be very different from that with magnons, which in a first approximation can be considered as excitations only in the magnetic subsystem, with usually only a relatively weak coupling to the lattice.

---

[7] Crystal field excitations always exist in TM ions, both in impurities and in concentrated systems. The optical study of such excitations is a standard tool used in inorganic chemistry to characterize such materials; it is known there as *ligand field spectroscopy*. The difference of orbitons as real excitations in crystals with orbital ordering from these ligand field excitations is that in a concentrated system these excitations should propagate through the crystal, that is they should acquire a certain dispersion. Thus to detect orbitons one really needs not only to find the corresponding excitations, for example in optical absorption, but also to measure their dispersion. To do this by optical means, as is most often done, is usually quite difficult.

Thus, in contrast to spins, orbitals are always quite strongly coupled to the lattice (JT coupling). In effect the excitations in these two subsystems, orbital and lattice, should intrinsically be rather strongly mixed. Therefore it may be difficult to disentangle orbital excitations from phonons. Experimentalists have already come across this problem: one of the first observations of orbitons in manganites (Saitoh *et al.*, 2001) was later challenged on the basis of results which seemed to point rather in the direction of phonons (Grüninger *et al.*, 2002).

In principle, one possibility to separate the orbital contributions from the mixed excitations may come from the spin–orbital coupling given by the KK interaction (6.14): one can think that for example by applying a magnetic field we could influence the spin subsystem, and then through the coupling (6.14) also the orbital one. If orbital degrees of freedom contribute to some excitations (which originally could be interpreted simply as phonons), the potential sensitivity of orbitals to the magnetic field would be transferred to the sensitivity of the corresponding "phonons" to the field. As the conventional phonons should not depend on the magnetic field, the observation that some phonon modes do depend on the external field could imply that there is a large contribution of orbital excitations in the corresponding mode.

As to orbiton–phonon coupling, the situation could be clearer in t_{2g} systems: the t_{2g}-electrons have a much weaker interaction with the lattice. However, for t_{2g} systems we have another complication – in this case the spin–orbit interaction may not be quenched, and it can also influence the orbital excitations, see the next section.

In any case, the story of orbital excitations is definitely very interesting, but there are as yet not many solid results in this field.

6.5 Orbital effects for t_{2g}-electrons

Until now in this chapter we have discussed mainly the case of doubly degenerate e_g levels. It is in this case that we expect strong JT coupling and large effects in all properties. But, as already mentioned in Section 3.2, orbital degeneracy may also exist for the triply degenerate t_{2g} levels. If we have a regular MO_6 octahedron with a cubic crystal field, we will have such degeneracy for quite a few TM ions: $Ti^{3+}(t_{2g}^1)$ and $V^{4+}(t_{2g}^1)$, $V^{3+}(t_{2g}^2)$ and $Cr^{4+}(t_{2g}^2)$, but also formally for $Fe^{2+}(t_{2g}^4 e_g^2)$, $Co^{2+}(t_{2g}^5 e_g^2)$, etc. There are three important factors which make the situation with the triple t_{2g} degeneracy different from that in the double e_g-degenerate case. The first is that the JT coupling of t_{2g}-electrons to the lattice is weaker than that for e_g-electrons; this is due to the fact that the lobes of t_{2g} orbitals point not toward the ligands (e.g., oxygens) but between them, see Fig. 3.3. Consequently, if the JT splitting of e_g-electrons can reach ~ 0.8–1 eV, for the t_{2g} case it is typically ~ 0.2–0.3 eV. The weaker coupling of t_{2g}-electrons to the lattice may be, in particular, beneficial for the observation of orbitons in such systems – see Section 6.4; it can also lead to stronger quantum effects (Section 6.6).

The second point is that, as discussed in Section 3.1, see Fig. 3.11, t_{2g} levels can be split not only by E_g distortions (tetragonal or orthorhombic) but also by T_{2g} ones (triply degenerate trigonal distortions, such as elongation or contraction of MO_6 octahedra along one of four possible [111] axes). As a result, the JT distortion for t_{2g} systems can lead both to tetragonal and trigonal symmetry.

Whether the distortion occurring in a given material with t_{2g} degeneracy will be tetragonal or trigonal is probably determined by which coupling – to the E_g mode or the T_{2g} mode – is stronger. But which will be stronger in which case, and why, is not always clear. An interesting empirical observation, already mentioned in Section 5.4, is that for typical materials of this type those with $Co^{2+}(t_{2g}^5 e_g^2)$, such as CoO or KCoF$_3$, usually undergo tetragonal distortions, with the spin-easy axis in, say, the [001] direction, whereas those with $Fe^{2+}(t_{2g}^4 e_g^2)$ (e.g., FeO, KFeF$_3$) typically have trigonal distortions with spins parallel to the [111] axis. How general this tendency is, and what causes it, is still an open question.

Probably the most important difference of t_{2g} levels, stressed already many times above, is that, in contrast to e_g levels, for t_{2g} levels the real relativistic spin–orbit coupling $\lambda \boldsymbol{L} \cdot \boldsymbol{S}$ is not quenched, but acts in the lowest order. We have discussed the consequences of this in much detail in Section 5.4; here we will, for completeness, repeat briefly the results of that treatment.

It turns out that for t_{2g} systems the spin–orbit coupling in a sense competes with the JT effect. As discussed in Section 3.4, spin–orbit coupling stabilizes complex orbitals with effective orbital moment $\tilde{\boldsymbol{l}}_{\text{eff}}$ parallel or antiparallel to the spin $\boldsymbol{S}$. For spins along the z-axis $S^z \neq 0$, and we have $\tilde{l}^z = \pm 1$. The corresponding wavefunctions are

$$|\tilde{l}^z = \pm 1\rangle \sim \frac{1}{\sqrt{2}}\big(|xz\rangle \pm i|yz\rangle\big) , \qquad (6.18)$$

and the shape of the wavefunction is that of an xz orbital rotated around the z-axis, that is a hollow cone, see Fig. 3.4. If we have one electron on such an orbital, it will "push away" the apical oxygens, so there will appear a tetragonal elongation, with $c/a > 1$. The crystal field-level scheme in this case will look as shown in Fig. 6.19(a). We see that for this distortion without the spin–orbit splitting we would have a degenerate doublet ($|xz\rangle$, $|yz\rangle$), or ($|l^z = \pm 1\rangle$) as the lowest state, with one electron on this doublet. Spin–orbit coupling splits this doublet further, creating a nondegenerate ground state. The energy of this state is

$$E_{(a)} = -\frac{E_{\text{JT}}}{2} - \frac{\lambda}{2} . \qquad (6.19)$$

However, if we would not have any spin–orbit coupling, then the triply degenerate states with one electron would rather split as shown in Fig. 6.19(b), that is the xy singlet would go down in energy and would be occupied by this one electron. This state would then be stabilized by the JT interaction, the corresponding distortion would not be a tetragonal elongation but a compression $c/a < 1$, and the energy gain would be

$$E_{(b)} = -E_{\text{JT}} . \qquad (6.20)$$

Comparing the energies $E_{(a)}$ and $E_{(b)}$, we see that the system would evolve according to the "JT scenario" of Fig. 6.19(b) if $E_{\text{JT}} > \lambda$; if however the spin–orbit interaction is strong

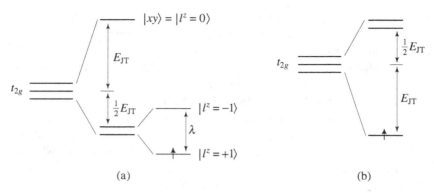

Figure 6.19 Competition between Jahn–Teller-driven and spin–orbit-driven distortions for the t_{2g}^1 configuration, for (a) tetragonal elongation and (b) tetragonal compression of MO_6 octahedra.

enough, $\lambda > E_{JT}$, then the spin–orbit energy will dominate, and the system would evolve according to the "spin–orbit scenario." The corresponding distortion will be opposite to what one would expect for the JT effect; it would occur simultaneously with magnetic ordering and would take the form of giant magnetostriction.

As the t_{2g} orbitals can be split both by tetragonal and trigonal distortions, the resulting distortions, which follow the spin direction (and maybe even determine it) can also be trigonal, with the same competition between spin–orbit and JT coupling and with very similar consequences.

As mentioned above, spin–orbit coupling strongly increases with the atomic number Z, and spin–orbit interaction is usually less important for the early $3d$ metals such as Ti, V but starts to dominate for heavier $3d$ elements (Fe, Co). Of course it plays a very important role for $4d$ and $5d$ elements.

This indeed agrees with the experimental observations: the eventual distortions at the beginning of the $3d$ series, for example in Ti or V compounds, are usually determined by the Jahn–Teller effect, whereas those of Co or Fe systems are determined by the spin–orbit coupling.

Concluding this section, we should stress once again that, as discussed above, the presence of the t_{2g} degeneracy, with its strong spin–orbit interaction, leads to a strong coupling of lattice distortions with spin ordering, in particular with spin orientation. Consequently, there usually exists a very strong magnetoelastic coupling and strong magnetostriction in such materials – typically it is one to two orders of magnitude stronger in systems with Co^{2+} compared with those without t_{2g} degeneracy (e.g., containing $Fe^{3+}(t_{2g}^3 e_g^2)$, $Mn^{2+}(t_{2g}^3 e_g^2)$, or $Ni^{2+}(t_{2g}^6 e_g^2)$). The magnetic anisotropy for t_{2g}-degenerate systems is also much stronger – this in fact is due to the same strong coupling with the lattice via the spin–orbit interaction. Thus these features, well known in the magnetism of cobaltites for example, are in fact a consequence of the original orbital degeneracy. We may say that they are caused by the "hidden JT effect": hidden because with strong spin–orbit coupling the degeneracy is already lifted by this interaction, and in such cases the distortions are of "anti-JT" type, see above. Still, all these effects – very strong magnetostriction, strong magnetoelastic coupling, and magnetic anisotropy – are in these cases intrinsically due to the original t_{2g} degeneracy.

6.6 Quantum effects in orbitals

Until now, when discussing orbital ordering, we always had in mind a quasiclassical picture, treating the orbitals and corresponding lattice distortions classically, or in a mean-field approximation. In contrast, we have mapped the description of the orbital subsystem for e_g orbitals onto an effective pseudospin-$\frac{1}{2}$ system. From the treatment of the corresponding spin systems we know that spins are quantum objects, and quantum effects are especially strong for spin $\frac{1}{2}$ (and especially for low-dimensional systems). One may suspect that this is also the case here, with pseudospin $\frac{1}{2}$ and with the directional character of orbitals often reducing the effective dimensionality of the system, see Section 6.2.

There are factors opposing that. One of them was mentioned in Section 6.4: the coupling to the lattice, always present in such systems, may suppress quantum effects. Indeed, especially for large systems we always treat the lattice classically.[8] But one may expect that at least for t_{2g} systems the electron–phonon interaction is weak enough so that it will not suppress eventual quantum effects. And indeed there were interesting suggestions that quantum effects may be important in some t_{2g} systems (Khaliullin and Maekawa, 2000) and some experiments, for example in LaTiO$_3$, YVO$_3$, and similar materials were interpreted from this point of view, see Khaliullin (2005). For some frustrated materials, such as LiNiO$_2$, one also notes the possible existence of orbital or rather spin-orbital liquid (Reynaud *et al.*, 2001).

One can think, by analogy with the conventional spins, of various quantum effects which may exist for orbitals. For spin systems there are, for example, the formation of spin singlets, valence bonds localized at particular bonds in crystals – we can call them valence bond crystals. Peierls or spin-Peierls transitions leading for example to dimerization of 1d chains, with singlets at every other bond, is one example of such states. Possible states at magnetization plateaux in frustrated magnets, such as those at the $\frac{1}{3}$ plateau in Kagome lattices (Zhitomirsky and Tsunetsugu, 2005) (cf. Fig. 5.42), are another example.

More general states stabilized by quantum effects in spin systems are the different types of spin liquids, discussed for example in Section 5.7. The resonating valence bond (RVB) states are typical examples thereof. Experience in spin systems shows that in low-dimensional or in frustrated systems one can expect the formation of much more exotic quantum states than the conventional long-range ordering.

---

[8] Note that, as discussed in Section 3.2, for isolated Jahn–Teller ions or for JT molecules one may need to treat not only electrons, but also the lattice vibrations quantum-mechanically, which constitutes the well-developed field of vibronic physics. However in concentrated systems, such as JT solids, we practically always ignore these effects, although exactly when we are allowed to do so and what would be the consequences of including the vibronic effects in concentrated solids are still open and interesting questions.

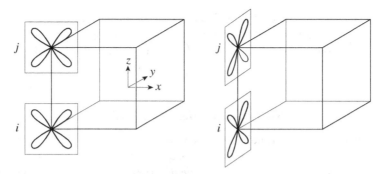

Figure 6.20 Two active t_{2g} orbitals for a transition metal pair in the c-direction.

Suggestions about possible manifestations of similar quantum effects for orbitals were also put forward for similar situations. Here, low dimensionality can follow not only from low-dimensional crystal structure but also, as explained in Section 6.2, from the directional nature of orbitals. This can also reduce the effective degeneracy. Thus, if we consider triply degenerate t_{2g} levels, we see that even in a cubic lattice only two orbitals, xz and yz, overlap in the z-direction, see Fig. 6.20. Thus for a pair $\{ij\}$ oriented along the z-direction we can keep only these two orbitals, and for this pair the effective model becomes equivalent to that of Section 6.1.2 for the doubly degenerate electrons, which here even takes a simple symmetric form with the hoppings $t^{11} = t^{22} = t$, $t^{12} = 0$ (with $1 \Longleftrightarrow xz$, $2 \Longleftrightarrow yz$), as is clear from the figure. As discussed, this symmetric model gives for $t \ll U$ the effective spin–orbit (KK) model (6.15), which for $J_H = 0$ has a very high symmetry, SU(4), and even for $J_H \neq 0$ this is a "double Heisenberg model," SU(2)$\times$SU(2). Consequently, we can expect very strong quantum fluctuations in this case (at least if we ignore coupling to the lattice).

In applications to real systems such as LaTiO$_3$ or YVO$_3$ with perovskite lattices the situation is more complicated, since one has to consider also Ti or V pairs in other directions, for which other pairs of orbitals would enter. Nevertheless quantum effects may still be present here, and they were invoked to explain certain experimental features of these systems. Thus for example the opening of a gap in the spin-wave spectrum in the c-direction in YVO$_3$ was explained by the formation of "orbital singlets" in the c chains (Ulrich *et al.*, 2002), very much like spin singlets in spin-Peierls systems.

Another system for which a possibly important role of quantum effects has been proposed is the quasi-two-dimensional material with triangular lattice LiNiO$_2$ (Reynaud *et al.*, 2001). The structure of this and similar systems is shown in Fig. 5.39. The Ni^{3+} ions in this material are in the low-spin state, $t_{2g}^6 e_g^1$, that is they have spin $\frac{1}{2}$ and double orbital degeneracy, or pseudospin $\tau = \frac{1}{2}$. As discussed above, for e_g-electrons one expects a strong coupling to the lattice, which should make the behavior of orbitals classical and lead to a long-range orbital ordering, with a structural phase transition. However, this system has a triangular lattice, which potentially may cause frustrations (this depends also on the details of the interactions: e.g. for ferro. exchange there will be, of course, no frustrations).

These expectations of "classical" behavior are indeed realized in a material with the same crystal structure, $NaNiO_2$: this material shows a structural transition at $T_{str} = 400$ K from a rhombohedral to a monoclinic structure, with all occupied e_g orbitals being the same (ferro. orbital ordering), and at a lower temperature $T_N = 20$ K it has the standard magnetic ordering – A-type antiferromagnetism, with 2d triangular layers ordered ferromagnetically, neighboring layers being oriented antiparallel (Chappel *et al.*, 2000).

But $LiNiO_2$ turns out to be a "bad actor." It does not show any structural phase transition down to the lowest temperatures, that is it does not experience cooperative orbital ordering, and it also does not have the usual magnetic ordering. Only at very low temperatures, ~ 8 K, do some samples of $LiNiO_2$ show a spin-glass transition, apparently connected with some disorder in the system: either some nonstoichiometry, or a certain interchange of Li and Ni, after which the magnetic Ni ions appear between the magnetic sublattices, which gives a strong frustrating interlayer interaction. But why $LiNiO_2$ itself shows neither the normal magnetic ordering nor orbital ordering remains a puzzle. Ideas have been proposed that maybe it is a rare example of a combined spin and orbital liquid (Reynaud *et al.*, 2001) (remember that the first candidate for the RVB state was a triangular system with $S = \frac{1}{2}$ (Anderson, 1973), and this system is just that). But, once again, for orbitals, despite a formally similar situation ($\tau = \frac{1}{2}$), the interactions are usually very anisotropic and often long-range, which should suppress quantum effects at least in the orbital sector. Also, the system $NaNiO_2$ is very similar and behaves "quite decently," with long-range order both in spins and orbitals. Thus what happens in $LiNiO_2$ is still a puzzle (Mostovoy and Khomskii, 2002). The seemingly unavoidable interchange of Li and Ni ions leading to magnetic frustrations can explain the absence of the ordinary magnetic ordering, but it is hardly sufficient to suppress the expected orbital ordering (the interaction of orbitals via the lattice is long-range and not very sensitive to weak disorder). Further experiments are needed to clarify the situation in $LiNiO_2$.

S.6 Summary of Chapter 6

As we have seen in previous chapters, orbital degrees of freedom play a very important role in the physics of TM compounds. Especially strong effects are met in symmetric situations, for example for regular MO_6 octahedra, where there exists an orbital degeneracy. It may be a double degeneracy for partially filled e_g levels (high-spin ions $Mn^{3+}(t_{2g}^3 e_g^1)$, $Cr^{2+}(t_{2g}^3 e_g^1)$, $Cu^{2+}(t_{2g}^6 e_g^3)$; low-spin $Ni^{3+}(t_{2g}^6 e_g^1)$) or a triple degeneracy for partially filled t_{2g} levels ($V^{4+}(t_{2g}^1)$, $V^{3+}(t_{2g}^2)$, $Co^{2+}(t_{2g}^5 e_g^2)$, etc.). As discussed in Chapter 3, such situations are unstable with respect to a decrease of symmetry due to lattice distortions, which would split the corresponding levels and lift the orbital degeneracy; this is the essence of the Jahn–Teller (JT) effect. Whereas for isolated JT ions this results in a modification of the properties of the corresponding centers (the vibronic effects, see Section 3.2), in concentrated systems with such ions this usually leads to a cooperative effect (cooperative Jahn–Teller effect, CJTE or orbital ordering, O.O.) – a structural phase transition with a reduction of symmetry, which leads to a static splitting of the degenerate crystal field levels, after which a particular orbital state is occupied at each site (that is orbital ordering occurs).

Such cooperative phenomena usually take the form of a real phase transition, often of II order, but sometimes I order. This is a rare case of structural phase transitions for which the microscopic nature is actually known: given the chemical formula of an (insulating) compound, one can check whether there would be an orbital degeneracy in a high-symmetry crystal structure, and if so, one can conclude that there should occur in such a system a structural phase transition with a lowering of symmetry. There are not many other solids for which such predictions are possible and for which the real microscopic origin of structural transitions is known.

There are two main mechanisms of CJTE and O.O. in TM compounds with orbital degeneracy. The first is just the JT interaction of the degenerate electrons with the lattice distortions. For concentrated systems such distortions around each JT center, dictated by the JT theorem, are not independent, but are coupled for example due to the presence of common oxygens belonging to two MO_6 octahedra; consequently, the distortion of one such octahedron influences possible distortions of neighboring ones. As a result there appears an interaction between such distortions, or between the corresponding orbital occupations, and the process takes a cooperative character. In fact, one does not even need to have common oxygens for two JT centers to give an interaction between them: in a lattice a local distortion around one center causes a long-range strain, which is "felt" by other JT sites, even if they are not nearest neighbors of the first one.

In any case, there appears an effective coupling between distortions and the corresponding orbital states of different ions, which leads to a CJTE and to O.O. A convenient way to describe this is to characterize the doubly degenerate e_g states by a special quantum number, which is similar to spin $\frac{1}{2}$ for the electronic spin. Such *pseudospin* $\tau = \frac{1}{2}$ can describe the orbital state of an ion: $\tau^z = +\frac{1}{2}$ would correspond for example to the orbital $|3z^2 - r^2\rangle \equiv |z^2\rangle$, and $\tau^z = -\frac{1}{2}$ to $|x^2 - y^2\rangle$. In effect the interaction of different JT centers (in this case via the lattice, or via phonons) can be written as some form of pseudo-spin exchange, $\sum_{ij} J_{ij} \tau_i \tau_j$ (which, generally speaking, can be long range and anisotropic). Thus we can treat orbital ordering by analogy with spin ordering, and depending on the particular situation we can have for example a ferro. orbital ordering (all $\langle \tau_i \rangle$ are the same, i.e. the same orbital is occupied at each site) or an antiferro. orbital ordering (alternation of different orbitals), or more complicated types of orbital ordering.

Another possible mechanism of orbital ordering which does not require coupling to the lattice is the superexchange mechanism (sometimes called the Kugel–Khomskii (KK) mechanism). This is analogous to the standard superexchange for nondegenerate electrons, Chapter 1, and is due to a tendency toward electron delocalization (formally due to virtual hopping of electrons from one site to another), which depends on the relative spin orientations on neighboring sites, but in this case also on the relative occupation of the respective orbitals. This extra energy gain due to virtual hopping may be larger for particular spin and orbital occupations, which favor the corresponding orbital ordering.

In contrast to the JT orbital interaction via the lattice, this mechanism leads simultaneously to both spin and orbital ordering in the ground state, which turn out to be coupled. (However the critical temperatures for these two types of ordering may be, and usually are, different.) The corresponding interaction has schematically the form $J_1 \sum \mathbf{S}_i \cdot \mathbf{S}_j + J_2 \sum \tau_i \tau_j + J_3 \sum (\mathbf{S}_i \cdot \mathbf{S}_j)(\tau_i \tau_j)$, where $J_\alpha \sim t^2 / U$ and where the interaction in the orbital variables $\boldsymbol{\tau}$ may be anisotropic.

In systems with 180° metal–oxygen–metal bonds and with e_g degeneracy this spin–orbital interaction often gives the "opposite" ordering on spin and orbital sectors: ferro. orbital ordering gives antiferromagnetic spin one and, vice versa, antiferro. orbital ordering gives spin ferromagnetism. This also agrees with the first and second Goodenough–Kanamori–Anderson rules, formulated and explained in Chapter 5. However this is not a general rule: it does not apply to e_g systems with edge-sharing octahedra and with 90° TM–O–TM bonds – in this case the exchange may be ferromagnetic irrespective of orbital occupation. Also, the situation is more complicated for t_{2g}-electrons.

Though the description of orbital ordering often looks rather similar to that of magnetic (spin) ordering, there are very important differences between them. The most important one comes from the directional character of orbitals. Because of that, the electronic structure of many orbitally ordered systems may effectively become low-dimensional (1d or 2d in 3d systems, e.g. cubic, etc.). Thus for example the practically cubic perovskite $KCuF_3$ becomes magnetically strongly one-dimensional because of the O.O. in this material; it is one of the best quasi-one-dimensional antiferromagnets known. Sometimes such modifications of the exchange interaction due to O.O. lead to the formation of spin singlet states: singlet dimers in VO_2, $TiOCl$, $NaTiSi_2O_6$, $La_4Ru_2O_{10}$ or singlet trimers in $LiVO_2$ and $LiVS_2$, or even larger singlet clusters, for example singlet heptamers – 7-site clusters in AlV_2O_4.

The directional character of orbitals can also lead to strong frustrations in the orbital sector even for simple lattices such as square or cubic ones. The resulting situation (the "compass" model (6.17)) resembles that of systems with dipole–dipole interactions:

dipoles ("magnetic needles") want to order "head to tail" but, for example on a square lattice in the xy-plane the interaction of the neighbors in the x-direction wants to order the dipoles parallel to the x-axis, but that of the neighbors in the y-direction parallel to y. As a result the system is frustrated and may have quite complicated properties. The situation with orbital ordering may sometimes resemble this, although in real systems the long-range interaction of orbitals via the lattice strain, and also higher-order terms in the interaction usually choose one particular ordered state even in these frustrated situations.

In solids, usually every type of ordering leads to the appearance of corresponding collective excitations. These are for example phonons in crystals, spin waves or magnons in magnetically ordered states, etc. Consequently, one could think that in systems with orbital ordering there should also appear special excitations – orbital waves, or orbitons. Indeed, in principle such excitations should exist, and there are some reports of their observation in the literature. However, several factors make the situation with orbitons more complicated than for example that with magnons. The spin waves or magnons are excitations in a magnetic subsystem, which is usually rather independent and interacts only weakly with other degrees of freedom such as phonons. The orbitals, however, have an intrinsically strong coupling with the lattice, thus it could be very difficult to disentangle orbital excitations from conventional phonons. Orbital excitations are also very closely related to crystal field excitations – actually they *are* just such crystal field excitations with a certain dispersion. It is this dispersion which could make them the real propagating excitations in a crystal. But it is not easy to measure this dispersion by the methods usually applied, for example by optics. Thus the situation with orbital excitations is at the moment rather uncertain, although in principle they should exist of course.

Yet another rather controversial question connected with orbital excitations is the possible importance of quantum effects in orbital physics. Again, the analogy with spin systems is very helpful, but can also sometimes be misleading. In magnetic systems we know that quantum effects may be very important, especially for small spins, for example $S = \frac{1}{2}$, and for low-dimensional or frustrated systems, even up to a point where quantum fluctuations can destroy the conventional long-range magnetic ordering completely, see Chapter 5. Similar situations can also be met in some orbital systems, especially in those with t_{2g}-electrons. Certain experiments, for example in $LaTiO_3$ and YVO_3, were interpreted from this point of view. However the situation here is also far from clear. Again, in contrast to spins, strong coupling of orbitals to the lattice can make the situation "more classical." In this sense there is not much hope of seeing significant quantum effects in e_g systems, for which this electron–lattice coupling is especially strong. The t_{2g} systems with weaker such coupling give a better chance of that.

For t_{2g} systems, in contrast, another factor can play an important role: the unquenched orbital moment and nonzero spin–orbit coupling $\lambda \boldsymbol{L} \cdot \boldsymbol{S}$, see also Chapters 3 and 5. This coupling usually competes with the JT effect and can also lift the t_{2g} orbital degeneracy, but in a way different from that due to the JT effect. Whereas the JT effect stabilizes real orbitals, for example the xy orbital for one t_{2g}-electron, the spin–orbit interaction rather stabilizes complex combinations such as $\frac{1}{\sqrt{2}}(|xz\rangle \pm i|yz\rangle)$, which correspond to $|l^z = \pm 1\rangle$. The localization of one electron on such an orbital also causes lattice distortion, but opposite to that due to the JT effect: the electron on the $(|xz\rangle \pm i|yz\rangle)$ orbital would cause a *local elongation* of the MO_6 octahedron in the z-direction, whereas the JT effect which puts the electron on a nondegenerate xy orbital would correspond to a *local compression* of such an octahedron. Thus typically for systems with t_{2g} degeneracy the JT effect and spin–orbit coupling compete with each other (see also Section 3.4) and, depending on which is stronger (i.e., whether the Jahn–Teller energy E_{JT} or the spin–orbit energy $\sim \lambda$ is larger), the system evolves either according to the JT or the spin–orbit route. As the spin–orbit interaction $\lambda \sim Z^4$ (or Z^2, see footnote 5 on p. 34), where Z is the atomic number, the JT effect usually dominates for the early $3d$ metals (Ti, V, Cr), and the spin–orbit interaction determines the properties of t_{2g} systems for heavier $3d$ ions (Fe^{2+}, Co^{2+}) and often for $4d$ and $5d$ elements.

In effect, orbital degrees of freedom can lead to a number of quite special phenomena in TM compounds, especially those with JT ions with orbital degeneracy. These are special types of ordering – structural transitions due to the cooperative Jahn–Teller effect with orbital ordering. Orbital ordering has a strong influence on the magnetic properties of the system and often leads to quite nontrivial magnetic structures, and sometimes to nonmagnetic (e.g., singlet) states. Detailed orbital features largely determine the properties of many important TM compounds.

7

Charge ordering in transition metal compounds

When dealing with transition metal compounds one has to look at the different degrees of freedom involved and their interplay. These degrees of freedom are charge, spin, and orbitals. And of course all electronic phenomena occur on the background of the lattice, that is one always has to think about the role of the interaction with the lattice, or with phonons.

The electron spins are responsible for different types of magnetic ordering. The orbitals, especially in the case of orbital (or Jahn–Teller) degeneracy, also lead to a particular type of ordering, and the type of orbital occupation largely determines the character of magnetic exchange and of the resulting magnetic structures.

As to charges, the first question to ask is whether the electrons have to be treated as localized or itinerant. We actually started this book by discussing two possible cases: a band description of electrons in solids, in which the electrons are treated as delocalized, and the picture of Mott insulators, with localized electrons.

But even for localized electrons there still exists some freedom, which has to do with charges. In some systems charges may be disordered in one state, for example at high temperatures, and become ordered at low temperatures. This charge ordering (CO) will be the main topic of this chapter. But, to put it in perspective, we will start by discussing different possible types of ordering, connected with charge degrees of freedom.

In charge ordering the charge density itself becomes modulated. Thus the corresponding order parameter in this case is the charge density $\rho(r)$ (or its Fourier transform) – a *scalar*; one can talk here of *"monopole ordering"* (charge monopoles, of course). Then, in certain cases, the ions and electronic clouds may shift in such a way as to produce *electric dipoles*; the resulting state could be ferroelectric (or, with some reservations, antiferroelectric – see below). This is also connected with an ordering in the charge sector, but the order parameter is a *vector* – the electric dipole moment d or the polarization P. These phenomena will be discussed in detail in Chapter 8. Finally, as discussed above, orbital ordering also deals with charge degrees of freedom. But in this case the charges at sites do not change, and electric dipole moments are not formed. What happens from the point of view of charge distribution is that this distribution, which was spherically symmetric in the orbitally disordered state at $T > T_{\mathrm{JT}}$ (or, rather, which had the cubic symmetry of undistorted MO_6 octahedra), becomes asymmetric in the ordered state below T_{JT}, see for example Fig. 7.1. With the

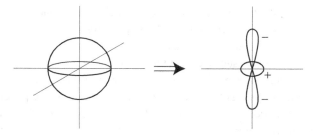

Figure 7.1 Transition from the spherical charge density to the uniaxial density distribution of quadrupolar type, occurring at orbital ordering.

orbital ordering shown in the figure, there appears a *quadrupolar* distribution of electron density in the orbitally ordered state. Thus, formally, in charge ordering the order parameter is a scalar – the density of electrons itself. In ferroelectricity it is a vector – the electric polarization. And in orbital ordering it is a second-rank tensor, the electric quadrupole.[1]

Now let us discuss charge ordering in TM compounds in different situations. The most typical situations in which one meets charge ordering in TM compounds are in systems in which there is a noninteger number of electrons per site. Such materials can be called mixed valence compounds (although this terminology is more often reserved for rare earth materials, see Chapter 11). There are many examples of such materials. Many TM elements can exist in different valence states, for example $Ti^{2+}(d^2)$, $Ti^{3+}(d^1)$, $Ti^{4+}(d^0)$; V can have valences from 2+ to 5+, Cr from 2+ to 6+, etc – see Periodic Table at the end of the book. And in some compounds the average valence can be intermediate, for example in Mn_3O_4 with, on average, $Mn^{8/3+}$; or in the so-called Magnéli phases V_nO_{2n-1}, which for $n = 2$ gives $V_2^{3+}O_3$, for $n \to \infty$ gives $V^{4+}O_2$, and for all intermediate values of n (which can be 3, 4, 5, ...) gives average valence of V between 3+ and 4+.

Another very typical case occurs when we dope a material with electrons or holes, as for example in $La_{1-x}Ca_xMnO_3$; for $x = 0$ all Mn ions are in the 3+ state, but with increased doping there appear more and more holes in the system, or formally there appears one Mn^{4+} per each Ca. And these extra holes, or different valence states of Mn, may order at low temperatures.

In some cases there exist different crystallographic positions in a crystal, each one occupied by the ion with a particular valence. Thus, Mn_3O_4 mentioned above (its formula can be written conveniently as $Mn^{2+}O \cdot Mn_2^{3+}O_3$)[2] has the structure of a spinel, see

---

[1] As discussed in Chapter 6, in many cases one can reduce the description of orbital ordering to a simpler one, characterizing it for example by an analog of a vector – a director, as in liquid crystals – which is nothing else but a symmetric tensor. But strictly speaking the order parameter here is the electric quadrupole, and in particular the similar phenomenon in rare earth compounds is often even called not Jahn–Teller or orbital ordering, but quadrupole ordering.

[2] This is in general a very useful way to present the compositions of complex compounds, decomposing them formally into simple components with well-defined valences. Such decomposition does not, of course, mean that these complex materials really consist of a mixture of such components, but it often simplifies the valence count and sometimes indeed suggests possible localization of different ions in the crystal.

Section 5.6.3, with tetrahedral A sites occupied by Mn^{2+} and octahedral B sites by Mn^{3+}. Often, however, all sites are equivalent crystallographically, nevertheless we formally have on them TM ions with different valence states.

There are several options in such a case. One is that the extra electrons, formally existing on the ions with smaller valence, are in fact delocalized in the crystal, hopping from one TM to another. If this hopping is fast enough, all TM ions would be equivalent, and in fact we would have metallic conductivity – the system will maybe be a bad metal, but still a metal. This is indeed what often happens in such situations.

Another possibility is that these extra electrons, or the corresponding valence states, are actually localized in the crystal, but in a random way. This can be caused for example by the random potential created by some defects in the crystal or, in doped systems such as $La_{1-x}Ca_xMnO_3$, by the Coulomb potential of dopands, here Ca. Or there may occur spontaneous self-trapping of the extra electrons, for example by local lattice deformations. In any case, the result would be a random system with a certain average valence of TM, which would be insulating (but for which local probes, such as NMR, would show the presence of inequivalent TM ions).

The third possibility, which is often realized at low temperatures, is the formation of a regular (super)structure with charge ordering. The system can go to this state from both states described above: from the "metallic" state with equivalent TM ions, and from the insulating state with a random distribution of TM ions with different valences. In the first case this ordering would typically be a metal–insulator transition, discussed in more detail in Chapter 10. In the second case such charge ordering would rather be an insulator–insulator transition. But in both cases there should appear a certain superstructure in the system, which could be detected by structural studies (X-ray, neutron scattering, electron microscopy).

7.1 Charge ordering in half-doped systems

A typical situation in which one has charge ordering is that with one extra electron per two sites, or a 50:50 concentration of two different valence states of an ion, in a bipartite lattice (recall that bipartite lattices are lattices which can be subdivided into two sublattices so that the nearest neighbors of one sublattice belong to the other; examples are simple square or cubic lattices, but also the honeycomb lattice, etc.). This situation is well studied in many examples, for example half-doped manganites such as $La_{0.5}Ca_{0.5}MnO_3$ (possibly with Sr instead of Ca, or other rare earths such as Pr or Nd instead of La), or half-doped "214" layered perovskites such as $La_{0.5}Sr_{1.5}MnO_4$. In both examples the average valence of Mn ions is 3.5+; that is, formally, there is an equal number of Mn^{3+} and Mn^{4+}. Typically in these cases at low temperatures we have charge ordering of checkerboard type in the basal plane (Fig. 7.2), where we denote Mn^{3+} by crosses and Mn^{4+} by circles. (Note that this convention, used throughout this chapter, is different from that used in other parts of the book, where circles typically denote anions – oxygens, fluorines, etc.) This seems to be the most natural type of ordering, which allows us to minimize the long-range Coulomb

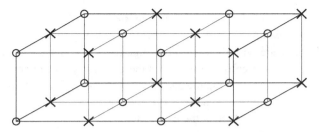

Figure 7.2 Typical charge ordering in half-doped manganites such as $La_{0.5}Ca_{0.5}MnO_3$. Here crosses denote Mn^{3+} and circles denote Mn^{4+}.

repulsion: the extra electrons on Mn^{3+} tend to stay as far away from each other as possible, which would give the checkerboard structure of the basal plane in Fig. 7.2. The situation is then very similar to the formation of a Wigner crystal in low-density electron systems, see for example Khomskii (2010).

However when we look at the experimental situation, for example in $La_{0.5}Ca_{0.5}MnO_3$, shown in Fig. 7.2, we realize immediately that this is not the whole truth, and that the Coulomb forces cannot be the only, or maybe even the main, driving force of charge ordering. If charge ordering were caused by the Coulomb interaction, we would expect that the charges, or valence states of Mn, would alternate in the third direction as well. Experimentally however this is not the case: as is seen from Fig. 7.2, the next plane is in fact *in phase* with the first one, so that Mn^{3+} is on top of Mn^{3+} and Mn^{4+} is on top of Mn^{4+}. This shows that other interactions contribute to this charge ordering as well. Most probably this extra interaction is the coupling via the lattice. We will see similar examples in other systems with charge ordering, for example in magnetite Fe_3O_4 (Section 7.4).

The conclusion that the simple Coulomb forces often do not explain the features of charge ordering in real TM compounds follows also from the fact that the actual charge difference of different ions, that is the degree of charge transfer, is usually much less than 1, much less than what would follow from the formal valences of Mn^{3+} and Mn^{4+}. Both experiment and theoretical calculations show that the actual charges in the CO state are $3.5 \pm \delta$, with $\delta \lesssim 0.2$. This is well known in inorganic chemistry – see for example Pauling (1998), where it is stressed that for example the real charge on O^{2-} ions is never really -2, but much closer to -1, or even smaller. Experimentally one can determine this value by carefully measuring the Mn–O distances for different Mn ions and calculating what in inorganic chemistry is known as the *bond valence sum* – the sum of all Mn–O bond lengths for a given Mn ion. There exists an empirical, but rather well-tested rule specifying how to find the valences of the corresponding ions from such a bond valence sum, and such experiments give the values presented above.

Theoretically one can estimate the degree of charge difference either from specific models or, more reliably, from *ab initio* band structure calculations. Such calculations also show that the actual charges at different Mn sites in CO states differ very little. Nevertheless, one can still use the terminology of ordering for Mn^{3+} and Mn^{4+} states: despite the fact that the actual charges of different Mn sites may be very different from those of Mn^{3+} and Mn^{4+}, the *quantum numbers* of respective states, such as their spin, orbital degeneracy, etc. actually coincide with those of the formal oxidation states 3+ and 4+. One can also rationalize this if one recalls that in oxides there is always rather strong hybridization between the *d* states of TM ions and the $2p$ states of oxygens. Owing to this hybridization the charges "leak out" to the surrounding oxygens, especially in the case of small charge-transfer energy (Chapter 4), which is the case at least for Mn^{4+}. Therefore the charges *on Mn ions* do not change much – the charges in any case stay largely on the oxygens. But the total quantum numbers of these hybridized (Mn–O) states are either those of Mn^{3+}, or of Mn^{4+}. The fact that the actual wavefunctions of the corresponding states are not those of pure *d* states but have a large covalency contribution and are more extended does change, for example, the magnetic form factor of the corresponding ions.[3]

In charge-ordered half-doped manganites there is yet another interesting aspect, which connects the phenomena in these systems with the discussion in the previous chapter. As mentioned above, despite a small charge difference between different Mn ions in charge-ordered states, the quantum numbers of these ions coincide with those of Mn^{4+} and Mn^{3+}. In particular, Mn^{3+} has double orbital degeneracy: its formal state is $t_{2g}^3 e_g^1$. Consequently, one can expect a certain orbital ordering in the charge-ordered state. This is indeed what happens. The orbital and spin structure in the basal plane of $La_{0.5}Ca_{0.5}MnO_3$ is shown in Fig. 7.3. The orbitals $|x^2\rangle$, $|y^2\rangle$ are ordered in "stripes" (often this picture is shown rotated by 45°).

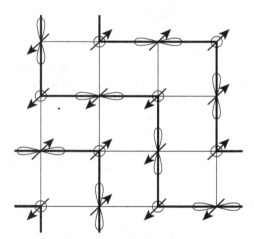

Figure 7.3 Charge, orbital, and spin ordering of CE type in the basal plane of half-doped manganites. Here circles are Mn^{4+} ions and the orbital occupation is shown for the Mn^{3+} ions.

---

[3] As mentioned in Chapter 4, the corresponding modification of the magnetic form factors has to be taken into account for example in the interpretation of magnetic neutron scattering. The deviations (in most cases reductions) of the magnetic moments of TM ions from those expected for a given formal valence, often observed in neutron scattering, can very often be connected with this factor.

One can explain this orbital pattern by the tendency to minimize the total strain in the system, the mechanism described in Section 6.1.1 (Khomskii and Kugel, 2003). Interestingly, as mentioned above, *ab initio* calculations (Anisimov *et al.*, 1997) give almost the same total charges at different Mn sites, but they reproduce the orbital ordering: the electron density around different Mn ions is very anisotropic, and it corresponds quite well to the picture of Fig. 7.3, which follows from experiment.

The checkerboard charge ordering and the corresponding orbital ordering shown in Fig. 7.3 occur in $La_{0.5}Ca_{0.5}MnO_3$ simultaneously, at $T_c \simeq 240$ K. The resulting states have localized spins which order antiferromagnetically at $T_N = 120$ K. This magnetic ordering, shown in Fig. 7.3, is called *CE* ordering; nowadays one often calls the charge ordering and the orbital ordering of Fig. 7.3 "*CE* ordering" also. This magnetic structure can be understood as ferromagnetic zigzags (shown in Fig. 7.3 by bold lines), stacked antiferromagnetically. This structure follows directly from the GKA rules formulated in Section 5.1: along these zigzags the occupied e_g orbitals of the edge-site Mn^{3+} point toward the corresponding corner-site Mn^{4+} ions with empty e_g levels. The virtual hopping of these e_g-electrons onto similar empty e_g states of Mn^{4+} would give rise to a ferromagnetic exchange along the zigzags. Antiferromagnetic interaction between zigzags is most probably caused by the antiferromagnetic exchange of t_{2g}-electrons.

Above, when talking about charge ordering, in particular in half-doped manganites, we had in mind that we should deal with the ordering of, formally, Mn^{3+} and Mn^{4+} ionic species, or that the extra electrons present here on top of the t_{2g}^3 configuration of Mn^{4+} would localize on certain particular Mn sites (in this case forming a checkerboard structure). However this is not the only possibility for localizing such electrons. Instead of being on certain particular ions, that is on *sites*, they may be localized on particular *bonds*. This possibility is actually not new: when we discussed the valence bond or RVB states in Section 5.7.2 we in fact already used this picture. The electrons were localized there (in singlet pairs) on particular bonds, not sites. If we have two electrons with opposite spins (in a singlet state) on such a bond, it would be just what chemists call a valence bond, as in the H_2 molecule. But we can also have just one electron on such a bond; it would then be the analog of the H_2^+ ion.

Such a possibility was proposed recently for half-doped manganites on experimental grounds by Daoud-Aladine *et al.* (2002) and theoretically by Efremov *et al.* (2004). The picture is that, instead of being localized on particular sites, as in Figs 7.2, 7.3, the extra electron from the x^2 orbital of Mn^{3+} starts to hop back and forth to one of its neighbors in the x-direction, say to its right neighbor (or from the y^2 orbital to its down-neighbor). The resulting structure, instead of Fig. 7.3, would look as shown in Fig. 7.4. The state with one electron on a bond connecting two sites is also the analog of a two-site polaron; sometimes it is called a Zener polaron, keeping in mind the analogy with Zener double exchange (one electron hopping between two sites with localized spins would order these spins ferromagnetically, which is also the case here).

The choice between the site-centered charge ordering of Figs 7.2, 7.3 and the bond-centered ordering of Fig. 7.4 is not a simple one. Theoretical treatment (Efremov

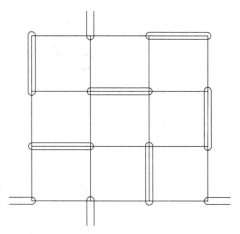

Figure 7.4 A possible alternative ordering in half-doped manganites (Zener polarons, with one electron delocalized over two neighboring sites), after Daoud-Aladine *et al.* (2002).

et al., 2004) shows that at $x = 0.5$ the site-centered structure wins, but if charge ordering survives for $x < 0.5$, as it does in $Pr_{1-x}Ca_xMnO_3$, see the next section, then the bond-centered state starts to compete with the site-centered one. It turns out that, at this doping range the situation may be intermediate, site- and bond-centered charge ordering may coexist, so that both the sites and the bonds become inequivalent. Most interestingly, this mixed state turns out to be ferroelectric, which will be discussed in detail in Chapter 8.

7.2 Charge ordering away from half-filling

7.2.1 Doped manganites

The situation for other relative concentrations of different ionic species may be more complicated. We will consider it again mainly for the example of doped manganites such as $La_{1-x}Ca_xMnO_3$ (LCMO) and $Pr_{1-x}Ca_xMnO_3$ (PCMO). The phase diagram of LCMO is presented in Fig. 5.16, and that of PCMO in Fig. 7.5. In both cases we have a well-defined charge and orbital ordering of *CE* type at $x = 0.5$. However, away from half-filling these systems behave differently. LCMO becomes metallic and ferromagnetic for $x < 0.5$; it is in this phase that one observes colossal magnetoresistance. These phenomena were described partially in Section 5.2, and will be treated in more detail in Section 9.2.1. PCMO, however, remains insulating for all x, including the region $x < 0.5$. And in a large part of the phase diagram, $0.3 \lesssim x \leqslant 0.5$, it remains charge ordered with the same $\sqrt{2} \times \sqrt{2}$ periodicity as the checkerboard CO state at $x = 0.5$, see Fig. 7.2.

For $x > 0.5$ the situation is different. Both systems remain insulating, with a certain superstructure resembling charge ordering, which, however, is probably better described in the itinerant picture as a state with a charge density wave (CDW) (Loudon *et al.*, 2005), see Section 7.3.

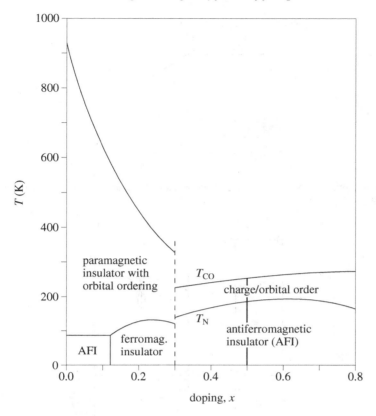

Figure 7.5 Schematic phase diagram of $Pr_{1-x}Ca_x MnO_3$.

The periodicity of the observed superstructure in LCMO and PCMO for $x > 0.5$ changes with doping: crudely, the wave vector of this superstructure, Q, which for $x = 0.5$ is $Q = (\frac{1}{2}, \frac{1}{2}, 0)$, for $x > 0.5$ becomes $Q = (\frac{1}{2}, \frac{1}{2} - \delta, 0)$, with $\delta \sim (x - 0.5)$. The behavior of this superstructure vector for PCMO is shown in Fig. 7.6.

An interesting situation occurs for $x = \frac{2}{3}$. In this case we would formally have Mn^{3+} and Mn^{4+} ions in the ratio 1:2. The ordered structure in this case is similar to that of the half-doped case (Fig. 7.3) but with an extra row (stripe) of Mn^{4+} (Fig. 7.7). In other words, we again have the stripes of Mn^{3+} with $|x^2\rangle$ and $|y^2\rangle$ orbitals, with such density as to give the average valence $Mn^{3.67+}$. Most probably the nature of this ordering, that is the formation of such stripes as shown in Fig. 7.7, can again be understood along the lines discussed in Section 6.1.1, as being due to an interaction of "impurities" (here extra electrons, or Mn^{3+} configurations on the background of Mn^{4+}) caused by strain coupling (Khomskii and Kugel, 2003).

The situation with the magnetic coupling in this case is somewhat more intricate. Again the spins of both Mn^{4+} ions to which the orbitals of Mn^{3+} in Fig. 7.7 point would be

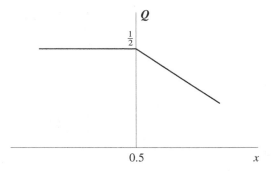

Figure 7.6 Behavior of the superstructure wave vector for underdoped ($x < 0.5$) and overdoped ($x > 0.5$) $Pr_{1-x}Ca_xMnO_3$.

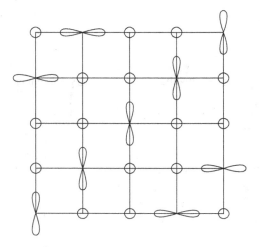

Figure 7.7 Possible stripe ordering in perovskite manganites such as $La_{1/3}Ca_{2/3}MnO_3$. Here circles denote Mn^{4+} and the occupied orbitals are shown for Mn^{3+}.

parallel to the spin of this Mn^{3+} ion. But we can connect these ferromagnetic (Mn^{4+}–Mn^{3+}–Mn^{4+}) blocks into a long-range-ordered structure in different ways, see Fig. 7.8. (We have shown here two different but equivalent structures with magnetic zigzags having "steps" of four sites, but equally well one could make 3×5 zigzags, or 5×3 zigzags, or any combination thereof.)[4] Therefore one should expect that, whereas charge and spin ordering at $x = \frac{2}{3}$ could form the well-defined stripe structure of Fig. 7.7, its magnetic ordering can have a lot of defects, magnetic stacking faults, and the corresponding peaks in magnetic neutron scattering could be rather broad.

---

[4] Notice that we have a similar problem at $x = 0.5$ in the "Zener polaron" structure of Fig. 7.4 compared with the simple site-centered ordering of Fig. 7.3: whereas in the latter the structure of ferromagnetic zigzags is unique, in the structure of Fig. 7.4 we can connect the ferromagnetic dimers into zigzags in different ways. However, in the mixed site- and bond-centered solution of Efremov *et al.* (2004) the magnetic structure is again unique.

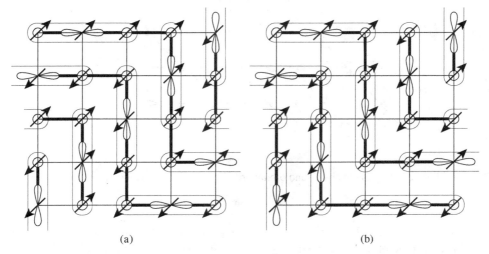

Figure 7.8 Two equivalent magnetic structures for the $\frac{2}{3}$-doped manganite $La_{1/3}Ca_{2/3}MnO_3$, with the same charge and orbital order; cf. Fig. 7.7.

When discussing charge ordering at arbitrary $x > 0.5$, we can mention two possibilities (this applies in fact to many other situations with periodic structures with periods incommensurate with the lattice period): either we have a harmonic (sinusoidal) density wave with wave vector Q which changes continuously with doping, as shown in Fig. 7.6 for $x > 0.5$, or this is only an average picture and for example for values of x between $\frac{1}{2}$ and $\frac{2}{3}$ we have a state which consists of alternating domains with structure similar to that of $x = \frac{1}{2}$ (Figs 7.2, 7.3) and $x = \frac{2}{3}$ (Fig. 7.7). In the second case the X-ray or neutron scatterings, probing the periodicity at large length scales, would show the average period but probes which can check the state very locally, such as transmission electron microscopy, could show the coexistence of domains with different CO patterns, corresponding to different periods. What is the real situation in overdoped manganites is still not clearly established; different experiments lead to different conclusions.

If the situation with charge ordering for $x > 0.5$ is at least more or less clear in principle (the average period of the resulting superstructure is determined by the electron concentration, as is often the case if the superstructure is caused by specific features of the Fermi surface, such as nesting, see Section 7.3), then what happens in PCMO for $x < 0.5$ is much less clear. It looks as though the superstructure typical for $x = 0.5$ is preserved for $x < 0.5$ as well, see Fig. 7.6. But in this part of the phase diagram we have more electrons than at $x = 0.5$, and we have to put these $(0.5 - x)$ extra electrons somewhere. These electrons, or holes, could become delocalized, and the system could become metallic. This is what happens in $La_{1-x}Ca_xMnO_3$ for $x < 0.5$. But PCMO in this doping range remains insulating, see Fig. 7.5. This means that in this system the electrons are localized in some way.

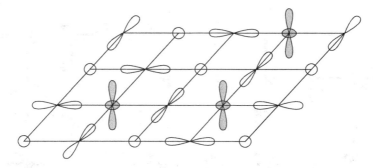

Figure 7.9 Possible charge and orbital ordering in $Pr_{1-x}Ca_xMnO_3$ for $x < 0.5$.

There exist in principle two possibilities. One possibility is that these extra electrons are localized at some of the former Mn^{4+} sites in the checkerboard structure of Figs 7.2, 7.3 at random; in this case most probably the e_g orbitals which these extra electrons should occupy would be z^2 orbitals, which point perpendicular to the basal planes shown in Figs 7.2, 7.3. This at least would not strain the lattice strongly in the xy or ab-plane: the radius of Mn^{3+} ions with z^2 orbitals in the xy-plane is also relatively small, comparable with that of Mn^{4+}. The resulting picture may then look as shown in Fig. 7.9: some of the former Mn^{4+} positions (circles) would be occupied by Mn^{3+} with z^2 orbitals.

Another option could be a larger-scale phase separation, in which the extra Mn^{3+} ions form clusters of certain sizes, the remaining part of the sample being more or less stoichiometric with $x = 0.5$ and with ordering of *CE* type. The actual situation is not really clear.

One should stress once again that, in contrast to naive expectations, there is no electron–hole symmetry in doped manganites, and the situation in $R_{1-x}Ca_xMnO_3$ for underdoped ($x < 0.5$) and overdoped ($x > 0.5$) cases is highly asymmetric, see the phase diagrams of Figs 5.16, 7.5, and 7.6. Whereas for $x < 0.5$ one can have a ferromagnetic metallic state with colossal magnetoresistance (e.g., for $La_{1-x}Ca_xMnO_3$), for $x > 0.5$ this is practically never the case, the overdoped manganites typically remain insulating, with one or the other type of charge ordering. Also with respect to CO in cases when it is observed for $x < 0.5$, for example in $Pr_{1-x}Ca_xMnO_3$, the character of this CO is very different from that for $x > 0.5$: whereas for $x > 0.5$ the period of the charge superstructure (CO or CDW) follows the doping, for $x < 0.5$ in PCMO the main periodicity of the superstructure remains the same as for $x = 0.5$, that is it is predominantly a checkerboard structure, with some "defects" (extra electrons, or extra Mn^{3+} ions) placed in the lattice in some random fashion. It is not really clear *why* there exists such marked asymmetry between hole-doped ($x < 0.5$) and electron-doped ($x > 0.5$) manganites. Most probably this is connected with the orbital degrees of freedom. The strong Jahn–Teller ions Mn^{3+} with one e_g-electron, when localized, have to distort the lattice, and whereas there is "a lot of space" for placing such Mn^{3+} ions for $x > 0.5$, for $x < 0.5$ there are too many of them, and their interaction may change the behavior drastically. This may be an important factor leading to such a strong electron–hole asymmetry in manganites. Some theoretical models agree with these ideas (van den Brink and Khomskii, 1999). However this question is not yet clarified completely.

There is one very interesting consequence of the presence of the charge-ordering state in manganites, in particular of this strange charge-ordered state for $x < 0.5$ in PCMO. We have two competing states: one is an insulating state with charge ordering (and orbital ordering), which at low temperatures is antiferromagnetic, and the alternative state without charge ordering, which would be metallic and, due to the double-exchange mechanism (see Section 5.2), also ferromagnetic. Then, if we apply an external magnetic field to the antiferromagnetic charge-ordered insulator, we can shift the equilibrium in favor of the ferromagnetic metallic (FM) state and cause the corresponding transition. And indeed, such experiments have been done by the group of Y. Tokura in Tokyo (Tomioka *et al.*, 1996; Tokunaga *et al.*, 1997), and they have observed a very sharp drop in resistivity at a certain critical field which is caused by a sharp I order transition from the charge-ordered insulator to the ferromagnetic metal. This drop, that is negative magnetoresistance, is the strongest here in all manganites.

But, interestingly enough, by increasing the temperature in some PCMO samples with $x < 0.5$ one could cause an inverse transition, from a ferromagnetic metal to a charge-ordered insulator. Typical phase diagrams for PCMO at different doping levels are shown in Fig. 7.10 (the shaded area is the region of hysteresis at these I order transitions). We see that, indeed, the magnetic field causes a transition to a FM state. And for $x = 0.5$ the charge-ordered state is rather robust – one needs fields $\gtrsim 20$ T to transform it to a ferromagnetic metal. We also see that for $x = 0.5$ the phase diagram looks "normal": with decreasing temperatures we go from a disordered to a charge-ordered state.

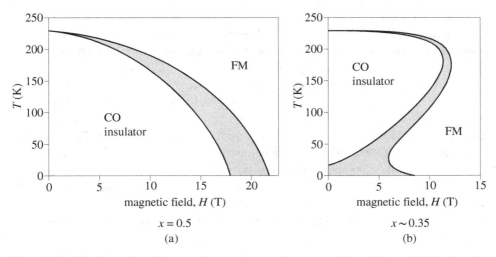

Figure 7.10 The H–T phase diagram of two representative compositions of $Pr_{1-x}Ca_xMnO_3$, after Tomioka *et al.* (1996) and Tokunaga *et al.* (1997). The shaded regions are regions of hysteresis at the sharp I order transitions.

The situation for $x < 0.5$ is different. First of all, the charge-ordered state is much less stable: we need much smaller fields to suppress it. But, most surprisingly, there is a re-entrant transition: as the temperature decreases, first the charge-ordered insulator appears, but at still lower temperatures it can disappear again, and the system transforms to a ferromagnetic metal. Why do we have such strange behavior?

To explain this we recall that in statistical mechanics the state of a system at a given temperature is determined by the condition of the minimum of the free energy

$$F = E - TS . \tag{7.1}$$

For low temperatures, for example for $T = 0$, this condition coincides with the condition of the minimum of the energy of the system, E. As the interaction energy is usually minimized for a certain ordered state, for example the exchange energy is minimal for the ferro- or antiferromagnetic state, we usually have a certain type of ordering at $T = 0$.

With increasing temperature, however, the second term in eq. (7.1), containing the entropy S, starts to play a more and more important role. Since the entropy, as a measure of disorder, is larger for disordered states, sooner or later, at certain temperatures, the second term in eq. (7.1) dominates, and the system enters a disordered state. Thus the general rule, which we will also meet below in discussing insulator–metal transitions in Chapter 10, is that with increasing temperature the system *always* goes to a state with higher entropy.

Why then does it look as if an order (charge ordering) which existed in PCMO at, say, $x = 0.35$ (Fig. 7.10(b)), disappears at low temperatures or, vice versa, why does this ordering appear with *increasing* temperature?

The answer is that apparently the entropy of this charge-ordered state at $x = 0.35$ is *higher* than that of a FM state at $T = 0$. Indeed, the ferromagnetic metallic state, though it does not have any superstructure such as charge ordering, is still a unique state, with a Fermi surface of electrons and with spin polarization. Consequently, it has low (nominally zero) entropy at $T = 0$. With increasing temperature this system seems to go to an even better-ordered, charge-ordered state. But, as we have discussed above, this charge-ordered state has some unavoidable intrinsic disorder (for $x < 0.5$): the period of the observed superstructure corresponds to that of $x = 0.5$, but in fact here $x = 0.35$, that is we have to put the remaining extra 0.15 electrons somewhere. And apparently we put them at random sites, as shown in Fig. 7.9, or there occurs a larger-scale but still random phase separation. Thus at $x = 0.35$ (and for all $x < 0.5$) this charge-ordered state is actually a "partially disordered charge-ordered state" with an intrinsic disorder and consequently with nonzero entropy. Apparently it is this disorder entropy which causes a transition to this "partially disordered CO state" with increasing temperature. For $x = 0.5$, for which there should be no such disorder, the phase diagram has the more "normal" form of Fig. 7.10(a) without re-entrance, and charge ordering, after it starts to develop, survives down to $T = 0$.

7.2.2 Stripes

One of the possible types of charge ordering in TM compounds is a quasi-one-dimensional ordering in the form of *stripes*. One can interpret the charge-ordering pattern in overdoped manganites such as $La_{1-x}Ca_xMnO_3$, for example that for $x = \frac{2}{3}$ (Fig. 7.7), as stripes. Stripe ordering was proposed especially for cuprates, for example $La_{2-x}Sr_xCuO_4$, and for similar layered materials such as $La_{2-x}Sr_xNiO_4$. The original idea (Zaanen and Gunnarsson, 1989) was that in doped antiferromagnetic materials it might be favorable to create antiferromagnetic domain walls, and localize doped charges (here holes) at such domain walls. The physics of this phenomenon is very similar to that discussed in Section 1.4. We saw there that the background antiferromagnetic ordering hinders the motion of charge carriers, for example holes, and that it might be energetically favorable to destroy antiferromagnetic ordering in some parts of the sample and put the holes there. However it is not clear in this picture why it would not be better to form small spherical (or circular in 2d case) hole droplets instead of stripes.

It is also not evident that such stripes, in which the holes are located along a line of TM ions, would indeed induce an antiferromagnetic domain wall. This may be the case in cuprates: the hole on Cu^{2+} would make this ion nonmagnetic (nominally $Cu^{3+}(t_{2g}^6(z^2)^2)$ with $S = 0$, or a Zhang–Rice singlet, see Section 4.4). Then indeed one may expect that the weaker, but still present antiferromagnetic interaction "across the hole" would make spins "to the right" and "to the left" of the stripe antiparallel, see Fig. 7.11(a), where we denote spins that are now reversed by double arrows.

But this picture would not work for example for manganites, where both Mn^{3+} ions existing in the undoped material and Mn^{4+} ions created when we hole-dope the system are magnetic. In this case, as we see from Fig. 7.11(b,c), even independently of the sign of the exchange between Mn^{4+} in a stripe (thick arrows) and the neighboring Mn^{3+} (thin arrows) the resulting antiferromagnetic ordering "to the left" and "to the right" of the stripe would not change, that is in this case such stripes would not be antiferromagnetic domain walls.

The situation may change and the stripes may become more favorable if they are oxygen-centered, that is, if the doped holes reside predominantly on oxygens (which is the case in cuprates and to some extent in nickelates). The resulting situation would then look as shown in Fig. 7.12: when we put localized holes on oxygens (circles in Fig. 7.12), these oxygen ions will themselves become magnetic, with spin $\frac{1}{2}$ and with certain exchange with neighboring TM ions, for example with Cu^{2+}. This exchange will be rather strong,

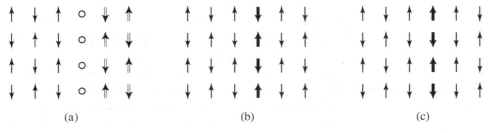

(a)   (b)   (c)

Figure 7.11 (a) Metal-centered stripes in systems such as cuprates (nonmagnetic sites are shown by circles); (b, c) Metal-centered stripes in systems such as manganites or nickelates; the magnetic ions with valence and spin different from the bulk are shown by bold arrows.

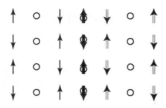

Figure 7.12 Oxygen-centered stripes. Here the arrows show the magnetic ions (spins) and the circles are oxygen ions: empty circles are nonmagnetic O_2^- and circles with arrows are oxygens with p-holes trapped on them. Double arrows mark spins reversed after doping, with doped holes located on oxygens.

$\sim t_{pd}^2/\Delta_{CT}$, much larger than the Cu–Cu exchange $\sim \dfrac{t_{pd}^4}{\Delta_{CT}^2}\left(\dfrac{1}{U_{dd}} + \dfrac{1}{\Delta_{pp}+U_{pp}/2}\right)$, cf. eq. (4.13). Even independently of the sign of this exchange (which is most probably antiferromagnetic, but may depend on the corresponding orbital occupation) such oxygen holes/spins (thick arrows in Fig. 7.12) would induce a strong frustration in the TM subsystem: the TM ions on two sides of such O^- ions would have *the same* spins, rather than opposite ones, as in the original antiferromagnetic structure of the undoped material. As a result, a chain of such oxygen holes would induce a reorientation of TM (e.g., Cu or Mn) spins say on the right of this stripe, as shown in Fig. 7.12 by double arrows. That is, such oxygen-centered stripes would indeed induce an antiferromagnetic domain wall.

There exists however another mechanism of stripe formation in doped oxides, independent of the magnetic structure (Khomskii and Kugel, 2001, 2003); this has already been mentioned briefly in Section 6.1.1. When we dope the TM oxide, we formally create ions with a different valence (e.g., Mn^{4+} in $La_{1-x}Ca_xMnO_3$), which also have a different size (different ionic radius). Consequently, such ions, if localized, would create a certain strain in the crystal. As is shown in Fig. 6.6 and in eqs (6.10)–(6.12), such strain for example in a cubic crystal has different signs depending on the direction in the crystal (the "windmill" of Fig. 6.6). Such strain also decays slowly, as $1/r^3$, and is not screened. In effect another such ion (doped hole) put somewhere else in the crystal feels this strain, which provides an effective hole–hole interaction, the sign of which depends on the relative orientation in the crystal; it is always attractive in some direction. Then the doped holes would order along these directions, which may be a natural mechanism of stripe formation in doped TM oxides such as manganites and nickelates, and possibly also cuprates. Note once again that this mechanism is independent of the magnetic structure, thus we do not have here the problems with antiferromagnetic domain walls which have been mentioned above in connection with metal-centered stripes in manganites (but it is possible that such stripes, formed by the strain mechanism, would then create or pin magnetic domain walls of some kind). This also agrees with the fact that experimentally charge ordering in manganites, which for example for $x = \frac{1}{3}$ or $\frac{3}{2}$ can be interpreted as stripes, occurs at T_{CO} which is much higher (typically twice higher) than the Néel temperature. Apparently this means that the magnetic mechanism is not the main driving force for stripe formation, at least in these materials.

7.3 Charge ordering vs charge density waves

As already mentioned in the previous section, in doped manganites such as $La_{1-x}Ca_xMnO_3$ and $Pr_{1-x}Ca_xMnO_3$ the periodicity of the superstructure for $x > 0.5$ follows directly the doping level (Fig. 7.6). This is rather typical for CDW systems, in which the charge density wave is caused by particular features of the Fermi surface such as nesting. The examples are CDW states in TM dichalcogenides such as $NbSe_2$ or TaS_2. These are layered materials in which the TM ions form triangular layers similar to those in $LiNiO_2$ or $LiVO_2$, cf. Section 5.7.1.

At high temperatures these materials are reasonably good metals, and one can describe them using the band picture. In this description the electrons occupy the Fermi surface and for a specific shape of this Fermi surface the normal metallic state may become unstable, and the system may develop a CDW or a spin density wave (SDW). This usually happens in case of the so-called *nesting* of the Fermi surface, when either the Fermi surface has flat parts which overlap when shifted by a certain wave vector (see the schematic picture in Fig. 7.13(a)), or the Fermi surface may have no flat parts but still have the property that finite parts of it are on top of one another when shifted by some vector Q (Fig. 7.13(b)). It can also be that the Fermi surface has several pockets, for example electron and hole pockets, and again if there is nesting between these pockets (Fig. 7.13(c)), that is if they can be superimposed by shifting by some wave vector, the system will be unstable with respect to CDW or SDW (see the detailed discussion in Khomskii, 2010).

The description of the resulting state, for example of a CDW, is that the electron density acquires a periodic modulation with the wave vector Q coinciding with the nesting vector,

$$\rho(r) = \rho_0 + \eta \, \mathrm{Re}(e^{i\,Q \cdot r + \varphi}) = \rho_0 + \eta \cos(Q \cdot r + \varphi), \tag{7.2}$$

where η is the amplitude of the CDW and φ is its phase. Thus in effect in CDW, similarly to charge ordering, the charge density varies periodically in space.

One can easily illustrate in this picture both the situation of site-centered and bond-centered charge ordering, discussed in the previous section. Thus, for example, for the one-dimensional case with one electron per site the CDW instability will correspond to

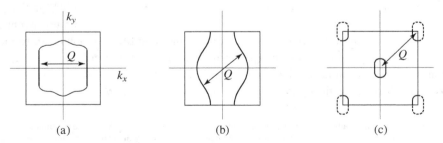

(a)   (b)   (c)

Figure 7.13 Schematic illustration of nesting of Fermi surfaces in metals, which could lead to instability and to the formation of CDW or SDW states.

(a)   (b)

Figure 7.14 Site-centered vs bond-centered charge density wave.

dimerization, that is the CDW or CO wave vector is $Q = \pi$ (the lattice constant is taken as 1) and the CDW can look as shown in Fig. 7.14. The case of Fig. 7.14(a) corresponds to a site-centered CDW, or the standard charge ordering, and the case of Fig. 7.14(b) corresponds to a bond-centered CDW, in which the sites are equivalent but the bonds are not.

The amplitude η of this CDW may in principle be quite small (and it is small for metallic systems with broad bands and weak coupling). But if we reduce the bandwidth and/or increase the coupling strength, this amplitude will increase, and if it becomes of order 1, we can talk of charge ordering. Thus these phenomena, CDW and charge ordering, are in fact the same except that one uses the terminology of CDW if the amplitude of charge modulation is weak (and has arbitrary period $d = 2\pi/Q$, which in particular can be incommensurate with the period of the lattice), and one speaks of charge ordering if such charge modulation is relatively large, and usually the period is commensurate with the lattice (dimerization, trimerization, etc.).

From eq. (7.2) and Figs 7.14(a, b) we see also that depending on the phase φ of the CDW this wave can be centered either on sites (Fig. 7.14(a)) or on bonds (Fig. 7.14(b)).[5] Thus from this point of view the site-centered and bond-centered charge-ordered states discussed in the previous section can be understood as being very similar, differing mainly by the phase of the corresponding density wave. But of course if we think which distortion accompanies one or the other state and contributes to their relative stability, we see that these states indeed look different, and they may have different energies: in real materials, in contrast to the simplified model illustrated in Fig. 7.14, site- or bond-centered charge-ordered states are accompanied (and may be caused) by different distortions. In manganites with site-centered charge ordering the effective sizes of different Mn ions become different, and the oxygens surrounding Mn^{4+} come closer to these ions, while those surrounding Mn^{3+} move away (breathing-type distortion). The distance between Mn^{4+} and Mn^{3+} in that process does not change in the first approximation. On the contrary, for the bond-centered states it is this distance that changes (so that short and long Mn–Mn bonds are formed), but the Mn–O distance does not change much. Thus in the charge-ordering picture, with a relatively large amplitude of charge variation, one can really discriminate between these two cases, which is practically impossible for weak-coupling CDW.

---

[5] In principle, for weak-coupling CDW the phase can be arbitrary, that is such CDW can be continuously shifted with respect to the lattice. Such *sliding CDW* can contribute to conductivity of such states – this is the so-called *Fröhlich conductivity*. For strong coupling and for a wave with large amplitude, as in the charge ordering we have been dealing with until now, nonlinear effects start to play a role, which can lead in particular to a *pinning* of the CDW or SDW to the lattice, and in the case of strong pinning this conductivity is not efficient anymore.

To what limit do real TM compounds with charge ordering belong? We have already mentioned that for example in TM dichalcogenides such as $NbSe_2$ or TaS_2 the picture of weak-coupling CDW is applicable. On the contrary, there are TM compounds with very strong charge ordering: thus for example for Fe_2OBO_3 it was shown that the amplitude of charge modulation is very close to 1 (Attfield *et al.*, 1998), so that in this system we are really dealing in practice with the ions Fe^{2+} and Fe^{3+}.

In most TM oxides with charge ordering the situation is somewhere in between, see for example Attfield (2006). As mentioned above, for half-doped manganites such as $La_{1-x}Ca_xMnO_3$ or $Pr_{1-x}Ca_xMnO_3$ the real charge difference of different Mn ions is $\sim 0.2e$. One should expect that away from half-filling the charge ordering should be even weaker (the extra electrons which order are further away from each other, and the interactions leading to charge ordering are weaker). And indeed this agrees with experimental observations. Whether one should still treat the resulting state as a charge-ordered one, or rather as a CDW, is an open question. At least some experimentalists prefer to interpret their results on overdoped manganites in terms of CDW, and not as charge ordering (Loudon *et al.*, 2005).

7.4 Charge ordering in frustrated systems: Fe_3O_4 and similar

Similarly to magnetic ordering, charge ordering has very special features in systems with geometric frustrations. Probably the most famous such example is the Verwey transition in magnetite (lodestone) Fe_3O_4. We have already mentioned that magnetite was the first magnetic material known to mankind, which gave us the very word "magnetism." In 1939 the Dutch physicist Verwey discovered in it what was apparently the first metal–insulator transition in TM oxides (Verwey, 1939; Verwey and Haaymann, 1941; Verwey *et al.*, 1947): with decreasing temperature at $T_V = 120\,K$ there occurs in Fe_3O_4 a first-order transition, in which the resistivity jumps by approximately 10^2, see the schematic picture in Fig. 7.15. The low-temperature phase is definitely an insulator, and above T_V Fe_3O_4 becomes relatively well conducting. Nevertheless, as can be seen from Fig. 7.15, it is still not really a metal: for $120\,K < T \lesssim 450\,K$ the conductivity still increases with T, whereas for normal metals it should decrease (or the resistivity should increase). And only above $\sim 450\,K$, after passing a broad maximum, does $\sigma(T)$ start to decrease. Nevertheless, with these reservations, one usually still calls this Verwey transition a metal–insulator transition. We will discuss qualitatively the nature of the intermediate phase above T_V later; first, however, we consider the low-temperate insulating state, the nature of which turns out to be quite nontrivial.

Fe_3O_4 has a spinel structure, see Section 5.6.3, in which Fe^{3+} ions are at the A sites and the B sites are occupied by, formally, Fe ions in a mixed valence state $Fe^{2.5+}$; we may say that there is an equal amount of Fe^{2+} and Fe^{3+} at the B sites. The first idea proposed to explain this transition was that whereas above T_V the extra electrons of "Fe^{2+}" ions hop more or less freely through the crystal, giving metallic-like conductivity, below T_V there occurs a charge ordering in the system, in which the ionic states $Fe^{2+}(d^6)$ and

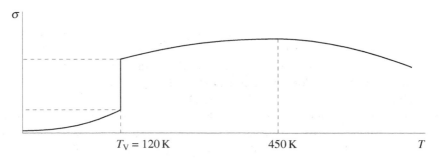

Figure 7.15 Schematic behavior of conductivity in magnetite Fe_3O_4. $T_V = 120\,K$ is the Verwey transition.

$Fe^{3+}(d^5)$ order in some fashion, after which the system becomes insulating. Such ordering should also cause a lattice distortion; and indeed it was found that the Verwey transition is accompanied by a structural distortion, from the high-temperature cubic phase typical for spinels, to a lower-symmetry phase.

But here the frustrated nature of the B sublattice of a spinel (forming a pyrochlore-like lattice of corner-sharing tetrahedra, cf. Sections 5.6.3, 5.7) comes into play. If we look at this sublattice, reproduced again in Fig. 7.16, we realize immediately that there is no unique way to order Fe^{2+} and Fe^{3+} ions (or the extra electrons) in this lattice so as to minimize the Coulomb repulsion between these electrons. Indeed, we have here one electron (of Fe^{2+} ion on the background of Fe^{3+}) per two B sites, and even if we put two extra electrons on each tetrahedron, which seems at first glance to be the best option from the point of view of Coulomb repulsion (this is known as the Anderson criterion, or Anderson rule; Anderson, 1956), already at the level of one tetrahedron we have several equivalent states, four of which (not all!) are shown in Fig. 7.16(a). If we should try to combine such tetrahedra into a pyrochlore-type lattice with each two tetrahedra having only one common corner, (Fig. 7.16(b)), we would have an enormous amount of equivalent states. We would then have finite entropy at $T = 0$.

There are of course other interactions in the system, such as long-range Coulomb interaction or interaction with the lattice. One also has to remember that Fe^{2+} ions in a cubic crystal field have orbital degeneracy (electronic configuration $t_{2g}^4 e_g^2$), so that the orbital effects, and also the spin–orbit interaction on Fe^{2+}, can potentially play some role. In effect, one may expect that some particular type of long-range charge ordering will still be realized in Fe_3O_4 despite the strong frustration. But such ordering can in principle be rather complicated.

The very first suggestion, made by Verwey himself (Verwey, 1939; Verwey and Haaymann, 1941; Verwey *et al.*, 1947) was an ordering shown in Fig. 7.16(b), in which we denote Fe^{3+} ions by circles and Fe^{2+} by squares. This ordering satisfies the Anderson criterion: there are two Fe^{2+} and two Fe^{3+} ions per tetrahedron. This charge ordering may

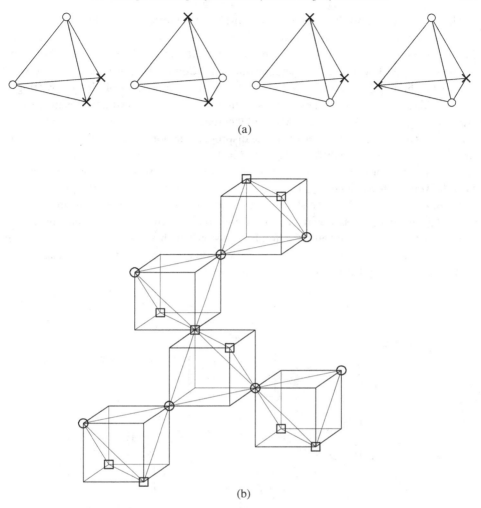

(a)

(b)

Figure 7.16 (a) Some configurations of charge ordering in Fe tetrahedra of the B sites of Fe_3O_4. The states (though not all) satisfying the Anderson rule (two Fe^{2+}, two Fe^{3+} per tetrahedron) are shown. (b) A possible simple type of charge ordering in Fe_3O_4 below the Verwey transition (apparently not the state realized in practice, see text). Circles are Fe^{3+}, squares are Fe^{2+}.

be visualized as consisting of alternating chains running in xy- and $\bar{x}y$-directions in the crystal (in fact they form whole [001] planes of Fe^{2+} and Fe^{3+}); of course there may exist different domains, in which such chains run in xz- or yz-directions. However the later, more detailed experiments have shown that the actual charge pattern in Fe_3O_4 below T_V is much more complicated. This story is indeed very interesting: the more accurate and more detailed experiments were made, the more complicated charge pattern one had to invoke to explain the results. Probably the most reliable structure was obtained by Wright *et al.* (2002); this structure is shown in Fig. 7.17.[6] Note first of all that in this structure the Anderson rule is strongly violated: in each tetrahedron the ratio of Fe^{2+} and Fe^{3+} ions is either 3:1 or 1:3. This tells us that apparently the Coulomb repulsion, at least between nearest neighbors, does not play a dominant role in the formation of charge ordering in Fe_3O_4. This also agrees with our general statement above, and also with the results of some experiments (Garcia and Subias, 2004) and theoretical calculations (Leonov *et al.*, 2004), which show that the actual degree of charge disproportion in Fe_3O_4 is rather small: the charge difference between different ions, nominally Fe^{2+} and Fe^{3+}, is at most $\sim 0.2e$. This is actually not surprising, bearing in mind that above the Verwey transition the magnetite has relatively high conductivity. This means that there exists an efficient hopping of electrons from site to site. And one should not expect this hopping to disappear immediately below T_V. However, it has a tendency to "average out" the charges at different sites.

Thus apparently it is not just Coulomb repulsion which stabilizes charge ordering in Fe_3O_4. Most probably the electron–lattice interaction is again an important mechanism, in which the orbital degrees of freedom can play some role.

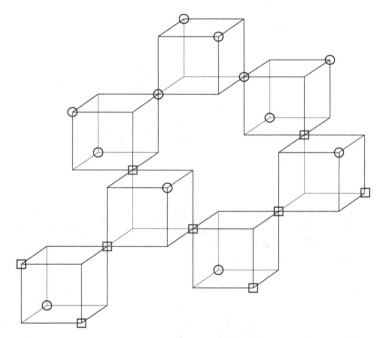

Figure 7.17 Alternative charge ordering in Fe_3O_4, after Wright *et al.* (2002). Circles are Fe^{3+} and squares are Fe^{2+}.

---

[6] Some authors of this investigation later proposed yet another, even more complicated structure (Senn *et al.*, 2011). Thus apparently this story is still not finished.

An interesting question is the nature of the state of Fe_3O_4 just above the Verwey transition and in the whole region $T_V \leqslant T \lesssim 450\,K$, in which, as we have already discussed, this material does not behave like a normal metal, but has the conductivity $\sigma(T)$ still increasing with temperature, see Fig. 7.15. Two qualitative pictures were proposed to explain this behavior. One attributes it to a strong electron–lattice interaction, and the transport properties of Fe_3O_4 in this temperature interval are explained in terms of the motion of mobile polarons – electrons heavily "dressed" by lattice polarization. Another picture uses the analogy of crystals vs liquids: if we could at least qualitatively consider the low-temperature charge-ordered state of Fe_3O_4 as some sort of electronic (Wigner) crystal, then above T_V it can behave as a "Wigner liquid" – with intersite correlations still relatively well developed, but random and with slow dynamics (and for really slow dynamics we should perhaps call this state not "Wigner liquid" but rather "Wigner glass"). Which of these pictures is closer to the reality in Fe_3O_4 is difficult to say at present.

It is worthwhile mentioning here that, apparently, charge ordering in Fe_3O_4 is such that it breaks inversion symmetry, so that in effect magnetite below the Verwey transition becomes ferroelectric, see for example van den Brink and Khomskii (2008) and the references cited therein; that is, it is multiferroic (simultaneously ferroelectric and magnetically ordered). These questions will be addressed in Chapter 8.

There are a few other (not so many) compounds with frustrated lattices and with charge ordering, for example Eu_3S_4; these are less well studied, but apparently the problems that we encounter in Fe_3O_4 are present to some extent in these systems as well.

7.5 Spontaneous charge disproportionation

A very interesting phenomenon connected with charge ordering occurs in some materials with nominally integer valence, but with a valence state which is intrinsically unstable. In some such systems there occurs spontaneous charge segregation, or charge disproportionation: the corresponding ions, for example with nominal configuration d^n, segregate into ions with one electron more and one less,

$$d^n + d^n \longrightarrow d^{n-1} + d^{n+1} .\tag{7.3}$$

Such phenomena occur not only in TM compounds: one of the best-known such materials is $BaBiO_3$, in which at low temperatures there occurs a structural phase transition with charge disproportionation, nominally

$$2Bi^{4+}(6s^1) \longrightarrow Bi^{3+}(6s^2) + Bi^{5+}(6s^0) .\tag{7.4}$$

Of course, similarly to the case of charge ordering discussed in the previous section, this charge disproportionation is never that strong – it is never one whole electron that is transferred from one site to another: the amplitude of charge variation is here rather weak, so this transition should probably be described rather as the formation of a charge density wave, cf. Section 7.3. But the intrinsic instability of the Bi^{4+} ionic state (as well as of some other ionic states such as Tl^{2+}, Pb^{3+}), well known in chemistry, definitely plays an important role here, being largely the driving force of this transition.

In transition metal compounds similar phenomena are observed in a number of materials, notably those formally containing $Fe^{4+}(t_{2g}^6 e_g^1)$, $Pt^{3+}(t_{2g}^6 e_g^1)$, $Au^{2+}(t_{2g}^6 e_g^3)$, and sometimes low-spin $Ni^{3+}(t_{2g}^6 e_g^1)$; we have already mentioned some of these in Chapter 4. One example is $CaFeO_3$ (Takano *et al.*, 1981), in which Fe^{4+} disproportionates into Fe^{3+} and Fe^{5+}; this is observed structurally, and also by the Mössbauer effect. There are other compounds containing Fe^{4+} which show similar charge disproportionation, for example $La_{2/3}Sr_{1/3}FeO_3$ (Takano *et al.*, 1981). These transitions are often simultaneously metal–insulator transitions: apparently the formation of Fe^{3+}–Fe^{5+} superstructures leads to opening of the gap in the electronic spectrum.

However, not all materials with Fe^{4+} exhibit this behavior: thus $SrFeO_3$ remains homogeneous and metallic. Most probably this is caused by the bigger size of Sr: the orthorhombic ($GdFeO_3$-type) distortion of $CaFeO_3$ is not present in cubic $SrFeO_3$, so the Fe–O–Fe angle is larger (180°); the effective d–d hopping is also larger and the bandwidth is broader. For this better metal, charge disproportionation does not occur.

A similar situation is observed in rare earth nickelates $RNiO_3$. The phase diagram of this family of compounds, presented in Fig. 3.47, shows that in most of them there occurs a metal–insulator transition with decreasing temperature, accompanied (or rather caused by) charge disproportionation, formally $2Ni^{3+} \longrightarrow Ni^{2+} + Ni^{4+}$ (of course again the actual degree of charge disproportionation is much smaller than 1). In this transition the orthorhombic high-temperature structure is transformed into the monoclinic structure, and there appear two inequivalent Ni sites (Alonso *et al.*, 1999; Mizokawa *et al.*, 2000), which can be associated with Ni^{2+} and Ni^{4+}. For smaller rare earths the tilting of NiO_6 octahedra ($GdFeO_3$-type distortion) is larger, consequently the d–d hopping t and the d bandwidth are smaller, and therefore the tendency toward charge disproportionation and the corresponding value of T_c increases. For larger rare earths the d bands are broader, which gradually suppresses charge disproportionation and with it the metal–insulator transition; and for the largest rare earth ion, La, the compound $LaNiO_3$ remains metallic, with all Ni ions equivalent. The situation here is thus similar to what we observe when we go from $CaFeO_3$ (with the small ion Ca and with charge disproportionation) to $SrFeO_3$ (with larger Sr, which remains homogeneous and metallic).

What are the conditions for spontaneous charge disproportionation in different materials? Apparently they are determined largely by local properties, even on the atomic level. C. M. Varma proposed the idea of "skipped valence" (Varma, 1988): by analyzing the systematics of atomic energies, in particular ionization potentials for different ionic states, he explained the tendency, well known in chemistry, that certain valences are much less stable than others. Thus in chemistry it is known that for example Bi^{4+} or Pb^{3+} practically do not exist, and in corresponding compounds they usually disproportionate into Bi^{3+} and Bi^{5+}, or Pb^{2+} and Pb^{4+}. Other examples of such "missing" or "skipped" valences are In^{2+} and Sb^{4+}. For TM elements, similarly, Pt^{3+} always segregates into Pt^{2+} and Pt^{4+}, etc.

An interesting example is presented by the column in the Periodic Table containing Cu, Ag, and Au (see the Periodic Table at the end of the book). For copper the state Cu^{2+} is well known and is quite stable. For silver one can stabilize Ag^{2+}, but this is rather difficult: Ag prefers to be either Ag^{1+} or Ag^{3+}. And for gold the divalent state Au^{2+} practically never occurs, at least in oxides: Au always occurs as Au^{1+} or Au^{3+}. A classical example is the system $Cs_2Au_2Cl_6$, which can be understood as a "derivative" of the ordinary perovskite $CsAuCl_3$: formally, in the latter Au should be Au^{2+}, but in fact it disproportionates into Au^{1+} and Au^{3+}, which order in the cubic lattice of the perovskite in a checkerboard fashion, and there appears a very strong distortion around $Au^{3+}(d^8)$ ions which can be understood as a strong JT distortion (the resulting configuration is a singlet $t_{2g}^6(z^2)^2$). In effect these two ionic species, Au^{1+} and Au^{3+}, occupy different specific positions in the crystal, and that is why one usually writes the formula of this compound as $Cs_2Au_2Cl_6 = Cs_2Au^{1+}Au^{3+}Cl_6$. Further, as in nominal Au^{3+} both e_g-electrons occupy the same orbital ($z^2\uparrow z^2\downarrow$), the resulting material is nonmagnetic. It is interesting that under pressure this charge disproportionation seems to disappear, and $Cs_2Au_2Cl_6$ becomes a homogeneous metal (Kojima *et al.*, 1994).

Going back to the compounds of $Fe^{4+}(t_{2g}^3 e_g^1)$ and low-spin $Ni^{3+}(t_{2g}^6 e_g^1)$ showing charge disproportionation (CaFeO$_3$ – Takano *et al.*, 1981; YNiO$_3$ – Alonso *et al.*, 1989; Mizokawa *et al.*, 2000), one can notice that, had it not been for this disproportionation, the ions Fe^{4+} and Ni^{3+} would have been strong Jahn–Teller ions, with one electron on the doubly degenerate e_g orbital. However, in all these cases there occurs no JT distortion and no orbital ordering, but charge disproportionation instead. It looks as if the system wants to "get rid of" orbital degeneracy, but here it does so not by JT distortion, but by "getting rid of the degenerate electron" itself: the states appearing after charge disproportionation, for example $Fe^{3+}(t_{2g}^3 e_g^2)$ and $Fe^{5+}(t_{2g}^3 e_g^0)$, are already nondegenerate, and similarly for Ni^{3+}. It is not really clear whether this incipient JT tendency of Fe^{4+} and Ni^{3+} is really necessary for the eventual charge disproportionation, but at least empirically this seems to be the case, or at least this is very helpful for such charge disproportionation. One can put forth some real theoretical arguments why this could be the case (Khomskii, 2005; Mazin *et al.*, 2008a). In a nutshell it is the gain in the Hund's energy in transferring an electron from one Fe^{4+} ion to the other that can stabilize this process (Fig. 7.18). In the state with charge disproportionation we have lost the Hubbard's on-site energy U, but gained the Hund's energy J_H. This process is definitely not favorable for strong Mott insulators, for which $U > J_H$. But when we approach the itinerant regime, U is effectively screened, while the Hund's coupling remains unscreened, see Section 2.2, and close to the Mott transition such charge segregation may become favorable (calculations show that one also needs a certain lattice distortion to promote this process; without such distortion purely electronic effects are still not sufficient to cause the transition). In any case, one can draw the qualitative phase diagram for JT systems as in Fig. 7.19: initially with the increase in electronic hopping t, or ratio t/U, the critical temperature of the cooperative JT effect increases, see for example Section 6.1.2 (eqs (6.15), (6.16)), but on approaching the metallic state for large t/U the

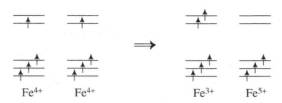

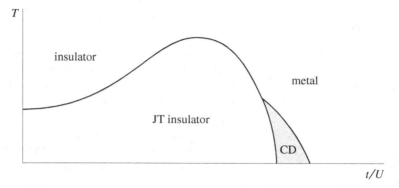

Figure 7.18 Schematic process of spontaneous charge disproportionation in systems such as $CaFeO_3$.

Figure 7.19 Possible form of the phase diagram of materials with Jahn–Teller ions and with small or negative charge-transfer gap close to the localized–itinerant crossover. CD denotes charge disproportionation.

JT distortion and orbital ordering should weaken and gradually disappear (nobody talks of orbital ordering in good metals, even those having different bands at the Fermi surface). And it turns out that close to this localized–itinerant crossover there may appear in such a system a novel state with spontaneous charge disproportionation, or with a charge density wave. This does not happen in every case, but it is a definite possibility.[7]

One more factor seems to be very important in these phenomena. One notices that in TM compounds charge disproportionation occurs in the case of late $3d$ elements with rather high nominal valence. As discussed in Section 4.1, this is just the situation in which we have very small or negative charge-transfer gaps and a lot of oxygen holes. Apparently this helps charge disproportionation and, maybe, it is even a necessary condition for

---

[7] These arguments work differently for Au^{2+}, for example in $Cs_2Au_2Cl_6$. The unstable $Au^{2+}(t_{2g}^6 e_g^3)$ would also be a strong JT ion, just as Cu^{2+}. But after charge disproportionation the state "$Au^{3+}(t_{2g}^6 e_g^2)$" is not in the high-spin state with $S = 1$, as for example Ni^{2+}, but in the low-spin state $t_{2g}^6 z^2\uparrow z^2\downarrow$ with $S = 0$. This state can be stabilized not by the Hund's rule energy J_H, but by a strongly enhanced JT distortion: as follows for example from eqs (3.21)–(3.23), when we put not one but n (here two) electrons on the lowest degenerate level, the energy (3.21) would become $E(n) = -gun + \frac{1}{2}Bu^2$, which, after minimization, would give $E_{JT} = -g^2n^2/2B$, that is it increases as n^2 – in this case 4 times. It may be just this extra energy gain which stabilizes charge disproportionation in $Cs_2Au_2Cl_6$.

its realization. Indeed, in this case we should not visualize this charge disproportionation as a real transfer of a d-electron from one TM ion to another – this would indeed be energetically very expensive. But with large participation of oxygen holes the corresponding charge redistribution may occur not so much on the d-shells themselves, but rather on the surrounding oxygens. That is, one can visualize the process

$$2Fe^{4+} \longrightarrow Fe^{3+} + Fe^{5+} \tag{7.5}$$

rather as

$$2(Fe^{3+}\underline{L}) \longrightarrow Fe^{3+} + (Fe^{3+}\underline{L}^2) \tag{7.6}$$

(here, as always, $\underline{L}$ denotes a ligand hole). Thus, in this process we transfer not the d-electrons themselves but rather move ligand holes, see Fig. 7.20. Apparently this would cost much less Coulomb energy: the oxygen orbitals are much more delocalized, and the state $\underline{L}$ is usually formed by a combination of six (or at least four) oxygens surrounding the TM ion (here Fe), so that the two holes in such a state can stay far from each other, on the "opposite sides" of Fe; this would strongly reduce the cost of charge disproportionation.

Concluding this discussion, we say a few words about a possible theoretical modeling of this situation. Of course in a complete theory one has to take into account all the factors discussed above, such as the role of ligands (oxygens), the details of the orbital structure, Hund's rule coupling, lattice relaxation, etc. But the main effect – the tendency toward charge disproportionation such as that of eq. (7.3) or eqs (7.5), (7.6) – can be described qualitatively by a model similar to the Hubbard model (1.6), but with effective on-site attraction instead of repulsion. Indeed, the process (7.3) means that there exists a tendency for two electrons to be at the same site, that is they effectively attract each other. Such a *negative-U Hubbard model* was often studied theoretically, in connection with charge disproportionation occurring for example in $BaBiO_3$ (Rice and Sneddon, 1981; Varma, 1988), but more often as a model for superconductivity. Indeed, the tendency of electrons to form a pair is the first step toward superconductivity. This tendency does not yet guarantee superconductivity: even after forming such pairs, their

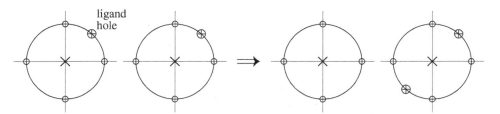

Figure 7.20 The process of spontaneous charge disproportionation in systems with small or negative charge-transfer gap; charge disproportionation involves predominantly the motion of ligand (e.g., oxygen) holes.

"fate" can be different. They may Bose-condense in momentum space, which would indeed give superconductivity; but they may also localize in the real space, forming some superstructure. This is apparently what happens in $CaFeO_3$ or in $YNiO_3$, and also in $BaBiO_3$, although doped $BaBiO_3$ ($Ba_{1-x}K_xBiO_3$) is indeed superconducting with a rather high $T_c \sim 30\,K$ (Cava *et al.*, 1988).

S.7 Summary of Chapter 7

The first degree of freedom we are dealing with in solids is the charge of the electrons. There are two general states of electrons in crystals: itinerant or band states, and localized electrons in the case of strong electron correlations. But there may occur extra types of ordering in solids, connected with the electron charge. The first one is the charge ordering, where the order parameter is the charge itself, or its modulation in space $\rho(r)$ (a scalar). Two other possible types of ordering connected with charge are ferroelectricity, where the average charge on a site or in a unit cell does not change, but there appears a dipole-like redistribution of charge leading to the appearance of nonzero average polarization $\boldsymbol{P}$ – a vector order parameter. The third type of ordering in the charge sector is orbital ordering, in which the electron distribution changes so as to preserve the inversion symmetry, that is the dipole moments do not appear but the electron density loses spherical symmetry and becomes anisotropic – there appear electric quadrupoles. Thus the real order parameters of orbital ordering are electric quadrupoles – second-rank tensors, although in many cases one can use a simpler description, see Chapter 6.

Charge ordering (CO) occurs typically in systems with noninteger number of electrons per site with, on average, noninteger valence of TM – for example in materials such as Fe_3O_4, or in doped systems such as $La_{1-x}Ca_xMnO_3$. In some such systems the extra electrons or holes may be delocalized, and the system would be metallic. But in certain cases, especially for commensurate electron concentration, say with one electron per two sites, there may occur charge ordering, after which the electrons become localized in an ordered fashion, either at particular sites or sometimes at particular bonds. Often such CO transitions are simultaneously also metal–insulator transitions: the CO states are typically insulating, whereas the disordered states may be (but not necessarily are!) metallic.

There are different possibilities of charge ordering in crystals. Probably the simplest is that with average electron density per site $n = \frac{1}{2}$ in a bipartite lattice, as for example in $La_{0.5}Ca_{0.5}MnO_3$. The most natural CO state in this case would be a checkerboard CO, with an alternation of occupied and empty sites (or an alternation of TM ions with different valences, in the case of half-doped manganites of Mn^{3+} and Mn^{4+}). Such a state could be favorable because it would at least minimize the energy of the Coulomb repulsion between the electrons, putting them as far as possible from each other. This state would thus be an analog of the Wigner crystal.

However Coulomb repulsion is not the only, and probably not the main mechanism of charge ordering in real crystals. Apparently electron–lattice or electron–phonon coupling plays a very important role here.

The second fact, related to the first, is that, although we usually describe such charge ordering as an ordering of, say, Mn^{3+} and Mn^{4+} ionic states, in reality the charge difference between different sites is almost always much smaller than the nominal one; it is typically $\sim 0.2e$. That is, one should rather speak about valences $Mn^{3.5\pm\delta}$, with $\delta \sim 0.1$–0.2. It is in most cases incorrect to visualize charge ordering as a real transfer of one whole electron: the actual charge difference is almost always rather small. This also explains why the Coulomb mechanism may not be very efficient for charge ordering. Physically this is connected with an important role of hybridization between TM and ligands, for example with oxygens. Because of this covalency the actual charges of ions in TM oxides are not those one would expect from the formal valences, but are usually smaller. Nevertheless, notation such as Mn^{3+}, Mn^{4+} has a definite meaning: though the actual charges *on the TM ions themselves* are not $+3$ or $+4$, the *quantum numbers* of the corresponding states, for example their spins, are those of the respective ionic states ($S = 2$ for Mn^{3+}, $S = \frac{3}{2}$ for Mn^{4+}, etc.).

The third complication, met already in half-doped manganites such as $La_{0.5}Ca_{0.5}MnO_3$ or $Pr_{0.5}Ca_{0.5}MnO_3$, is that there exists an alternative possibility instead of the simple checkerboard on-site charge ordering: the electrons may localize *not on TM sites*, but *on bonds* (such a state will be the analog of Peierls dimerization). Such a possibility was suggested experimentally for manganites, and a theoretical study confirmed that this may be the case in certain situations (and there may also exist "mixed" situations, in which the site- and bond-centered charge orderings coexist; such states, interestingly enough, would be ferroelectric – see Chapter 8 for more details).

The situation with CO for other electron concentrations, or other (e.g., frustrated) lattices, is even richer and more complicated. For manganites such as $R_{1-x}Ca_xMnO_3$ for $x \neq 0.5$ charge ordering still occurs often, especially in the overdoped regime $x > 0.5$. For $x < 0.5$ some manganites, such as $La_{1-x}Ca_xMnO_3$ or $La_{1-x}Sr_xMnO_3$, become metallic (with colossal magnetoresistance) but others, for example $Pr_{1-x}Ca_xMnO_3$, remain insulating and charge-ordered. But the character of charge ordering for $x > 0.5$ and (when present) for $x < 0.5$ is very different. For overdoped systems $x > 0.5$ the period of the resulting superstructure follows the doping. That is why sometimes one describes this state not as charge ordering, but rather as a state with charge density wave (CDW), by analogy with CDW states for example in transition metal dichalcogenides such as $NbSe_2$ or TaS_2. In those latter systems the formation of CDW is explained by the special features of the energy spectrum in the respective metallic states, with the so-called *nesting* of the Fermi surface (e.g., the presence of flat parts of the Fermi surface). The period of the wave vector of the resulting CDW superstructure is determined by the corresponding nesting wave vector $\boldsymbol{Q}$, and a change of band filling for example by doping leads to a change of this wave vector and, correspondingly, of the period of the CDW – which is observed in overdoped manganites. But charge ordering or CDW in $Pr_{1-x}Ca_xMnO_3$ for $x < 0.5$ behaves quite differently: the period of CO in this case remains the same as for $x = 0.5$, that is it corresponds on average to a doubling of the period in the *ab*-plane. The extra electrons, or extra Mn^{3+} ions, which should be present in this case are apparently distributed in this superlattice in some random fashion. This, in particular, makes this CO state not fully ordered, which may explain the re-entrant character of the corresponding phase transition in $Pr_{1-x}Ca_xMnO_3$ for $x < 0.5$.

In effect we see that the phase diagram of doped manganites $R_{1-x}A_xMnO_3$ ($R = $ rare earth, $A = $ Ca, Sr) is very asymmetric for the hole-doped ($x < 0.5$) and electron-doped ($x > 0.5$) cases: there may exist a ferromagnetic metallic (CMR) phase or the strange "partially disordered charge-ordered state" for $x < 0.5$, but for $x > 0.5$ we have typically an insulating state with a superstructure (CO or CDW) with the period determined by doping. The reasons for such strong electron–hole asymmetry are not completely clear: one possibility is that it can be connected with the orbital degrees of freedom and with the corresponding distortions around the Mn^{3+} sites.

One interesting possibility in partially doped TM compounds is ordering in the form of *stripes* – structures in which the doped charge carriers are ordered in chains. They are found in lightly hole-doped cuprates and nickelates, and also in overdoped manganites. Even more than charge ordering in half-doped manganites, the formation of stripes definitely cannot be explained by Coulomb forces. Usually one connects their origin with magnetic interactions: doped carriers may be accumulated at the antiferromagnetic domain walls (and they, in turn, may stimulate the formation of such domain walls). This mechanism can work for cuprates, and for oxygen-centered stripes in other materials. But there exists another quite general mechanism of stripe formation, not depending on the magnetism and caused by the inhomogeneous strain created in elastic media by "impurities" (ions with different valence). Probably stripes in overdoped manganites (which appear at much higher temperatures than magnetic ordering) are due to this mechanism, which in fact is universal and can also act in other materials with stripes.

Especially complicated may be charge ordering in frustrated lattices. The most famous example is the Verwey transition in magnetite Fe_3O_4 – probably the first metal–insulator transition observed in TM oxides. This transition, occurring in Fe_3O_4 at $T_V = 120\,K$, is usually explained by a charge ordering of Fe^{2+} and Fe^{3+} in the B-site sublattice of a spinel lattice. But this sublattice is heavily frustrated – it is formed by corner-shared tetrahedra (the pyrochlore lattice) and there is no unique way to order these ions (in a 50:50 ratio) in this sublattice. Actually, despite many years of research, the detailed character of charge ordering in Fe_3O_4 is still not established. The best picture, at present, is that shown in Fig. 7.17 – a structure which, in particular, strongly violates the so-called Anderson condition that there should be two Fe^{2+} and two Fe^{3+} ions in each tetrahedron (the system is still strongly frustrated even with this restriction!). In this structure the ratio of different valence states at each tetrahedron is not 2:2 but 3:1 and 1:3. This again demonstrates that the Coulomb interaction is not the main driving force of charge ordering in this system.

Most often, charge ordering is found in TM compounds with noninteger electron concentration (having formally TM ions with different valence). But there are some systems in which a similar phenomenon is seen in the situation with integer electron count, that is with nominally the same valence of TM ions. Examples are some systems with Fe^{4+}, for example $CaFeO_3$, or with Ni^{3+} ($YNiO_3$). In these systems there occurs *spontaneous charge disproportionation*, which may be represented formally as a "reaction" $d^n + d^n \rightarrow d^{n+1} + d^{n-1}$, say $2Fe^{4+}(d^4) \rightarrow Fe^{3+}(d^5) + Fe^{5+}(d^3)$. This phenomenon is apparently connected with the atomic properties. It is known in chemistry that some elements have "unfavorable" valence states, for example Au is never Au^{2+}, but can only be Au^{1+} or Au^{3+}, Pb can only be Pb^{2+} or Pb^{4+} but not Pb^{3+}, etc. Thus if in the average structure of a material the metal would be in such a "skipped" valence state, there may occur spontaneous charge disproportionation. Another option could be that the material would be metallic, in which case the notion of valence does not have a strict meaning, and all metallic ions may remain equivalent. Thus in some TM compounds, for example in $YNiO_3$, there occurs a phase transition from a homogeneous metallic state in which all Ni are equivalent to an insulating state with charge disproportion, $2Ni^{3+} \rightarrow Ni^{2+} + Ni^{4+}$, but for example $LaNiO_3$ remains homogeneous and metallic. Similarly, $CsAuCl_3$ at ambient pressure is an insulator with a superstructure with two inequivalent Au ions, Au^{1+} and Au^{3+} (usually its formula is even written as $Cs_2Au_2Cl_6$), but at high pressure it apparently transforms to a homogeneous metal.

The phenomenon of spontaneous charge disproportionation in TM compounds is usually met for TM ions which, in the average valence state, would be strong Jahn–Teller ions ($Fe^{4+}(t_{2g}^3 e_g^1)$; the low-spin $Ni^{3+}(t_{2g}^6 e_g^1)$; $Au^{2+}(t_{2g}^6 e_g^3)$). It seems that this orbital degeneracy plays some role and makes such ionic states less stable and more susceptible to charge disproportionation. But more important seems to be that usually this phenomenon is found for TM ions with an unusually high oxidation state and with small or negative charge-transfer gaps, see Chapter 4. Apparently the large contribution of ligand (oxygen) states, specifically oxygen holes, is very important for charge disproportionation. For negative charge transfer gaps one would have, for example, not $Fe^{4+}(d^4)$ but rather $Fe^{3+}(d^5)\underline{L}$, where $\underline{L}$ denotes a ligand (oxygen) hole. Consequently the "reaction" $2Fe^{4+}(d^4) \rightarrow Fe^{3+}(d^5) + Fe^{5+}(d^3)$ should rather be written $2Fe^{3+}(d^5)\underline{L} \rightarrow Fe^{3+}(d^5) + Fe^{3+}(d^5)\underline{L}^2$, so it is not so much the d-electrons which move from site to site but rather it is the ligand (oxygen) holes which redistribute between sites. Consequently, such a redistribution would cost much less Coulomb (Hubbard) energy, and would be facilitated. Once again, though the actual charges may be mainly on the oxygens, the *quantum numbers* of the resulting states will be those of Fe^{3+} and Fe^{5+}. In this sense one can still use this terminology – but remember that the real charges on the TM ions do not change much, the main changes occur on the oxygens.

8

Ferroelectrics, magnetoelectrics, and multiferroics

As discussed at the beginning of Chapter 7, there can exist different types of ordering phenomena connected with charge degrees of freedom. Examples are ordering of charges themselves (charge "monopoles"); ordering of electric dipoles, giving ferroelectricity (FE); or ordering of electric quadrupoles, which happens in orbital ordering. We have already discussed the first and third possibilities; we now turn to the second.

Ferroelectricity is a broad phenomenon, in no way restricted to TM compounds. There are ferroelectrics among organic compounds, in some molecular crystals, in systems with hydrogen bonds. But the best, and most important in practice, are ferroelectrics on the basis of TM compounds such as the famous $BaTiO_3$, or the widely used $Pb(ZrTi)O_3$ ("PZT"). And it is in these compounds that one also sometimes meets a very interesting interplay of ferroelectricity and magnetism – the field now known mostly as multiferroicity. By multiferroics (Schmid, 1994) we refer to materials which are simultaneously ferroelectric and magnetic – possibly ferro- or ferrimagnetic, although such cases are rather rare, most of the known multiferroics being antiferromagnetic. (Sometimes ferroelastic systems are also included in this class.) In this chapter we discuss these classes of compounds, paying attention mostly to the microscopic mechanisms of ferroelectricity and its eventual coupling to magnetism.

A general treatment of ferroelectricity, dealing mainly with the macroscopic aspects of ferroelectrics and their phenomenological description, with special attention paid to practical applications, may be found in many books, for example Megaw (1957), Lines and Glass (1977), Scott (2000), Gonzalo (2006), Blinc (2011).

8.1 Different types of ferroelectrics

When we discuss the microscopic origin of magnetism in strong magnets, at least conceptually the situation is simple: due to strong electron correlations the electrons in solids become localized, there appear localized spins or localized magnetic moments in the system, and their exchange interaction leads to one or other type of magnetic ordering. All the variety of behavior of different magnetic materials is connected with particular details of this general picture.

In contrast to magnetism, the situation with ferroelectricity is much more diverse and complicated. There exist several quite different types of ferroelectrics, with different mechanisms responsible for ferroelectricity. This makes the phenomenon more difficult to explain.

Ferroelectricity is, first of all, the appearance of nonzero electric polarization in a system. But having nonzero polarization does not yet guarantee that the material will be ferroelectric. For real ferroelectricity we need to have the possibility of switching the polarization by applying an appropriate electric field, which would produce the famous polarization loop – the same hysteresis loop as observed in ferromagnets in a magnetic field. If polarization always points in one direction and cannot be reversed, we have *pyroelectrics*. But if one can reverse the polarization direction by applying the opposite electric field, the system would be truly ferroelectric.

Without trying to be complete, we will discuss here several typical classes of ferroelectrics and several mechanisms of this ordering. Conceptually the simplest case could be, for example, materials made of molecules with intrinsic dipole moments, such as H_2O or HCl. If these molecules are allowed to orient in such a way that all local dipoles point in the same direction, we would get net electric polarization P. Such local dipoles may be disordered at high temperatures – we would then have the paraelectric phase. And they can order at a certain temperature, below which we would have the ferroelectric phase. The classical example of ferroelectrics of this type is $NaNO_2$.

In other cases such dipole moments can only *appear* below T_c: above T_c there may be no dipole moments at all. These two different types of ferroelectrics are called, respectively, order–disorder ferroelectrics and ferroelectrics of the displacement type. To elucidate whether a particular material belongs to one or the other class is not always easy; often we have an intermediate situation where there is no long-range ordered ferroelectricity above T_c, but in a certain temperature interval above T_c the system behaves as having local disordered dipoles.

From the phenomenological point of view, a very useful concept in the field of ferroelectricity is that of *Born effective charges*. When we move the ions in a solid, the polarization thus created contains, first of all, the contribution from the ionic charges themselves,

$$P_{\text{ion}} = \sum_i (r_i - r_{i0})Z_i , \qquad (8.1)$$

where the sum goes over all ions with charges Z_i, with initial positions r_{i0} changed by the displacement to r_i. However by shifting the ions we also change the details of the chemical bonds between them, as a result of which there occurs electron redistribution, that is a shift of the electron density, which also contributes to the polarization. Thus for example if we start from the regular chain of H^+ and Cl^- ions, and then shift these ions in such a way as to form HCl molecules (see Fig. 8.1), the original distribution of the electron density (the valence electrons in the ionic picture are located on the Cl^- ions) changes as shown in Fig. 8.1(b) (the electron density is shaded in this figure): the electrons at each Cl ion move "to the left" to form a chemical bond with the H^+ ion approaching it.

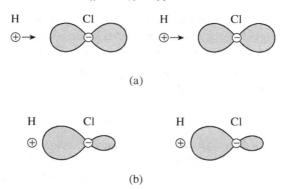

Figure 8.1 (a) Regular chain of H^+ and Cl^- ions, with the density of the valence electrons shaded. (b) HCl molecules formed after shifts of ions shown in (a) by arrows. One sees that the valence electrons also shift during this process.

In effect we have moved the positive H^+ ions (i.e., protons) to the right and simultaneously the negative electron density to the left, thus increasing the total polarization appearing during this process. That is, the total polarization will be given by the sum of the ionic polarization (8.1) and the electronic polarization

$$P_{\text{total}} = P_{\text{ion}} + P_{\text{el.}} = \sum_i (r_0 - r_{i0})Z_i + P_{\text{el.}} . \tag{8.2}$$

One can describe this increase in polarization appearing with the shift of ions by saying that it looks as if the effective charge of the moving ions increases: instead of the charges Z_i in (8.1), (8.2) of, say, positive ions (in our example $Z_i = +1$ for H^+) we have an increased charge Z_i^* such that the total polarization, eq. (8.2), could be represented as

$$P_{\text{total}} = \sum_i (r_i - r_{i0})Z_i^* \tag{8.3}$$

(the electronic polarization $P_{\text{el.}}$ is in the first, linear approximation also proportional to the atomic displacements $r_i - r_{i0}$). The value Z_i^* is called the Born effective charge; the formal definition of Z_i^* is thus (see, e.g., Spaldin, 2012)

$$Z_i^* = \frac{v}{e} \frac{\partial P_i}{\partial r_i}, \tag{8.4}$$

where v is the unit cell volume and e is the electron charge.[1]

Born effective charges may sometimes be much larger than the nominal ionic charges corresponding to the formal valence of the ions (remember that the actual charge at the ion is usually less than that corresponding to the formal valence). Thus the Born effective charge of Ti^{4+} in $BaTiO_3$ is not $+4$ but $+7.1$, and the Born effective charge of the oxygen

---

[1] In principle the Born effective charge is a tensor, $(Z_i^*)_{kl} \sim \partial(P_i)_k / \partial(R_i)_l$ (Spaldin, 2012), because the shift in electron density following the displacement of an ion may be in a different direction. Often this factor is ignored, at least in the simplified treatment.

is here -5.8 instead of the formal charge -2 of O^{2-} (Ghoses *et al.*, 1998). Large Born effective charges often indicate a tendency toward ferroelectricity. They also have important manifestations in many other properties of the corresponding solids, in particular in their optical properties, increasing strongly the oscillator strength of some phonons; one even talks sometimes about "charged phonons", see for example Rice and Choi (1992).

Going back to microscopic mechanisms of ferroelectricity, one can specify several different mechanisms and different types of ferroelectrics.

1. *Ferroelectrics from structural units with permanent dipole moments.* We have already mentioned the first class of ferroelectrics – materials made of building blocks with permanent dipole moments, such as dipole molecules, which order below a certain temperature. An example is the already mentioned $NaNO_2$, of Fig. 8.2(a): below T_c, in the ferroelectric phase, the "molecules" NO_2 with dipole moments are all oriented in the same direction, and above T_c this orientation is random.

2. *Ferroelectrics with hydrogen bonds.* In these systems the H ions have two positions, for example close to one or the other oxygen. At high temperatures such hydrogens occupy these positions at random, but at low temperatures they may order in particular positions, which may give ferroelectricity (see Fig. 8.2(b)). One example of such ferroelectrics is the Rochelle salt – actually the first material in which ferroelectricity was discovered. Another well-known example is KH_2PO_4 ("KDP"). Even one form of ice – the so-called ice XI – is ferroelectric (and this ferroelectric ice was even found recently in the universe; Fukazawa *et al.*, 2006).

3. *Ferroelectricity due to lone pairs.* Some elements, such as Bi or Pb, can exist in particular valence states such as $Bi^{3+}(6s^2)$ and $Bi^{5+}(6s^0)$, or $Pb^{2+}(6s^2)$ and $Pb^{4+}(6s^0)$. We have already mentioned these elements in Section 7.5 as "valence skippers." Here we are interested in another feature: often in the materials containing for example Bi^{3+}, the two outer electrons do not participate in the formation of chemical bonds, but form what is known in chemistry as *lone pairs* or *dangling bonds*. Such lone pairs (not pure $6s^2$, but hybridized with $6p$ states on the same Bi, and often also with p states of ligands) can have a certain orientation and point in a particular direction (Fig. 8.2(c)). There may be several equivalent such directions. If so, these lone pairs may have a cooperative ordering below a certain temperature, and as any such lone pair forms in fact a dipole moment, their ordering can produce ferroelectricity. Indeed, many materials containing Bi^{3+} and Pb^{2+} are ferroelectric.

Usually the positive influence of Bi^{3+} and Pb^{2+} on ferroelectricity is explained classically by saying that these ions have large polarizability, and this can lead to the well-known polarization catastrophe, which is the traditional classical explanation of ferroelectricity. However the microscopic picture is that of easily oriented lone pairs, described above. The examples of ferroelectrics with this mechanism are $BiFeO_3$ – one of the best multiferroics; or the many so-called Aurivillius phases – layered materials containing Bi_2O_2 blocks alternating with perovskite-like blocks, with the general formula $Bi_{2m}A_{(n-m)}B_nO_{3(n+m)}$, for

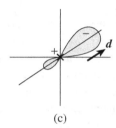

(a)

(b)

(c)

Figure 8.2 (a) Possible formation of a ferroelectric state in systems with structural units (e.g., molecules) with permanent dipole moments. (b) Ferroelectricity due to hydrogen bonds. (c) Ferroelectricity driven by ordering of lone pairs in ions such as Bi^{3+} or Pb^{2+}.

example $SrBi_2Ta_2O_9$. There are quite good ferroelectrics among materials of this class, which nowadays attract much attention because of their good properties, in particular low fatigue. Some of them, containing magnetic TM ions, could also be multiferroics, see below.

4. *"Geometric" ferroelectrics.* In some inorganic materials the origin of ferroelectricity can be traced back to some crystallographic distortions, which by themselves are not ferroelectric, or at least their driving force is not connected with ferroelectricity, but in which ferroelectricity appears as a "byproduct." Such is the nature of ferroelectricity in hexagonal systems such as $YMnO_3$. The structure of these compounds is shown schematically in Fig. 8.3. Similar to perovskites, the size mismatch between different ions leads to a structural distortion, consisting mainly of rotation of rigid building blocks – the MO_6 octahedra in perovskites, and MO_5 trigonal bipyramids here, see Fig. 8.3(b). This distortion is caused by the tendency toward close packing. In perovskites ABO_3 the resulting rotation and tilting of the BO_6 octahedra, leading for example to the orthorhombic $Pbnm$ structure, see Section 3.5, gives a quite short distance between the A ion and one of the apical

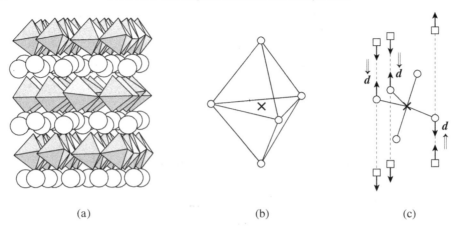

(a)　　　　　　　　　(b)　　　　　　　　　(c)

Figure 8.3 The mechanism of "geometric" ferroelectricity in YMnO$_3$. (a) Schematic structure of YMnO$_3$. (b) Coordination of Mn^{3+} in YMnO$_3$ (the cross is the Mn^{3+} ion, circles are oxygens). (c) Distortions in YMnO$_3$ due to tilting of MnO$_5$ trigonal bipyramids, leading to the formation of a net dipole moment (the squares are Y ions, local dipoles are shown by double arrows).

oxygens, oxygen O1 of Fig. 3.45. But the resulting A–O dipole moments in different unit cells of perovskites point in opposite directions and cancel out, and thus do not lead to a net ferroelectricity. However a similar phenomenon in hexagonal manganites RMnO$_3$, such as YMnO$_3$, does produce net uncompensated dipole moments due to nonlinear coupling with the polar mode (van Aken *et al.*, 2004), see Fig. 8.3(c), where we denote Y ions by squares and Mn and O by crosses and circles, respectively, as usual; the resulting dipole moments are shown by double arrows. As a result, such rotation and tilting of MO$_5$ blocks makes these hexagonal manganites ferroelectric at rather high temperatures, with $T_c \sim 800$–1000 K. At lower temperatures ~ 70–80 K there appears a magnetic ordering in these systems, so these materials are multiferroic.

5. *Ferroelectricity due to charge ordering.* Yet another mechanism which can lead to ferroelectricity is charge ordering (CO), often observed in TM compounds, especially those formally containing TM ions with different valence. Charge ordering occurring in this situation usually takes the form of localization of "extra" electrons at particular sites in an ordered fashion, see for example Fig. 8.4(a). In the case of Fig. 8.4(a) there will be no net polarization: this structure has an inversion symmetry with an inversion center for example at each "+" site; it has mirror planes, etc. But nonzero polarization $P \neq 0$ (a vector) requires that the inversion symmetry should be broken, see eq. (8.11) below; this is not the case in the structure of Fig. 8.4(a).

However in principle such an "extra" electron can be delocalized over two sites (forming an analogue of the H$_2^+$ molecular ion), Fig. 8.4(b); this is the bond-centered charge ordering discussed in Section 7.1, see Fig. 7.4. In this case the sites would have the same charge, for example $+\frac{1}{2}$, and be equivalent, but the bonds would become inequivalent: typically the bond on which such an electron is localized would become shorter, see Fig. 8.4(b).

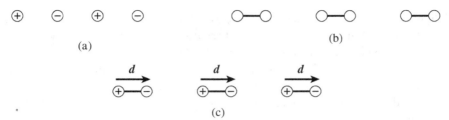

Figure 8.4 Formation of a ferroelectric state due to combination of site- and bond-centered charge ordering, after Efremov *et al.* (2004).

For example, as discussed in Section 7.1, in half-doped manganites such as $Pr_{1/2}Ca_{1/2}MnO_3$ there appears charge ordering which is usually treated as site-centered (checkerboard) ordering. But there exists also an alternative picture – with the bond-centered ordering, the so-called Zener polarons (Daoud-Aladine *et al.*, 2002). In any case, it is clear that there will be no dipole moments in this case either; this structure is also centrosymmetric, with the inversion centers in the middle of the bonds.

If, however, we form a "combined" ordering, with ordered ions of different charges, but also with alternating short and long bonds (Fig. 8.4(c)), we would create "molecules" with dipole moments and all such moments would be parallel. As a result, this combination of inequivalent sites and inequivalent bonds would give ferroelectricity. From the symmetry point of view we see that in the structure of Fig. 8.4(c) we lose the inversion symmetry, which was present in cases (a) and (b).

Interestingly enough, theoretical treatment (Efremov *et al.*, 2004) shows that this situation, intermediate between site- and bond-centered charge ordering, may indeed be realized in certain situations. The corresponding mechanism can be at work in systems such as $Pr_{1/2}Ca_{1/2}MnO_3$ or in nickelates $RNiO_3$ (van den Brink and Khomskii, 2008), in which there occurs spontaneous charge disproportionation (formally the "reaction" $2Ni^{3+} \longrightarrow Ni^{2+} + Ni^{4+}$). But more often we meet a situation in which there are ions with different charges (often simply different elements), and dimerization occurs on top of that; such examples will be treated in the next section, devoted to multiferroics. Apparently $TbMn_2O_5$ (Hur *et al.*, 2004) and the quasi-one-dimensional multiferroic Ca_3CoMnO_6 (Choi *et al.*, 2008) belong to this class, see below. Similarly, there exist organic ferroelectrics with this mechanism, for example the so-called charge-transfer (donor–acceptor) molecular systems, in which the "sites" (molecules) are inequivalent due to charge transfer (or due to neutral–ionic transitions), and the bonds become inequivalent (e.g., due to Peierls-like dimerization; see Horiuchi and Tokura, 2008). Another similar possibility is when the bonds are inequivalent from the very beginning, simply because of the particular structure of the material, and site-centered charge ordering appears on top of that. This is apparently the case in some organic systems (Monceaux *et al.*, 2001), and the same mechanism may also work in $LuFe_2O_4$ (Ikeda *et al.*, 2005). In the latter there exist bilayers containing Fe ions with mixed valence $Fe^{2.5+}$, and with decreasing temperature there appears charge ordering in this system, after which the two "partner" layers of each bilayer could become charged differently. As a result we would have a situation exactly similar to that of Fig. 8.4(c): inequivalent "bonds" (different interlayer distances within and between bilayers), with the average charges of each layer also different due to charge ordering. Depending on the ordering of the resulting dipoles (parallel or antiparallel), $LuFe_2O_4$ would become either ferroelectric or antiferroelectric. Later experiments, however, have cast some doubt on this picture: according to de Groot *et al.* (2012), the charge ordering in $LuFe_2O_4$ does not produce dipoles in bilayers. Thus the question of ferroelectricity in $LuFe_2O_4$ is still open.

Ferroelectricity due to charge ordering, however, seems to be firmly established in the magnetite Fe_3O_4 below the Verwey transition (Section 7.4), see the discussion and references in van den Brink and Khomskii (2008) and Senn *et al.* (2011). Note also that this phenomenon occurs in the magnetically ordered state, thus Fe_3O_4 below the Verwey transition is multiferroic.

6. *Ferroelectricity in transition metal perovskites; d^0 vs d^n problem.* The most common ferroelectrics among TM compounds are the materials similar to $BaTiO_3$. There are quite a lot of ferroelectrics among TM perovskites: besides $BaTiO_3$, these are for example $KNbO_3$, $KTaO_3$, etc. We have put these materials at the end of the list, but actually these are the most numerous and the most important ferroelectrics.

When one looks at perovskite ferroelectrics with TM ions, one immediately notices one fact: whereas magnetic perovskites such as $LaMnO_3$, $GdFeO_3$, etc. have partially filled d-shells d^n, $0 < n < 10$, practically all perovskite ferroelectrics ABO_3 have B ions with an empty d-shell, with configuration d^0. Such are $BaTiO_3$ and $PbTiO_3$ with $Ti^{4+}(3d^0)$, $KNbO_3$ with $Nb^{5+}(4d^0)$, $KTaO_3$ with $Ta^{5+}(5d^0)$, etc. It looks as if there is mutual exclusion of ferroelectricity and magnetism, at least in perovskites: perovskites with TM with empty d-shells may be ferroelectric (though not necessarily are!), but as soon as TM ions have at least one real d-electron, the material could be magnetic, but definitely not ferroelectric. There are hundreds of magnetic perovskites, see for example the tables in Goodenough and Longo (1970), and hundreds of ferroelectric perovskites (even more extensive data on them are collected in Mitsui *et al.*, 1981), but the comparison of these two sets of extensive tables shows that there is practically no overlap between them! The question is, why? This question is especially important in the modern quest for multiferroics – materials combining magnetism and ferroelectricity, see below.

There appear, at first glance, a few exceptions from this rule. The best known is $BiFeO_3$ with "magnetic" $Fe(3d^5)$, which is a very good ferroelectric with $T_c \sim 1100\,K$ and with quite large polarization reaching 60–$90\,\mu C/cm^2$ (in $BaTiO_3$ it is $\sim 60\,\mu C/cm^2$), and which becomes magnetically ordered at $T_N = 640\,K$. Other apparent exceptions could include the ferromagnetic insulator $BiMnO_3$ (though it is not really clear whether it is ferroelectric or antiferroelectric) and the recently synthesized $PbVO_3$ (Belik *et al.*, 2005) and $BiCoO_3$ (Belik *et al.*, 2006) (which are rather not ferroelectrics, but pyroelectrics: the polar distortion in them is so strong that one cannot switch the polarization by any reasonable electric field). But all these materials contain either Bi^{3+} or Pb^{2+} and, as we have discussed above, these elements often produce ferroelectricity due to their lone pairs. Apparently this is what really happens in all these systems: ferroelectricity is indeed caused by Bi or Pb and has nothing to do with the TM ions such as Fe, Mn, V, or Co, whereas in most ferroelectric perovskites such as $BaTiO_3$ or $KNbO_3$ it is these TM ions with empty d-shells which are responsible for the appearance of ferroelectricity. Thus the problem of this mutual exclusion of ferroelectricity and magnetism in this class of materials is still a very important problem, and a better understanding of the nature of this mutual exclusion could open ways toward creating new multiferroics.

The present explanation of the nature of ferroelectricity in compounds of this class, such as $BaTiO_3$, is (in a simplified form) the following: the TM ions, for example $Ti^{4+}(3d^0)$, may remain in the centers of the respective TiO_6 octahedra, Fig. 8.5(a), or may shift from the center toward one oxygen, Fig. 8.5(b). (In reality the Ti ions in $BaTiO_3$ at low temperatures shift along the [1 1 1] directions, but at higher temperatures there exists a tetragonal phase in which the average displacement of Ti ions is indeed along the x-, y-, or z-direction, i.e. toward one oxygen.) In the simplified picture of Fig. 8.5, the Ti ion in this process, Fig. 8.5(b), forms a strong covalent bond with one oxygen instead of having weaker bonds with two oxygens as in Fig. 8.5(a).[2] And if such a process leads to energy gain, that is if the energy gain by strengthening one bond (in Fig. 8.5(b) – with the oxygen to the right) would exceed the energy loss at the other bond (and the elastic energy loss of the corresponding distortion), then the process of shifting the Ti ions from the centers of O_6 octahedra would be favorable and occur spontaneously. If all such displacements, at every unit cell, are in phase, that is occur in the same direction, we get ferroelectricity. (And if they alternate from one unit cell to the next, we can talk of antiferroelectricity.[3]) Thus the question is: when is such a displacement favorable, and why does one need TM ions with empty d-shells for that?

$$O \!-\! Ti \!-\! O \qquad \Longrightarrow \qquad O\text{-----}Ti\!-\!O$$

$$\text{(a)} \qquad\qquad\qquad\qquad \text{(b)}$$

Figure 8.5 Qualitative explanation of the formation of a ferroelectric state in systems such as $BaTiO_3$ due to the formation of strong covalent bonds of Ti with one (or three) neighboring oxygens.

---

[2] This covalent bond is established between the oxygen p orbitals directed toward the TM ion (e.g., Ti^{4+}) and the corresponding d orbitals (the strongest covalency will be with one of the e_g orbitals directed toward this oxygen). One can say that the O^{2-} ion "donates" an electron which it initially "took" from Ti, back to the empty d states of Ti^{4+} ions.

[3] The notion of antiferroelectricity, strictly speaking, is not well defined, whereas ferroelectricity is: ferroelectricity is a spontaneous breaking of inversion symmetry, but inversion is not broken in antiferroelectrics, that is, formally an antiferroelectric transition cannot be discriminated from any other structural phase transition in which inversion is not broken. Indeed, any structural transition is accompanied by a certain charge redistribution, for example within the unit cell, and we can say that there appear corresponding local dipole moments. But their definition depends on how we subdivide the unit cell, so this is not a unique characterization of the transition. Also in antiferroelectrics in the new doubled unit cell the polarization in one half of it (in the "old" unit cell) points for example to the right, but in the other half to the left, so that these cancel. Thus, rigorously speaking, antiferroelectric transitions are not different from other structural transitions, whereas ferroelectricity is indeed special: in a transition to the ferroelectric state the inversion symmetry is spontaneously broken. Nevertheless, sometimes it is useful to refer to antiferroelectricity as of a special subclass of general structural transitions: we expect that the anomalies, for example in the dielectric constant at the transition would be quite strong, whereas they are weak (although still present) for most other structural transitions.

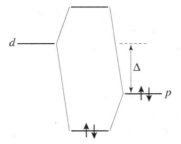

Figure 8.6 Possible energy gain due to the formation of a strong Ti–O covalent bond.

Very crudely, one can argue as follows (and real *ab initio* calculations support this picture; see Cohen, 1992): by shifting for example Ti^{4+} toward one oxygen (the one to the right in Fig. 8.5), one increases the p–d hybridization by a certain amount, $t_{pd} \longrightarrow t_{pd} + \delta t_{pd}$. Then the corresponding bonding (dp) orbital would go down, Fig. 8.6, and the corresponding energy would be

$$E_r \sim -\frac{(t_{pd} + \delta t_{pd})^2}{\Delta} = -\frac{t_{pd}^2}{\Delta} - \frac{2t_{pd}\,\delta t_{pd}}{\Delta} - \frac{(\delta t_{pd})^2}{\Delta}. \tag{8.5}$$

However the p–d hopping matrix element to the other oxygen, the one to the left in Fig. 8.5, would decrease, $t_{pd} \longrightarrow t_{pd} - \delta t_{pd}$, which would give the bonding energy with that oxygen

$$E_l \sim -\frac{(t_{pd} - \delta t_{pd})^2}{\Delta} = -\frac{t_{pd}^2}{\Delta} + \frac{2t_{pd}\,\delta t_{pd}}{\Delta} - \frac{(\delta t_{pd})^2}{\Delta}. \tag{8.6}$$

In the resulting energy (the sum of expressions (8.5) and (8.6)) the terms linear in δt_{pd} (or in the displacement u, if we take $\delta t_{pd} \sim gu$) would cancel, but we still get an energy gain

$$\Delta E_{\text{total}} \sim -\frac{2(\delta t_{pd})^2}{\Delta} = -\frac{2g^2 u^2}{\Delta}, \tag{8.7}$$

that is such off-center displacement of Ti^{4+} would give a net energy gain $\sim (\delta t_{pd})^2 \sim u^2$ (quadratic in the displacement). The elastic energy loss is also $\sim u^2$ (it is $\frac{1}{2}Bu^2$, where B is the bulk modulus). Thus if the energy gain (8.7) exceeds the energy loss $\frac{1}{2}Bu^2$, such distortions would occur spontaneously and the material would become ferroelectric. But if we compare this situation with for example that in the systems with the Jahn–Teller effect (Section 3.2), we see that they are different: in the JT systems the electronic energy gain is linear in distortion, and for small u it always exceeds the elastic energy loss, see eq. (3.21). That is why JT systems are always unstable with respect to distortions reducing the symmetry. In contrast, here both energies are quadratic in distortion. That is, we need the electron–lattice coupling g to exceed a certain critical value so as to make the ferroelectric distortion favorable. That is why not all TM perovskites, even those containing d^0 ions, are ferroelectrics. Thus $BaTiO_3$ is, but $SrTiO_3$ is not (although it is very close to becoming ferroelectric, and it does become ferroelectric under strain, and even after isotope substitution $^{16}O \to {}^{18}O$). In that sense the Jahn–Teller effect is a "theorem," but the appearance of ferroelectricity in TM compounds is a "game of numbers."

In fact, the comparison of the origins of ferroelectricity and the JT effect can be developed much further (Bersuker, 2006). One can consider ferroelectric instability as a consequence of a *pseudo-JT effect*, or second-order JT effect: lattice distortions (breaking inversion) mix the valence band and the conduction band states of different symmetries (with different parities, e.g. *s* and *p*, or *p* and *d*), and the corresponding energy gain can be the reason for the ferroelectric instability. It is clear in this picture that one needs some critical strength of the electron–lattice coupling to overcome the elastic energy loss.

The simple qualitative explanation of the appearance of ferroelectricity in materials such as $BaTiO_3$, described above, on the one hand tells us that TM ions such as Ti in $BaTiO_3$ or Ta in $KTaO_3$ are indeed instrumental in providing the mechanism of ferroelectricity in these compounds, and on the other hand allows us to understand the conditions for the appearance of ferroelectricity in these systems and, in particular, gives us at least some hints why indeed the d^0 configurations are especially favorable for ferroelectricity. One such factor can be illustrated in Fig. 8.7 (cf. Fig. 8.6): if the *d* level is not empty but there is for example one electron on it, then after establishing the strong covalent bond with one oxygen (the one on the right in Fig. 8.5) we would occupy not only the bonding orbitals (by two electrons) but also the antibonding orbital (by one electron). Thus we would lose on this antibonding orbital half of the energy gain (8.7) which we had in the case of an empty *d*-shell. This energy loss would make this process less likely.

However it seems that this effect is not the only one, and maybe not even the main one by which real electrons in *d*-shells seem to suppress ferroelectricity. Thus, it would not explain why for example the perovskites $RCrO_3$ (where *R* is a rare earth) containing $Cr^{3+}(t_{2g}^3 e_g^0)$ (so-called orthochromates) are not ferroelectric in the paramagnetic phase (the situation may be different in a magnetically ordered phase): the e_g orbitals, which have the strongest hybridization with the oxygens, are empty in Cr^{3+}, that is it looks as if we are dealing here with the situation of Fig. 8.6 and not Fig. 8.7 (if by *d* levels there we mean the e_g orbitals). Another possible mechanism, confirmed by numerical calculations, is presented in Khomskii (2006). Qualitatively this mechanism may be explained as

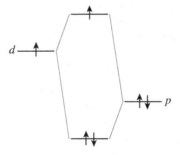

Figure 8.7 Illustration of one mechanism explaining how the presence of "real" *d*-electrons on a transition metal ion can weaken the tendency to form a covalent bond with the oxygen and hence suppress the tendency toward ferroelectricity.

follows: the strong covalent bond between the TM ion and one oxygen means the formation of a singlet state

$$\frac{1}{\sqrt{2}}(d\uparrow p\downarrow - d\downarrow p\uparrow)\,. \tag{8.8}$$

Now, suppose there are other localized d-electrons at this TM ion (e.g., on the t_{2g} orbitals of $Cr^{3+}(t_{2g}^3 e_g^0)$ in the example mentioned above). These localized electrons would interact with the d-electron (e.g., the e_g-electron) participating in the valence bond (8.8) by the Hund's rule exchange. Suppose these localized spins are say $\uparrow$. Then the Hund's coupling would be favorable for the first half of the wavefunction (8.8), but not for the second half. In other words, localized spins "do not like" the singlet state (8.8), and they tend to act as "valence bond breakers" – similarly to magnetic impurities which act as pair breakers for the singlet Cooper pairs in ordinary superconductors. This may be another mechanism by which the real d-electrons can suppress ferroelectricity, which can contribute to the mutual exclusion we are discussing.

Note that all these mechanisms do not really forbid ferroelectricity for d^n configurations with $n \neq 0$; they only make it much less likely. In that sense it seems that the mutual exclusion of magnetism and ferroelectricity is indeed not a "theorem," but rather that everything depends on specific parameters such as the strength of the electron–lattice coupling g in (8.7), the elastic moduli of the crystal, the occupation of d-shells, the strength of the Hund's rule coupling, etc. One cannot exclude that there may exist situations in which TM ions with partially filled d levels would still give ferroelectricity. This seems not very likely but, strictly speaking, not forbidden, and such cases have indeed been discovered experimentally – see for example Sakai *et al.* (2011).

All these considerations will be very important for us when we discuss multiferroics in Section 8.3.

7. *Electronic contribution to ferroelectricity.* Up till now we mostly talked about ferroelectricity induced by the shifts of ions in the crystal. However, as always, lattice distortions are accompanied (or caused) by the corresponding modifications of the electronic wavefunctions and by a redistribution of the electron density; the question of which of these effects is primary is, in a sense, a chicken-and-egg problem. Nevertheless the question of what is the relative role of electronic and ionic contributions to the polarization is not an idle one. One can also ask whether there exists a purely electronic mechanism of ferroelectricity not depending crucially on the lattice (of course even in this case the lattice would respond, adding to the resulting polarization).

Recent theoretical calculations demonstrate that indeed electronic contributions to ferroelectricity do exist, and in some cases they may be at least as important as the ionic ones, in particular in some multiferroic compounds. The situation, however, may differ significantly from system to system: if in some case, for example $TmMnO_3$ (Picozzi *et al.*, 2006), the electronic contribution either dominates or at least gives $\sim 50\%$ of the total polarization, in other cases, for example $TbMnO_3$ (Malashevich and Vanderbilt, 2008), the ionic contribution dominates, giving $\sim 80\text{--}90\%$ of the polarization.

8. *Ferroelectricity caused by magnetic ordering.* In all types of ferroelectrics discussed above (cases 1–7 in this section), ferroelectricity, even if it occurs in materials which are simultaneously magnetic, has a microscopic nature independent of magnetism. In fact, the opposite is rather often true: as we have seen above in the discussion of ferroelectricity in TM perovskites, the presence of magnetic ions (or of localized "magnetic" electrons) was harmful for ferroelectricity. There are however materials in which ferroelectricity is observed only in some particular magnetically ordered states, so that the very mechanism of ferroelectricity is intrinsically connected to magnetism: in these systems ferroelectricity is *caused* by magnetic ordering.

These systems and the corresponding phenomena will be discussed in more detail in Section 8.3; here we only note that there are several possible classes of such multiferroics. In all of them the magnetic ordering should be such that it breaks inversion symmetry. But the microscopic mechanisms which lead to the appearance of polarization can be different. Thus, in one class of materials ferroelectricity is caused by magnetostriction, that is by the dependence of the exchange constants J_{ij} in the exchange interaction $J_{ij} S_i \cdot S_j$ on the atomic coordinates (on the distances R_{ij}, or on the angles M_i–O–M_j if we have superexchange via the oxygens, etc.). This mechanism has essentially the same nature as the mechanism described in case 5 above, cf. Fig. 8.4: if we have alternating magnetic ions with different charges, then for some magnetic structures, for example with ↑↑↓↓ spin ordering, the magnetostriction in principle would be different for ferro. and antiferro. bonds. Because of that these bonds would become inequivalent, for example the ferro bonds would become shorter. We would then end up in the same situation of inequivalent sites and bonds as that shown in Fig. 8.4(c), that is we would get ferroelectricity, which in this case will be caused by magnetostriction in this particular magnetic structure.

Another microscopic mechanism in the magnetically driven ferroelectrics is connected to the role of spin–orbit coupling and is met in magnetic systems with particular types of spiral magnetic ordering. Yet a third mechanism can be connected to the modifications of the p–d hybridization due to spin–orbit interaction. These situations will be described in more detail in Section 8.3.

When considering the appearance of ferroelectricity in different situations, one interesting question arises. We have seen above that there are many ferroelectrics among perovskites, either driven by the valence-bond formation in d^0 systems such as $BaTiO_3$, or magnetically induced as in $TbMnO_3$. There exists however another class of transition metal systems, probably as big as that of perovskites – the spinels AB_2O_4, see Sections 3.5 and 5.6.3. But there are practically no ferroelectrics among them. (The only exception seems to be Fe_3O_4, which is apparently ferroelectric below the Verwey transition; see van den Brink and Khomskii, 2008 and the references cited therein.)

Similarly, in the big class of pyrochlores there are only one or two ferroelectrics – the only well-known such system is $Cd_2Nb_2O_7$. The question is, why is that so? A possible answer could be that both spinels (the B-site sublattices) and pyrochlores have lattices with geometric frustration (consisting of corner-shared tetrahedra). As discussed in Section 5.7, such frustrations can modify the magnetic properties of these systems drastically, often strongly suppressing long-range magnetic ordering. It could be that the same frustrations also hinder ferroelectric ordering (Seshadri, 2006; McQueen *et al.*, 2008b) – the ordering of electric dipoles instead of magnetic dipoles (spins). The question of the conditions for ferroelectricity in systems with geometric frustrations definitely deserves further study.

8.2 Magnetoelectric effect

Electricity and magnetism were combined into one common discipline in the 19th century, culminating in the Maxwell equations. But electric and magnetic ordering in solids, and their electric and magnetic response, are most often considered separately. And usually with good reason: the electric charges of electrons and ions are responsible for "electric" effects, whereas it is electron spins which are responsible for magnetic properties.

However there are cases where these degrees of freedom couple strongly. This, in particular, gave rise to a very big field of spintronics. In this field one deals mainly with the effects of spins on the transport properties of solids and studies the possibility of controlling such transport for example by magnetic field, or vice versa, controlling magnetic properties by currents, etc.

Starting from the 1960s insulating systems were also discovered in which magnetic and electric degrees of freedom are strongly coupled. Although the history of this field can be traced back to Pierre Curie (Curie, 1894), the real beginning of this field occurred in 1959 with a remark by Landau and Lifshitz in one of the volumes of their famous *Course of Theoretical Physics* (Landau and Lifshitz, 1960), in which they wrote: "Let us point two more phenomena which in principle could exist. One is piezomagnetism, which consists of linear coupling between magnetic field in a solid and a deformation (analogous to that which exists in piezoelectrics). The other consists of linear coupling between magnetic and electric fields in media, which would result e.g. in the appearance in an electric field of a magnetization, proportional to this field. Both these phenomena could exist for certain classes of magnetocrystalline symmetry. We will not however discuss these phenomena in more details because it seems that till present, presumably, they have not been observed in any substance."

Very soon after that Dzyaloshinskii (1959) predicted and Astrov (1961) observed such an effect, which became known as the *linear magnetoelectric (ME) effect*, in the well-known antiferromagnet Cr_2O_3. This was rapidly followed by the discovery of many other compounds of this class, and by a rather complete symmetry analysis and classification of possible groups allowing for this effect, see for example Schmid (2008). The magnetoelectric effect was studied actively both experimentally and theoretically, first in the 1970s–1980s and then again in the 2000s, after a revival of interest in magnetoelectric phenomena.

Phenomenologically the magnetoelectric effect, which is in essence a linear coupling between electric and magnetic degrees of freedom, may be described by terms in the free energy of type

$$F_{\mathrm{ME}} = -\alpha_{ij} E_i H_j \ , \tag{8.9}$$

where here (and in the rest of this chapter) we assume summation over repeated indices. From this expression we get

$$
\begin{aligned}
P_i &= -\frac{\partial F}{\partial E_i} = \alpha_{ij} H_j \ , \\
M_j &= -\frac{\partial F}{\partial H_j} = \alpha_{ij} E_i \ ,
\end{aligned}
\tag{8.10}
$$

that is we indeed have polarization which depends linearly on the magnetic field, and magnetization which depends linearly on the electric field. The tensor α_{ij} is called the magnetoelectric tensor. It can in general have both symmetric and antisymmetric parts; its detailed form is determined by the symmetry of the crystal lattice and the type of magnetic ordering.

Let us first say a few words about symmetry relations, which turn out to be very important for the study of magnetoelectric and multiferroic effects. Magnetic and electric characteristics of the system have different transformation properties with respect to different transformations, the most important of which are the spatial inversion $\mathcal{I}$ and time reversal $\mathcal{T}$. Ordinary vectors such as coordinates r and polarization P are odd with respect to spatial inversion, but even with respect to time reversal, for example

$$
\begin{aligned}
\mathcal{I} P &= -P \ , \\
\mathcal{T} P &= P \ .
\end{aligned}
\tag{8.11}
$$

In contrast, magnetization (a pseudovector) does not change under inversion but changes sign under time reversal,

$$
\begin{aligned}
\mathcal{I} M &= M \ , \\
\mathcal{T} M &= -M \ .
\end{aligned}
\tag{8.12}
$$

One can explain this if we recall that magnetization is produced by currents, see Fig. 8.8, and spatial inversion would leave the magnetic moment M unchanged, Fig. 8.8(b). However, reversing the direction of time would reverse the direction of the current $j = ev = e\,dr/dt$, that is the magnetic moment would also be reversed. For electric polarization, as is clear from Fig. 8.8(a), the situation is just the opposite.

One can use similar arguments to understand how P and M change under other transformations, for example under mirror reflections. As is clear from Fig. 8.9(a, b), when reflected in a mirror plane the polarization P changes sign when it is perpendicular to the mirror plane (Fig. 8.9(a)) but remains unchanged when it is parallel to the mirror plane (Fig. 8.9(b)). For magnetic moments the situation is the opposite: moments perpendicular to the mirror plane remain unchanged under reflections (Fig. 8.9(c)) but if the moment is parallel to the mirror plane, it changes sign under reflections (Fig. 8.9(d)). These rules

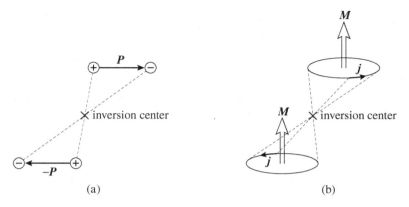

Figure 8.8 Transformation of (a) a vector and (b) a pseudovector (axial vector) under spatial inversion.

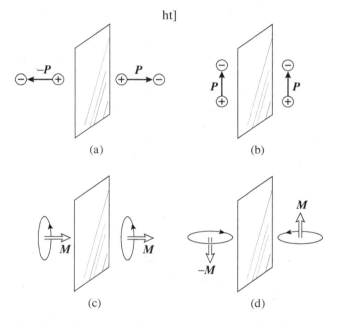

Figure 8.9 Transformation of a vector (a, b) and a pseudovector (c, d) under mirror reflections.

are indeed quite clear if we represent magnetic moments by small current loops; but they remain essentially the same also for spins of electrons (which may be understood as originating from "internal currents"). These rules, together with those for inversion and time reversal, eqs (8.11) and (8.12), are very important for understanding the conditions for the existence of the linear magnetoelectric effect, and they will also be very useful for interpreting the properties of multiferroics, Section 8.3.

In particular, from the relations (8.9), (8.10) with the transformation properties (8.11), (8.12), one sees that the magnetoelectric response function α_{ij} should be odd under both spatial inversion and time reversal. The first condition gives some restrictions on the symmetry groups, allowing for the linear magnetoelectric effect; in particular it tells us that the inversion symmetry in the material should be broken, for example by the corresponding magnetic ordering, and the second condition tells us that the corresponding materials should be magnetic, for instance antiferromagnetic. The classical example of a magnetoelectric material is Cr_2O_3, and the comparison of its situation with that of Fe_2O_3, which has the same crystal structure but does not exhibit the linear magnetoelectric effect, illustrates this. Cr_2O_3 and Fe_2O_3 have four TM ions in a unit cell of the corundum structure, see Fig. 8.10. We see that, while they have the same crystal structure, the magnetic orderings in Cr_2O_3 and Fe_2O_3 are different. In Fe_2O_3 the magnetic structure goes into itself under inversion (the inversion center is marked by a star in Fig. 8.10), that is the inversion symmetry in Fe_2O_3 is not broken: for example site 2 goes under inversion to 3, and in Fe_2O_3 they have the same spin directions. On the contrary, the antiferromagnetic structure of Cr_2O_3 breaks the inversion symmetry: under inversion site 2 goes to 3, but in Cr_2O_3 these sites have opposite spins. Thus the necessary conditions for the existence of the linear magnetoelectric effect are met in Cr_2O_3 but not in Fe_2O_3. Indeed, Cr_2O_3 is a magnetoelectric material whereas Fe_2O_3 is not. (But, on the contrary, there exists weak ferromagnetism in Fe_2O_3 but not in Cr_2O_3. One can prove rigorously (Turov, 1994) that, if in the paramagnetic phase the inversion symmetry is not broken, then in the antiferromagnetic phase the material can have *either* the linear magnetoelectric effect *or* weak ferromagnetism, but never both (of course it may also have neither!): that is, in this situation these two phenomena cannot

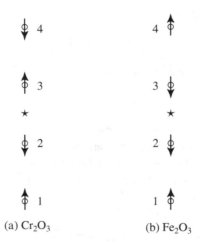

(a) Cr_2O_3 (b) Fe_2O_3

Figure 8.10 Different inversion symmetries of Cr_2O_3 and Fe_2O_3 caused by different types of magnetic ordering.

coexist. The case of Cr_2O_3 compared with Fe_2O_3 gives a good illustration of this general rule.)[4]

The magnetoelectric response function, the tensor α_{ij}, may have both symmetric and antisymmetric components. The symmetric part of α_{ij} can always be diagonalized. One can show that in Cr_2O_3, in the antiferromagnetic state shown in Fig. 8.10(a), α_{ij} is diagonal:

$$\alpha_{ij} = \begin{pmatrix} \alpha_{xx} & 0 & 0 \\ 0 & \alpha_{yy} & 0 \\ 0 & 0 & \alpha_{zz} \end{pmatrix}$$

(and, by symmetry, $\alpha_{xx} = \alpha_{yy}$). In this case, according to eq. (8.10), the electric field E will cause polarization $P \parallel E$ if E is directed along the principal axes of the tensor α_{ij}. But for other systems, and for other magnetic structures, the magnetoelectric tensor α_{ij} may also contain antisymmetric nondiagonal components $\alpha_{ij} = -\alpha_{ji}$. Such antisymmetric tensors can be represented by the dual vector

$$T_i = \varepsilon_{ijk}\alpha_{jk}, \tag{8.13}$$

where ε_{ijk} is the fully antisymmetric tensor (again we assume summation over repeated indices). The (pseudo)vector T is called the *toroidal moment* (or sometimes *anapole moment*), see for example Spaldin *et al.* (2008) (in general the relation between the toroidal moment and the magnetoelectric tensor may be more complicated). If the material is such that it has the magnetoelectric effect with such an antisymmetric magnetoelectric tensor α_{ij}, then it can also be characterized by having a nonzero toroidal moment T, pointing in a particular direction. The toroidal moment is odd with respect to both spatial inversion $\mathcal{I}$ and time reversal $\mathcal{T}$. If we have such a situation, we can see that the polarization in a magnetic field can be expressed as

$$P = T \times H, \tag{8.14}$$

and similarly the magnetization in an electric field is

$$M = -T \times E, \tag{8.15}$$

that is the induced moments will be *perpendicular* to the respective fields. Magnetic structures giving nonzero toroidal moments are of "magnetic vortex" type, see Fig. 8.11.

From the general expression for the toroidal moment which can be obtained in electrodynamics (we do not present it here; see, e.g., Spaldin *et al.*, 2008) one can obtain that for

---

[4] However this general rule may not be true if the inversion symmetry is already broken in the paramagnetic phase: if so, the weak ferromagnetism and linear magnetoelectric effect can coexist. Apparently this is what happens in $BiFeO_3$: this material becomes ferroelectric (because of Bi lone pairs, see Section 8.1) below $\sim 1100\,K$, and in this phase the inversion symmetry is already broken. Then in the magnetically ordered phase of $BiFeO_3$ the weak ferromagnetism and linear magnetoelectric effect can coexist, and indeed they have both been observed in $BiFeO_3$ under certain conditions.

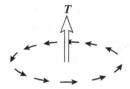

Figure 8.11 A "magnetic vortex" and the corresponding toroidal magnetic moment.

systems of localized spins (with zero net magnetization, $\sum_i S_i = 0$) the toroidal moment is given by the expression

$$T = \frac{g\mu_B}{2} \sum_i r_i \times S_i, \tag{8.16}$$

where g is the g-factor and μ_B is the Bohr magneton. (Note that if $\sum_i S_i \neq 0$, then the toroidal moment defined in this way would depend on the choice of the coordinate origin, as it would change under $r \rightarrow r + r_0$.) The spin vortex shown in Fig. 8.11 has, according to eq. (8.16), a nonzero toroidal moment along the axis of this spin vortex. According to eqs (8.14), (8.15) we then expect that such a spin vortex would exhibit the linear magnetoelectric effect, so that for example the in-plane magnetic field would create an electric polarization $P \perp H$ and $\perp T$ (see, e.g., Delaney et al., 2009).

The magnetoelectric effect attracts considerable attention not only because of its scientific interest: potentially it can serve as a basis for many applications. Thus in magnetoelectric systems one can control magnetic properties by electric field, which might be very useful for many applications, including memory devices. One can, for example, write on magnetic hard discs electrically, not by using electric currents but by applying an electric voltage, which is very desirable: one can hope to reduce dramatically the dimensions of the corresponding devices, and avoid Joule heating. One can also use the inverse effect and employ such materials for electrical measurements of small magnetic fields. From that point of view, the search for materials with larger magnetoelectric coefficients presents an important problem.

As to the possible values of the magnetoelectric response α_{ij}, there exists an important limitation:

$$|\alpha_{ij}| \leqslant \frac{1}{4\pi} \sqrt{\epsilon_{ii}\mu_{jj}}, \tag{8.17}$$

where ϵ_{ij} is the (tensorial) dielectric constant and μ_{ij} is the magnetic permeability of the material. This condition, or even a stronger one,

$$|\alpha_{ij}| \leqslant \sqrt{\chi_{ii}^e \chi_{jj}^m}, \tag{8.18}$$

where $\chi_{ij}^{e,m}$ are the electric and magnetic susceptibilities, follows from the condition of stability of the system (Brown et al., 1968).

New possibilities in the search for materials with nontrivial coupling of electric and magnetic properties can be opened up by materials in which not only is there linear coupling between electric and magnetic degrees of freedom, but also magnetic and ferroelectric ordering are really combined. These so-called multiferroics will be discussed in the next section.

8.3 Multiferroics: materials with a unique combination of magnetic and electric properties

As discussed in Section 8.2, in 1958–1960 a nontrivial interplay of electric and magnetic properties was discovered in some materials, when the linear magnetoelectric effect was predicted theoretically (Dzyaloshinskii, 1959) and observed experimentally in Cr_2O_3 (Astrov, 1960), after which many other materials of this class were found. A new twist in this problem was the idea that there may exist not only a strong cross-coupling of *responses* (e.g., the appearance of magnetization M in an electric field E, or the inverse effect of electric polarization P generated by application of a magnetic field H), but also systems in which the corresponding two types of ordering, (ferro)magnetism and ferroelectricity (FE), coexist in one material in the absence of external electric and magnetic fields. Following the suggestion of Schmid (1994), such systems are now called multiferroics (MF).[5]

Probably the first such materials found were boracites (Ascher *et al.*, 1964, 1966), in particular the Ni–I boracite $Ni_3B_7O_{13}I$, and very soon after that several other multiferroics were either found in nature, or synthesized artificially, see for example Smolenskii and Chupis (1982). The most active study of this field was carried out at that time by two groups in the Soviet Union, those of Smolenskii in Leningrad (St. Petersburg) and Venevtsev in Moscow.

Somehow, after relatively intense studies in the 1960s and 1970s, the activity in this field decreased. This field got new life at the beginning of this century, and one can single out three important developments which led to that. The first was a theoretical development, in which the general problem of why the coexistence of magnetism and ferroelectricity is so rare was raised, see for example Hill (2000), Khomskii (2001). This attracted new attention to the field, but most important of course were two experimental achievements. One was the rapid progress in growing thin films, and in particular the successful growing of films of one of the most popular multiferroics – $BiFeO_3$, with greatly improved properties (Wang *et al.*, 2003). This gave rise to enormous developments in the field, aimed mainly at practical applications.

The second experimental development which really moved this field to the forefront of active research was the discovery in 2003–4 of a novel class of multiferroics, in which magnetism and ferroelectricity do not just coexist, but in which ferroelectricity is *caused* by

---

[5] In principle one often includes in this category also a third type of ordering: spontaneous deformation, leading to ferroelasticity. This is however a somewhat separate field, and most often nowadays the term multiferroics refers predominantly to the coexistence of magnetism and ferroelectricity.

magnetism: the group of Y. Tokura and T. Kimura discovered this phenomenon in $TbMnO_3$ (Kimura *et al.*, 2003a), and the group of S.-W. Cheong in $TbMn_2O_5$ (Hur *et al.*, 2004). These so-called 113 and 125 multiferroics and their analogs with other rare earths till present serve as the "Rosetta stones," the test grounds of this whole field, on which the main physical ideas are tested – although since then many other multiferroics of this type have been discovered.

The discovery of these systems, with their very rich and interesting physics, together with the realization of the great potential importance of these materials for practical applications – such as the possibility of addressing magnetic memory electrically (and without currents!), the creation of novel types of four-state logic (up and down polarization, up and down magnetization), novel types of sensors, etc. – led to an extremely rapid development of this field; see, for example, the extensive reviews of multiferroics in Fiebig (2005), Ehrenstein *et al.* (2006), Khomskii (2006), Cheong and Mostovoy (2007), Ramesh and Spaldin (2007), Khomskii (2009). Probably the most complete coverage of different aspects of this field can be found in the collection of short review articles published in the special issue of the *Journal of Physics: Condensed Matter* (2008).

As discussed in Section 8.1, whereas the nature of strong magnetism is more or less the same in different materials, there exist many different mechanisms and different types of ferroelectricity. Consequently, one can think of different types of multiferroics, which differ by the nature of their ferroelectricity. Generally speaking, one can subdivide all multiferroics into two groups.[6]

The first group, which we call *type-I multiferroics*, are those multiferroics in which ferroelectricity and magnetism have different sources and appear largely independent of each other (though of course with some coupling between them, which is the primary interest in this field). Typically (though not always) ferroelectricity in these systems appears at higher temperatures, and spontaneous polarization is often rather large (of order $10–100 \,\mu C/cm^2$). Such are for example the materials $BiFeO_3$, very popular nowadays ($T_{FE} \sim 1100 \, K$, $T_N = 623 \, K$, $P \sim 90 \,\mu C/cm^2$), and $YMnO_3$ ($T_{FE} \sim 910 \, K$, $T_N = 75 \, K$, $P \sim 6 \,\mu C/cm^2$).

The second group, which we call *type-II multiferroics*, are the relatively recently discovered systems in which ferroelectricity is *due to* magnetism (i.e., they are also improper ferroelectrics – although there may be improper ferroelectrics among type-I systems as well). As mentioned above, it is just the discovery of this class of systems which led to an upsurge of interest in this field in fundamental science; these materials are now studied very intensively. In this section we will devote our primary attention to such magnetically driven ferroelectricity, that is to type-II multiferroics, although at the beginning we also discuss briefly type-I systems.

---

[6] This classification differs from the division of ferroelectrics into *proper* and *improper* ones, used sometimes (see, e.g., Levanyuk and Sannikov, 1974), in which one pays most attention to whether the polarization P is the primary order parameter or whether it arises as a secondary effect, for example due to a coupling with some other order parameter η, maybe of type $\eta^2 P^2$ – see Appendix C.

As to practical applications, *composite multiferroics* consisting of known magnets and ferroelectrics, both in the form of multilayers and self-organized nanostructures, are also widely studied now. I will not discuss this direction of research below, but one has to say that from the point of view of potential applications these composite systems look the most promising. Nevertheless, even for these applied studies the understanding of the basic physics determining multiferroic properties, the very conditions for coexistence of magnetism and ferroelectricity, and the factors determining the coupling of these degrees of freedom is indispensable. To these basic problems we will now turn.

8.3.1 Type-I multiferroics

The "older" and more numerous are type-I multiferroics, in which ferroelectricity and magnetism are due to different mechanisms. As mentioned above, these are often good ferroelectrics, and critical temperatures of magnetic and ferroelectric transitions can be well above room temperature. However, the coupling between magnetism and ferroelectricity in them is usually rather weak. The challenge for these systems is to enhance this coupling, keeping all their other positive features. As we will see below, the problems in type-II multiferroics are just the opposite.

One can single out at least four different subclasses of type-I multiferroics, depending on the mechanism of ferroelectricity in them; this classification follows closely that of Section 8.1.

1. *Multiferroics with asymmetric chemical units.* Conceptually the simplest are multiferroics in which different chemical groups in a crystal are responsible for magnetism and for ferroelectricity. The first natural multiferroics – boracites such as $Ni_3B_7O_{13}I$ – are of this type. Boracites form a major group of compounds, many of which are well-known minerals. They can contain different TM ions such as Ni, Co, Mn. Their magnetic properties are due to these ions, and their ferroelectricity is apparently connected with the BO groups. Many boracites containing for example BO_3 groups without inversion symmetry are multiferroics, or at least exhibit a linear magnetoelectric effect. However, other type-I multiferroics are more important.

2. *Multiferroic perovskites.* Probably the best-known ferroelectrics are perovskites such as $BaTiO_3$ or the widely used $Pb(ZrTi)O_3$ ("PZT"). There are many magnetic materials among perovskites, and also many ferroelectrics. As discussed in Section 8.2, magnetism and ferroelectricity in these materials are usually mutually exclusive: magnetism requires partially filled d-shells, whereas ferroelectricity is found in perovskites with empty d-shells of TM ions. Apparently the realization of this tendency suggested one possible strategy to get around this problem: creating "mixed" perovskites, so as to combine in one material both "ferroelectric" d^0 and "magnetic" d^n ions. Indeed several such mixed perovskites have been synthesized successfully, for example $PbFe_{1/2}Nb_{1/2}O_3$ and $PbFe_{2/3}W_{1/3}O_3$ (see, e.g., Smolenskii and Chupis, 1982). Some of these multiferroic systems are ferrimagnetic

and have nonzero net magnetization. Unfortunately, the coupling between magnetism and ferroelectricity in them is rather weak.

As mentioned in Section 8.2, the mutual exclusion of magnetism and ferroelectricity in perovskites is not a universal law of nature. Indeed, recent *ab initio* calculations (Bhattacharjee *et al.*, 2009) have also shown that the classical magnetic perovskite $CaMnO_3$ could become ferroelectric (or antiferroelectric) when subjected to negative pressure or tensile strain. This prediction was confirmed experimentally on the system $Ba_{1-x}Sr_xMnO_3$ (Sakai *et al.*, 2011). Thus in principle one cannot exclude that other type-I multiferroic perovskites could be found or synthesized.

3. *Multiferroics with lone pair ferroelectricity.* Another way to make multiferroics, in particular among perovskites, is to use the other mechanism of ferroelectricity – that due to lone pairs. Indeed, many mixed d^0–d^n perovskites such as the ones cited above also contain Pb^{2+} ions, which in themselves may be active in providing the mechanism of ferroelectricity. Also $BiFeO_3$, the most popular multiferroic material, is of this type: the ferroelectricity in this system is predominantly due to Bi, and Fe gives rise to magnetic ordering. The same class contains also $BiMnO_3$, which is a rare case of a ferromagnetic insulator: ferromagnetism in this system is due to a very specific type of orbital ordering. However, as mentioned above, it is not really clear whether it is ferroelectric or antiferroelectric. As mentioned in Section 8.2, two other similar materials have also been synthesized: $PbVO_3$ and $BiCoO_3$,[7] but in both of them the ferroelectric-type distortion is so strong that one cannot switch polarization by any reasonable electric field, so they should rather be classified as pyroelectrics. (Also, Co^{3+} in $BiCoO_3$ is in the low-spin state, i.e. it is nonmagnetic, thus this material, strictly speaking, should not be called multiferroic.)

There are many other compounds of Bi and Pb in which ferroelectricity is due to this mechanism and which could be multiferroic. Among these are the above-mentioned Aurivillius phases such as $Bi_4Ti_3O_{12}$ – well-known ferroelectrics, layered materials containing Bi_2O_2 layers alternating with perovskite blocks. There are also among them compounds containing magnetic ions, for example $Bi_5FeTi_3O_{15}$, with rather high ferroelectric transition temperature of 1050 K. This class of potential multiferroics is practically unexplored yet; first reports are just beginning to appear, see for example Sharma *et al.* (2008).

4. *"Geometric" multiferroics.* A special class of multiferroics are the systems $RMnO_3$, where R can be Y or small rare earths. As discussed above, despite the formula which looks similar to that of perovskites, these systems are not perovskites at all: they are layered hexagonal systems containing fivefold-coordinated Mn ions at the centers of trigonal bipyramids, see Fig. 8.3 above. They are good ferroelectrics with $T_{FE} \sim$ 800–1000 K and with the antiferromagnetic transition at around 70–80 K. Owing to the

---

[7] In $PbVO_3$ the lone pair of Pb plays an important role in the formation of the observed pyroelectric structure. But V^{4+} is also known in chemistry for its tendency to shift away from the center of symmetry of ligand octahedra and for forming a strong covalent bond with one oxygen – this even carries a special name, vanadile bond.

frustrated nature, their magnetic properties are rather rich. In particular, these materials provide a good opportunity for studying the interplay of magnetic and ferroelectric domains. But the magnetic–ferroelectric coupling in these materials, though definitely present, is not strong enough, and they do not seem to be very promising for practical applications.

5. *Multiferroics with ferroelectricity due to charge ordering.* Materials with ferroelectricity due to the coexistence of site-centered and bond-centered charge ordering can also be multiferroic. The examples given in Section 8.1 – $(PrCa)MnO_3$ and $LuFe_2O_4$ – are of this class. They are both claimed to be ferroelectric (although ferroelectricity in both these systems is not yet established with certainty), and both are magnetic. The coupling between electric and magnetic degrees of freedom in these materials has not been studied in practice.

As discussed above, ferroelectricity in such systems may either be caused by spontaneous charge ordering intermediate between site- and bond-centered (Efremov *et al.*, 2007) or, more likely, it may occur in systems in which sites or bonds are inequivalent from the beginning and the second ordering occurs spontaneously. In particular, inequivalent bonds can appear in mixed TM oxides, containing TM ions with different charges, due to magnetostriction. Such is apparently the situation in Ca_3CoMnO_6 (Choi *et al.*, 2008) and in $TbMn_2O_5$ (Hur *et al.*, 2004). But these materials belong to type-II multiferroics, and we will discuss them in more detail below.

8.3.2 Type-II (magnetic) multiferroics

As we have already stated above, a special new type of multiferroics are those in which ferroelectricity exists only in the magnetically ordered state and is caused completely by a particular type of magnetism. Thus, for example, in $TbMnO_3$ the magnetic ordering appears at $T_{N1} = 41$ K, and at a lower temperature $T_{N2} = 23$ K the magnetic structure changes. In this low-temperature phase there appears nonzero polarization, that is the material becomes multiferroic (Kimura *et al.*, 2003a). Similarly, in $TbMn_2O_5$ ferroelectricity appears below the magnetic ordering temperature in a particularly ordered phase (Hur *et al.*, 2004). In these two papers it was shown that in these cases one can strongly influence the polarization by magnetic field: in $TbMnO_3$ at $H = 0$ the polarization is directed along the c-axis, but in a field (e.g., in the a-direction) for $H_a \sim 5$ T there occurs a spin flop accompanied by a polarization flop, that is the polarization rotates by $90°$, and for $H_b > 5$ T it is oriented in the a-direction. In $TbMn_2O_5$ the influence of the external field is even stronger: the polarization charges sign with the field, so that for example the field alternating between $+5$ T and -5 T leads to the corresponding oscillations of the polarization. Such a strong sensitivity of the polarization to the magnetic field immediately attracted considerable attention, and since then these magnetic or type-II multiferroics have been investigated very intensively. A large part of the activity in the fundamental research of multiferroics is devoted to this class of materials.

Already, many such type-II multiferroics have been discovered besides those mentioned above: $Ni_3V_2O_6$, $MnWO_4$, $LiCu_2O_2$, CuO, $CuFeO_2$, $(LiNa)CrO_2$, $NaFeSi_2O_6$, and new systems of this class are constantly being discovered. These systems and their properties, both static and dynamic, are studied experimentally using various techniques. Further, significant theoretical progress has been reached in their understanding.

From the point of view of the mechanism of the multiferroic behavior of these systems, one can divide type-II multiferroics into two groups: those in which, as is now established, ferroelectricity is caused by a particular type of magnetic spirals (mostly of cycloidal character) and those in which ferroelectricity can appear in collinear magnetic structures. It turns out that microscopically, in the first group (type IIa) a crucial role is played by the relativistic spin–orbit interaction, acting mostly via the antisymmetric (Dzyaloshinskii–Moriya) exchange, whereas in the second group (type IIb) this is not required, and ferroelectricity is caused by the exchange striction.

1. *Spiral type-II multiferroics.* Most type-II multiferroics known to date belong to this subgroup. As has been established, mostly by neutron scattering, ferroelectricity appears in them in a particular type of magnetic structures – magnetic spirals (Fig. 8.12), mostly of cycloidal type (Fig. 8.12(c)). This is for example the case in $TbMnO_3$, in $Ni_3V_2O_6$, and in $MnWO_4$. Thus, in $TbMnO_3$ below $T_{N1} = 41$ K the magnetic structure is incommensurate; however it is not formed by rotating spins, but is rather a sinusoidal spin-density wave with all spins in the same direction (with the average ordered spin component changing in magnitude, e.g. $\langle S_x \rangle = \langle S_y \rangle = 0$, $\langle S_z \rangle \sim |S| \cos(\boldsymbol{Q} \cdot \boldsymbol{r})$), Fig. 8.12(a). However, below $T_{N2} = 23$ K the structure changes to a cycloidal one, of type $\langle S_x \rangle \sim |S| \cos(\boldsymbol{Q} \cdot \boldsymbol{r})$, $\langle S_z \rangle \sim |S| \sin(\boldsymbol{Q} \cdot \boldsymbol{r})$, with the wave vector of the spiral $\boldsymbol{Q} = (Q_x, 0, 0)$. And it is in this state that one finds nonzero polarization, in this case $P_z \neq 0$ (Fig. 8.12(c)).

Theoretical treatment using a microscopic approach and the phenomenological Landau treatment lead to the same conclusion. According to Mostovoy (2006), for a cycloidal spiral there appears polarization given by the expression

$$P = c(\boldsymbol{Q} \times \boldsymbol{e}), \tag{8.19}$$

where c is a certain coefficient, $\boldsymbol{Q}$ is the wave vector of the spiral, and $\boldsymbol{e} \sim \boldsymbol{S}_1 \times \boldsymbol{S}_2$ is the spin rotation axis. In the microscopic approach of Katsura *et al.* (2005), the same expression is written as

$$\boldsymbol{P}_{ij} = c'\boldsymbol{r}_{ij} \times (\boldsymbol{S}_i \times \boldsymbol{S}_j), \tag{8.20}$$

where $\boldsymbol{P}_{ij}$ is the polarization of a pair of spins $\boldsymbol{S}_i$ and $\boldsymbol{S}_j$. The authors called this mechanism the "spin-current mechanism"; this term is often used nowadays. One should probably better call it "spin-supercurrent," so as not to mix it up with the real dissipative spin currents in metallic systems, for example those used in spin torque experiments and devices. This term comes from the analogy with superconductors: it is well known that the supercurrent in a superconductor is proportional to the gradient of the phase of the superconducting order parameter, $j \sim \mathrm{grad}\,\varphi$. Similarly here we can characterize the cycloidal structure of

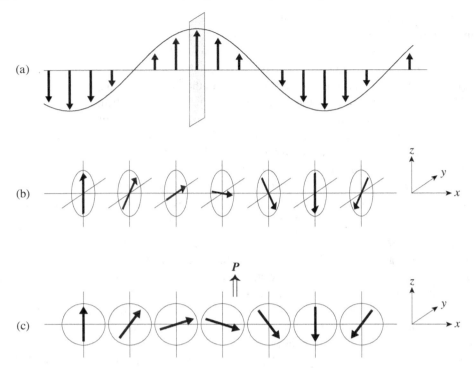

Figure 8.12 (a) Sinusoidal spin density wave; a mirror plane is shown. (b) Helicoidal spiral (proper screw). (c) Cycloidal spiral. The electric polarization is shown by a double arrow.

Fig. 8.12(c) by the angle φ_i, made by the spin S_i for example with the z-axis. And if the spins rotate as shown in Fig. 8.12(c), this phase changes from site to site, that is in analogy with supercurrents in superconductors we can talk here about spin supercurrent, and the term $S_i \times S_j$ entering (8.20) is a measure of it.

Microscopic treatment shows that the coefficients c, c' in (8.19), (8.20) are proportional to the strength λ of the spin–orbit coupling $\lambda L \cdot S$. Being a relativistic effect, it has the expected relativistic smallness, which explains why the polarization created by this mechanism is usually small: typical values of spontaneous polarization in cycloidal multiferroics are $\sim 10^{-2}$–$10^{-3}\,\mu C/cm^2$, that is 2–3 orders of magnitude smaller than in good ferroelectrics such as $BaTiO_3$ or $BiFeO_3$.

The microscopic origin of multiferroicity in cycloidal magnets has much in common with the phenomenon of weak ferromagnetism, discussed in Section 5.3. As shown by Sergienko and Dagotto (2006), one can explain the appearance of ferroelectricity in systems such as $TbMnO_3$ by the inverse Dzyaloshinskii–Moriya effect. According to Dzyaloshinskii (1958), in the low-symmetry magnets there can exist an antisymmetric exchange of the type $D_{ij} \cdot (S_i \times S_j)$, see (5.29), where the Dzyaloshinskii vector D is nonzero for certain symmetries. Crudely one can say that for the pair of TM ions i, j,

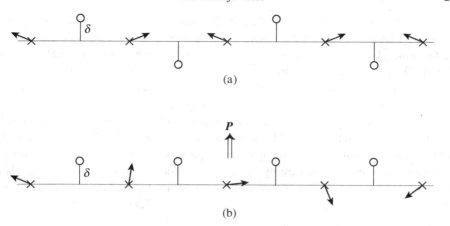

Figure 8.13 (a) Mechanism of weak ferromagnetism due to the Dzyaloshinskii–Moriya interaction, caused by shifts of oxygens from the centers of metal–metal bonds. (b) Shifts of oxygens for a cycloidal spiral, giving ferroelectric polarization. Crosses are transition metal ions, circles are oxygens.

with the oxygen between them, the Dzyaloshinskii, or Dzyaloshinskii – Moriya vector, is $D_{ij} \sim r_{ij} \times \delta$, see (5.32), where δ is the shift of the oxygen from the middle of the bond r_{ij}. If there exists such a shift, then $D_{ij} \neq 0$, and the spins S_i, S_j will cant and become non-collinear, see Fig. 8.13(a) (cf. Figs 5.21, 5.22). Vice versa, if for some reason the spins are noncollinear, as for example in a cycloidal spiral, then such oxygens would shift to make D_{ij} nonzero in order to gain the Dzyaloshinskii–Moriya energy, and in a cycloidal spiral they will all shift in one direction, Fig. 8.13(b), which gives finite polarization in cycloidal magnets (the "centers of gravity" of positively charged TM ions (crosses in Fig. 8.13) and negatively charged oxygen ions (circles) are now shifted in a direction perpendicular to the spiral axis).

To get a spiral magnetic ordering in an insulator one needs magnetic frustration; therefore this type-IIa multiferroicity is usually observed in frustrated systems. Thus, these multiferroics provide a good example of the situation when magnetic frustrations – a very active topic in itself (Lacroix et al., 2011), see Section 5.7 – lead to unexpected new consequences, with potentially important applications.

From the picture obtained above and described by eqs (8.19), (8.20), one can understand the possible role of the magnetic field. The application of an external field H to the standard spiral antiferromagnet can cause a phenomenon exactly analogous to the well-known spin-flop transition, see for example Coey (2010) or Khomskii (2010): in a field the sublattice moments prefer to lie in the perpendicular plane (of course with some canting in the direction of the applied field). Similarly, if in the appropriate field the plane of the cycloid should flop by 90°, the polarization P would also flop: according to (8.16), (8.19) P lies in the plane of rotating spins but perpendicular to Q.[8]

---

[8] In some cases, such as $TbMnO_3$, one should also take into account the reaction to H of other magnetic ions, here Tb; this can sometimes change the behavior of such systems in the external field.

Thus, due to the inverse Dzyaloshinskii–Moriya effect, cycloidal magnetism induces electric polarization. Interestingly, the opposite effect exists as well, appearing also in type-I multiferroics. For example, it leads to the formation of a long-periodic (~ 500 Å) spiral in $BiFeO_3$. The large polarization already present in this system below ~ 1100 K distorts the simple collinear antiferromagnetic ordering of G-type which would exist in the absence of ferroelectricity (and which exists in the non-ferroelectric orthoferrites $RFeO_3$). The resulting magnetic structure of $BiFeO_3$ is a cycloidal spiral, satisfying the relations (8.19). Indeed, the energy gain, described by a term of the type $-E \cdot [Q \times (S_i \times S_j)]$ in the free energy, is proportional to the pitch angle of the spiral, but the energy loss of the exchange interaction is quadratic in this tilting angle of neighboring spins. Thus the spiral should appear in the initially collinear (antiferro)magnet, if nonzero polarization of any nature is present in it. The appearance of a spiral ordering in $BiFeO_3$ can also be explained by the standard Dzyaloshinskii mechanism, see Section 5.3.2, with the Dzyaloshinskii vector appearing below the transition to the ferroelectric phase.

One word of caution here in connection with the expressions (8.19), (8.20). They have become very popular in the community of scientists studying multiferroics. However one should remember that these expressions were obtained for a particular situation within a particular microscopic model, and when conditions are different from those assumed in Katsura *et al.* (2005) and Mostovoy (2006), the conclusions may differ as well. Thus for example if the magnetic order is more complicated and cannot be described by an antiferromagnetic order parameter $L = M_1 - M_2$ slowly changing in space, the expression (8.19) should be modified (Mostovoy, 2008). Similarly, the derivation of Katsura *et al.* (2005) and Mostovoy (2006) assumed a particular symmetry of the crystal, for example simple cubic or tetragonal. In this case indeed the expressions (8.19), (8.20) are valid, and the polarization of for example a proper screw, in which the spins rotate in the plane perpendicular to the wave vector of the spiral Q, see Fig. 8.12(b), would be zero: the spin rotation axis e or $S_i \times S_j$ in this case is parallel to Q. However, as shown by Arima (2007), this may not be the case for other symmetries, in which case there may appear a nonzero polarization even for the proper screw.[9]

Experimentally this situation was observed in $RbFe(MoO_4)_3$, and more recently in other layered triangular systems, such as $CuFeO_2$, $LiCrO_2$, $NaCrO_2$, see for example Seki *et al.* (2008). Theoretically, the appearance of polarization in this case can be explained by the spin–orbit dependence of the p–d hybridization (Jia *et al.*, 2007).

An interesting case of such multiferroics can be provided by materials which, in Johnson *et al.* (2011), were called *ferroaxial*. These are systems with inversion symmetry, but which can exist in two modifications which are mirror images of each other. An example of such symmetry is the well-known symbol shown in Fig. 8.14(a, b): each such figure has an

---

[9] For a proper screw by symmetry the polarization can point only in the direction of Q (all perpendicular directions are equivalent). If there exists a twofold rotation axis perpendicular to Q, for example rotation along the z-direction in Fig. 8.12(b), the corresponding rotation, on the one hand, should leave the situation invariant, but, on the other hand, it would change P to $-P$. From this we see that in this case $P = 0$. However if such a twofold rotation axis is absent, polarization may appear; at least it is not forbidden. (One should consider similarly the other symmetry elements.)

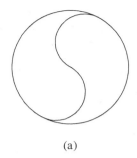 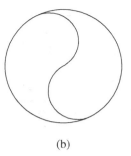

(a)   (b)

Figure 8.14 A symbol demonstrating the existence of two inequivalent forms of a figure (or a crystal), though each of them has an inversion symmetry. This is an example of a ferroaxial symmetry, characterized by the axial vector A.

inversion center (the center of the disc), and (a) is the mirror image of (b). Such symmetry is characterized by an axial vector, or pseudovector A (in this case pointing perpendicular to the plane of the figure, for example into the page (away from us) in (a) and out of the page (towards us) in (b)). In Johnson *et al.* (2012) the crystal classes having such a property (ferroaxial classes) are enumerated.

When we have such a system (characterized by a pseudovector A), and when there is in the system simultaneously a magnetic ordering of the proper screw type with the screw axis parallel to A (characterized by pseudoscalar $\kappa = r_{ij} \cdot (S_i \times S_j)$), then there can (and would) appear polarization $P \sim \kappa A$. Note that both κ and A are odd with respect to time inversion, hence their product κA is a normal vector ($\mathcal{I}$-odd and $\mathcal{T}$-even), as the polarization should be. Apparently, type-II multiferroics $RbFe(MO_4)_2$ (Kenzelmann *et al.*, 2007), $CaMn_7O_{12}$ (Bhattacharjee *et al.*, 2009), and some others belong to this class.

2. *Collinear type-II multiferroics.* The second group of magnetically driven ferroelectrics are those in which ferroelectricity appears in collinear structures without any necessary involvement of spin–orbit interaction. Polarization can appear as a consequence of exchange striction.

The simplest example (realized in Ca_3CoMnO_6; see Choi *et al.*, 2008), Fig. 8.15, is actually closely related to the mechanism discussed in Section 8.2 – the mechanism of the appearance of ferroelectricity due to a combination of site- and bond-centered charge ordering. If we have for example a chain made up of ions with different charges (such as Co^{2+} and Mn^{2+} in Ca_3CoMnO_6), Fig. 8.15(a), there is still a center of inversion in the lattice, for example at each site. But if the magnetic ordering is of $\uparrow \uparrow \downarrow \downarrow$ type, Fig. 8.15(b), then the inversion symmetry is broken, see Section 8.2, which can give rise to polarization (remember that in an inversion the spin directions do not change; the magnetic structure of Fig. 8.15(b) violates this symmetry). Physically the magnetostriction of ferro- and antiferromagnetic bonds would be different, so that for example the ferromagnetic bonds would become shorter than the antiferromagnetic ones, Fig. 8.15(c) (or vice versa). We then obtain the situation shown in Fig. 8.4(c), with nonzero net polarization, that is a ferroelectric state, in this case driven by magnetostriction in this particular magnetic structure.

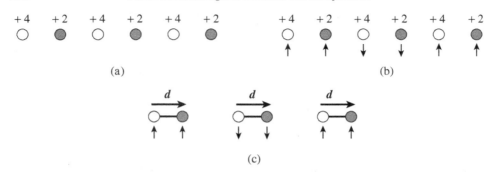

Figure 8.15 The chain Co^{2+}–Mn^{4+}–Co^{2+}–Mn^{4+} in Ca_3CoMnO_6, with ↑ ↑ ↓ ↓ magnetic structure. The inequivalent magnetostriction of the ↑ ↑ and ↑ ↓ pairs leads to the formation of electric polarization; this polarization was discovered by Choi *et al.* (2008).

Experimentally the situation in Ca_3CoMnO_6 is exactly like that. Ca_3CoMnO_6 consists of 1d chains in which the ions Co^{2+} and Mn^{4+} alternate, see Fig. 8.16. At high temperature the distances between them along the chains are the same, there exists a center of symmetry, and polarization is absent. Magnetic ordering at $T_N = 13$ K, however, breaks the inversion symmetry: the magnetic structure is of ↑ ↑ ↓ ↓ type, cf. Fig. 8.15. Owing to the mechanism described above (exchange striction) we get the situation of Fig. 8.15(c), that is the material becomes ferroelectric. Theoretical calculations (Wu *et al.*, 2009) confirm this picture.

Apparently the situation in the RMn_2O_5 systems is very similar, and ferroelectricity in them is also of the exchange striction type (Radaelli and Chapon, 2008). Recent experiments have shown that a certain cycloidal component in the Mn magnetic ordering is also present in these systems; but the most probable explanation is that it is of secondary nature, being not the cause, but rather a consequence of polarization, should polarization appear for any reason (in this case due to exchange striction). Indeed, as explained above for the example of $BiFeO_3$, the same coupling of polarization with the spin spiral which gives rise to the connection (8.19), (8.20) would also work in the opposite direction: if for some reason there appears spontaneous polarization in a system, its presence would turn a collinear magnetic structure into a cycloidal one. This is a possible origin of the weak spiral component in RMn_2O_5. Another possible explanation of this incommensurate component of magnetic ordering is that it may be connected with frustrations existing in this class of compounds; as discussed in Section 5.9, frustrations often lead to the formation of such incommensurate structures.

As mentioned above, the exchange striction can in principle give ferroelectricity in many systems, for example in perovskite nickelates of the type $YNiO_3$ in which there exists spontaneous charge disproportionation of checkerboard type (alternating [111] planes with different charges – formally Ni^{2+} and Ni^{4+}), see the discussion in Section 7.5. The usually considered magnetic structure corresponds to ↑ ↑ ↓ ↓ ordering in the same [111] direction (Alonso *et al.*, 1999), after which the situation along this direction becomes exactly equivalent to that of Ca_3CoMnO_6, or to that in Fig. 8.15. In this case these nickelates should

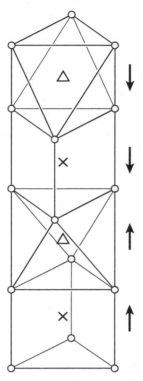

Figure 8.16 Crystal and magnetic structure of Ca_3CoMnO_6. Crosses are Co^{2+} ions, triangles are Mn^{4+} ions, circles are oxygens.

become multiferroic, with polarization in the [111] direction. In the alternative magnetic structure, proposed for example in Fernandez-Diaz (2001), the situation, however, would be different: the system would still be multiferroic, but with a different direction of the polarization (Giovanetti *et al.*, 2009).

   The exchange striction systems considered above contained TM ions with differ-ent valence (Co^{2+} and Mn^{4+} in Ca_3CoMnO_6, Mn^{3+} and Mn^{4+} in RMn_2O_5, formally Ni^{2+} and Ni^{4+} in $RNiO_3$). One can, however, get the same effect even for identical mag-netic ions, when one takes into account the fact that the exchange in TM oxides usually occurs via the intermediate oxygens and depends not only on the TM–TM distance, but also on the TM–O–TM angle. Thus in perovskites $RMnO_3$ with small rare earths the E-type magnetic structure is realized, which, along the Mn rows in the basal plane, is of the already mentioned type $\uparrow \uparrow \downarrow \downarrow$. As proposed by Sergienko *et al.* (2006), the exchange striction here can lead to a shift of oxygen ions perpendicular to the Mn–Mn bonds, and this shift will be such as to produce polarization in this direction. Indeed, the basic ele-ment of the crystal structure in such systems has the form shown in Fig. 8.17: due to the tilting of MO_6 octahedra there appear zigzags in the M–O chains (cf. Fig. 3.45). In the

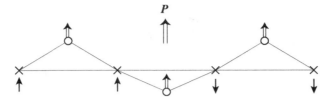

Figure 8.17 Illustration of the exchange striction mechanism leading to multiferroic behaviour in perovskite manganites $RMnO_3$ with small rare earth R and with E-type magnetic structure. Short double arrows show the shifts of oxygen ions (circles), leading to the polarization $\boldsymbol{P}$.

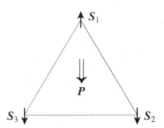

Figure 8.18 The appearance of polarization in a spin triangle, after Bulaevskii *et al.* (2008).

E-type magnetic structure the spins along such chains are ordered as $\uparrow\uparrow\downarrow\downarrow$. According to the Goodenough–Kanamori–Anderson rules (Section 5.1), the nearest-neighbor exchange would be "more ferromagnetic" for Mn–O–Mn angle closer to 90°, whereas for the antiferromagnetic coupling it is more favorable to have this angle closer to 180°. Then, with the $\uparrow\uparrow\downarrow\downarrow$ structure shown in Fig. 8.17, to gain this magnetic energy it is more favorable to move the oxygen ions as shown by double arrows; that is, all negatively charged oxygens shift for example upwards, and in effect there appears net polarization $\boldsymbol{P}$, marked in Fig. 8.17. Such polarization was indeed observed in Lorenz *et al.* (2007) and Feng *et al.* (2010), although its value was smaller than estimated theoretically (Picozzi *et al.*, 2006; Sergienko *et al.*, 2006). This mechanism can also work in other systems, for example in nickelates $RNiO_3$, see Giovanetti *et al.*, (2009).

3. *Electronic ferroelectricity in frustrated magnets.* Interestingly enough, there may exist a purely electronic mechanism which could give magnetically driven ferroelectricity in frustrated systems (Bulaevskii *et al.*, 2008). One can show that for example in a magnetic triangle with three electrons, described by the standard Hubbard model (1.6) with $t \ll U$, there could appear polarization directed for example from site 1 to the middle of the bond (23), see Fig. 8.18, given by the expression

$$\boldsymbol{P}_{1,23} = \frac{24t^3}{U^2}[\boldsymbol{S}_1 \cdot (\boldsymbol{S}_2 + \boldsymbol{S}_3) - 2\boldsymbol{S}_2 \cdot \boldsymbol{S}_3] . \tag{8.21}$$

If the magnetic structure is such that the combination of spin correlation functions entering eq. (8.21) is nonzero, then the system would develop electric polarization. Note that this mechanism requires neither spin–orbit coupling nor lattice distortion. Interestingly enough, the same treatment shows that for other spin textures – those with nonzero scalar spin chirality $S_1 \times [S_2 \times S_3]$ – there would appear real spontaneous orbital currents in the triangle. Thus the state with orbital currents is a counterpart of the state with electric polarization. This connection can be developed rather far, and it leads to quite nontrivial consequences, such as electrically driven electron spin resonance (ESR), etc. (Bulaevskii *et al.*, 2008).

The static properties of different classes of multiferroics, such as the type of ordering in the ground state, the change of magnetic and ferroelectric properties by magnetic and electric fields, the rich domain structure, etc., are actively studied, and many of their properties are already clear. But multiferroics display also quite nontrivial dynamic properties. Probably the most interesting is the appearance of *electromagnons* – magnetic excitations created by an a.c. electric field (Pimenov *et al.*, 2006). The existence of such excitations is based on the coupling between magnetization and electric polarization, intrinsic for multiferroics, but also present in systems with the linear magnetoelectric effect. Both the Dzyaloshinskii–Moriya interaction and the exchange striction can lead to the appearance of electromagnons, the second mechanism being apparently more important (Valdes-Aguilar *et al.*, 2009). One can find more details on these phenomena in the reviews of Pimenov *et al.* (2008) and Sushkov *et al.* (2008).

8.3.3 Multiferroics and symmetry

From the very beginning of the development of the whole field of magnetoelectricity and multiferroicity, starting from the remark of Landau and Lifshitz (1960) and from the seminal papers by Dzyaloshinskii (1958, 1959), symmetry considerations played a crucial role in this area; we have already touched upon this point in several places in this chapter. Indeed, as discussed in Section 8.1, from the symmetry point of view the presence of polarization breaks the inversion symmetry $\mathcal{I}$ (P is a normal vector changing sign under inversion, $\mathcal{I}P = -P$), and magnetic ordering breaks time reversal invariance, or time inversion $\mathcal{T}$ ($\mathcal{T}M = -M$). Among all crystallographic Shubnikov point groups there are 31 allowing for spontaneous polarization (the so-called pyroelectric groups). In magnetic systems one has to consider how the magnetic structure changes under crystallographic symmetry elements, and also add the time-reversal transformation.

As discussed in Section 8.2, for the presence of the linear magnetoelectric effect both spatial inversion and time-reversal invariance should be broken. This is evident for multiferroics (if both P and M are nonzero, both $\mathcal{I}$ and $\mathcal{T}$ are broken). For magnetoelectrics this is clear from the relations (8.10)–(8.12): in order for the linear magnetoelectric effect to be present, the magnetoelectric coefficients α_{ij} should be both $\mathcal{I}$- and $\mathcal{T}$-odd, so that left- and right-hand sides of this equation would remain invariant under inversion and time reversal.

In type-I multiferroics the polarization and magnetic order appear independently, so that $\mathcal{I}$ and $\mathcal{T}$ symmetries are broken separately, each at its own transition temperature. In type-II multiferroics the inversion symmetry could be broken by the corresponding magnetic order. Thus for example the sinusoidal spin density wave of Fig. 8.12(a) is centrosymmetric, and has mirror planes, and because of that in itself it does not give multiferroicity. But the inversion symmetry is broken for example in the spiral states of Fig 8.12(b, c), that is these structures may give ferroelectricity (as explained above, polarization may still be suppressed in the proper spiral of Fig. 8.12(c) if there exists a perpendicular twofold rotation axis). Similarly, in Ca_3CoMnO_6 (Choi *et al.*, 2008) the collinear magnetic order breaks the inversion symmetry, see Fig. 8.16, which makes this material multiferroic (remember that spins are axial vectors, which do not change sign under inversion, $\mathcal{I}S = S$).

For many multiferroics the symmetry considerations are indispensable, especially for type-II multiferroics. In this field they were applied successfully, for example, by Harris *et al.* (2008). A detailed classification of possible symmetries, allowing for multiferroicity and magnetoelectricity, is presented in the useful review of Schmid (2008).

An interesting group of questions which we come across in this field concerns the connection of multiferroicity with such notions as chirality and handedness. Vector chirality is defined as the outer product $S_i \times S_j$; it is a $\mathcal{T}$-even axial vector. From expressions (8.19), (8.20) we see that it is closely related to the phenomenon of multiferroicity, especially in type-II multiferroics. However, whereas the usually nonferroelectric proper screw of Fig. 8.12(b) is a chiral object, a cycloidal spiral (more favorable for ferroelectricity) may be nonchiral – this, in fact, depends on its commensurability with the lattice. By talking about chiral objects we refer not so much to the properties under spatial inversions, but rather those under mirror reflections: the definition of a chiral object, going back to Lord Kelvin, is this: "an object or a system is called chiral if it differs from its mirror image, and its mirror image cannot be superimposed on the original object," using all possible transformations (shifts, rotations). This is indeed the case for a proper screw: under mirror reflection a right screw transforms into a left screw, and everyone knows that one cannot superimpose one on the other.

The situation with magnetic moments is different, as they transform under mirror reflections not in the same way as standard vectors. As explained in Section 8.1, the mirror transformation of standard vectors is such that the components of a vector (say P) perpendicular to the mirror plane change sign, $P_\perp \rightarrow -P_\perp$, while the parallel components do not change, $P_\parallel \rightarrow P_\parallel$ (see Fig. 8.9(a, b)). But for magnetic moments the reflection rules are just the opposite: $M_\perp \rightarrow M_\perp$ and $M_\parallel \rightarrow -M_\parallel$ (Fig. 8.9(c, d)). By applying these rules to the proper screw of Fig. 8.12(b) we indeed see that under mirror reflection the right magnetic screw goes over to the left one, that is the proper screw is a chiral object. However some magnetic cycloids, for example with period four times the period of the lattice, change under reflection (with mirror planes both parallel and perpendicular to the spin rotation plane) in such a way that by rotating the image around the twofold axis and using corresponding shifts it can be superimposed on the original cycloid. On the contrary, this is not true for example for a cycloid with period three, or for incommensurate cycloids. Thus, the question of chirality in cycloidal spirals is not a trivial one.

8.4 "Multiferroic-like" effects in other situations

The lessons we learn by studying multiferroics, especially type-II ones, are very helpful for predicting or explaining a number of miscellaneous phenomena, some of which can be very useful for potential applications. Consider for example the standard domain wall in a ferromagnet. There are two types of such wall, shown schematically in Fig. 8.19. The wall shown in Fig. 8.19(a) is called a Bloch domain wall; the magnetization here rotates in the plane of the wall, that is perpendicular to the direction q from one domain to the other. In the other case, Fig. 8.19(b), which is called a Néel domain wall, spins rotate in the plane containing q. From comparison of Figs. 8.19 and 8.12(c) it is immediately clear that in the second case one can consider the Néel domain wall as part of a cycloidal spiral. According to the expressions (8.19), (8.20), we then expect that there will appear an electric polarization in such domain wall, localized in the wall and directed, according to eqs (8.19), (8.20), parallel to the domain wall and in the spin rotation plane (Mostovoy, 2006).

If we now put a system with such Néel domain wall in the gradient of an electric field, the wall, that is the electric dipole associated with it, will be either pulled inside the region of stronger field or pushed out of it, depending on the orientation of the dipole (here

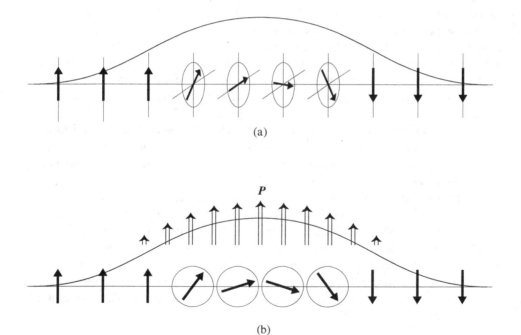

(a)

(b)

Figure 8.19 (a) Bloch domain wall; (b) Néel domain wall.

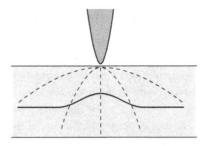

Figure 8.20 Bending of the Néel domain walls (thick line) in iron garnet films under the influence of inhomogeneous electric field (force lines are shown by dashed lines), after Logginov *et al.* (2007, 2008).

these dipoles cannot rotate freely as point dipoles do). This effect was actually discovered and studied experimentally by a group in Moscow (Logginov *et al.*, 2007, 2008). These authors performed a conceptually very simple experiment: using epitaxial films of iron garnet $(BiLu)_3(FeGa)_5O_{12}$, a well-known insulating ferromagnet with high T_c, they brought a sharpened copper wire close to the surface of the film and applied a voltage pulse to it. They detected that under the influence of the resulting inhomogeneous electric field the domain walls in the film, which they visualized magneto-optically, started to move (or rather bend), see Fig. 8.20. The analysis allowed them to conclude that this indeed happens because of the mechanism described above, and one could even estimate the resulting velocity of the domain walls. This observation, besides confirming our general understanding, may be potentially very important, as it allows us in principle to control the domain structure and consequently magnetic memory even in ordinary insulating ferromagnets by applying an electric field, that is voltage, without using electric currents.

One can also consider similar effects for other micromagnetic structures, for example magnetic vortices (Fig. 8.11), which are now widely studied experimentally. It seems that with some ingenious engineering one can modify and control these structures, potentially making them useful.

Another spin-off from the physics discussed above concerns magnetic structures of magnetic films on different substrates. An experiment carried out by Bode *et al.* (2007) and Ferriani *et al.* (2008) demonstrated that in an atomically thin layer of Mn on W there develops not a simple antiferromagnetic structure, as one would expect, but rather a long-periodic cycloidal spiral. And, most surprisingly, all such spirals have the same sense of spin rotation, for example the right one.[10] The same group made the observation that the Néel domain walls in a two-atom-thick layer of Fe on W also have a unique sense of spin rotation (Kubetska *et al.*, 2002; Vedmedenko *et al.*, 2004).

---

[10] Note that for 2d systems one can talk of chirality and handedness also for cycloidal structures.

At first glance this is a very unexpected phenomenon. However it becomes immediately obvious if we remember the physics of the formation of magnetic spirals in BiFeO$_3$ described in Section 8.3.1. According to the discussion there, in type-II multiferroics a cycloidal spiral causes nonzero polarization; but, vice versa, a nonzero polarization would create a cycloidal spiral. This is what happens in BiFeO$_3$; and this is also apparently what happens in the situations studied in Kubetska *et al.* (2002), Vedmedenko *et al.* (2004), Bode *et al.* (2007), and Ferriani *et al.* (2008). Indeed, close to the surface there exists a drop in the potential (i.e., the work function), that is, in a double layer at the surface there exists an electric field perpendicular to the surface. Consequently it can, in principle, create a local polarization P, for example pointing outwards. And, according to the arguments presented above, it would modify the collinear magnetic structure, transforming it into a cycloidal spiral, with the same sense of spin rotations everywhere, determined by eqs (8.19), (8.20).

Microscopically the corresponding mechanism is the same Dzyaloshinskii–Moriya interaction; this was indeed proposed in the original publications (Kubetska *et al.*, 2002; Vedmedenko *et al.*, 2004; Bode *et al.*, 2007; Ferriani *et al.*, 2008) as the explanation of the observed effect. (The inversion symmetry is always broken at the surface, which allows for the Dzyaloshinskii–Moriya interaction.) The knowledge acquired in the study of multiferroics makes the explanation very simple and transparent. It seems that this effect should be present to a certain extent and should be taken into account in all magnetic thin films, although its strength would of course depend on the particular situation: this effect requires strong spin–orbit interaction, for which one should work with heavy elements (for example W). Similar physics is also responsible for the appearance of skyrmion crystals in Fe monolayers on the (111) surface of Ir (Heinze *et al.*, 2012).

Yet one more effect can be proposed (Khomskii, 2009). Consider a standard insulating ferromagnet and a spin wave, or magnon, in it. One usually talks about the reversed spin propagating through the crystal, but the quasiclassical picture of a spin wave is that shown in Fig. 8.21(a), so the magnon is a spin deviation from the average magnetization M, with each spin precessing around M in time and with the phase shifting along the propagation direction. But from Fig. 8.21(a) we see that the instantaneous picture (snapshot) of the spin wave is a cone with a constant M and with the perpendicular component forming a cycloid. According to our general rules (8.19), (8.20), in such a spin configuration there should appear electric polarization perpendicular to the magnetization and to the propagation vector. And indeed such polarization was discovered for a static cone magnetic structure in CoCr$_2$O$_4$ (Yamasake *et al.*, 2006). But it should also work in dynamics: if we create a spin wave in the form of a wave packet, the wave packet would carry with it not only magnetization, but also an electric dipole moment, see Fig. 8.21(b).[11]

---

[11] If one represents this dipole moment as "+" and "−" charges on the opposite sides of the wave packet, the corresponding currents will be such as to create, in accordance with Faraday's law, a magnetic moment M_z, carried by the magnon.

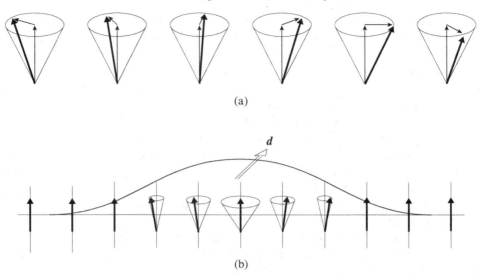

(a)

(b)

Figure 8.21 (a) Quasiclassical picture of a spin wave in a ferromagnet. Each spin has precession around the average magnetization, with a phase shift between different spins. (b) The magnon wave packet, with the corresponding dipole moment (shown by a double arrow) it carries.

S.8 Summary of Chapter 8

The study of ferroelectrics is a quite well-developed field. This class of materials is in no way confined to transition metal compounds. But the best known ferroelectrics, and those most important practically, are transition metal compounds, most often perovskites such as $BaTiO_3$ or $Pb(ZrTi)O_3$. From our point of view just such ferroelectrics are the most interesting – and especially those in which one can combine ferroelectricity and magnetism. Such systems are now known as *multiferroics*. They attract considerable attention, mostly in view of the possibilities of practical application, for example controlling magnetic memory electrically, etc.

Despite all the variety of magnetic structures, the origin of magnetism in "strong" magnets is conceptually simple: one needs the presence of localized electrons with their localized spins, and magnetic ordering is due to the exchange interaction of these spins. In contrast, there exist several different types of ferroelectrics with very different mechanisms of ferroelectric ordering. This makes ferroelectricity more difficult to understand and to describe microscopically.

One can single out several types of ferroelectrics; their general classification, with corresponding examples, is presented in Section 8.1. For our purposes, the following four types of ferroelectrics are the most important: those in which ferroelectricity is due to the presence of TM ions with empty d-shells (with configuration d^0), such as $BaTiO_3$; those with ferroelectricity due to the presence of lone pairs, or dangling bonds – typically materials containing Bi^{3+} or Pb^{2+}; "geometric" ferroelectrics such as $YMnO_3$, in which ferroelectricity is a "byproduct" of some distortions such as rotation and tilting of polyhedra building the crystal (MO_6 octahedra, MO_5 trigonal bipyramids); and materials with ferroelectricity due to a coexistence of site-centered and bond-centered charge ordering, or a simultaneous presence of inequivalent sites (e.g., with different ions) and inequivalent bonds between them. All of these ferroelectrics can be simultaneously magnetic, and then they may be classified as multiferroics; one can call these multiferroics in which ferroelectricity and magnetism in principle exist independently, though of course with some coupling between them, *type-I multiferroics*.

But there exists yet another class of ferroelectrics, or rather multiferroics: materials in which ferroelectricity appears only in certain magnetically ordered states, and thus is *driven by* magnetism. We call such systems *type-II multiferroics*.

The relation between magnetism and ferroelectricity is not simple. Thus, in perovskites, among which there is quite a lot of magnetic systems and which also include many ferroelectrics, there seems to exist a mutual exclusion of magnetism and ferroelectricity: for magnetism one needs partially filled d levels, d^n, $n \geqslant 1$, but practically all perovskite ferroelectrics are those with TM ions with empty d-shells, or with configuration d^0. There are special reasons for that, discussed in more detail in Section 8.2, but actually this question is not yet completely clarified. Nevertheless one can also combine magnetism and ferroelectricity in perovskites – for example by making a "mixture," a perovskite containing both magnetic ions d^n and ferroelectrically active TM ions with configuration d^0. Another option is to use ferroelectricity with other microscopic mechanisms. Such is for example the perovskite $BiFeO_3$ – one of the best multiferroics with large polarization $\sim 90\,\mu C/cm^2$, larger than in $BaTiO_3$, and with high ferroelectric and magnetic transition temperatures $T_{FE} \sim 1100\,K$, $T_N \sim 640\,K$. Unfortunately, in these type-I multiferroics, with good ferroelectric and magnetic properties, the coupling between ferroelectric and magnetic subsystems is usually relatively weak, which limits potential practical applications. In this sense type-II multiferroics, in which ferroelectricity is *due to* magnetism, may be more promising – though until now such materials mostly had weak polarization and low critical temperatures.

Type-II multiferroics can be subdivided into two groups. In the first, until now the more numerous group, ferroelectricity occurs in spiral magnetic structures, mostly cycloidal spirals, and is microscopically due to the inverse Dzyaloshinskii–Moriya (DM) effect, Section 5.3.2. In the weak DM ferromagnetism some structural distortions removing the inversion centers between two magnetic sites produce the interaction $\sim \boldsymbol{D} \cdot [\boldsymbol{S}_i \times \boldsymbol{S}_j]$, which leads to a canting of spins $\boldsymbol{S}_i$ and $\boldsymbol{S}_j$. In spiral magnets such spins are canted or become noncollinear because of the spiral structure itself (which, in turn, is usually caused by frustrated exchange interactions). And such spin canting may cause lattice distortions by the inverse DM effect, that is for example it will cause shifts of oxygens sitting between sites i and j, as a result of which there may appear local polarization $\boldsymbol{P}_{ij} \sim \boldsymbol{r}_{ij} \times [\boldsymbol{S}_i \times \boldsymbol{S}_j]$, where $\boldsymbol{r}_{ij}$ is the vector connecting sites i and j. In effect, in cycloidal magnetic structures there appears ferroelectric polarization $\boldsymbol{P} \sim \boldsymbol{Q} \times [\boldsymbol{S}_i \times \boldsymbol{S}_j]$, where $\boldsymbol{Q}$ is the wave vector of the cycloidal spiral. In other words, in this case there appears electric polarization lying in the plane of the rotating spins but perpendicular to the direction of the cycloid.[12] And if we modify the magnetic structure by external magnetic field, for example if we flop the spin plane, polarization would also rotate by 90°; this is what is indeed observed say in $TbMnO_3$ or $DyMnO_3$.

Note that the inverse DM mechanism responsible for ferroelectricity in this subclass of type-II multiferroics, similar to weak ferromagnetism caused by the DM interaction, is microscopically due to the relativistic spin–orbit interaction. Therefore it is not surprising that the value of the resulting polarization in these systems is usually rather small – typically $\sim 10^2$–10^3 times smaller than in good ferroelectrics such as $BaTiO_3$ or $Pb(ZrTi)O_3$. On the contrary, as we saw above, the *coupling* of magnetism and ferroelectricity in them is intrinsically strong, and we can change polarization by relatively weak magnetic fields.

Another subclass of type-II multiferroics are materials in which polarization appears due to magnetostriction, without requiring spin–orbit interaction. Therefore in such multiferroics the polarization is usually larger than in spiral type-II multiferroics, though still smaller than for example in $BaTiO_3$. Such ferroelectricity is observed for example in

---

[12] Note however that in some cases the proper screw structure can also produce polarization.

TbMn$_2$O$_5$ or in Ca$_3$CoMnO$_6$, but also in perovskites RMnO$_3$ with small rare earths such as Tm and with the so-called E-type magnetic ordering.

Owing to the coupling of electric and magnetic degrees of freedom, there may appear new types of excitation in multiferroics – for example magnons which can be excited by the electric field. Such excitations, called *electromagnons*, form an important field of investigation nowadays.

As mentioned above, multiferroics can be very promising for practical applications, for example due to the possibility of acting directly on magnetic states by electric fields or vice versa, which in principle could be used to write and read magnetic memory electrically (without using electric currents). But in principle one may not even need real multiferroics for that: it may be enough just to have strong coupling between magnetic and electric degrees of freedom, or strong magnetoelectric response, without necessarily having real ferroelectricity (nonzero spontaneous polarization). Materials with these properties are well known among TM compounds; they are called systems with *linear magnetoelectric effect*. This effect was predicted theoretically in Cr$_2$O$_3$ in 1959 by Dzyaloshinskii and discovered experimentally a year later by Astrov, and since then quite a lot of such magnetoelectric materials were found. In such systems the magnetization M can be caused by an electric field, $M_i = \sum_j \alpha_{ij} E_j$, or, vice versa, polarization may be caused by a magnetic field, $P_i = \sum_j \alpha_{ij} H_j$. The magnetoelectric tensor α_{ij} may be symmetric, in which case it can be diagonalized, $\alpha_{ij} = \alpha_i \delta_{ij}$, and for fields along the principal axes we will have $M \parallel E$ or $P \parallel H$. But α_{ij} may also contain an antisymmetric part, which is equivalent to the presence of a dual vector T, called the *toroidal moment*. In systems with $T \neq 0$ the induced magnetization would be perpendicular to the electric field, $M \sim T \times H$, and similarly $P \sim T \times E$. Such, in particular, may be the properties of some magnetic vortices.

The experience reached in studying multiferroic and magnetoelectric materials can be used to explain some interesting effects in ordinary magnetic materials and to suggest some new effects. Thus, one of the types of domain wall in ordinary ferromagnets, the Néel domain wall, in which spins rotate in the plane perpendicular to the plane of the domain wall, locally has the same magnetic structure as a cycloidal spiral, and as such it should have nonzero electric polarization localized at the domain wall. Consequently, one can act on it by electric field, for example it can be pulled inside the region of stronger field. Such experiments were indeed done, and they confirmed these expectations. One can also suggest several other similar effects, some of which are discussed in Section 8.3.3.

Summarizing, we can say that there appear very interesting connections between the phenomenon of ferroelectricity, which for many years was a separate subject, and magnetism. One sees especially interesting effects in multiferroic materials and in systems with the linear magnetoelectric effect, but also in such purely magnetic objects as magnetic vortices or domain walls. The coexistence and interplay of nontrivial electric and magnetic properties in one material, with the possibility of influencing one through the other, can open up prospects for practical applications of these interesting multifunctional materials.

9

Doping of correlated systems; correlated metals

Until now we have largely been discussing the properties of correlated systems with integer number of electrons; only in a few places, for example in the sections on charge ordering and on the double exchange, did we touch on some properties of doped correlated systems. But in principle the variety of phenomena which can occur in such systems with the change in electron concentration is quite broad – from a strong modification of magnetic properties up to a possibility of obtaining non-trivial, possibly high-temperature superconducting states.

A number of questions arises when we start thinking about doped strongly correlated systems. Would the system be metallic? And if so, would it be a normal metal described by the standard Fermi liquid theory? In effect, even with partially filled bands the electron correlations can still remain strong, with the Hubbard's U (much) bigger than the bandwidth; thus these questions are really nontrivial.

The other question is, what magnetic properties will result when we dope Mott insulators? As we have argued in Chapter 1 and Section 5.2, for partially filled bands the chances of ferromagnetic ordering are strongly enhanced, whereas Mott insulators with integer occupation of d-shells are typically antiferromagnetic.

One may also expect that some other, new features could appear in strongly correlated systems with partial occupation of d levels. One can ask the question whether such systems would remain homogeneous, or whether some kinds of inhomogeneity would appear, such as for example the stripes mentioned in Section 7.2.

The situation with noninteger electron concentration and with partially filled bands can be reached in several ways. The most straightforward is the doping of Mott insulators, by electrons or holes. Examples are the colossal magnetoresistance (CMR) manganites such as $La_{1-x}Ca_xMnO_3$, with the average valence of Mn being Mn^{3+x} (i.e., we have x holes per unit cell, or formally $(1-x)$ Mn^{3+} ions and x Mn^{4+} ions), or high-T_c cuprates $La_{2-x}Sr_xCuO_4$, with, formally, Cu^{2+x}, that is with x holes. Similar resulting configurations can also be met in some stoichiometric compounds such as Fe_3O_4, V_4O_7, or $YBaCo_4O_7$, in which the average valence of TM ions is intermediate just by the chemical composition. Thus, in the example mentioned above the vanadium ions are $V^{3.5+}$ and the cobalt ions are $Co^{2.25+}$ ($Y^{3+}Ba^{2+}Co_4^{2.25+}O_7^{2-}$). In doped systems the material is formally disordered (random occupation of e.g. Ca^{2+} or Sr^{2+} instead of La^{3+} in $La_{1-x}Ca_xMnO_3$ or $La_{2-x}Sr_xCuO_4$); sometimes this may be important, although in most cases one ignores this. In contrast, mixed oxides with, on average, intermediate valence of TM ions, such as V_4O_7, are regular systems with periodic lattices. In these materials there may exist inequivalent crystallographic positions for the TM ions just from the structural considerations – as for example the A-site (tetrahedral) and B-site (octahedral) positions in spinels such as Mn_3O_4 or Fe_3O_4. Thus in Mn_3O_4 one A site is occupied by Mn^{2+}, and two B sites by Mn^{3+}, that is all Mn ions have in fact integer valence (though different in different crystallographic positions) – despite the noninteger average valence. Consequently one should expect that for strong correlations such systems would behave as the more conventional Mott insulators with integer electron density. But in other, similar systems the situation may be different. Thus in the magnetite Fe_3O_4 the A sites are occupied by Fe^{3+}, and Fe ions on the B sites have mixed valence $Fe^{2.5+}$, which changes drastically the physical properties of this system.

Finally, there may exist situations in which both strongly correlated and band electrons coexist in the system. This is more typical for rare earth compounds, to be discussed in Chapter 11, but we can meet such a situation in TM compounds as well. And if the localized d levels or narrow d bands overlap with broader, less correlated bands, one can also have the situation with fractional occupation of correlated d states. This may be the situation for example in TM oxides with small or negative charge transfer gaps, see Section 4.3, in which case instead of the more localized d states the doped holes go predominantly into broader p bands. It can also happen that several d bands overlap. In all these cases we typically have a metallic state, although the correlations still play an important role.

All these and similar questions will be discussed in this chapter. We will mostly illustrate the resulting phenomena with three examples, the first being the colossal magnetoresistance manganites $R_{1-x}A_xMnO_3$ (where R is a rare earth, most often La, and A is Ca or Sr). The second system, which illustrates a possible change of spin states, is $La_{1-x}Sr_xCoO_3$. And the third, probably the most important example, is that of cuprates such as $La_{2-x}Sr_xCuO_4$, which form the basis of high-T_c superconductors. Eventually we will also mention other systems of this type.

9.1 Nondegenerate Hubbard model at arbitrary band filling

We start by considering the properties of doped systems described by the nondegenerate Hubbard model; it is the simplest case which nevertheless allows us to illustrate many fundamental properties of doped TM compounds. The material of this section partially overlaps and elaborates on the topics touched upon briefly in Chapter 1; for completeness we repeat part of this material here, or refer to the treatment of Chapter 1.

9.1.1 General features

We start by considering a general theoretical description of conceptually the simplest case, the nondegenerate Hubbard model (1.6),

$$\mathcal{H} = -t \sum c_{i\sigma}^\dagger c_{i\sigma} + U \sum n_{i\uparrow}n_{i\downarrow} , \qquad (9.1)$$

for different strengths of interaction U/t and for different values of the electron concentration $n = N_{el}/N$. We know the properties of this system for $n = 1$ and for $U \gg t$:

it is a Mott insulator with localized electrons and with the antiferromagnetic exchange interaction

$$\mathcal{H}_{\text{eff}} = \frac{2t^2}{U} \sum_{\langle ij \rangle} S_i \cdot S_j , \qquad (9.2)$$

see (1.12). In principle we expect that with decreasing interaction U/t and for $n = 1$ there should occur an insulator–metal transition (Mott transition); these transitions will be discussed in the next chapter. However for certain simple cases such as those with a square or cubic lattice with only nearest-neighbor hopping the energy spectrum

$$\varepsilon(k) = -2t(\cos k_x + \cos k_y + \cos k_z) \qquad (9.3)$$

has a special property called nesting (see also Section 7.3):

$$\varepsilon(k + Q) = -\varepsilon(k), \qquad (9.4)$$

with $Q = (\pi, \pi, \pi)$. One can show, see for example Khomskii (2010), that in this case the metallic state is unstable with respect to the formation of either charge or spin density waves (CDW, SDW), which open the gap at the Fermi energy ε_F and make the system insulating even for a weak interaction. For electron–electron repulsion, as in the Hubbard model, we expect the formation of SDW, which with increasing U/t will gradually go over to the standard checkerboard antiferromagnetic ordering (type-G antiferromagnetism), see the discussion in Section 1.3. Thus for this particular energy spectrum there would be no transition to a metallic state, and the system would remain insulating down to $U/t \ll 1$.

But if the electron density deviates from one electron per site, $n \neq 1$, the situation is different. (Note that often it is more convenient to talk about doping concentration, or about the deviation $\delta = |n - 1|$ of electron concentration from $n = 1$.) If we consider homogeneous states (see below), doped materials could have finite conductivity, that is behave as metals: in a periodic system the extra electrons ("doublons") or holes created by doping can be located at any site in a crystal; that is, in principle they could move without any activation energy.

However the actual situation is not that simple. We have to remember that even for a doped system the strong Hubbard repulsion $U \gg t$ is still present, and that the doped charge carriers would move on the background of other electrons, with which they would strongly interact. We have already seen in Section 1.4 that because of that, in particular because the undoped system for $U/t \gg 1$ is an antiferromagnet, the motion of doped charge carriers will be strongly affected by the underlying magnetic structure (the antiferromagnetic ordering will not disappear immediately at low doping).

One can easily see that for bipartite lattices such as square or cubic lattices, with nearest-neighbor hopping, there is an electron–hole symmetry in the Hubbard model: the transformation from electrons to holes corresponds to the interchange $c_{i\sigma}^{\dagger} \leftrightarrow c_{i\sigma}$, and we can compensate it by changing the sign of the hopping matrix element t, which for the spectrum (9.3) is simply equivalent to the change $k \rightarrow k + (\pi, \pi, \pi)$. Therefore in what follows we will mostly talk about hole doping (which is what we actually have in the three examples chosen – CMR manganites, doped cobaltites, and high-T_c cuprates). Formally, in the simple Hubbard model on bipartite lattices the results for electron doping would be the same as for hole doping. (This is however not the case for some other lattices such as triangular or fcc lattices, nor when we take into account further-neighbor hopping or orbital degeneracy.)

9.1.2 Magnetism in the partially filled Hubbard model

As argued in Section 1.4, the motion of doped holes is hindered by the antiferromagnetic background, so that the electron of hole bandwidth may be strongly reduced, or the doped carriers could become completely localized. Therefore we cannot gain the kinetic energy which we would get if the holes were allowed to move freely through the crystal. Mathematically this means that the localized electron has energy $\varepsilon_0 = 0$ (the middle of the band (9.3)). If, however, the electron or hole could move freely, its energy would correspond to the bottom of the band; that is, we would gain the energy $-zt$, where z is the number of nearest neighbors (for cubic lattices with the spectrum (9.3), $z = 6$). Thus it may be favorable to modify the antiferromagnetic ordering, for example making it ferromagnetic, so as to gain the kinetic (band) energy of the doped holes. For $U \to \infty$ this tendency was demonstrated by Nagaoka (1966), who has proven that for $U = \infty$ and for one extra electron or hole, $N_{el} = N \pm 1$, the ground state of the system is indeed ferromagnetic. For large but finite U and for finite *concentration* of holes $\delta = 1 - n$ one cannot prove anything rigorously, but the approximate treatment (Penn, 1966; Khomskii, 1970) and numerical calculations give the phase diagram which is shown schematically in Fig. 9.1 (already presented in Chapter 1). As we have discussed above, for $n = 1$ (or for doping $\delta = |1-n| = 0$) we have an insulator with antiferromagnetic ordering for $U/t \gg 1$, which, due to nesting (9.4), would remain formally insulating down to $U = 0$ (but with exponentially small gap $E_g \sim e^{-U/t}$), and with a weakened antiferromagnetic ordering taking the form of SDW (but with the same periodicity as antiferromagnetism at $U/t \gg 1$). For almost full ($n \simeq 1$) or almost empty ($n \simeq 0$) bands the system should be metallic, with the normal paramagnetic Pauli susceptibility. One can see this if one uses the so-called "gas" or low-density approximation valid for $n \ll 1$: in this case the Hubbard repulsion U is replaced by the scattering amplitude, or scattering length, which is equivalent to the transition to an effective U,

$$U_{\text{eff}} \simeq \frac{U}{1 + U/W} \sim \frac{U}{1 + U/2zt} \tag{9.5}$$

(mathematically this corresponds to the summation of the so-called ladder diagrams, see e.g. Khomskii, 2010). That is, in the low-density limit the effective interaction U_{eff} (9.5) remains finite even when $U \to \infty$. Therefore we can think that in this limit we would indeed have a normal metal, the standard Fermi-liquid state.[1]

---

[1] There may be some problems with these arguments though. The result (9.5) is obtained when $n \to 0$ for finite U, and only then do we take the limit $U \to \infty$. But in principle the results may depend on the order in which we take the limits $n \to 0$ and $U \to \infty$: if we reverse the order and first take $U \to \infty$ and only after that $n \to 0$, the results could in principle change. Nonetheless the existing numerical calculations seem to confirm the conclusion that we have the normal Fermi-liquid behavior for $n \ll 1$, though, again, such calculations may also suffer from the same problem of the order in which these limits are taken.

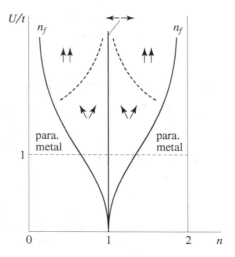

Figure 9.1 Qualitative phase diagram of the Hubbard model for arbitrary band filling and variable interaction strength.

Very interesting (and still rather controversial) is the situation for large U/t and for the intermediate doping range. Approximate treatment gives a ferromagnetic state (denoted by ↑↑ in Fig. 9.1) in a certain part of the phase diagram (Penn, 1966; Khomskii, 1970). This also follows from many numerical calculations. And the transition from the antiferromagnetic state at $n = 1$ to such a ferromagnetic state occurs in these treatments by continuously changing the antiferromagnetic structure so that it gradually approaches the ferromagnetic one. There are several possibilities for this crossover. One is the canting of antiferromagnetic sublattices, as occurs in weak ferromagnetism (see Section 5.3.2) or in double exchange (see Section 5.2, Fig. 5.18), with such canting becoming stronger and stronger with doping, until we reach the ferromagnetic state, see Fig. 9.2(a). Another possibility is that the magnetic ordering becomes a spiral with the wave vector $\boldsymbol{Q}$ which changes with doping: for the undoped case $n = 1$ (or $\delta = 0$) we have $\boldsymbol{Q} = (\pi, \pi, \pi)$, which corresponds to a simple two-sublattice antiferromagnetism (here we principally have in mind a simple cubic lattice, with the lattice parameter taken as 1). With increasing δ the wave vector $\boldsymbol{Q}$ of the spiral decreases, and at a certain critical doping $\delta_c(\sim t/U)$ we have $\boldsymbol{Q} \to 0$, that is the period of the spiral becomes infinite and the ordering becomes ferromagnetic, see Fig. 9.2(b). The corresponding "transitional" phase between the antiferromagnetic and ferromagnetic states is marked in Fig. 9.1 by ↖↗; whether this would be a canted antiferromagnetism (with the original checkerboard periodicity of the antiferromagnetic component), or a helicoidal state with the same pitch angle between neighboring spins, is difficult to say. In any case, the deviations from pure antiferromagnetic ordering start in this approach formally at arbitrarily small doping, and the critical line for the transition to a pure ferromagnetic state is approximately given by the condition (Khomskii, 1970)

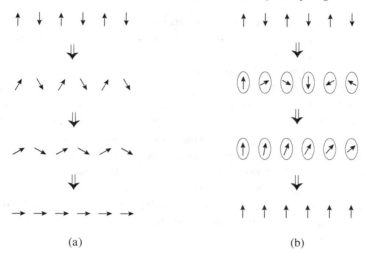

       (a)   (b)

Figure 9.2 Two possible routes to go from the homogeneous antiferromagnetic state at $n = 1$ (or doping $\delta = |n - 1| = 0$) to a ferromagnetic state at large doping and strong interaction: (a) via a canted state or (b) via a helicoidal state with changing wavelength. (The third route – the possibility of phase separation – is discussed in Section 9.7.)

$$0.25\,\delta_c = t/U \ . \tag{9.6}$$

This criterion for the creation of the ferromagnetic state in the nondegenerate Hubbard model agrees quite well with the criterion $0.246\,\delta_c = t/U$, obtained by Nagaoka (1966).

The most problematic is actually not so much the question of the detailed nature of any intermediate state, but rather the very question of the existence of the ferromagnetic phase, and the exact conditions it requires. The question of what is the "outer" boundary of the ferromagnetic state at large doping (large deviation from half-filling $n = 1$), marked as n_f in Fig. 9.1, that is at which n does this state go over to the nonmagnetic metal state realized for the almost empty or almost full bands ($n \sim 0$ and $n \sim 2$), is a matter of considerable debate. Different calculations give, apparently, different values for this critical concentration, for example $\delta_f = |1 - n_f| \sim 0.29$ for large values of U (von der Linden and Edwards, 1991), and it is rather discouraging that the more accurate calculations usually give a smaller region of existence of the ferromagnetic state. This raises a general question: does the nondegenerate Hubbard model on simple bipartite lattices *ever* give ferromagnetism? Maybe the only rigorous result in this field, the Nagaoka theorem, describes just one "point" in the phase diagram which is ferromagnetic, and the other results giving ferromagnetism in a finite region of the phase diagram of Fig. 9.1 are just artefacts of the approximations made.

For more complicated lattices, such as lattices with geometric frustrations, one can sometimes prove rigorously that the ferromagnetic state does exist. These are the lattices which give a very special type of electron energy spectrum, with one completely flat, dispersionless band. It is this flat band which is responsible for ferromagnetism in such lattices (Mielke and Tasaki, 1993). But this is a rather exotic situation. For more conventional lattices such proof is absent. And one often expresses the opinion that the real ferromagnetism can appear only in the presence of several levels, or bands, and is actually due to "atomic ferromagnetism" – the stabilization of the maximum spin of an atom or an ion due to the intra-atomic Hund's rule coupling, see Section 2.2. From this point of view one would not have ferromagnetism in the nondegenerate Hubbard model; one can get ferromagnetism only due to the presence of several orbitals, with the Hund's interaction, which appears in that case, being the main mechanism of ferromagnetism.

This question has, in a certain sense, a somewhat academic character: in all real correlated materials, in particular those which have ferromagnetic phases, one usually has the situation with several active orbitals. But from the conceptual point of view the question of the existence of ferromagnetism in the simple nondegenerate Hubbard model is definitely very interesting, and its clarification would contribute significantly to our understanding of the physics of systems with strongly correlated electrons.

9.1.3 Eventual phase separation in the doped Hubbard model

One more important factor has to be mentioned right away. In considering different possible magnetic states above we always had in mind a homogeneous state of the system – be it antiferromagnetic at $n = 1$, or a canted intermediate phase, or a ferromagnetic state. But there exists yet another possibility: there may appear, for $n \neq 1$, an inhomogeneous state, that is there can occur a *spontaneous phase separation* in such systems. Such a phase-separated state was first proposed for the partially filled Hubbard model by Visscher (1974), and later rediscovered in several other works, mostly in the context of high-T_c superconductors, see for example Khomskii (2010). Such an inhomogeneous state can, for example, consist of a ferromagnetic phase (ferromagnetic droplets) in which all doped electrons or holes would concentrate, the remaining undoped part of the sample being antiferromagnetic. One can show that in a simple approximation such an inhomogeneous state is a better candidate (having lower energy) for the intermediate phase between the antiferromagnetic state at $n = 1$ and the ferromagnetic state at larger doping (if it indeed exists). The antiferromagnetic–ferromagnetic crossover would then occur by the creation of such ferromagnetic metallic droplets, and with increasing doping their volume would increase until they occupy the entire sample, which would then be in a homogeneous ferromagnetic state.

The instability of doped Hubbard systems with partially filled bands with respect to phase separation can of course be a special property of this model, with only short-range (on-site) interaction: the long-range Coulomb interaction, which is not included in the Hubbard model, would strongly counteract such a phase separation and stabilize a state with charge neutrality. But if such intrinsic instability toward phase separation is indeed present in the short-range Hubbard model, the long-range Coulomb forces would not necessarily suppress it completely, but rather may just set a limit on the size of such charged droplets, the inhomogeneous state with small droplets being still more favorable than a completely homogeneous metallic state. Such an inhomogeneous state with small metallic droplets can still be insulating for direct current (d.c.) if the system is below the percolation threshold, whereas for a homogeneous solution with canted AF or ferromagnetic ordering we should expect metallic conductivity. Experimentally, phase separation with the formation of inhomogeneous states was indeed observed in many doped correlated systems; all these questions will be discussed in more detail in Section 9.7.

9.1.4 Hubbard (sub)bands vs usual bands; spectral weight transfer

There is yet another very important point which should be kept in mind when dealing with doped strongly correlated systems. In Chapter 1 we talked about the semiconductor analogy, often used to describe Mott insulators, with the lowest Hubbard (sub)band playing the role of the valence band in conventional semiconductors, and the upper Hubbard (sub)band playing the role of the conduction band, see Fig. 1.11. Quite often this picture is very helpful, and it is used in interpreting various properties of Mott insulators, for example optical properties. However, as already mentioned in that chapter, one has to be careful in using this analogy. The origin of these bands and of the energy gap ($\sim U$) that separates them is very different from that in ordinary insulators or semiconductors such as Ge or Si. If for the latter the band structure is formed by the interaction of the electrons with the periodic potential of the lattice, and one can use the one-electron description of the properties of such systems, in Mott insulators the energy gap appears because of the electron–electron interaction itself; thus the single-particle picture, which forms the basis of the standard band theory, may not be applicable for strongly correlated systems. One of the consequences of this, discussed in Sections 1.4 and 9.1.2, is the strong interplay of charge-carrying one-particle excitations (extra electrons, or "doublons," and holes) with the background magnetic structure. This interplay strongly affects the properties of such electron or hole excitations, and may even lead to a modification of the type of magnetic ordering in the ground state.

Another unusual and very important feature, especially for doped Mott insulators, is the fact that the total "capacity" (number of states) in the lower and upper Hubbard bands is not constant, but changes with doping (for ordinary band systems it is of course constant, equal to $2N$ for each band). This *spectral weight transfer* has very important implications for many properties of these systems.

The easiest way to understand this phenomenon is to consider the situation with localized electrons (ignoring the electron hopping t in the Hubbard model (1.6)), and look at the probability of electron removal and electron addition for a system with N_{el} electrons per N centers, that is with electron concentration $n = N_{el}/N$ (see Fig. 9.3), where one such typical state is shown. If we have N_{el} electrons, there are N_{el} ways to remove one electron from the system (wavy line in Fig. 9.3). Hence the number of occupied states at energy $E = 0$ is N_{el}.

Now let us add extra electrons to the system (dashed lines in Fig. 9.3). We can add N_{el} electrons to *occupied sites* – at each site we can add one electron with spin opposite to the spin of the electron already sitting there. The energy of the state to which we add

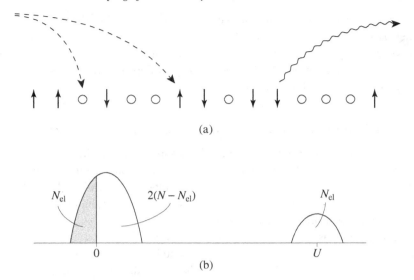

Figure 9.3 (a) The processes of removing (wavy line) and adding (dashed line) electrons to a Hubbard system with partial electron occupation (the number of electrons N_{el} is less than the number of sites N). (b) The corresponding density of states in the lower and upper Hubbard bands, demonstrating the process of spectral weight transfer. The occupied states in the lower Hubbard band are shown by shading.

these electrons is U – the energy of the on-site Hubbard repulsion. That is, the number of available states at energy U is also N_{el}.

Then we can put extra electrons at the *empty* sites, of which there are $N - N_{el}$. The energy of these states will be $E = 0$; however at each site we can put two electrons, with spin up and spin down. That is, the total number of such states will be $2(N - N_{el})$.

If we now include a small electron hopping t, which will somewhat broaden the atomic levels $\{0, U\}$, the resulting picture of the density of states of occupied (shaded) and unoccupied (empty) states will look as shown in Fig. 9.3(b). That is, for $N_{el} \neq N$ ($n \neq 1$) one of the Hubbard subbands, for example the lower one, will be partially filled, with the Fermi energy at zero, and there will be unoccupied states both in the lower ("valence") and in the upper ("conduction") bands. However, as one sees from this construction, the number of states in the lower and upper Hubbard bands, in contrast to the valence and conduction bands in the usual semiconductors, is not constant, but depends on the total number of electrons, and changes with the change in band filling, that is with doping.

The total number of states in both bands in Fig. 9.3(b) is of course constant, $N_{el} + 2(N - N_{el}) + N_{el} = 2N$, as it should be. But there occurs a very strong spectral weight redistribution when the total number of electrons changes. Thus, for example, for very small N_{el} the upper Hubbard band practically disappears – that is, the states at $E \sim U$ do exist, but the total weight of the upper Hubbard band goes to zero as $N_{el} \to 0$.

Such systems with arbitrary N_{el} have partially filled lower or upper Hubbard bands, thus formally they should be metallic. It is only at $N_{el} = N$ that we will have Hubbard bands with equal weight – N in each band – and then the lower band will be completely filled and the upper one empty. This will be the state of the Mott insulator of Fig. 1.11. But as soon as we start to dope the system, not only do we create electrons in the upper or holes in the lower Hubbard bands, but we also change the total number of states in these bands.

One consequence of this concerns the behavior of some transport coefficients such as the Hall constant; this will be discussed later, in Section 9.3. Another general feature, closely related to the spectral weight transfer, is that it is highly nontrivial to separate the low-energy sector from the high-energy states in strongly correlated systems. The standard approach used in treating many-particle systems is to "exclude" or "get rid of" high-energy states and describe low-temperature properties using only the ground state and low-energy excited states, with energies ω of the order of the temperature T. Such low-energy excitations may be strongly renormalized compared with the bare electrons or phonons, but, after such renormalization is done, we can operate only with these low-energy excitations, or quasiparticles. However, the spectral weight transfer described above demonstrates that in strongly correlated systems the low-energy excitations with $\omega \sim 0$ are strongly coupled to the high-energy excitations with $\omega \sim U$, and one cannot disentangle them, at least in the usual manner. This feature is seen directly in many experiments, for example in the modifications of the optical spectra across insulator–metal transitions in TM oxides, for example in V_2O_3 (Rosenberg *et al.*, 1995). One might think that at this transition, occurring in V_2O_3 at $T_c \sim 150$ K, with the closing of the energy gap $E_g \sim 0.6$ eV, only the features at energies of at most E_g would change. However, experiment shows that there is a strong change in the optical weight of this system, extending up to ~ 5 eV. This general feature has to be kept in mind in dealing with strongly correlated systems, especially with doped Mott insulators. To reduce the description of such a system to one making use only of the low-energy sector is quite a nontrivial problem.

9.2 Representative doped transition metal oxides

After this general discussion of the conceptually simplest case of the nondegenerate Hubbard model (formally the simplest one, but even for which we still do not have full answers to some fundamental questions), we go over to the discussion of the situation in real materials, for which qualitative considerations and approximate methods of treatment often give a reasonable description. These examples will serve as an illustration of the main physical concepts in the field of doped strongly correlated systems. We start with CMR manganites.

9.2.1 Colossal magnetoresistance manganites

Doped manganites $R_{1-x}A_x MnO_3$ (where R^{3+} is a rare earth such as La, Pr, etc. and A^{2+} is usually Ca or Sr) with the perovskite structure attract considerable attention due to their very rich properties, in particular colossal magnetoresistance, observed in many of these materials. They show an astonishing variety of different types of ordering: orbital,

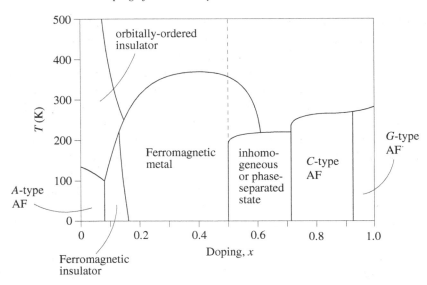

Figure 9.4 Simplified phase diagram of $La_{1-x}Sr_xMnO_3$, after Hemberger *et al.* (2002) (cf. a similar diagram for $La_{1-x}Ca_xMnO_3$ in Fig. 5.16).

magnetic, charge ordering; they have insulating and metallic phases, they are very sensitive to external influences such as electric and magnetic fields, irradiation, etc.

These systems have very complicated and rich phase diagrams. A typical phase diagram for $La_{1-x}Ca_xMnO_3$ was already presented in Fig. 5.16, and that of $Pr_{1-x}Ca_xMnO_3$ in Fig. 7.5; in Fig. 9.4 we present a somewhat simplified phase diagram of yet another manganite, $La_{1-x}Sr_xMnO_3$. (The detailed phase diagram for $La_{1-x}Sr_xMnO_3$ is in fact even more complicated; see Hemberger *et al.*, 2002.) Undoped $LaMnO_3$, which contains Jahn–Teller ions $Mn^{3+}(t_{2g}^3 e_g^1)$ with $S = 2$, is a Mott insulator (although at high temperatures it is a relatively good conductor). It experiences the cooperative JT transition with orbital ordering at $T_{O.O.} \sim 800$ K, and at a much lower temperature $T_N \simeq 140$ K it becomes antiferromagnetic. The orbital and magnetic structure (of type A, according to the standard classification, see Fig. 5.24 in Chapter 5) is shown in Fig. 9.5.[2]

With doping by Ca^{2+} or Sr^{2+}, that is hole doping, we formally introduce into the system ions $Mn^{4+}(t_{2g}^3 e_g^0)$ with $S = \frac{3}{2}$ and without orbital degeneracy, instead of the JT ions Mn^{3+}. The temperature of the JT transition or of orbital ordering, $T_{O.O.}$, decreases rapidly with doping x, and T_N goes down as well; the magnetic ordering changes into some other state, for example a spin-glass-like state, which is also rather inhomogeneous spatially, that is we have here a phase separation of some kind.

---

[2] In fact, as discussed in Chapter 6, the real orbital structure of $LaMnO_3$ is not exactly that shown in Fig. 9.5 (with the orbitals $|x^2\rangle$ and $|y^2\rangle$, which correspond to the orbital mixing angle $\theta = \pm\frac{2}{3}\pi$ in eq. (6.1), see Fig. 6.2); instead, this mixing angle is rather close to $\sim 108°$, which gives the orbital structure shown in Fig. 6.14(b). But this small difference does not play a crucial role, and for most purposes it is usually sufficient to deal with the orbitals shown in Fig. 9.5.

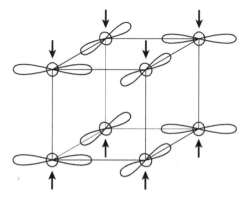

Figure 9.5 Orbital and spin structure of undoped LaMnO$_3$.

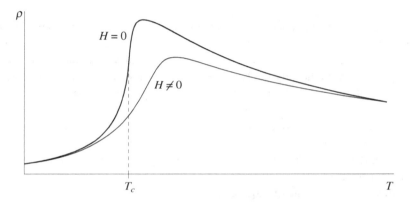

Figure 9.6 Schematic behavior of CMR manganites, typical for systems like La$_{1-x}$Ca$_x$MnO$_3$.

Probably the most interesting phase (or at least the one which has attracted most attention) is reached in both La$_{1-x}$Ca$_x$MnO$_3$ and La$_{1-x}$Sr$_x$MnO$_3$ at $x \sim 0.2$, and extends
typically up to $x = 0.5$; this is the FM phase in Figs 5.16 and 9.4. (See also a similar
phase in Fig. 9.9 later for the bilayer manganites La$_{2-2x}$Sr$_{1+2x}$Mn$_2$O$_7$.) It is in this phase
that one meets the phenomenon of colossal magnetoresistance, which gave its name to this
whole class of compounds – CMR manganites. The typical behavior of resistivity in this
phase at $H = 0$ and at nonzero H is shown in Fig. 9.6: the resistivity is semiconducting-
like at $T > T_c$, and at the transition to the ferromagnetic state it drops sharply and is
metallic for $T < T_c$ (this is the case in Ca-doped systems; in La$_{1-x}$Sr$_x$MnO$_3$ the resistivity above T_c is more metallic-like, see Fig. 9.7, although it also drops below T_c). The
magnetic field shifts the ferromagnetic transition to higher temperatures (and somewhat
broadens the transition), and in effect in the vicinity of T_c the resistivity decreases strongly
with H, that is we have here strong negative magnetoresistance, which can reach up
to 80%. Thus we see yet another manifestation of the tendency we have already mentioned

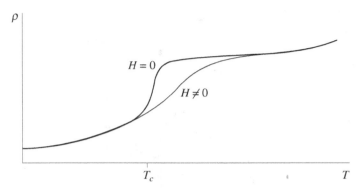

Figure 9.7 Schematic behavior of resistivity in CMR manganites such as $La_{1-x}Sr_xMnO_3$.

several times – that the undoped Mott insulators are typically antiferromagnetic, and ferromagnetism usually coexists with metallic conductivity: ferromagnetic ordering makes electron hopping easier, and this electron hopping, in its turn, promotes ferromagnetism, see Section 5.2. Only relatively rarely, in very specific situations (mostly in cases of specific orbital ordering) can we have an insulating ferromagnetic state, see for example Section 6.1.2.

The formation of inhomogeneous states due to phase separation, clearly seen in low-doped manganites, may still exist at higher doping as well. There are some arguments, see for example Dagotto (2003), that it can also be important for the very phenomenon of colossal magnetoresistance in these systems.

For higher doping the state of $La_{1-x}Ca_xMnO_3$ changes again, see Fig. 5.16. At $x = 0.5$ and at some higher doping levels there is charge ordering in the system: at $x = 0.5$ it is of the checkerboard type in the basal plane, but "in phase" in the third direction, see Section 7.1; in the overdoped regime it is incommensurate, in the form of CDW or stripes, see Sections 7.2, 7.3. The magnetic structure at $x = 0.5$ is of *CE*-type, Fig. 7.3; in the stripe phases at higher doping it is also of zigzag type, Fig. 7.7; for still higher doping it is of *C*-type (ferromagnetic chains stacked antiferromagnetically), until at $x \sim 1$, close to and in pure $CaMnO_3$ without any orbital freedom, it becomes a simple two-sublattice structure of *G*-type. All the phases for $x \geqslant 0.5$ are insulating, although maybe with not very high resistivity. The transition to a charge-ordered state is usually accompanied by an increase in resistivity, shown schematically in Fig. 9.8.

The sequence of phases for $La_{1-x}Ca_xMnO_3$ and $La_{1-x}Sr_xMnO_3$, shown in Figs 5.16 and 9.4, is more or less typical for other manganites as well. But there are also significant differences. Thus, as mentioned above, for $La_{1-x}Sr_xMnO_3$ the resistivity in the paramagnetic phase above T_c in the concentration range $0.25 \lesssim x \lesssim 0.5$, in which the ground state is also a ferromagnetic metal, behaves not as shown in Fig. 9.6 but remains more or less metallic (Fig. 9.7). Correspondingly, the negative magnetoresistance in this system close to T_c is smaller than in $(LaCa)MnO_3$. The charge ordering is also much less pronounced, or completely absent.

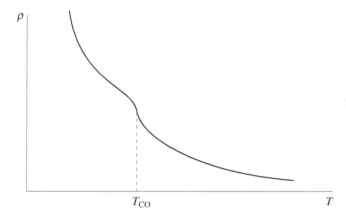

Figure 9.8 Typical behavior of resistivity in manganites with charge ordering at T_{CO}.

In $Pr_{1-x}Ca_xMnO_3$, however, the ferromagnetic metallic phase does not exist at all: as discussed in Section 7.2, this system remains insulating at all x, and for $0.3 \lesssim x \lesssim 0.5$ it has the same checkerboard charge ordering as for $x = 0.5$, with extra Mn^{3+} most probably occupying some sites at random (Fig. 7.9) or with a larger-scale phase separation.

The system $Nd_{1-x}Sr_xMnO_3$ has an interesting novel phase: immediately after the charge-ordered insulating phase at $x = 0.5$ it becomes metallic again, but with A-type magnetic ordering. Most probably the partially filled band in this system, responsible for metallic conductivity, is a two-dimensional band made of $x^2 - y^2$ orbitals (or $x^2 - z^2$, if the plane with parallel spins is not the xy- but the xz-plane).

Thus we see that CMR manganites present a large variety of different phases, all this diversity being determined by the interplay between different degrees of freedom: charge, spin, and orbital. In many such phases one also observes phase separation, especially close to (usually quite sharp) transitions between charge-ordered insulating antiferromagnets and ferromagnetic metallic states which lie close in energy; these transitions can be caused by temperature, by magnetic field, and even by isotope substitution (Babushkina *et al.*, 1998). At some such transitions, typical phase diagrams of which are presented in Fig. 7.10, the resistivity may drop by up to 5–6 orders of magnitude, that is these systems have even stronger negative magnetoresistance than $La_{1-x}Ca_xMnO_3$ shown in Fig. 9.6.[3]

Similarly to manganites of the type (LaCa)MnO_3, some layered perovskite-like manganites also show a large variety of properties. Thus, the bilayer "327" manganite $La_{2-2x}Sr_{1+2x}Mn_2O_7$ has a phase diagram with many phases, Fig. 9.9 (Zheng *et al.*, 2008) (cf. the diagram in Fig. 9.4), with a ferromagnetic metallic (CMR) phase FM, an A-type

---

[3] Yet another example of extremely large negative magnetoresistance is met in slightly nonstoichiometric EuO, see Section 10.3 for more details. With decreasing temperature EuO has a transition at $\sim 70\,K$ from a paramagnetic insulator to a ferromagnetic metal, accompanied by a huge jump in resistivity, up to $\sim 10^{10}$ (Shapira *et al.*, 1973). And when this transition is shifted to higher temperatures by a magnetic field, close to T_c the resistivity drops by the same amount, that is the negative magnetoresistance in EuO is really huge, much larger than in CMR manganites. Unfortunately, it occurs at lower temperatures, $\sim 70\,K$.

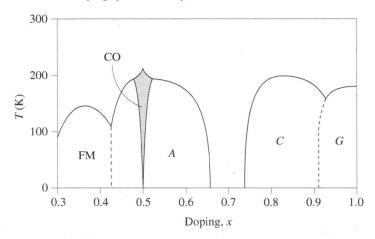

Figure 9.9 Simplified form of the phase diagram for the bilayer manganite $La_{2-2x}Sr_{1+2x}Mn_2O_7$.

phase which is also metallic, a phase with charge ordering (with a curious "ice-cream cone" shape – the shaded region in Fig. 9.9), etc. Note however that a similar single-layer "214" system $La_{2-x}Sr_xMnO_4$ remains insulating for all x. This is in general typical for many such systems: very often such single-layer compounds (with Mn, Co, Fe) have a much stronger tendency to remain insulating even when doped, in comparison with double-layer, and even more so three-dimensional "113" perovskite structures.

9.2.2 Doped cobaltites

The system $La_{1-x}Sr_xCoO_3$ demonstrates yet another possible effect of doping. As discussed in Section 3.3, $Co^{3+}(d^6)$ can exist in different spin configurations: in the high-spin (HS), intermediate-spin (IS), and low-spin (LS) states, Fig. 9.10 (see also the corresponding discussion in Section 5.9). In undoped $LaCoO_3$ at low temperatures the LS state of Co^{3+} is stable, and the material in this state is practically a nonmagnetic insulator (the LS state of Co^{3+} has spin $S = 0$). But those configurations that are magnetic, that is HS or IS, apparently lie close in energy to the LS state, and they become thermally populated with increasing temperature. As a result some magnetic states with $S = 2$ (HS) or $S = 1$ (IS) appear in the system, which one sees from the rapid increase of magnetic susceptibility, Fig. 9.11: it is small at low temperatures, but increases strongly at $T \sim 80$ K, and then starts to decrease, more or less following the Curie–Weiss law. There is another similar crossover at higher temperatures ~ 400–500 K, above which the material becomes a (bad) metal.

Similarly to the temperature-induced transition, the transition to magnetic and metallic state can be caused by (hole) doping. Replacing La^{3+} by Sr^{2+} leads to such a transition. The resulting phase diagram is shown schematically in Fig. 9.12. The hatched regions in this phase diagram are the regions of the crossover between states with magnetic behavior, shown for $x = 0$ in Fig. 9.11, and a ferromagnetic metallic state, appearing with increasing doping.

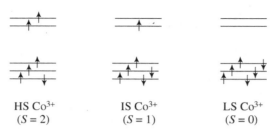

HS Co³⁺   IS Co³⁺   LS Co³⁺
$(S = 2)$ $(S = 1)$ $(S = 0)$

Figure 9.10 Different possible spin states of Co^{3+}: high-spin, intermediate-spin and low-spin states.

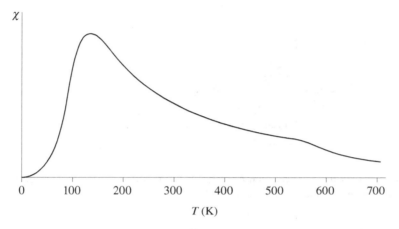

Figure 9.11 Qualitative behavior of magnetic susceptibility of $LaCoO_3$.

Again, as in the case of doped manganites, phase separation takes place here, at least for not very large doping (Wu and Leighton, 2003). Thus again the transition to a ferromagnetic metallic state with increasing x goes via inhomogeneous states. Their presence was established by different methods, including local probes such as NMR and ESR, and is confirmed especially by neutron scattering (Phelan *et al.*, 2006; Podlesnyak *et al.*, 2006, 2008). The microscopic factors leading to this behavior, in particular possible reasons for the stabilization of a ferromagnetic metallic state, will be discussed in Section 9.7.

9.2.3 Cuprates

Our third, and probably the most important example, are doped cuprates, which, in particular, may become high-T_c superconductors. After the discovery of this phenomenon in 1986 (Bednorz and Müller, 1986) they attracted enormous attention and have been studied using probably all existing methods, and quite a lot of information about them has already been accumulated. Nevertheless many important questions are still open, not least the detailed understanding of the microscopic mechanism of superconductivity, though quite a lot of theoretical ideas have been proposed.

This is already a huge field, and there are many good reviews and monographs devoted to these problems; see for example Anderson (1997), Norman and Pepin (2003),

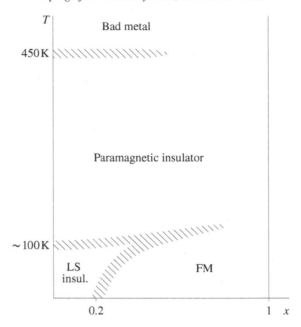

Figure 9.12 Schematic phase diagram of $La_{1-x}Sr_xCoO_3$. LS denotes the nonmagnetic low-spin state phase; FM denotes the ferromagnetic metal phase.

Plakida (2010). We will not of course be able to cover this whole field, but mostly try to present the main physical ideas, using cuprates as a good example of the general phenomena which may occur in doped TM compounds.

The basic prototype material for high-T_c cuprates, which was actually the first such superconductor discovered, is the "214" layered perovskite $La_{2-x}Sr_xCuO_4$. Its main building blocks, CuO_2 layers (Fig. 9.13), are present in most other high-T_c materials of this class.

Undoped La_2CuO_4 ($x = 0$) is a Mott insulator with $Cu^{2+}(t_{2g}^6 e_g^3)$ ions, that is with one hole in the d-shell, with spin $\frac{1}{2}$, arranged in CuO_2 layers in a simple square lattice. It has antiferromagnetic ordering of G-type (simple checkerboard alternations of spins ↑ and ↓) at $T_N = 317$ K (or ~ 200–220 K in less stoichiometric samples). Cu^{2+} is a famous Jahn–Teller ion with one hole in the doubly degenerate e_g orbitals. It always causes strong tetragonal distortion (elongation of the surrounding ligand octahedra), which puts two e_g-electrons on the z^2 orbital, the hole remaining in the $(x^2 - y^2)$ orbital, see Fig. 9.14. The splitting of e_g levels $2E_{JT}$ is usually quite large, typically $\gtrsim 1$ eV. In the process of this JT distortion the two apical oxygens move away from the basal plane, see Fig. 9.15(a). This distortion can be so strong that one of these oxygens "can move to infinity," leaving Cu^{2+} in a fivefold coordination, Fig. 9.15(b). In fact, even both apex oxygens can move away, and Cu will remain square-coordinated, Fig. 9.15(c). These situations are quite typical for Cu^{2+}. In particular in "YBCO" – the high-T_c superconductor $YBa_2Cu_3O_{7-\delta}$ with T_c

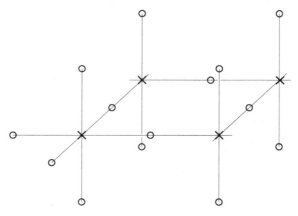

Figure 9.13 Basic elements of the crystal structure of La_2CuO_4. Here crosses are Cu ions and circles are oxygens.

reaching up to ~ 90 K – the Cu ions in CuO_2 planes are in square pyramids, that is are fivefold coordinated as in Fig. 9.15(b) (and there are also predominantly Cu^{1+} ions in linear chains), and there exist also superconductors with the square coordination of Fig. 9.15(c), in particular electron-doped superconductors $Nd_{2-x}Ce_xCuO_4$.

The hole occupation of the $(x^2 - y^2)$ orbital is very common for most known materials with Cu^{2+}. But the axes of these distorted CoO_6 octahedra (or pyramids or squares) need not be in the z-direction. As we have seen for example in Section 6.1.3 for the example of $KCuF_3$, the same "cross-shaped" hole orbitals may also alternate, with local tetragonal axes lying in x- and y-directions, Fig. 6.10. Similarly, the material K_2CuF_4 with, nominally, the same crystal structure as the prototype material for high-T_c superconductors La_2CuO_4 (known as K_2NiF_4-type structure) has alternating $(x^2 - z^2)$ and $(y^2 - z^2)$ hole orbitals in the basal plane, Fig. 6.11. But in the oxide La_2CuO_4 the situation in this sense is simpler: all long axes of the CuO_6 octahedra are parallel to the z-axis, that is the hole orbitals are all of $(x^2 - y^2)$ type. And, as the JT splitting of $(x^2 - y^2)$ and z^2 orbitals is quite large, one usually considers only these $(x^2 - y^2)$ orbitals, ignoring the z^2 bands. Thus in this and similar systems the holes are in nondegenerate $(x^2 - y^2)$ levels or bands, and these cuprates can be to a first approximation described by the simple nondegenerate Hubbard model (1.6), or, more accurately, including also the oxygen p orbitals, by the three-band (or $d-p$) model (4.1), again for nondegenerate d-electrons.[4]

---

[4] There are however some theories which rather stress the original degeneracy of e_g-electrons and the corresponding JT physics, see for example Aoki and Kamimura (1987). Note that for K. A. Müller, who together with G. Bednorz discovered superconductivity in cuprates (Bednorz and Müller, 1986), the notion of the JT effect, in particular of strong electron–lattice interaction in this case, was the guiding principle which in fact led him to the discovery of these compounds. And many scientists, including Müller himself, still believe that these effects are important for the very phenomenon of high-T_c superconductivity, and there exist some experimental results supporting this point of view, though the main tendency now is to attribute this superconductivity to other factors – see more details in Section 9.6.

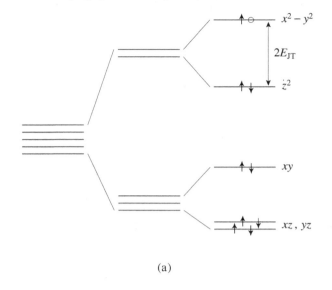

(a)

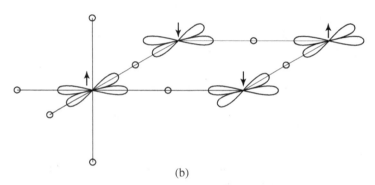

(b)

Figure 9.14 (a) Crystal field splitting of d levels of Cu^{2+} in La_2CuO_4; (b) $x^2 - y^2$ hole orbitals occupied in La_2CuO_4.

One can argue (Zhang and Rice, 1988) that for cuprates one can reduce this three-band or $d-p$ model to the nondegenerate Hubbard model. As we have discussed in Section 4.3, the state Cu^{3+}, which one would get formally by hole doping, corresponds to a very small or negative charge-transfer gap, see Figs 4.7, 4.8. This means that the doped holes would predominantly go to the p orbitals of oxygens, that is instead of the state $Cu^{3+}(d^8)$ we would rather have $Cu^{2+}(d^9)\underline{L}$, where $\underline{L}$ is the ligand (oxygen) hole. The "core" state $Cu^{2+}(d^9)$ has spin $\frac{1}{2}$, the ligand hole $\underline{L}$ around it also has an unpaired spin $\frac{1}{2}$, and one can show that due to the strong $d-p$ hybridization this ligand hole would form a bound singlet state with Cu^{2+}. This state is known as the *Zhang–Rice singlet*, see the detailed discussion in

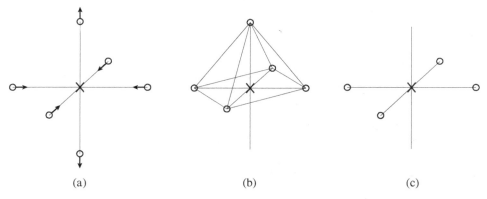

(a)   (b)   (c)

Figure 9.15 Possible coordination of Cu in high-T_c cuprates: (a) elongated octahedron; (b) square pyramid (fivefold coordination); (c) square (fourfold coordination).

Section 4.4. That is, when we introduce a hole into, say, La$_2$CuO$_4$, such a hole would form a singlet state centered at a given Cu site – exactly the same as for the hole doping (removal of an electron) in the simple Hubbard model. Thus we can map this situation formally onto the Hubbard model (1.6) (or onto the so-called $t-J$ model, see eq. (9.22) below), only keeping in mind that such holes are not really localized on d states (Cu), but are the Zhang–Rice singlets with the hole delocalized over four oxygens surrounding a given Cu ion (and strongly hybridized with the $(x^2 - y^2)$ state on it). This approximation is used widely in the theoretical description of high-T_c cuprates, although there are also some arguments (Emery and Reiter, 1988) that this approximation can sometimes break down.

When we dope the Mott insulator La$_2$CuO$_4$, increasing the Sr content x, first of all the antiferromagnetic order is suppressed rapidly, and the material becomes metallic. And this metallic phase is also a high-T_c superconductor. The typical behavior of T_c in these systems is dome-shaped: T_c increases initially with doping, passes through a maximum, and then starts to decrease and disappears in the overdoped regime (which however is very difficult to reach experimentally – for doping that is too large, $x \gtrsim 0.35$, the material is not homogeneous anymore and has a tendency to decompose). A typical phase diagram for high-T_c superconductors is shown in Fig. 9.16, where we mark different phases: the AF insulator; an intermediate phase which behaves as spin glass (though this phase is not observed in all cases); and the superconducting phase (HTSC). For overdoped systems we usually have a more or less normal metal state described by the standard Fermi-liquid theory. But in the intermediate concentration range, close to optimal doping, the properties of the normal state above T_c are unusual: the resistivity does not behave as in conventional metals, etc. This is the phase of an anomalous metal. And yet another special state is observed typically for smaller x: the so-called *pseudogap phase*, in which both transport and magnetic properties are anomalous. There is significant controversy about the nature of this pseudogap phase. One point of view is that it is a precursor to the superconducting phase, in

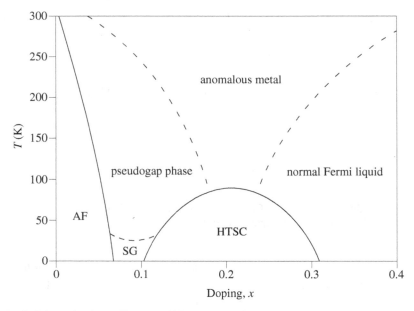

Figure 9.16 Schematic phase diagram of high-T_c superconducting cuprates. Here, AF denotes the antiferromagnetic insulator phase; SG denotes the probable spin-glass phase; HTSC denotes the phase of high-temperature superconductivity.

which the electrons (or rather holes) are already bound in Cooper pairs, but there is no phase coherence between them, required for real superconductivity. The alternative point of view is that this pseudogap phase is due not to preformed superconducting pairs, but rather due to some other instability, for example charge ordering, which rather competes with superconductivity.

The other high-T_c cuprates such as "YBCO" ($YBa_2Cu_3O_7$), "BISCCO" (e.g., $Bi_2Sr_2CaCu_2O_{8+x}$), or similar materials with Hg or Tl, in which maximum superconducting transition temperatures up to 150 K are reached, all have basically the same main building blocks – CuO_2 planes with ($x^2 - y^2$) hole orbitals, as shown in Fig. 9.14. The difference lies in the nature of the "charge reservoirs" which provide doping in these CuO_2 planes.

We have described briefly above three typical classes of doped TM compounds: CMR manganites; doped cobaltites with spin-state transitions; and high-T_c cuprates. We will use these examples to illustrate the general effects which one meets in doped Mott insulators or, in general, in materials with significant electron correlations and with partially filled bands. We now discuss these different phenomena, stressing general trends, and also formulating problems that are still open.

9.3 Doped Mott insulators: ordinary metals?

As we have already mentioned above, the first question which arises when one considers strongly correlated systems with partially filled bands is whether these systems behave as conventional metals. For systems with integer numbers of electrons per site, for example $n = 1$ in the Hubbard model, the ground state for strong interaction $U \gg t$ is a Mott insulator with localized electrons. When we dope such systems, we introduce charge carriers, electrons or holes, which in principle can be mobile: all positions of these extra electrons or holes are equivalent, and formally we do not have to spend any energy by moving them from site to site. But, on the contrary, there are still strong correlations present; these electrons or holes move on the background of other electrons, with their magnetic and eventually orbital ordering, etc. Thus the properties of the resulting state can in principle be very different from those of a normal metal with weakly interacting electrons.

The description of normal metals is given in many monographs and textbooks; see, for example, Ziman (1964, 2000), Ashcroft and Mermin (1976), Harrison (1989), Khomskii (2010). Essentially we treat them as consisting of almost free electrons. The electrons occupy the lowest states in the corresponding energy bands, up to a maximum energy ε_F – the Fermi energy. In momentum space the electrons occupy all states below ε_F, and the boundary between the occupied and empty states is the Fermi surface. This can have a simple form, as for example in the metal Na, or it can be rather complicated, consisting of several electron and hole pockets, etc.

All thermodynamic and transport properties of the standard metals are determined by elementary excitations close to the Fermi surface. The resistivity and optical properties are described by the Drude theory, which gives a specific temperature and frequency dependence of conductivity $\sigma(\omega, T)$:

$$\sigma_{\mathrm{dc}} = \sigma(\omega = 0, T) = \frac{ne^2\tau}{m}, \qquad \sigma(\omega, T) = \sigma(0, T)\frac{1 + i\omega\tau}{1 + \omega^2\tau^2}, \qquad (9.7)$$

where $\tau(T)$ is the relaxation time. The conductivity $\sigma(\omega, T)$ can be measured by the standard d.c. transport measurements and by optics. The typical behavior of resistivity $\rho = 1/\sigma$ as a function of temperature is shown schematically in Fig. 9.17: starting from the residual resistivity ρ_0, determined by impurities and defects in the crystal, $\rho(T)$ increases, but finally should reach some saturation. For normal metals with broad bands, such as Na or Al, the main mechanism of scattering that leads to the increase of resistivity is electron–phonon scattering, which at low temperatures gives $\rho \sim T^5$.

The saturation of resistivity at high temperature is reached when the electron mean free path $l = v_F\tau$ (where v_F is the Fermi velocity and τ is the relaxation time), which decreases with increasing T, becomes of the order of the lattice constant, $l \sim a$. After that the standard approach used in describing conductivity by the kinetic theory (Boltzmann equation) is not applicable anymore, and the conductivity process will have a diffusion character. The limit $l \sim a$ is called the *Yoffe–Regel limit*, and the corresponding conductivity $\sim ne^2a/mv_F$ (see (9.7)) is *Mott's minimum metallic conductivity*.

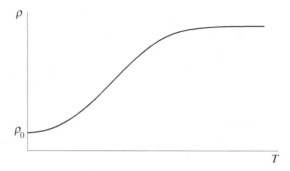

Figure 9.17 Typical temperature dependence of resistivity in normal metals (Fermi liquids): at low temperatures $\rho(t) \sim \rho_0 + AT^2$; then the resistivity grows with T approximately linearly, until reaching saturation at a value close to the Mott's minimum metallic conductivity (or maximum metallic resistivity).

For systems with narrow bands and stronger electron correlations, when the electron–electron scattering starts to play a role, the low-temperature dependence of resistivity is

$$\rho = \rho_0 + AT^2 \; ; \tag{9.8}$$

this is sometimes called Baber's law. Such dependence is seen in some transition metals or in metallic TM compounds, and more clearly in rare earth heavy-fermion compounds, see Chapter 11.

The behavior of many TM compounds in the metallic phase is often rather different from that of conventional metals, although some of them may indeed behave as normal Fermi liquids. Thus, the ferromagnetic metallic phase of manganites (the CMR phase) behaves apparently as a normal metal. However the situation in cuprates is more complicated. Overdoped cuprates, see Fig. 9.16, also behave apparently as Fermi liquids. But, as already mentioned, at optimal doping the normal state properties of cuprates are different, for example the resistivity behaves more or less linearly in temperature, $\rho \sim T$ (or in some cases differently, e.g. as $\rho \sim T^{1.5}$), without any apparent tendency toward saturation. Also their optical properties are different from those of normal metals: the standard Drude dependence (9.7) is rarely observed, or if it is seen, the intensity of the Drude peak in $\sigma(\omega)$ at low temperatures is usually much less than expected for normal metals. This is typical, in general, for many doped Mott insulators.

Similarly, a very characteristic feature of such systems is the appearance of the so-called mid-infrared peak in the optical absorption. Typically this peak is observed at $\omega \sim 0.3$–0.5 eV. The nature of this mid-IR peak is still somewhat controversial: most often one ascribes it to some kind of excitations connected with the Hubbard's energy (e.g., the excitations across the Hubbard gap or pseudogap).

One sees a very special behavior in high-energy spectroscopy of correlated systems, especially in (angle-resolved) photoemission (PES or ARPES) (or, less frequently, inverse photoemission (IPES or BIS)). Photoemission is the process in which an electron is ejected

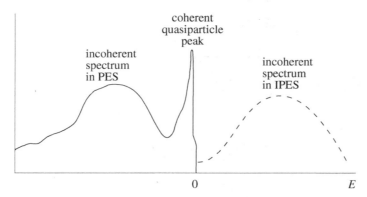

Figure 9.18 Schematic form of the spectrum of photoemission (solid line) and inverse photoemission (dashed line) for metallic systems with correlated electrons. Zero energy is the Fermi energy.

from the solid under illumination (by ultraviolet light or X-rays). By measuring its energy and momentum (or the related angle under which the electron leaves the crystal), one can obtain information about the state of the electron in the solid, about its energy, dispersion relation, and density of states.

For noninteracting or weakly interacting electrons, their excitations have the form of well-defined coherent quasiparticles with a particular dispersion $\varepsilon(\mathbf{k})$. But for strongly interacting electrons, incoherent excitations are also present: they are seen in ARPES as broad peaks. In particular, for strong electron correlations there may exist well-defined quasiparticles at and close to the Fermi energy, but a large part of the spectral weight is in the incoherent parts of the spectrum, which can be associated with the lower (and upper) Hubbard bands, see the schematic picture in Fig. 9.18, where the typical structure of the photoemission spectrum of strongly correlated systems is shown.[5]

---

[5] One can show that what is measured in photoemission and inverse photoemission is the spectral function of electrons, which is connected to the one-electron Green function $G(\mathbf{k}, \omega)$, see for example Khomskii (2010). The intensity of (AR)PES and IPES is proportional to the spectral function $A(\mathbf{k}, \omega)$ given by the relation

$$G(\mathbf{k}, \omega) = \int \frac{A(\mathbf{k}, \omega)\, d\omega'}{\omega - \omega' + i\omega'\delta} \,,$$

this spectral function being proportional to $\operatorname{Im} G(\mathbf{k}, \omega)$, with

$$\operatorname{Im} G = \begin{cases} -\pi A(\mathbf{k}, \omega) & \text{for } \omega > 0, \\ \pi A(\mathbf{k}, \omega) & \text{for } \omega < 0 \,. \end{cases}$$

For weakly interacting electrons the Green function $G(\mathbf{k}, \omega)$ has a structure with the poles at the values of excitation spectra,

$$G(\mathbf{k}, \omega) \sim \frac{Z_{\mathbf{k}}}{\omega - \varepsilon(\mathbf{k}) + i\Gamma} \,,$$

and the spectral function is $A(\mathbf{k}, \omega) \sim Z_{\mathbf{k}}\delta(\omega - \varepsilon(\mathbf{k}))$ (actually this is a slightly broadened delta-function, with the strength of the peak $Z_{\mathbf{k}}$ and the width of the peak Γ determined by the quasiparticle lifetime, $\Gamma \sim 1/\tau$). But for strong interaction such Green functions have small quasiparticle peaks, $Z \ll 1$, and large incoherent parts not having the structure with well-defined poles; this is seen in ARPES as broad peaks away from the Fermi energy (taken in Fig. 9.18 as the zero energy).

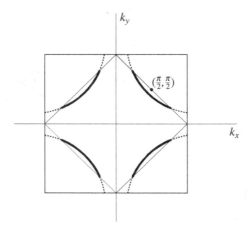

Figure 9.19 "Fermi arcs" in high-T_c cuprates.

In cases where one sees, in correlated systems, well-defined sharp quasiparticle peaks close to the Fermi energy ε_F (zero in Fig. 9.18), one can say that we are still dealing with Fermi-liquid-like systems (maybe having strongly renormalized parameters, e.g. increased effective mass m^*, or strongly reduced weight if the intensity of such quasiparticle peaks is low), even though a large part of the spectral weight is transferred to the incoherent parts of the spectrum. But there may be cases where even in nominally conducting materials the real quasiparticles are destroyed; or maybe they are destroyed not at the whole Fermi surface, but only at some parts of it. Then we would deal with really unusual states. This seems to be the case in many high-T_c cuprates. Often one finds that only parts of the Fermi surface in them are still preserved – usually in the form of Fermi-surface *arcs*, see Fig. 9.19. That is, the quasiparticles in these systems are more or less well-defined in the vicinity of $(\pm\frac{\pi}{2}, \pm\frac{\pi}{2})$ points, or in $[\pm 1, \pm 1]$ directions, but not on other parts of the original Fermi surface (which for noninteracting electrons with this lattice would have the form shown in Fig. 9.19 by dashed lines). It may also be significant that in the superconducting state, which is realized in these systems at low temperatures, we have the so-called d-wave pairing, with the energy gap having zeros just in these directions, that is in the vicinity of these $(\pm\frac{\pi}{2}, \pm\frac{\pi}{2})$ points. Thus this structure of the energy spectrum shows that high-T_c cuprates at this doping range are indeed not the standard Fermi-liquid metals. Apparently the existence of this unconventional normal state is also very important for the very phenomenon of high-T_c superconductivity.

It is interesting to note that this structure of the electronic spectrum may also be related to factors discussed in detail in Chapter 1 – notably to the interplay of electron or hole motion with the magnetic background in correlated systems. The magnetic structure of underdoped cuprates, for example of La_2CuO_4, is a simple two-sublattice (G-type) antiferromagnetism, see Fig. 9.14(b). Consequently, as discussed in Section 1.4, for the Hubbard model with only nearest-neighbor hopping t the coherent motion of doped carriers on such a background will be either completely impossible or at least strongly suppressed. But if

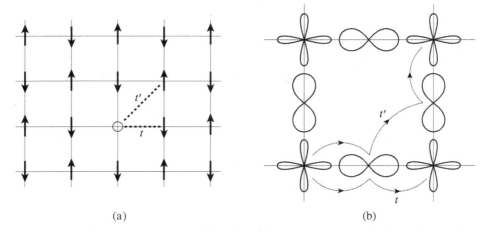

(a)   (b)

Figure 9.20 (a) Motion of a hole on an antiferromagnetic background, with nearest-neighbor and next-nearest-neighbor hoppings t, t'. (b) Possible mechanisms giving nearest-neighbor and next-nearest-neighbor hoppings t, t' in a CuO_2 plane.

we allow hopping t' to next-nearest neighbors, that is along the diagonals in Fig. 9.20(a), then this hopping would be possible on the same magnetic sublattice, and the antiferromagnetic correlations would not hinder such motion. In effect the electrons would be able to move freely and coherently in such directions, that is in [11] and in [$\bar{1}$11], but not in other directions. And we may expect the existence of well-defined coherent quasiparticles, or parts of the Fermi surface, just in these directions. This is what we indeed see in Fig. 9.19: the Fermi arcs exist precisely in these directions.

In real cuprates such next-nearest-neighbor hopping t' may indeed exist, for example due to the overlap of the respective p orbitals of two oxygens. Estimates show that $t' \sim 0.15$–0.18 eV, whereas the nearest-neighbor hopping is $t \sim -0.4$ eV (the signs are important!). The real antiferromagnetic ordering present for example in undoped La_2CuO_4 is suppressed by doping, see Fig. 9.16, but the local antiferromagnetic correlations can still survive even in doped systems, and they can be responsible for the "arcs" shown in Fig. 9.19.

If the long-range antiferromagnetic order had survived in this doping range, the situation could be even simpler, see a schematic illustration in Fig. 9.21: the new antiferromagnetic Brillouin zone (small inner square in Fig. 9.21) would cut the original (e.g., spherical) Fermi surface and create four small hole pockets (shaded in Fig. 9.21) just around the points ($\pm\frac{\pi}{2}, \pm\frac{\pi}{2}$). And if for some reason the matrix elements for photoemission should be much weaker at the outer parts of these pockets of the resulting Fermi surface, then in ARPES one would see the picture of arcs of Fig. 9.19. There are some indications in the de Haas–van Alfven experiments supporting the existence of such small hole pockets. But in real situations the long-range antiferromagnetic ordering at these dopings already disappears, and only some short-range magnetic correlations of this type remain; this would strongly broaden the features shown in Fig. 9.21. Similarly, the applicability of mean-field

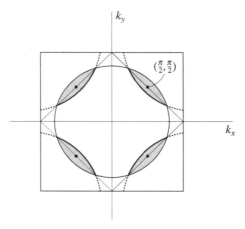

Figure 9.21 Possible origin of the Fermi arcs due to (short-range) antiferromagnetic correlations.

treatment of the electronic spectrum, which would lead to the situation shown in Fig. 9.21, is questionable.

In fact, the two pictures described above, the next-nearest-neighbor hopping in the simple Hubbard model and the mean-field-like description leading to Fig. 9.21, correspond essentially to the same physics: the strong electron correlations resulting in an antiferromagnetic state, and the strong interplay of the motion of electrons or holes with magnetic ordering or magnetic correlations, can lead to a significant modification of the electron spectrum, which, in particular, can have the form of "Fermi arcs" of Fig. 9.19 or Fig. 9.21, with coherent quasiparticles well-defined only in particular directions.

One can see that doped Mott insulators may also differ from normal metals in other respects. One example is the behavior of these systems at low doping. Consider for simplicity the case of a nondegenerate Hubbard system. When we deal with normal metals with electron density $n \simeq 1$ (i.e., for the Hubbard interaction $U \to 0$), we have a large Fermi surface like that shown in Fig. 9.19 by the dashed lines. For the normal metals (normal Fermi liquids) there exists a famous theorem – the Luttinger theorem – which states that the volume of the Fermi surface, even for (weakly) interacting electrons, is equal to that of free electron gas with the same density. Thus for example for weakly hole-doped, weakly interacting systems with $n = 1 - \delta$, with $\delta \ll 1$, we would have the concentration n of charge carriers, and these charge carriers would be electrons (i.e., they would have a positive mass); this would correspond to a slightly less-than-half-filled band, see Fig. 9.22(a). Correspondingly for example the Hall effect coefficient

$$R_{\mathrm{H}} = \frac{1}{nec} \qquad (9.9)$$

would have sign corresponding to the electrons, that is it would be negative and small ($e < 0$, $n \simeq 1$). But if we were to start from the Mott insulator ($n = 1$, $U \gg t$) and

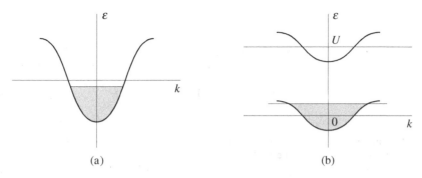

Figure 9.22 Partially filled free electron band (a) and lower Hubbard subband (b) for electron concentration $n \lesssim 1$.

dope it slightly with holes, the actual charge carriers would not be electrons but doped holes with concentration $\delta \ll 1$, Fig. 9.22(b). Indeed, in a simple picture we would have an almost filled *lower Hubbard subband*, with a small amount δ of holes close to the top of this subband. Thus in this case we should expect a positive and large Hall effect coefficient

$$R_{\mathrm{H}} = +\frac{1}{\delta |e| c} \qquad (9.10)$$

(that is, we would have $\delta \ll 1$ instead of $n \sim 1$ in the denominator, compared with (9.9)). And with increasing doping, for example with increasing Sr concentration x in $La_{2-x}Sr_xCuO_4$ (with hole concentration $\delta = x$) this Hall coefficient, remaining positive, should decrease with x as $\sim 1/x$. This is indeed what is observed experimentally in $La_{2-x}Sr_xCuO_4$. In this sense this system at low doping levels behaves very differently from a Fermi liquid, which for this concentration should give, according to the Luttinger theorem, a large Fermi surface and a small and negative Hall coefficient. And only in the overdoped regime $x \gtrsim 0.3$ do we gradually go over to this Fermi-liquid behavior: the Hall coefficient indeed becomes small and negative. Thus there seems to be a contradiction, for $x \ll 1$, between the behavior of many transport properties such as the Hall effect, which can qualitatively be explained quite reasonably as being due to a small number of positively charged holes, and for example photoemission, which seems to give in many cases a large electron-like Fermi surface – although strong modifications of this Fermi surface in cuprates, when only the arcs remain as well-defined quasiparticles (Fig. 9.19), may help to resolve this contradiction.

This would indeed be clear if the picture of Fig. 9.21 applied: in the presence of long-range antiferromagnetism the electron spectrum treated in the mean-field approximation would give, instead of a large Fermi surface (dashed lines in Fig. 9.19), only four small hole pockets around $(\pm\frac{\pi}{2}, \pm\frac{\pi}{2})$ (shaded pockets in Fig. 9.21). But, as we have mentioned above, real long-range antiferromagnetism at these doping levels in cuprates already disappears, and the applicability of the mean-field approximation is very questionable. Thus, in principle, this problem still exists, and its complete solution is still not known: how do

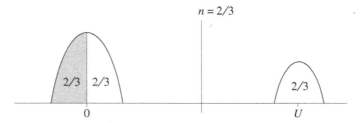

Figure 9.23 Illustration of the qualitative arguments that the effective mass of charge carriers in the lower Hubbard band, and with it the Hall coefficient, may change sign at the electron concentration $n = N_{\mathrm{el}}/N \simeq \frac{2}{3}$, or hole doping $x = 1 - n = \frac{1}{3}$ (cf. Fig. 9.3, and the phase diagram of doped cuprate high-T_c superconductors, Fig. 9.16).

we reconcile the apparent metallic behavior of many doped Mott insulators, which would imply the presence of a large electron-like Fermi surface, with the simple intuitive physical picture in which we ascribe all properties to a small number of doped holes?

A related question is the following: at what doping, or what band filling, does the Hall effect, dominated by this small number of holes, go over to the Hall effect of a large Fermi surface with high concentration of electrons, and how does this crossover occur? For non-interacting electrons with the tight-binding spectrum of Fig. 1.8 or Fig. 9.22(a) it occurs for the half-filled free-electron band, that is for $n = 1$. Where will it occur for strongly correlated electrons? The arguments of the spectral weight transfer, Section 9.1 (see Figs. 9.3 and 9.23), may be used to argue that in this case such a change from electron-like to hole-like charge carriers should occur at $n \sim \frac{2}{3}$: it is precisely at $n = \frac{2}{3}$ that the number of states in all three parts of the spectral function shown in Fig. 9.23 would be equal, $\frac{2}{3}$ each. In particular, the number of occupied and empty states in the lower part of the subband, $\epsilon \sim 0$, would be equal. Thus one could then think that for $n < \frac{2}{3}$ the carriers would be electron-like, while for $n > \frac{2}{3}$ they would be hole-like. This also agrees with the approximate treatment which can be carried out using the so-called Hubbard I approximation (Hubbard, 1963) (see also Phillips, 2009). All these are very interesting and important questions, which definitely deserve further study.

9.4 Magnetic properties of doped strongly correlated systems

As we have discussed above, undoped Mott insulators have a certain magnetic ordering, usually antiferromagnetic. When we change the electron density, we may expect the magnetic properties to also be strongly modified. The material may go to a metallic state, which may be a simple Pauli paramagnet, as are normal metals. But, once again, the presence of strong electron correlations may have a manifestation in the magnetic properties, in particular making the doped systems still magnetically ordered but with a different type of magnetic structure.

On a theoretical level these problems were discussed in Section 9.1 for the example of the nondegenerate Hubbard model, and in Section 5.2 for two-component systems. We have presented there some arguments that, instead of antiferromagnetism, in doped materials there may appear a ferromagnetic ordering stabilized by the gain in kinetic (band) energy, possibly with an intermediate phase of canted antiferromagnetic or spiral type. And only at very high doping, or with bands close to being empty or completely filled, could one have a nonmagnetic metal. But, as described in those sections, in fact the question of the presence of a ferromagnetic state in the nondegenerate Hubbard model on simple lattices is still open: the alternative possibility is that to get the ferromagnetic state one in fact needs degenerate levels or bands, with the intra-atomic "ferromagnetism" due to Hund's rule playing a crucial role. Here we will briefly repeat and elaborate these arguments, paying most attention to the "more metallic" situations.

In real materials we almost always deal with the situation in which different d levels or d bands play a role simultaneously. This is for example the situation in CMR manganites, described in Section 9.2.1. Undoped $LaMnO_3$ with Jahn–Teller ions Mn^{3+} ($t_{2g}^3 e_g^1$) is an orbitally ordered antiferromagnetic Mott insulator. When we replace La^{3+} by Ca^{2+} or Sr^{2+}, we introduce some e_g-holes into the system, or, formally, Mn^{4+} states with the configuration $t_{2g}^3 e_g^0$.

The e_g-electrons in perovskite lattices have a relatively strong hybridization with neighboring oxygens, and the corresponding large $t_{pd\sigma}$ hopping leads finally to the formation of e_g bands, with the bandwidth ~ 1–2 eV. It is still smaller or of the same order as the Hubbard energy $U_{dd} \sim 3$–4 eV, that is these e_g-electrons should be treated as correlated, but forming narrow partially filled e_g bands. Correspondingly they can give rise to metallic conductivity in $La_{1-x}Ca_xMnO_3$ or $La_{1-x}Sr_xMnO_3$ at a certain doping level, typically 0.2–$0.3 \lesssim x \leqslant 0.5$, see Figs 5.16, 9.4, 9.9. At the same time t_{2g}-electrons can still be treated as localized, that is we would have a situation with antiferromagnetically coupled, localized t_{2g} spins $S = \frac{3}{2}$ at every site, with partially filled narrow e_g bands coexisting with them. These e_g-electrons will interact with the localized t_{2g}-electrons by the Hund's rule coupling, which can be written schematically as

$$- J_H S_i \cdot c_{i\sigma}^{\dagger} \sigma_i c_{i\sigma}, \tag{9.11}$$

where S_i is the localized spin $\frac{3}{2}$ of three t_{2g}-electrons, and σ is the spin of e_g-electrons – see a more detailed discussion of the Hund's rule coupling, and the possibility of describing it by such intra-atomic exchange, in Section 2.2. The typical values of Hund's coupling for one electron pair for 3d elements are ~ 0.8–0.9 eV, for example for Mn ions with three localized t_{2g}-electrons the e_g–t_{2g} interaction J_H in (9.11) would be ~ 2.4–2.7 eV – comparable with or even larger (but not much larger) than the e_g bandwidth $W \sim 1$–2 eV. In theoretical treatment of manganites, however, one usually treats Hund's coupling as strong, larger than the bandwidth. In this limit the resulting situation would be that of the *double exchange*, described in detail in Section 5.2. Remember that in this model, assuming strong Hund's coupling, we would have in the ground state the spins of conduction (here e_g) electrons parallel to those of the localized (t_{2g}) ones. And the motion of conduction electrons, their hopping from one site to another, would "pull up" the localized spins of the surrounding centers, making the system ferromagnetic. This double-exchange picture is used as an explanation of the appearance of ferromagnetic metallic state in CMR manganites.

A similar mechanism is invoked to explain the appearance of ferromagnetism in metallic states in many other TM compounds, which either become metallic with doping, or are metallic because the intrinsic energy bands are broad enough. For example, this seems to be the mechanism of the ferromagnetic metallic state in CrO_2 – the material used for many years in "Cr tapes" for tape recorders. In this material, according to *ab initio* calculations of Korotin *et al.* (1997), one of two t_{2g}-electrons of $Cr^{4+}(t_{2g}^2)$ is practically localized and has a localized magnetic moment. The second electron forms a metallic band and provides a mechanism of ferromagnetic ordering due to the same double exchange.

One uses this mechanism as an explanation of ferromagnetism in many other materials, including even transition metals themselves, such as metallic iron, cobalt and nickel. In all these systems we are also dealing not with one type of d-electrons, but with electrons of different bands. The Hund's rule exchange, together with Hubbard-like correlations, are both important, despite the metallic character of these systems. In the approach describing these systems as itinerant ones, in particular in *ab initio* band structure calculations using density functional methods, such as local density approximation (LDA), these interactions take the form of exchange splitting; in this field it is usually denoted by I, and the conditions for magnetic ordering (in particular for ferromagnetism) take the form of a generalized Stoner criterion

$$I\chi(k) > 1 \,, \tag{9.12}$$

where $\chi(k)$ is the magnetic susceptibility for the corresponding wave vector. The instability for a certain $k = Q$ would then mean that the magnetic ordering with the corresponding wave vector Q or with the respective period $\hbar/Q$, for example the helicoidal structure with this Q, would be realized. For $Q = 0$ this would be the criterion of instability toward ferromagnetism, which can be rewritten as

$$IN(\varepsilon_F) > 1 \,, \tag{9.13}$$

where $N(\varepsilon_F)$ is the density of states at the Fermi energy. This is the usual Stoner criterion in the narrow sense. And the exchange splitting I introduced in this approach is connected intrinsically with U and J_H in a localized description.

These considerations also show that in metallic regimes there may occur situations in which the susceptibility $\chi(k)$ is maximal not for $k = 0$, which would mean the tendency toward ferromagnetism, nor for $k = (\pi, \pi)$ or (π, π, π), which would signal a tendency toward antiferromagnetic ordering in 2d square or 3d cubic lattices, but rather for certain intermediate values of $k = Q$, which, if the condition (9.12) is satisfied, would mean the instability toward a spiral state, or spin density wave with the respective period. Usually this situation can be met in two typical cases. First, this happens for more localized electrons when there exist competing interactions between nearest-neighbor and further neighboring sites, see Section 5.7. But in metallic systems we meet this situation more often in the case of particular energy spectra, or particular shapes of the Fermi surface with nesting, see eq. (9.4). This nesting can occur in correlated bands themselves, as is for example the case in metallic Cr or in transition metal dichalcogenides (in which, however, we rather have a charge density wave and not a spin density wave as in Cr). Or the nesting can occur in relatively broad conduction bands coexisting with localized electrons, which is often the case in rare earth metals and compounds.

In any case, we see that metallic systems with strong electron correlations, be it doped Mott insulators or more itinerant systems which are "on the metallic side of Mott transition," often show rather interesting magnetic properties – in quite a few cases with ferromagnetic ordering, but also with more complicated magnetic structures, such as magnetic spirals. And even if such metallic systems remain nonmagnetic (if the interactions are not strong enough so that the Stoner criteria (9.12), (9.13) are not satisfied), often the presence of strong correlations leads to unusual magnetic response of such systems – for example Curie-like susceptibility $\chi \sim 1/T$ instead of Pauli susceptibility $\chi \sim$ const. typical for normal metals. And the proximity to one or other magnetic instability (9.12), (9.13) may also give rise to an interaction of electrons via the corresponding spin fluctuations, ferro- or antiferromagnetic, which, in particular, can provide the mechanism of superconducting pairing – see Section 9.6.

9.5 Other specific phenomena in doped strongly correlated systems

When dealing with Mott insulators we saw that, depending on the particular situation, they could have different types of ordering. And all of them could be modified if we change the electron concentration, especially if as a result of doping the material would become conducting. In the previous section we saw how a magnetic ordering is changed by that. What about other possible types of ordering?

Of course, ferroelectricity would disappear: electric polarization cannot exist in conducting materials. Nevertheless there can exist metals without inversion symmetry, and even those with the symmetry of a pyroelectric class, that is with one particular "vector." What the properties of such "ferroelectric metals" would be (Anderson and Blount, 1965) is a very interesting, but practically unexplored question.

Other types of ordering, such as orbital ordering, or the spin state of particular ions, can also be changed by doping. On the contrary, new types of ordering can appear. One of the most common is charge ordering of different types, already considered in Chapter 7. And of course the possible appearance of superconductivity in some such systems is very interesting.

9.5.1 Modification or suppression of orbital ordering in metallic systems

Systems with localized electrons and with orbital degeneracy usually display some kind of orbital ordering, together with the corresponding structural (Jahn–Teller) distortion. However when the electrons become itinerant, such orbital ordering usually disappears. Thus there exist many metallic systems in which different orbitals contribute to band formation. And usually in these cases we do not talk about orbital ordering, and there is no corresponding distortion. There may be several bands of different origin, crossing the Fermi level; there may also exist different pockets of the Fermi surface; sometimes one can even say in which of them one or the other orbital contributes more strongly, be it d-electrons, or p- or s-electrons. But for broad enough bands this is only an approximate language, and usually in these cases one does not talk about orbital ordering in the usual sense.[6] Strictly speaking, for band electrons one cannot refer to an orbital state: only at the Γ-point (at $k = 0$) can the band states be classified by the point symmetry of the respective ions, that is one can say which states originate from, say, e_g-electrons and which from t_{2g}-ones. At an arbitrary k-point in the Brillouin zone this classification is not valid. Nevertheless often, when the bands are narrow enough, one can still say which band originates from which d level. In this case, if the occupation of one band is different from the other, one can still talk about something like orbital ordering, but one has to be rather careful with this terminology; one should always mention that one speaks about *bands*, not localized electrons. And the features of respective distortions for band electrons may differ from those met in the localized case. Thus, we stressed in Chapter 6 that for the normal Jahn–Teller systems with e_g degeneracy, local distortions practically always correspond to the *elongation* of the corresponding MO_6 octahedra: out of hundreds of known insulators with cooperative Jahn–Teller effect and with orbital ordering there are actually none (or at most very few) with locally compressed octahedra. But this need not be the case for partially filled e_g *bands* having predominantly one or the other orbital character. Thus, in doped manganites of the type $R_{1-x}A_x MnO_3$, Section 9.2.1, the resulting bands are relatively narrow, and often one can still talk about the bands having predominantly $(x^2 - y^2)$ or z^2 character. (This is also the case for spinels with degenerate t_{2g}-electrons, such as $MgTi_2O_4$ or ZnV_2O_4, Section 6.2, and in many other systems.) But for example if the electrons occupy predominantly the $(x^2 - y^2)$ band, which may be the case in layered (but also in 3d perovskite[7]) manganites, such band occupation may give a tetragonal distortion with $c/a < 1$, that is corresponding to a local *compression* of the MnO_6 octahedra. But, once again, this is possible for partially filled e_g *bands*, but practically never for *localized e_g-electrons*.

When we are really talking about orbital ordering of localized electrons, this ordering is usually suppressed when we dope the system. This we can see from the phase diagram of manganites of Figs 5.16, 9.4: the normal orbital ordering, present in undoped $LaMnO_3$, is suppressed with doping, and at least in the usual form it is absent in the ferromagnetic metallic (CMR) state. Detailed structural studies show that for the doping range $0.3 \lesssim x \leqslant 0.5$ at temperatures above T_c of the ferromagnetic transition there still exist short-range charge and orbital correlations, reminiscent of the charge and orbital ordering

---

[6] Sometimes one even takes the presence or absence of orbital ordering and of the corresponding lattice distortions in cases where, by electron count, we would expect orbital degeneracy, as a "fingerprint" of whether we are dealing with localized or itinerant electrons (Goodenough, 1963). The electrons should be treated as localized if there is a JT distortion in the system, and if for nominally orbitally degenerate systems it is absent, it can be taken as an indication that the corresponding electrons should be treated as itinerant.

[7] For example, in the A phase of $Nd_{1-x}Sr_x MnO_3$ for $x > 0.5$.

present at $x = 0.5$, but these correlations disappear rapidly upon entering the ferromagnetic state (Vasiliu-Doloc *et al.*, 1999; Campbell *et al.*, 2001). Thus it looks as though the orbital ordering disappears completely at low temperatures in the FM phase.[8] Similarly, some metal–insulator transitions in TM compounds, for example in VO_2 and V_2O_3, are accompanied by a certain orbital ordering in the insulating phase, whereas orbitals are more or less equally populated in the high-temperature metallic state, see below and Chapter 10. In some cases such orbital ordering leads to an effective reduction of dimensionality of the system, after which an energy gap may appear due to the Peierls phenomenon. This seems to be the case in VO_2 (Haverkort *et al.*, 2005), and it can explain the metal–insulator transition with the formation of spin singlets in spinels $MgTi_2O_4$ and $CuIr_2S_4$ (Khomskii and Mizokawa, 2005). Thus, once again, orbital ordering is rarely seen in metallic phases, but it has a large chance of being realized in insulating states.

9.5.2 *Spin blockade and spin-state transitions in doped Mott insulators*

Doping can significantly modify other properties of correlated electron systems. As discussed in Sections 3.3, 5.9, and 9.2.2, in several situations, especially in materials containing Co^{3+} and Fe^{2+} with the d^6 configuration, there exists a possibility of different spin states of corresponding ions, for example the high-spin $Co^{3+}(t_{2g}^4 e_g^2)$, $S = 2$; intermediate-spin $Co^{3+}(t_{2g}^5 e_g^1)$, $S = 1$; and nonmagnetic low-spin $Co^{3+}(t_{2g}^6 e_g^0)$, $S = 0$. In some cases, for example in $LaCoO_3$, with increasing temperature there occurs a transition (or rather a crossover) from the nonmagnetic LS state to a magnetic (HS or IS) state, see Section 9.2.2. Similar transitions can also be caused by doping, for example in $La_{1-x}Sr_xCoO_3$. What are the physical phenomena occurring at such transitions, and what are the factors causing these transitions?

One important effect determining the spin state is the phenomenon of a *spin blockade*, proposed by Maignan *et al.* (2004). It essentially consists of the following. Suppose we start from the situation such as that in $LaCoO_3$, with the low-spin Co^{3+}, Fig. 9.24. If we add an extra electron to this system, it would form the $Co^{2+}(d^7)$ state. Co^{2+} ions practically always exist in the high-spin state $t_{2g}^5 e_g^2$ with spin $S = \frac{3}{2}$, Fig. 9.24; only for

---

[8] There exists yet another possibility, besides just the equal statistical population of both e_g orbitals in the metallic state: if the correlations are still strong enough, as they seem to be in the CMR state of manganites, in doped systems there may appear orbital ordering of a completely different type from the one realized in the usual cases and described by the wavefunctions (6.1). In this case there may occur an ordering of *complex* orbitals – complex linear combinations of e_g orbitals of the type $\frac{1}{\sqrt{2}}(|z^2\rangle \pm i|x^2 - y^2\rangle)$ (van den Brink and Khomskii, 2001). One can show that such orbitals give a symmetric electron density, that is unlike the usual orbitally ordered states they would not produce any electric quadrupole moment, and would not lead to structural distortions. On the contrary, such states (actually eigenstates of the pseudospin operator τ^y, see Section 6.1, whereas the usual states (6.1) are eigenstates of τ^x and τ^z) would break time-reversal invariance and, as all complex states, would be magnetic. However they do not have a nonzero magnetic dipole moment: what is nonzero for such states is the *magnetic octupole* moment. Similar states with ordering of higher multipoles are sometimes involved in different situations, most often for rare earth compounds such as CeB_6 (Kuramoto *et al.*, 2009) or URu_2Si_2 (Kiss and Fazekas, 2005; Haule and Kotliar, 2009), but also for some TM compounds (Jackeli and Khaliullin, 2009). In our case one can show that such states may indeed be favorable for the doped doubly degenerate Hubbard model (6.13) (van den Brink and Khomskii, 2001). Whether such states are actually realized in a ferromagnetic state of doped manganites is still an open question.

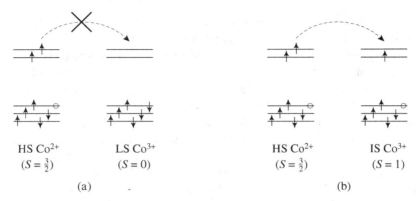

HS Co^{2+} LS Co^{3+} HS Co^{2+} IS Co^{3+}

$(S = \frac{3}{2})$ $(S = 0)$ $(S = \frac{3}{2})$ $(S = 1)$

(a)   (b)

Figure 9.24 (a) Spin blockade for electron hopping between high-spin Co^{2+} and low-spin Co^{3+}. (b) Hopping is allowed for high-spin Co^{2+} and intermediate-spin Co^{3+}.

very strong crystal fields, created by very strong ligands such as CO or NO$_2$, can they become low-spin. Then from Fig. 9.24(a) we see that the hopping of this extra electron, or the exchange of the Co^{2+} and Co^{3+} states, is forbidden: if we transfer an e_g-electron from the HS Co^{2+} to the LS Co^{3+}, the resulting Co^{3+} state "on the left" would be in the "wrong" state – it would be an IS Co^{3+}, not the original LS Co^{3+}. Similarly, the Co^{2+} thus created "on the right" would be "wrong" – it would be the LS Co^{2+}($t_{2g}^6 e_g^1$). In other words, we cannot interchange HS Co^{2+} and LS Co^{3+} by only moving one electron. One can also explain this in a different way: one cannot interchange the state with $S = \frac{3}{2}$ (the HS Co^{2+} on the left in Fig. 9.24(a)) with the ion with $S = 0$ (the LS Co^{3+} on the right) by only moving one electron: such hopping can only change the spins of the corresponding ions by $\pm\frac{1}{2}$. Therefore this situation leads to a *spin blockade*: the ionic state Co^{2+} $(S = \frac{3}{2})$ cannot propagate freely through the background of Co^{3+} $(S = 0)$ ions. Therefore we do not gain in kinetic energy, as we would if such hopping was allowed.

The way out of this difficulty is similar to that we have met before, in Sections 3.3, 5.9, and 9.2.2. One can promote neighboring Co^{3+} ions from the LS to the IS state with $S = 1$, Fig. 9.24(b): we see that in this case one can easily move a second e_g-electron from Co^{2+} to this site, and we recover the initial state with Co^{3+} in the IS state and Co^{2+} in the HS state, only interchanged on the lattice. Therefore after such promotion the doped electron would be free to move through these excited IS Co^{3+}, and we would gain a certain kinetic energy. If this energy gain exceeds the energy loss of promoting the LS Co^{3+} to the excited IS state, this process would occur spontaneously.

We meet a similar situation if we hole-dope such cobaltites, as is actually the case in La$_{1-x}$Sr$_x$CoO$_3$. Formally in this case we create Co^{4+} ions, which are usually in the LS state. The corresponding situation is shown in Fig. 9.25. As we see from Fig. 9.25(a), in principle there exists a possibility of interchanging an LS Co^{3+} $(S = 0)$ and an LS Co^{4+} $(S = \frac{1}{2})$ by moving one t_{2g}-electron. But the hopping of t_{2g}-electrons is much less efficient, and the corresponding gain in kinetic energy is much smaller than that for e_g-electrons: we

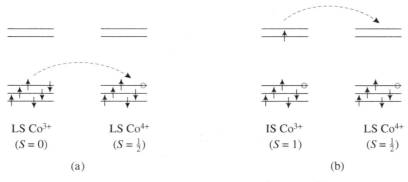

LS Co^{3+} LS Co^{4+} IS Co^{3+} LS Co^{4+}

$(S = 0)$ $(S = \frac{1}{2})$ $(S = 1)$ $(S = \frac{1}{2})$

(a)   (b)

Figure 9.25 Possible t_{2g}–t_{2g} and e_g–e_g hoppings between Co^{3+} and Co^{4+} for different spin states.

have seen this with many examples, such as CMR manganites, in which the t_{2g}-electrons are practically localized, whereas the e_g-electrons form partially filled bands. We can have a similar situation here, but to gain the kinetic energy of e_g-electrons, we first have to promote the LS Co^{3+} to the IS state, see Fig. 9.25(b). For that we have to spend the energy of this promotion, but the e_g-electron thus created can then hop freely to the neighboring LS Co^{4+}, Fig. 9.25(b). Again, if the corresponding energy gain exceeds the energy loss required to promote (a certain number of) LS Co^{3+} ions to the IS state, this process would occur spontaneously. Note that the spins of the neighboring Co ions between which this e_g-electron hops are parallel, that is the process would stabilize the ferromagnetic ordering.

This is apparently what happens in hole-doped LaCoO$_3$. As shown in Fig. 9.12, La$_{1-x}$Sr$_x$CoO$_3$ becomes a ferromagnetic metal for $x \gtrsim 0.2$; and in this phase not only these "x" Co^{4+} ions formally created by doping have nonzero spins, but in fact all Co ions are magnetic – whether closer to IS or HS states is still a matter of debate.[9] Most probably a better description is in terms of IS states (although this language may not be very appropriate for metallic states with itinerant electrons). Experiment shows (Podlesnyak *et al.*, 2008) that the same promotion of some Co^{3+} ions to the IS state in La$_{1-x}$Sr$_x$CoO$_3$ also occurs locally for very low doping $x \sim 0.2\,\%$. What happens in this case is that the six Co^{3+} ions surrounding the Co^{4+} state are promoted to the IS states, see Fig. 9.26, that is each doped hole creates a local object – a magnetic polaron, or rather a spin-state polaron, with the extra hole delocalized over these seven sites. As we see from Fig. 9.25(b), for the free motion of a hole we also need the spins of all sites to be parallel. Thus in this picture each hole (LS Co^{4+}, $S = \frac{1}{2}$) creates in its vicinity six magnetic IS Co^{3+} ions with $S = 1$ each, so that the total spin of this object would be $6 + \frac{1}{2} = \frac{13}{2}$, that is, the total magnetic moment would be $M = 13\mu_B$. This is indeed seen in the magnetic susceptibility of very low-doped LaCoO$_3$ (Yamaguchi *et al.*, 1996), and is confirmed by neutron scattering (Podlesnyak *et al.*, 2008) and by ESR (Noguchi *et al.*, 2002).

---

[9] Note that the free hopping of e_g-electrons can occur not only between LS Co^{4+} ($S = \frac{1}{2}$) and IS Co^{3+} ($S = 1$), but also between IS Co^{4+} ($t_{2g}^4 e_g^1$) ($S = \frac{3}{2}$) and HS Co^{3+} ($t_{2g}^4 e_g^2$) ($S = 2$).

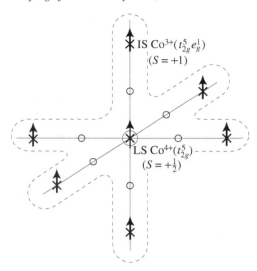

Figure 9.26 The formation of a magnetic cluster in slightly Sr-doped LaCoO$_3$, after Podlesnyak *et al.* (2008). Here crosses are Co^{3+} ions (and the cross at the center is Co^{4+}), and circles are oxygens.

Somewhat surprisingly, such promotion of low-spin Co^{3+} to a magnetic state is not observed in electron-doped cobaltites, neither in the low-doping region nor at higher concentration of doped electrons. Thus for example the half-doped layered cobaltite La$_{1.5}$Sr$_{0.5}$CoO$_4$, which nominally has half of Co ions in 2+ and half in 3+ states, shows Co^{2+}/Co^{3+} charge ordering of the checkerboard type, as in Fig. 9.27, and the Co^{2+} ions (crosses in Fig. 9.27) are magnetic with $S = \frac{3}{2}$, as they should be (HS Co^{2+}), but the Co^{3+} ions remain nonmagnetic, LS Co^{3+} (Zaliznyak *et al.*, 2000, 2001). The magnetic ordering in this system, involving only Co^{2+}, is also shown in Fig. 9.27; see also Cwik *et al.* (2009). Why the mechanism of spin blockade described above (e.g., Fig. 9.24) does not work here, that is why the Co^{3+} ions remain in the nonmagnetic LS state despite being surrounded by Co^{2+}, is an interesting and open question. Probably two factors contribute to this apparent asymmetry between electron- and hole-doping in cobaltites. The first is just the lattice effects: Co^{2+} ions have much bigger ionic radii than Co^{3+}. Therefore, when we create such Co^{2+} states by electron doping, the neighboring Co^{3+} sites would be locally compressed, which stabilizes the LS state of Co^{3+}. Indeed, as explained in Section 3.3, LS Co^{3+} have much smaller ionic radii than for example HS Co^{3+} (by almost 15%). Therefore compression gives extra stability to the LS state of Co^{3+}, which agrees with experiment (Lengsdorf *et al.*, 2004). This may be an important factor preventing the transition of Co^{3+} adjacent to Co^{2+} from LS to HS state in the case of electron doping. (And the same factor may contribute strongly to the very formation of the checkerboard charge ordering in La$_{1.5}$Sr$_{0.5}$CoO$_4$, which consists of alternation of large HS Co^{2+} and small LS Co^{3+}.) On the contrary, Co^{4+} ions created during hole doping, for example in La$_{1-x}$Sr$_x$CoO$_3$, are

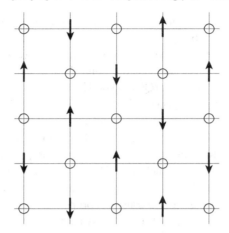

Figure 9.27 Charge and magnetic ordering for half-doped single-layer cobaltite $La_{1.5}Sr_{0.5}CoO_4$, after Zaliznyak *et al.* (2000, 2001). Here circles are low-spin Co^{3+} ions with $S = 0$; the magnetic ordering (arrow) is shown for the Co^{2+} ions.

relatively small; the neighboring Co^{3+} will have more space and can more easily transform to the bigger IS or HS states.

Another factor may be that the extra electrons really go to d states, creating normal Co^{2+} ions. However the "Co^{4+}" states formally created by hole doping are known to be relatively unstable: as discussed in Chapter 4, these states have a very small or even negative charge-transfer gap, so that a large fraction of these holes would actually go to the neighboring oxygens, and instead of $Co^{4+}(d^5)$ we would then have $Co^{3+}(d^6)\underline{L}$ or even $Co^{2+}(d^7)\underline{L}^2$, where $\underline{L}$ denotes a ligand hole. This would reduce the corresponding crystal field splitting and therefore would facilitate LS–HS or LS–IS promotion for hole-doped $LaCoO_3$ (remember that the relative stability of LS vs HS or IS states is determined by the ratio of crystal field splitting and Hund's rule coupling: large crystal field splitting stabilizes the LS states, but reduction of this splitting would favor HS or IS states, see eqs (3.33), (3.36), (3.38)).

9.5.3 Quantum critical points and non-Fermi-liquid behavior in correlated systems

We have seen already in Section 9.3 that in some cases doped Mott insulators, even when they show metallic conductivity, may behave differently from conventional metals of Fermi-liquid type. Thus the resistivity of optimally doped cuprates is typically linear in temperature, $\rho \sim T$, in a broad temperature interval (in some systems it behaves as $\rho \sim T^{1.5}$), and the absolute value of resistivity at high temperatures may be larger, or conductivity smaller than Mott's minimal metallic conductivity. One explanation of such unconventional metallic behavior is that the dopand concentration x, at which this behavior is seen most clearly, more or less coincides with the concentration at which some ordering in the system disappears. This situation, when by changing a certain parameter q (pressure, or magnetic field, or doping concentration) we suppress some ordering, so that the corresponding critical temperature $T_c(q)$ goes to zero for $q \rightarrow q_c$ (e.g., pressure $P \rightarrow P_c$), is known as a *quantum critical* regime, and the point $q = q_c$ in the (q, T) phase diagram is known as a *quantum critical point* (QCP) (von Löhneisen *et al.*, 2007; Sachdev, 2011), see Fig. 9.28. Indeed, close to the critical point T_c, especially if we are dealing with a second-order (continuous) phase transition, there are usually strong fluctuations present. For finite T_c these fluctuations are mainly thermal fluctuations, which are essentially classical in nature. But if by changing some external parameter q we can suppress T_c so that $T_c \rightarrow 0$ for $q \rightarrow q_c$, then close to such quantum critical point not classical but quantum fluctuations would play the main role. And this can change the behavior of many properties of the system, both thermodynamic and transport properties.

In particular, if we are dealing with metallic systems, or at least if one of the phases (e.g., the disordered phase in Fig. 9.28) is a metal, then often close to the quantum critical point and in a certain vicinity of it (shaded area in Fig. 9.28) the behavior of the system can differ markedly from the normal Fermi-liquid behavior away from this point. We talk then of a *non-Fermi-liquid* state.

The behavior of different systems in these cases is not always the same, and it may depend on a particular situation and on the detailed type of quantum critical point. Thus for example the resistivity may be linear, $\rho \sim T$, or it may behave as $\rho \sim T^\alpha$, with a certain exponent α (e.g., $\alpha = 1.5$, etc.) Also such quantities as specific heat or magnetic susceptibility may be different from the normal Fermi-liquid behavior $c \sim \gamma T$, $\chi \sim$ const.: they may become nonanalytic, for example $c \sim T^\zeta$, etc.

There are now many examples of such quantum critical point and of the corresponding non-Fermi-liquid behavior in different systems.[10] Most clearly one sees these effects in

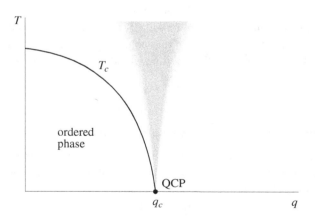

Figure 9.28 Schematic phase diagram with a quantum critical point.

---

[10] Note that non-Fermi-liquid behavior may exist not only close to quantum critical points, but also in some other situations – see for example von Löhneisen *et al.* (2007).

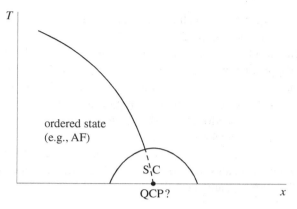

Figure 9.29 Schematic phase diagram with the appearance of a dome-shaped superconducting phase close to a quantum critical point; this is one of the possible scenarios in high-T_c superconducting cuprates.

rare earth compounds, such as in heavy-fermion systems, see Chapter 11. However, some TM compounds also show such behavior. In particular, one explanation of unconventional normal state properties of high-T_c cuprates close to optimal doping attributes this behavior to the proximity of this phase to a "hidden" quantum critical point at about $x \sim 0.22$, just inside the superconducting dome, see Fig. 9.29 (cf. Fig. 9.16). What the detailed nature of this QCP is (if it indeed exists) and what kind of ordering it is connected to is not completely clear. It may be connected to the pseudogap phase in Fig. 9.16 (the nature of which, however, is also a matter of debate).

The extra argument in favor of this interpretation in high-T_c cuprates is that very often, in several different systems, one observes the appearance of superconductivity close to the quantum critical point, at which a certain ordering (most often magnetic) disappears, see Fig. 9.29. Such a phenomenon is seen in CePd$_2$Si$_2$ and CeIn$_3$ (Mathur *et al.*, 1998), in UGe$_2$ (Saxena *et al.*, 2000), and in several organic compounds (Kagoshima *et al.*, 2006); see also von Löhneisen *et al.* (2007). Typically it is the antiferromagnetic ordering which is suppressed (but in UGe$_2$ it is the ferromagnetic one), and one usually attributes the appearance of superconductivity in these cases to electron pairing due to interaction with antiferromagnetic spin fluctuations. This is also the most popular explanation of the nature of pairing in high-T_c cuprates.

One can expect that the superconductivity in this case would be unconventional, not of s-wave type (with the superconducting order parameter or gap $\Delta(\boldsymbol{k}) \sim$ const.) but of d-wave type ($\Delta(\boldsymbol{k}) \sim (\cos k_x - \cos k_y)$ or $\sim (k_x^2 - k_y^2)$). For coupling via the ferromagnetic spin fluctuations we could expect p-wave pairing – for more details see Section 9.6.

There are also situations, more rare, in which non-Fermi-liquid behavior is observed not in the vicinity of special *points* (QCP), but in a finite region of the phase diagram. This is the situation met for example in MnSi: the spiral magnetic order is suppressed by pressure, but the non-Fermi-liquid state is observed in a large region of pressures $P > P_c$

(Pfleiderer *et al.*, 2004). Another such example are the nickelates such as $PrNiO_3$ (Zhou *et al.*, 2005b). As mentioned in Sections 3.5 and 7.5 (see Fig. 3.47), in these materials there exists a metal–insulator transition driven by temperature, and the corresponding T_c may be strongly suppressed by pressure. It was observed by Zhou *et al.* (2005b) that for $P > P_c \sim 10$ kbar, for which $T_c \to 0$, $PrNiO_3$ becomes metallic, but of non-Fermi-liquid type: the resistivity in the resulting metallic phase behaves as $\rho \sim T^\alpha$, with $\alpha \sim 1.3$ for $P_c < P \lesssim 16$ kbar and $\alpha \sim 1.6$ for $P > 16$ kbar. That is, in these systems we have not just a quantum critical point giving a non-Fermi-liquid behavior but again a whole non-Fermi-liquid region, and with different properties at different pressures. What the nature of this behavior is, and which mechanisms lead to the non-Fermi-liquid state in a large region of pressures in MnSi or $PrNiO_3$, are not really clear.

9.6 Superconductivity in strongly correlated systems

In Section 9.2.3 we discussed briefly some basic aspects of the structure and properties of high-T_c cuprate superconductors, mainly using $La_{2-x}Sr_xCuO_4$, the first high-T_c superconductor discovered, as a prototype example (Bednorz and Müller, 1986; note though that the original discovery was rather made on $La_{2-x}Ba_xCu_2O_4$). In this section we describe some other high-T_c superconductors based on transition metals, and devote our attention mostly to conceptual problems such as the type and mechanism of pairing, that is to the very nature of this phenomenon itself.

First a few extra words about cuprates. There are now several groups of high-T_c cuprate superconductors known. Besides the so-called "214" or LSCO system of the type $La_{2-x}Sr_xCu_2O_4$, discussed in Section 9.2.3, with maximum critical temperature $T_c \sim 40$ K, there are also the systems $YBa_2Cu_3O_7$ (YBCO); materials on the basis of Bi, for example $Bi_2Sr_2CaCu_2O_8$ (BSCCO 2212); and similar systems based on Hg or Tl. The maximum values of T_c are as follows: YBCO and BSCCO have $T_c^{max} \sim 90$ K, while mercury and thallium-based materials have $T_c^{max} \sim 120$–130 K (or, under pressure, even up to ~ 140–150 K).

All these systems have the same basic building blocks – two-dimensional CuO_2 layers shown for example in Fig. 9.13. In some cases the in-plane Cu ions are in elongated octahedra, for example in $La_{2-x}Sr_xCuO_4$ as shown in Fig. 9.13; in other systems, for example in $YBa_2Cu_3O_7$, the "active" layers contain fivefold-coordinated Cu in CuO_5 square pyramids, Fig. 9.15(b); and in certain cases Cu ions are square-coordinated (note that these are typical coordinations of Cu^{2+} in general, due to the very strong Jahn–Teller nature of these ions).

The blocks sitting between such CuO_2 layers (sometimes bilayers or trilayers) play the role of charge reservoirs (but sometimes also a more active role). In some such spacer layers there are also Cu ions, but usually in different coordination, for example linearly coordinated Cu^{1+} (with the dumbbell configuration).

A change in composition, for example the substitution of Sr^{2+} for La^{3+} in $La_{2-x}Sr_xCu_2O_4$, or a change of oxygen stoichiometry in $YBa_2Cu_3O_{6+x}$, leads to hole

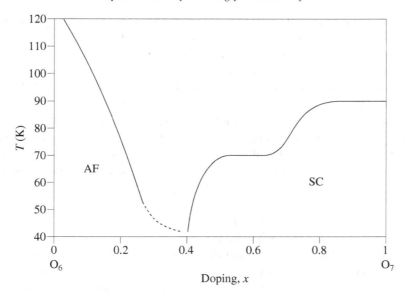

Figure 9.30 Schematic phase diagram of $YBa_2Cu_3O_{6+x}$.

doping of CuO_2 planes, and starting from a certain hole concentration superconductivity appears. The schematic phase diagram of high-T_c cuprate superconductors has already been shown in Fig. 9.16 for the example of $La_{2-x}Sr_xCu_2O_4$, but its general structure is very similar (with some minor modifications) in other systems of this class as well. For example in Fig. 9.30 we show schematically the phase diagram of $YBa_2Cu_3O_{6+x}$, where oxygen content x plays the same role as Sr concentration does in $La_{2-x}Sr_xCu_2O_4$. For $x = 0$ we have a Mott insulator with magnetic $Cu^{2+}(d^9)$ ions with $S = \frac{1}{2}$, which order magnetically. Hole doping in both cases leads to the appearance of the superconducting phase. In all cases of high-T_c cuprates the superconducting phase lies in close proximity to such magnetic phases and is dome-shaped, reaching the maximum at a certain doping level and decreasing in the overdoped regime. In most theories one relies on this proximity to the magnetic phase and attributes the very phenomenon of high-T_c superconductivity to the interaction of electrons with magnetic degrees of freedom, although the details often differ wildly. There are, however, still some theoretical approaches which ascribe superconductivity to the coupling of electrons to some charge excitations (remember that in the "classical" superconductors the main mechanism of superconductivity is the electron–phonon interaction, see below.)

In 2008 a second big class of superconductors with transition metal elements and with high transition temperatures, reaching ~ 55 K, was discovered (Kamihara *et al.*, 2008): the iron-based high-T_c superconductors. The known variety of these systems is already more extensive than that of cuprates. There are several families of these superconductors, usually denoted by short codes: 1111 systems of the type $LaFeAs(O_{1-x}F_x)$; 122 systems, for example $BaFe_2As_2$; 111 systems, $LiFeAs$; 11 systems, α-$FeSe$.

There are similarities and differences between this class of materials and high-T_c cuprates. Both are layered (quasi-two-dimensional) systems. In both there is usually a magnetic state close to the superconducting one. The general shape of the phase diagram is also similar, see below. But there are also important differences between them. One is that, in contrast to cuprates, here the doping may be isovalent, and it can be replaced by pressure. Also, whereas the undoped cuprates are Mott insulators, with strong electron correlations which apparently persist to the superconducting composition, the Fe-based systems are practically always metallic (with only a few exceptions). The electron correlations in them are weaker – probably they are still present, but definitely not as strong as in cuprates. Magnetism in them is presumably mostly of itinerant character, resembling spin density waves.

We do not describe all these systems in detail; this extensive field is already well covered in review articles – see for example Sadovskii (2008), Mazin and Schmalian (2009), Hirschfeld *et al.* (2011), Stewart (2011). We only show here the main structural block of these systems (Fig. 9.31), and the typical phase diagram (Fig. 9.32) (as always we denote TM ions, here Fe, by crosses and anions – As, or Se, etc. – by circles). The Fe ions form a square lattice, and they are surrounded by As or Se tetrahedra, and the pnictogene (As) or chalcogene (Se) ions also form square lattices above and below the Fe layer. Other components (LaO layers in 1111 systems, Ba or Li in 122 and 111) lie between these FeAs layers, and in 11 systems such as FeSe these layers connect directly, without any spacer in between.

The typical phase diagram of Fe-based superconductors (Fig. 9.32), has many similarities with that of cuprates, cf. Fig. 9.16. In both cases there is usually a magnetic phase close to the superconducting one; in both cases this superconducting phase is dome-shaped, with T_c reaching the maximum at a certain doping and disappearing both in underdoped and overdoped systems. In some such systems the magnetic and superconducting phases overlap slightly, in others they do not; in some cases the antiferromagnetic (or SDW) phase disappears and goes over to the superconducting one abruptly.

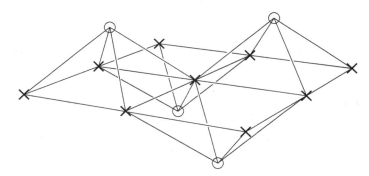

Figure 9.31 Basic elements of the structure of Fe-based superconductors, formed by the square lattice of Fe ions in tetrahedral coordination by As or Se. Here crosses are Fe ions, and circles are pnictogene (e.g., As) or chalcogene (Se) anions.

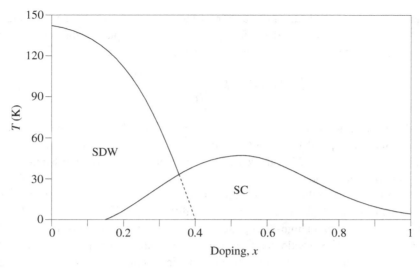

Figure 9.32 Typical phase diagram of Fe-based superconductors, as exemplified by $Ba_{1-x}K_xFe_2As_2$.

The most important question, of course, is what the mechanism of superconductivity is in these systems. We see that both big groups of systems discussed above are based on TM elements with, at least potentially, strong electron correlations – and the values of T_c in them are (much) higher than in the older, more familiar superconductors such as Al or Sn. At first glance this looks quite unexpected and surprising. Indeed, in normal metals super-conductivity is well described by the standard Bardeen–Cooper–Schrieffer (BCS) theory, and the superconducting pairing is explained by the effective attraction between electrons, provided by the electron–phonon interaction, see for example Schrieffer (1964). The stan-dard pairing in this case is in the singlet state and occurs in the *s*-wave channel, that is the order parameter is

$$\Delta(\mathbf{k}) \sim \langle c^{\dagger}_{\mathbf{k}\uparrow} c^{\dagger}_{-\mathbf{k}\downarrow} \rangle = \Delta_0 = \text{const.} \tag{9.14}$$

From this point of view the very appearance of superconductivity in systems with strong Coulomb *repulsion* seems very surprising: in the standard approach, in order to form a Cooper pair of two electrons one needs *an attraction* between these electrons.

The answer is that in the case when electron repulsion dominates, the superconduct-ing pairing of electrons is still possible, but typically this pairing would not be of *s*-wave, but of *p*-wave or *d*-wave type; and it would not be hindered by electron repulsion, as in conventional superconductors such as Sn or Al, but would rather be *promoted* by it. Thus it is now established rather firmly that superconductivity in high-T_c cuprates is of the *d*-wave type,

$$\Delta_d(\mathbf{k}) \sim \Delta_0 \cdot (\cos k_x - \cos k_y) \sim \Delta_0 \cdot (k_x^2 - k_y^2). \tag{9.15}$$

We cannot go into a too detailed discussion of these questions here; for more details see Schrieffer (2007) or Plakida (2010). Suffice it to say that typically the self-consistency equation for the superconducting gap has schematically the form

$$\Delta(\mathbf{k}) = - \int d^3 k' \, \Gamma(\mathbf{k} - \mathbf{k}') \frac{\Delta(\mathbf{k}')}{\sqrt{\left(\varepsilon(\mathbf{k}') - \varepsilon_{\mathrm{F}}\right)^2 + |\Delta(\mathbf{k}')|^2}} \, . \tag{9.16}$$

If the effective interaction $\Gamma(\mathbf{k} - \mathbf{k}')$ is attractive, $\Gamma < 0$, then there may exist a solution of this equation of the type (9.14), with $\mathbf{k}$-independent gap Δ_0. If however we have dominating repulsion between electrons, the kernel is $\Gamma > 0$ and such a solution would not exist: the left- and right-hand sides of this equation would have opposite signs.

But there may exist in this case a solution in which the gap function $\Delta(\mathbf{k})$ changes sign as a function of $\mathbf{k}$, that is it has different signs at different parts of the Fermi surface. This seems to be the situation in high-T_c cuprates. The Fermi surface then has schematically the form shown in Fig. 9.33 (we do not discuss here the details of the spectrum of cuprates in the normal state; see Section 9.3 for details). If the interaction $\Gamma(\mathbf{k} - \mathbf{k}')$ in eq. (9.16) is a repulsion, $\Gamma > 0$, but it is nonzero predominantly for $\mathbf{k} - \mathbf{k}' \sim (\pi, \pi)$, that is if it connects the sectors of the Fermi surface related by the wave vector $\mathbf{Q} = (\pi, \pi)$, shown in Fig. 9.33 by arrows, then there would exist a nontrivial solution of (9.16) which changes sign for $\mathbf{k} \to \mathbf{k} + \mathbf{Q}$; for example $\Delta(\mathbf{k})$ is positive in the unshaded sectors in Fig. 9.33 and negative in the shaded ones. The d-wave solution (9.15) is of just this type.[11]

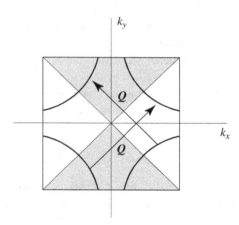

Figure 9.33 Origin of d-wave pairing in high-T_c cuprates.

[11] Sometimes one explains the preference for d-wave pairing in the case of electron repulsion compared with s-wave pairing by arguing that, whereas for the s-wave Cooper pair there is a large probability of two electrons being close to each other, which is very unfavorable for electron repulsion, for the Cooper pair with nonzero orbital momentum l ($l = 1$ for p-wave, $l = 2$ for d-wave, etc.) the wave function behaves as $\psi(r) \sim r^l$ (Landau and Lifshitz, 1965), that is the probability of having two electrons close to each other (with relative coordinate $r \to 0$) is suppressed. Although this factor can play some role, the argument is not completely correct: even for s-wave pairs this probability can be reduced strongly for strong on-site repulsion. The treatment presented above is more rigorous and indeed describes the main physics of d-wave pairing for the repulsive interaction.

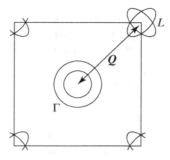

Figure 9.34 Schematic form of the Fermi surface and a possible origin of $s_{\pm}$-pairing in Fe-based superconductors.

This is not the only possible situation which can give superconductivity under repulsion. Thus, in Fe-based superconductors (see above) there exist two (or rather four) pockets of the Fermi surface at Γ- and L-points of the Brillouin zone, see Fig. 9.34. If there is an effective repulsion between electrons, $\Gamma(k - k') > 0$, which is peaked at momentum close to $k - k' \sim Q \sim (\pi, \pi)$, there may exist a nonzero solution of the superconductivity equation (9.16) with the energy gaps having opposite signs at these different pockets of the Fermi surface, for example $\Delta(k) > 0$ for $k \sim 0$ (close to the Γ-point of the Brillouin zone) and $\Delta(k) < 0$ for $k \sim Q$ (close to the L-point). The sign of Δ in each pocket can remain the same, that is it can be of the s-wave type, but it will be opposite on these two pockets. This type of pairing is called pairing of $s_{\pm}$ type (Mazin *et al.*, 2008b); most probably it is realized in the iron-based superconductors.

Why can the effective electron–electron interaction $\Gamma(k - k')$ be concentrated close to particular values of the momentum? The usual approach is to ascribe this interaction to the exchange of spin fluctuations, in analogy with the standard electron–electron attraction in the BCS theory, which is explained as being due to the exchange of phonons. We expect that in strongly correlated systems with dominant electron repulsion we either have some magnetic ordering, or the system is close to such ordering – for example as happens in cuprates, see Figs 9.16, 9.29. In this case we may expect that even if the magnetic long-range order is suppressed (and it seems that for superconductivity to be realized it should be suppressed), still the well-developed spin fluctuations would remain. And in this case, crudely speaking, the effective electron–electron interaction $\Gamma(k - k')$ entering eq. (9.16) would be proportional to

$$\Gamma(q) \sim \frac{\chi(q)}{1 - g\,\chi(q)}, \tag{9.17}$$

where g is a certain coupling constant and $\chi(q)$ is the corresponding magnetic susceptibility. We see that this interaction $\Gamma(q)$ would be maximal at values of q for which $\chi(q)$ is maximal.

If the system is close to antiferromagnetism, as is the case for example for high-T_c cuprates or Fe-based superconductors, one should expect that the strongest spin fluctuations would be those with $q \sim Q \sim (\pi, \pi)$, in which case we are in the situation of Figs 9.33, 9.34 (usually $\chi(q)$ is maximal at q values corresponding to the eventual magnetic ordering). In this case such interaction would indeed give singlet but d-wave superconductivity in cuprates (Fig. 9.33) and $s_\pm$-type superconductivity in iron-based superconductors (Fig. 9.34). If, however, the strongest spin fluctuations were ferromagnetic, that is with $q \sim 0$, one would also get unconventional superconductivity, however not with the singlet Cooper pairs as in the d- or $s_\pm$-wave case, but with triplet p-wave pairing. (According to quantum-mechanical rules the singlet pairing with $S = 0$ (antisymmetric spin part) can coexist with $l = 0$ (s-wave) or $l = 2$ (d-wave) (symmetric coordinate wave function), and the triplet pairing with $S = 1$ can only be of p-wave type ($l = 1$) – or maybe f-wave ($l = 3$), etc.) Such triplet p-wave pairing is most probably realized in Sr_2RuO_4 – the material with the same K_2NiF_4-type crystal structure as $La_{2-x}Sr_xCuO_4$, but with much lower $T_c \sim 0.95$ K and, as mentioned above, most probably with triplet p-wave superconductivity.

Many very important questions in high-T_c superconductivity, notably in cuprates, remain open. In Section 9.2.3 we already mentioned the controversy in the literature regarding the nature of the pseudogap phase: whether it is a precursor of superconductivity, that is due to preformed superconducting pairs, only without phase coherence required for true superconductivity, or whether it is due to some other type of ordering competing with superconductivity.

Another such open question is what the energy gain actually is in creating the superconducting state. In conventional low-T_c superconductors with electron–phonon mechanism of pairing this is the energy gain of interaction. But for cuprates there are theoretical arguments (Leggett, 2006) and experimental indications (Molegraaf *et al.*, 2002) that this may be rather the gain in kinetic energy – in particular the energy of interlayer tunneling. One of the arguments is that in strongly correlated systems, to which high-T_c cuprates definitely belong, the character of elementary excitations may be different from the conventional electrons in metals: as discussed above, see Chapter 1 and Section 5.7.2, these may be for example spin-carrying neutral excitations – spinons and spinless charge excitations, holons. As such, they cannot, for example, tunnel from one CuO_2 plane to the next – they have first to combine into a real hole with charge $+e$ and spin $\frac{1}{2}$. But in the superconducting state the character of the state and of the excitations in it change and become "more normal," which can help gain extra kinetic energy, and this energy gain can be the driving force of superconductivity in cuprates – see Anderson (1997). Once again, this is only one suggestion; the more traditional approach, described at the beginning of this section, is at the moment more popular. But in principle the question of the main driving force of superconducting pairing in high-T_c cuprates is still not finally solved; the same is true for the iron-based superconductors.

As mentioned in Section 9.5.3, there are also suggestions that in high-T_c cuprates there may exist a hidden quantum critical point, see Fig. 9.29. It is still an open question whether such a quantum critical point indeed exists, and if so, what is its nature and what could be its role in establishing high-T_c superconductivity. But one should note that there are now many different systems with correlated electrons, in which indeed superconductivity appears in the vicinity of a quantum critical point. These are, for example, the already mentioned rare earth systems $CePd_2Si_2$, $CeRh_2Si_2$, $CeNi_3$, $CeCoIn_5$, UGe_2, $URhGe$; the system Cs_3C_{60}; and also some organic materials in which one has correlated electrons and magnetically ordered states, where magnetism can be suppressed for example by pressure, see von Löhneisen *et al.* (2007).

9.7 Phase separation and inhomogeneous states

In systems with fractional occupation of bands, in particular in doped Mott insulators, one very often gets different types of inhomogeneous states. We have already encountered some examples above: different types of charge ordering (CO), stripes, or charge density waves (CDW). These states are inhomogeneous, but still ordered. There may also appear random inhomogeneous states, with different phases mixed at random, which often leads to percolation phenomena. Such states can usually occur due to phase separation in the system.

The instability of a homogeneous state is often met, in particular, in doped Mott insulators. One gets such instability in the standard theoretical models describing strongly correlated systems, when the concentration of charge carriers differs from some simple commensurate values such as $n = 1$ or $n = \frac{1}{2}$. There are also many experimental indications of such phase separation in different real systems.

The simplest model showing this phenomenon is the standard nondegenerate Hubbard model (1.6). We have already mentioned in Section 9.1 that for large U/t and $n \neq 1$ the crossover from the antiferromagnetic insulating state at $n = 1$ to the eventual ferromagnetic metal for a large deviation from half-filling can occur either via canted or via spiral states shown in Figs 9.1 and 9.2, but also via a phase separation into ferromagnetic regions which contain all doped electrons or holes, and undoped antiferromagnetic insulating regions with $n = 1$; such separation can take the form of ferromagnetic droplets of finite size, if we take into account long-range Coulomb interaction which prevents large-scale charge inhomogeneities. (We recall that the holes gain maximum kinetic energy if they move on the ferromagnetic background; it is this factor which can promote phase separation with the formation of ferromagnetic droplets.) One can see this tendency toward phase separation in the Hubbard model away from half-filling as follows (for more detailed discussion see Khomskii, 2010). Suppose we have such a charge separation, and let us compare the energy of the resulting inhomogeneous state with the energy of the best alternative homogeneous state. If the average concentration of doped holes is $\delta = N_{holes}/V = (N - N_{el})/V$, where V is the total volume of the system, and out of the total volume V the part V_F is ferromagnetic, containing all the doped holes, then the actual hole concentration in this ferromagnetic part of the sample would be $\delta_F = N_{holes}/V_F = (V/V_F)\delta$. (Here for concreteness we talk about holes, although, as explained above, for simple lattices we have electron–hole symmetry.) The remaining part of the sample $V_{AF} = V - V_F$ would be undoped and antiferromagnetic. The energy of this antiferromagnetic region would then be

$$E_{\text{AF}} \sim -J V_{\text{AF}} = -J V \left(1 - \frac{V_{\text{F}}}{V} \right), \tag{9.18}$$

where J is the exchange constant in the Hubbard model, $J = 2t^2/U$, see (1.12).

The energy of the ferromagnetic region with volume V_{F} consists of magnetic energy $+J V_{\text{F}}$ (we lose the exchange energy in this volume), and of the energy of holes occupying the corresponding band. This energy is $-t z \delta_{\text{F}} V_{\text{F}} + t V_{\text{F}} \delta_{\text{F}}^{5/3}$: the holes would move freely in the ferromagnetic background, with the hopping matrix element t and with the bandwidth $2zt$, that is these δ_{F} holes would be at the bottom of the corresponding band, $-tz$, and the finite filling of this band would give the second term $\sim \delta_{\text{F}}^{5/3}$. In effect the total energy of the ferromagnetic part of the sample would be

$$E_{\text{F}} = J V_{\text{F}} - t z \delta_{\text{F}} V_{\text{F}} + t \delta_{\text{F}}^{5/3} V_{\text{F}} . \tag{9.19}$$

Minimizing the total energy $E_{\text{AF}} + E_{\text{F}}$, (9.18) and (9.19), with respect to the fraction V_{F}/V, we finally find

$$\left(\frac{V_{\text{F}}}{V} \right)_0 = \left(\frac{U}{t} \right)^{3/5} \delta , \tag{9.20}$$

and the minimal energy of such a phase-separated state is

$$\frac{E_0}{V} = -\frac{2t^2}{U} - t z \delta + t \delta \left(\frac{t}{U} \right)^{2/5} . \tag{9.21}$$

Comparing this energy with that of the homogeneous states such as for example the canted state of Section 9.4 (which for small δ changes quadratically, $\sim -(2t^2/U) - c\delta^2$), we see that the energy (9.21) of the inhomogeneous state is lower, at least for small δ. Thus according to this treatment the partially filled Hubbard model with $U/t \gg 1$ should phase-separate into a ferromagnetic metallic part containing all the doped charge carriers, and the remaining part which would be an undoped antiferromagnetic Mott insulator. As follows from eq. (9.20), this inhomogeneous state would exist if $V_{\text{F}}/V < 1$, that is for the total doping $\delta < \delta_c \sim (t/U)^{3/5}$; above this doping level the ferromagnetic phase would occupy the whole sample, and we would have a homogeneous ferromagnetic state. The full behavior of the energies of different states in the Hubbard model is illustrated in Fig. 9.35.

This treatment is of course rather crude. It does not take into account for example the surface energy of the ferromagnetic phase in the antiferromagnetic matrix. But most important is that quantum fluctuations are not taken into account. Also, for real materials, of course one has to take into account the long-range Coulomb interaction, which would prevent large-scale phase separation and could instead produce ferromagnetic droplets of finite size, for which the surface energy would be even more important. Nevertheless, the intrinsic general tendency of instability of the homogeneous state in this regime seems definitely to be present, and it can have many manifestations in real systems.

The same tendency toward phase separation is seen even more clearly in the so-called t–J model, which is nowadays often used to describe strongly interacting Hubbard systems, in particular high-T_c cuprates. In this model, instead of the Hubbard model (1.6), one maps, for $U \gg t$, the electronic system to the model

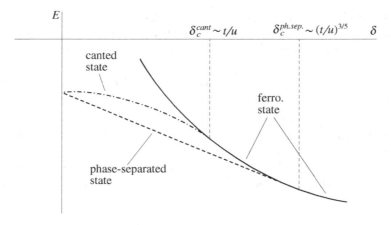

Figure 9.35 Phase separation in the doped Hubbard model.

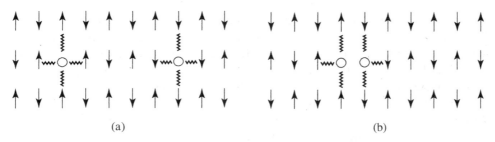

(a)   (b)

Figure 9.36 Illustration of the tendency toward phase separation in the $t-J$ model for strong exchange interaction $J > t$.

$$\mathcal{H}_{t-J} = -t \sum_{\langle ij\rangle,\sigma} \tilde{c}_{i\sigma}^{\dagger}\tilde{c}_{j\sigma} + J\sum_{\langle ij\rangle} \mathbf{S}_i \cdot \mathbf{S}_j \,, \qquad (9.22)$$

where the second term is the antiferromagnetic exchange of localized spins and the first term describes the motion of doped electrons or holes (note that the effective operators $\tilde{c}^{\dagger}$, $\tilde{c}$ are similar, but not identical to the original electronic operators $c^{\dagger}$, c). In this $t-J$ model (for $n < 1$) one effectively excludes doubly occupied sites, keeping only the states with $\langle n_i \rangle = 0$ and 1 (but the "price" for this is the difference between $\tilde{c}$ and c operators). One can easily see that in this case, if we put a certain number of holes into such an antiferromagnetic system, then at least for $J > t$ the holes would prefer to cluster, see Fig. 9.36: if in the case of two separate holes, Fig. 9.36(a), we lose z antiferromagnetic bonds per hole (here four bonds per hole, i.e. altogether eight bonds), in the situation when these holes come close together, Fig. 9.36(b), we lose only seven bonds, that is with such clustering we gain energy J. And if $J > t$, this process would definitely be favorable.

Of course if we go back to the original Hubbard model, this limit of $J > t$ does not seem realistic: in the Hubbard model with strong interaction $J = 2t^2/U \ll t$, and if

$t \sim U$, the perturbation theory in t/U used to obtain the antiferromagnetic exchange $J S_i \cdot S_j$ in (9.22) is strictly speaking not valid. But qualitatively the arguments illustrated in Fig. 9.36 may work also in more realistic cases. Different treatments using mostly the $t-J$ model indeed confirm the tendency toward phase separation, at least for $J > t$ (see, e.g., Carlson *et al.*, 1998), although some results differ slightly (Emery *et al.*, 1991; Hellberg and Manousakis, 1997; White and Scalapino, 2000). The situation in the original Hubbard model however is not so clear: different calculations give very different results in this respect. Nevertheless it seems that the possibility of phase separation is a real and important factor, which one always has to take into account.

It turns out that other situations and other models may also show the same tendency. One such model is the double exchange (DE) model, described in Section 5.2. The Hamiltonian of this model is (see (5.13))

$$\mathcal{H}_{\mathrm{DE}} = -t \sum c_{i\sigma}^{\dagger} c_{j\sigma} + J \sum S_i \cdot S_j - J_{\mathrm{H}} \sum_i S_i \cdot c_{i\sigma}^{\dagger} \sigma c_{j\sigma}, \tag{9.23}$$

where $c^{\dagger}$, c describe conduction electrons, interacting by the Hund's exchange J_{H} with localized spins S_i, which also have antiferromagnetic exchange interaction J between themselves. In this model one usually considers the case of narrow bands, or small hopping $t \ll J_{\mathrm{H}}$, with the exchange of localized electrons still smaller, $J < t$. As argued in Section 5.2, for large J_{H} the spins of conduction electrons should always be parallel to the localized spins, and because of that the effective hopping t_{ij} depends on the angle θ between spins S_i and S_j, $t_{\mathrm{eff}} = t \cos(\theta_{ij}/2)$, see (5.14). As a result, as shown in that chapter, to gain the kinetic energy of mobile electrons the localized spins start to cant with doping x, $\cos(\frac{1}{2}\theta) = tx/4JS^2$, see (5.16), and if we assume that the system is homogeneous, then with doping the antiferromagnetic structure of the localized spins starts to change – first we would get a canted state, and finally, for doping $x > x_c = 4JS^2/t$ (5.17), we would get a ferromagnetic metallic state (de Gennes, 1960).

However there exists another possibility: the system can become inhomogeneous, so that for example instead of homogeneous canting we would get a phase-separated state with the doped electrons all clustered in a ferromagnetic part of the sample, and the remaining part would be an undoped antiferromagnet. We see that this possibility is very similar to the one described for the partially filled Hubbard model.

Indeed, one can easily see that this is what happens in the double exchange model (9.23) (Kagan *et al.*, 1999). With the canting angle θ given by $\cos(\frac{1}{2}\theta) = tx/4JS^2$ (5.16), the energy of the system (5.15) becomes

$$E_{\mathrm{min}}^{\mathrm{canted}} = -JS^2 z - \frac{z}{8} \frac{(xt)^2}{(JS^2)^2}. \tag{9.24}$$

We see from this expression that the energy of such a homogeneous canted state is a concave function of electron density $x = N_{\mathrm{el}}/V$, that is $d^2 E/dx^2 < 0$. But this is actually the inverse compressibility of the system, and its negative value means that this state is absolutely unstable. The dependence of the energy on the electron density has the form shown

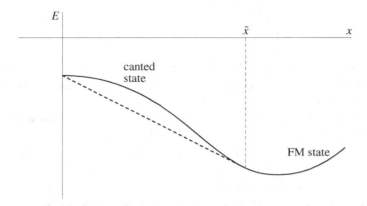

Figure 9.37 Phase separation in double exchange magnets. The dashed line shows the Maxwell construction for the phase-separate state.

in Fig. 9.37, resembling the corresponding dependence of the Hubbard model, Fig. 9.35: the homogeneous canted state of de Gennes is absolutely unstable, and one has to use the Maxwell construction to find the two-phase state: for average electron concentrations $0 \lesssim x < \tilde{x}$ the system decomposes into two phases with $x = 0$ (undoped antiferromagnet) and $x = \tilde{x}$ (ferromagnetic metallic state), with relative volumes determined by the conventional Maxwell rule.

Again, the inclusion of long-range Coulomb interaction would limit the sizes of the resulting ferromagnetic droplets, which would be formed instead of one big ferromagnetic "puddle." The bulk properties of the resulting state, for example d.c. conductivity, would then have a percolating character: below the percolation threshold, when these metallic droplets do not form an infinite cluster, the system would behave as insulating for direct current, despite the presence of a certain fraction of ferromagnetic metallic phase. Only above the percolation threshold will the d.c. conductivity become metallic.

One can easily estimate the average size of the magnetic microregion which will be created by doped electrons or holes (one calls such ferromagnetic microregions in an antiferromagnetic matrix *magnetic polarons*, or *ferrons*). If we assume that each charge carrier creates a spherical ferromagnetic microregion with radius R (measured in units of the lattice constant), its energy would be

$$E(R) \sim \frac{4\pi}{3} R^3 J S^2 - tz + \frac{t}{R^2} , \qquad (9.25)$$

where the first term is the energy loss of the exchange interaction of localized spins, which are "forced" to be ferromagnetic, and the second and third terms represent the energy of the charge carrier moving in a spherical potential well of radius R (the electron or hole can move freely inside such a ferron, but cannot go outside into the antiferromagnetic matrix). Minimizing this expression in R we obtain

$$R_0 \sim \left(\frac{t}{J}\right)^{1/5}, \qquad E_0 = E(R_0) \sim -tz + J^{2/5} t^{3/5} \qquad (9.26)$$

(where we have omitted numerical factors ~ 1). This estimate is, of course, valid only for $t \gg J$, in which case the ferron radius is large enough, bigger than the lattice constant (remember that we measure R in units of a).

One can show that in other situations and models we have a similar tendency toward phase separation, when we dope the ordered system and when this ordering suppresses the motion of doped carriers: in all these cases it may be favorable to "deform" the ordering or to destroy it completely so as to facilitate electron motion and therefore gain kinetic energy. One can show that such situations can exist in charge-ordered systems such as $La_{1-x}Ca_xMnO_3$ slightly away from the commensurate doping $x = 0.5$ (Kagan *et al.*, 2001); similarly, doped systems with spin-state transitions can become unstable toward phase separation (Kugel *et al.*, 2008; Sboychakov *et al.*, 2009), etc. And in all these cases inhomogeneous states can be formed, with the percolation picture of conductivity, and with many other specific properties.

One often sees manifestations of this tendency toward phase separation in real materials. This was observed in doped manganites, so that even the very phenomenon of colossal magnetoresistance in them is sometimes connected to the tendency toward such phase separation (Dagotto, 2003). Indications of phase separation are also seen in cobaltites (Wu and Leighton, 2003; Hoch *et al.*, 2004; Aarborgh *et al.*, 2006) and in many other systems. And to interpret the properties of these materials one often has to use the picture of percolation. Thus for example in $La_{1-x}Sr_xCoO_3$ the crossover to a ferromagnetic metallic state, which occurs at the values of $x \sim 0.18$–0.2, see Fig. 9.12, can be explained by such a percolation picture (remember that the percolation threshold of three-dimensional systems is ~ 0.18, see e.g. Stauffer and Aharony, 1994). One can see the formation of inhomogeneous states especially clearly by local probes such as NMR: they often give an indication of the presence of different phases, for example of two different local configurations of TM ions in manganites or in cobaltites.

In relation to real systems, we have to make one important remark. Often in correlated systems (and not only in them, of course) one deals with sharp I order transitions. Such are for example transitions from the charge-ordered state to the ferromagnetic metallic state in some manganites, cf. Fig. 7.10. In this case such transitions can be accompanied by a strong hysteresis, and in the vicinity of such phase transition the system may become inhomogeneous simply because of that: this is similar to the formation of fog in supercooled vapor. In such cases the inhomogeneous states thus created have a nonequilibrium nature: their existence and specific features depend for example on the presence of nucleation centers, etc. The supercooled or superheated states would relax with time to a certain homogeneous equilibrium state. The size of inhomogeneous regions close to such I order transitions can in principle be quite large, almost macroscopic. Thus for example at the crossover from the charge-ordered insulating to the ferromagnetic phase in $Pr_{1-x}Ca_xMnO_3$ (Fig. 7.10) the coexisting phases can differ by the type of ordering, by conductivity, they may have different lattice parameters and different band structure, but they do not differ in charge density, that is these coexisting different phases are not charged: their chemical, or rather electrochemical potentials should be the same. In this case the condition of electroneutrality does not put any limits on the sizes of these regions, in contrast to the case of intrinsic phase separation considered above using the example of the Hubbard and double exchange models. Thus one has to discriminate between the macroscopic and nonequilibrium phase separation often seen close to I order phase transitions, and the intrinsic instability in some cases, which would lead to a microscopic phase separation. Thus in fact one often uses the same term, "phase separation," for two distinct phenomena. Many experimental observations of phase separation for example in CMR manganites are in fact the detection of large-scale inhomogeneities close to I order phase transitions (Khomskii, 2000). However some experiments mentioned above, for example Hennion and Moussa (2005), give indications of an intrinsic tendency toward phase separation, often present in doped Mott insulators. One can even say that this tendency is not an exception, but rather the rule in such cases, and one always has to keep this possibility in mind when analyzing the behavior of such systems.

9.8 Films, surfaces, and interfaces

Until now we always dealt with bulk transition metal compounds – real three-dimensional solids. Some of them could have layered (quasi-two-dimensional) or chain-like (quasi-one-dimensional) structures; but, in principle, even those ultimately form three-dimensional crystals. There are, however, particular systems in which the main specific effects are due to their two-dimensional character. Such are for example thin films, or surfaces of bulk materials, or interfaces of different systems, for example in multilayered materials. Such systems are studied very actively and used widely in practice: after all, all our modern electronic devices are largely based on the phenomena occurring at interfaces, mainly those of semiconductors.

Very specific effects can occur in such situations in transition metal compounds. Thus, because the inversion symmetry is broken at surfaces and interfaces, there may appear nontrivial magnetic and ferroelectric states on them, see for example Section 8.4. Similarly, the lattice periodicity can change because of the lattice mismatch between the film and the substrate. As we discuss below, the electronic structure and different types of ordering can all change at surfaces and interfaces.

Transition metal compounds present a special case due to the very specific effects that can occur in such situations. This field is studied very actively at present, and we will not be able to cover it in its entirety. In this short section we rather attempt to describe qualitatively some interesting effects that may occur, the physics of which is related to the main topics of this book. One can find more details for example in the following books and review articles: Heinrich and Cox (1994), Freund *et al.* (1996), Noguera (1996), Mannhart *et al.* (2008), Mannhart and Schlom (2010), Hwang *et al.* (2012). Note that many basic phenomena important for both surfaces and interfaces of transition metal compounds – such as, for example, band bending, the formation of (Shottky) barriers, etc. – are basically similar to those in semiconductors. Some phenomena, however, are indeed specific to transition metal compounds; and it is these that will be the main focus of the discussion in this section.

When dealing with surfaces or interfaces of TM compounds, one of the first effects we meet is a modification of the characteristic parameters of the system. For systems with correlated electrons these are, first of all, the effective electron hopping t and the resulting bandwidth $W \sim t$, and the Hubbard interaction U, see the simplest Hubbard model (1.6). The effective electron hopping t may change due to structural transformations often occurring at the surface, such as structural relaxation and eventual surface reconstruction. These can occur already at free surfaces. At the interface of different materials the matching of respective lattice parameters is crucial, as it can change the interatomic distances and therefore the values of t.

But the simplest and most important fact is that for a surface layer the number of nearest neighbors z^*, to which the d-electrons can hop, is reduced compared with the number of nearest neighbors in the bulk, z. Consequently the effective bandwidth $W^* = 2z^*t$, which is a measure of the kinetic energy of electrons, is reduced at the surface in comparison with its bulk value $W = 2zt$. This suppression of the kinetic energy can be very important for TM surfaces and interfaces. It leads to a tendency which makes surfaces or interfaces more insulating.

At the same time the value of the Coulomb (Hubbard) energy would also change in this situation. The screening of the Coulomb interaction by the creation of charge excitations, for example by the process $d^n d^n \rightarrow d^{n-1} d^{n+1}$, would be different at the surface or in thin films compared with the bulk, especially for films on metallic substrates. A simple way to describe this effect is to use the classical notion of image charges, well known from regular courses on electricity and magnetism. Positive and negative charges created in Mott insulators by electron transfer from site to site would attract to their image charges, and in effect the energy required to create such an electron–hole pair (which, by definition, is the Hubbard's U) would decrease, $U^* < U$. How strong this effect would be depends on many factors. In principle it could be so strong that the surface or a thin film of a Mott insulator (in which in the bulk $U > W = 2zt$) could even become metallic, $U^* < W^* = 2zt^*$, or at least the energy gap could decrease strongly at the surface, even though the effective bandwidth W^* itself is reduced in comparison with the bulk. Such situations were discussed theoretically, and there are some indications that they may be realized in some cases: for example the energy gap of NiO on a metallic substrate is 1.2 eV, smaller than the corresponding gap of 3 eV in the bulk material (Tjeng *et al.*, 1989).[12]

However, as mentioned above, the opposite effect – the reduction of the kinetic energy (suppression of electron hopping in one direction) – can also occur, leading to a more insulating character of the surface or interface layers; and it seems to be observed even more

---

[12] The long-range part of the Coulomb interaction can also be strongly modified in thin films, depending on the film thickness and the ratio of the dielectric constants of the film and the substrate (Keldysh, 1979).

frequently than the metallization of the surface due to the reduction of the Hubbard interaction. Thus, in many cases one sees that the material, which is metallic in the bulk, turns out to be insulating when made in the form of a very thin film – one, two, or three layers thick. This is for example a typical situation in thin films or multilayers of nickelates $RNiO_3$. As is seen for example from the phase diagram of Fig. 3.47, whereas the ground state of most of these nickelates is insulating, $LaNiO_3$ in the bulk is metallic. But very thin films or multilayers of $LaNiO_3$ with number of layers $\lesssim 3$ are insulating (Thiel *et al.*, 2006), even on compressive substrates which act as an external pressure and which usually make a system more metallic. Most probably this is explained by the above-mentioned suppression of the effective kinetic energy and the bandwidth W^* due to the reduced number of nearest neighbors and the corresponding suppression of electron hopping in the perpendicular direction.

One should also notice that, as discussed in Chapter 4, there exist two types of insulators with strong electron correlations: Mott–Hubbard insulators and charge-transfer insulators. One may expect that the modifications of the properties at surfaces and interfaces could be even stronger in charge-transfer systems, as they are very sensitive to the distribution of charge and spin density across the binding oxygens at the interface. The charge-transfer energy Δ_{CT} can change in these situations even more strongly than the Hubbard repulsion U.

When we go to a more realistic situation and include in the description such factors as orbital effects, etc., the situation may become even more rich and more complicated. Especially in systems with a particular orbital ordering the orbital structure can change at the surface or at the interface. This may occur due to several factors. First of all, lattice distortions accompanying a particular orbital ordering may be suppressed by the strain imposed by the substrate. Also the superexchange contribution to orbital ordering ("Kugel–Khomskii" contribution, see Section 6.1) may strongly change. In effect, orbital structure at the interface can be very different from that in the bulk, and can depend on the substrate. One can even use this to control the orbital structure by choosing a particular substrate (Konishi *et al.*, 1999). Orbital reconstruction at the interface was seen experimentally for example on a contact of the CMR manganite $La_{1-x}Ca_xMnO_3$ with the high-T_c superconductor $YBa_2Cu_3O_7$ (Chakhalian *et al.*, 2006, 2007). Other effects – for example the change of the crystal field splitting in anisotropic situations, see Wu *et al.* (2011) – can also play some role in the modification of the orbital occupation in surfaces and interfaces.

The change in strength of electron correlations at surfaces or in thin films and in multilayers, and also the eventual change of the orbital structure, can strongly modify the exchange interaction and change the type of magnetic ordering. Thus for example if the surface of a Mott insulator should become metallic, one could in general expect that, instead of the bulk antiferromagnetism, such a surface may become ferromagnetic. This would agree with the general tendency mentioned several times elsewhere in this book, that the states of Mott insulators are most often antiferromagnetic whereas metallic systems with strong correlations have a good chance of becoming ferromagnetic. If however both states, in the bulk and at the surface/interface, remain insulating, then the possible change in orbital structure, discussed above, could in principle lead to the opposite effect. Thus for example one could speculate that, whereas the orbital ordering in K_2CuF_4 (shown in Fig. 6.11) makes this material ferromagnetic, if the orbital ordering at the interface should change (due to compressive strain) to that typical say for La_2CuO_4, with the $(x^2 - y^2)$ hole occupied at every site, see Fig. 6.12, then such a surface layer could become antiferromagnetic.

Yet another very important class of phenomena, studied very actively especially since 2004 (though many basic ideas were formulated much earlier) are the effects connected with the possible existence of charged layers in films, and especially on interfaces, and with the related phenomena of the polarization catastrophe and the resulting charge redistribution – see for example Noguera (1996) and Goniakowski *et al.* (2008).

The fact that the presence of charged layers can lead to very nontrivial effects was actually realized in the physics of semiconductors long ago, see for example Harrison *et al.* (1978), and in the present context in the most clear form in Hesper *et al.* (2000). Let us look for example at the [001] surface of a typical perovskite such as $LaMnO_3$. We have in this case alternating layers $Mn^{3+}O_2^{2-}$ with total charge -1 per unit cell, and $La^{3+}O^{2-}$ with charge $+1$, see Fig. 9.38.

If, using classical electrostatic theory, we now look at the effective electric field $\mathcal{E}$ and the electric potential V in this situation, we see that they would look as shown in Fig. 9.39. That is, the potential would grow with the thickness of such a film, $V(n) = V_0 n$, where n is the number of layers. In effect such a potential would grow linearly with thickness, and at a certain thickness the potential drop would exceed the energy gap of the system E_g, which would imply an instability of the system.

Simple estimates show that the nonzero electric field in Fig. 9.39 would be $\mathcal{E}_0 \sim 10^7$ V/cm, the potential drop V_0 at one unit cell with thickness $d \sim 5$ Å would be $V_0 \sim \mathcal{E}_0 d \sim 0.5$ V, and the energy of moving an electron across such a unit cell would be $E_0 \sim eV_0 \sim 0.5$ eV. In effect already for four unit cells the potential drop would be ~ 2 V, that is the corresponding energy would be $E \sim 2$ eV – of the order of or larger than the typical energy gap of the system. For thicker films it would increase still further and would exceed even much larger energy gaps! This situation is called *polarization catastrophe*: such a state, with such a large potential difference between the lower and upper layers, is definitely unstable.

Figure 9.38 Charged [001] layers in perovskite systems such as $LaMnO_3$ or $LaAlO_3$.

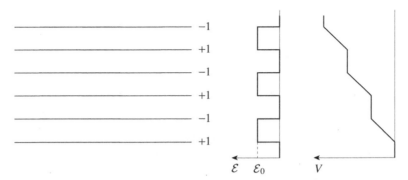

Figure 9.39 Effective electric field and potential in a stacking of charged layers, demonstrating the phenomenon of polarization catastrophe (unlimited increase of potential with thickness).

There may be different resolutions of this problem. In real situations there may occur some adsorption of charged ions from the surrounding media (liquid, air) on the surface of the film. Another option is the change in stoichiometry of boundary layers, for example the creation of oxygen vacancies. Indeed, the large unresolved potential difference between the charged layers may result in a strong electric field pushing the atoms out of the already metastable equilibrium positions (due to the broken symmetry at the interface), leading to the formation of vacancies. In this case instead of the upper $Mn^{3+}O_2^{2-}$ layer in Figs 9.38, 9.39, with charge -1, we may have an oxygen-deficient layer $Mn^{3+}O_{2-x}^{2-}$, with charge $+3 - 2(2 - x) = -1 + 2x$; thus for $x = 0.5$ ($\frac{1}{4}$ oxygen vacancies in this layer), this layer would be charge-neutral. This is a very realistic scenario, and apparently that is what actually happens in many cases. Simply put, in the process of film growth the first few layers could grow stoichiometrically, but at a certain stage, after reaching the situation with a large enough potential drop, it would no longer be favorable to grow stoichiometric layers, and nonstoichiometry would appear to make the surface layer neutral.

Another case that occurs often is the intermixing of ions at the interface during growth, for example the polarization catastrophe at the $SrTiO_3/LaTiO_3$ interface would be "healed" if the interface layer is intermixed.

But there exists also a more interesting, intrinsic possibility of overcoming polarization catastrophe. It is *electron reconstruction* – a transfer of some electrons from for example the upper layer to the lower layer. Indeed, in the situation with the potential drop between the upper and lower layers being $V = V_0 n$ and with the corresponding energy $E(n) = eV(n) = eV_0 n$ exceeding the energy gap of the system, E_g, it is favorable to move an electron from the upper surface to the bottom layer, or to create a (spatially separated) electron–hole pair: one loses the energy E_g in this process, but gains a bigger energy $E(n)$. That is, in this case there should occur electron reconstruction – spontaneous charge redistribution between the upper and lower layers of the film.

In our model case, shown in Fig. 9.39, one should move $\frac{1}{2}$ electrons per unit cell from the upper (MnO_2) layer to the lower (LaO) layer. Then the electrostatic situation would look

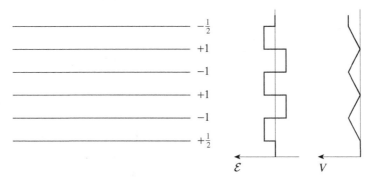

Figure 9.40 Internal electric field and potential after electron reconstruction (transfer of $\frac{1}{2}$ electrons per unit cell from the upper to the lower layer).

as shown in Fig. 9.40. We see that in this case the electric field alternates in sign between consecutive layers (electric "capacitors"), and the total voltage drop does not grow with the thickness of the film, but has a sawtooth shape. As mentioned above, such electron reconstruction was proposed for example by Harrison *et al.* (1978) and Hesper *et al.* (2000), and by Ohtomo and Hwang (2004) in connection with the study of multilayers of $SrTiO_3$ (STO) and $LaAlO_3$ (LAO).

Some quite nontrivial consequences of the process of electron reconstruction shown in Fig. 9.40 can be anticipated right away. After such charge redistribution the upper boundary layer, which had integer valence before the redistribution, would now be strongly doped: there would be for example ~ 0.5 holes per MnO_2 unit cells in the upper layer of Fig. 9.40. In this case one can expect that such a surface, or such an interface of two insulating materials, would become metallic. And this is indeed what was observed by Ohtomo and Hwang (2004): whereas the bulk materials they have used, $SrTiO_3$ and $LaAlO_3$, are both very good insulators, their interface was metallic! The physical mechanism discussed above – the polarization catastrophe and the resulting electron reconstruction – would give a very natural explanation of the observed phenomenon. Moreover, as mentioned above, in the presence of strong electron correlations one can expect that such partially filled bands could lead to ferromagnetism. Experimental indications for that will be discussed below.

Of course one has to be prudent and keep in mind the other possibilities mentioned above, especially the possible eventual change in stoichiometry, first of all the formation of oxygen vacancies. Probably in some experimental conditions these effects indeed play a crucial role. But there are also arguments in favor of the intrinsic mechanism of the observed effect.[13] Probably the strongest one is the observation (Thiel *et al.*, 2006) that in

---

[13] Note that in certain cases oxygen vacancies have to be created even in the intrinsic charge redistribution due to the polarization catastrophe. When we remove an electron (actually $\frac{1}{2}$ electron) from the upper MnO_2 surface layer of $LaMnO_3$, Fig. 9.38, the Mn ions in the upper layer can change valence and become Mn^{4+}, instead of the original Mn^{3+}. But this would not work so simply for example for $LaAlO_3$. In that case we again remove $\frac{1}{2}$ electron from the upper layer $Al^{3+}O_2^{2-}$, but Al^{3+} cannot change its valence. We would then have

the STO/LAO interface the metallic conductivity appears starting from a particular critical thickness of the LAO layer. In $SrTiO_3$ the respective layers, $Ti^{4+}O_2^{2-}$ and $Sr^{2+}O^{2-}$, are neutral. But in $LaAlO_3$, as in our example of $LaMnO_3$ considered above, we are dealing with charged [001] layers: $Al^{3+}O_2^{2-}$ layers with charge $q = -1$, and $La^{3+}O^{2-}$ layers with charge $q = +1$. And, as mentioned above, the polarization catastrophe in this case starts from a certain critical thickness of LAO, such that the corresponding energy difference $E(n) \sim eV_0 n$ exceeds the energy gap E_g. In the experiment of Thiel *et al.* (2006) the one-, two-, and three-layer systems remained insulating, but starting from $n = 4$ the interface became metallic. A similar critical thickness was observed later in several other combinations of materials. The intrinsic mechanism of the polarization catastrophe and the resulting electron reconstruction gives a natural explanation of this phenomenon, including the existence of a critical thickness for the appearance of metallic conductivity.

Yet another argument in support of the intrinsic mechanism of this phenomenon is that the metallic layers thus formed have rather high electron mobility. If the main mechanism were connected with nonstoichiometry, for example with the formation of oxygen vacancies at this interface, one would expect that such disorder could strongly reduce the electron mobility. Still, once again, one always has to keep in mind the possibility of such nonstoichiometry (which could also be driven by the same polarization catastrophe, and consequently could also depend on the thickness of the film). The mobility in this case can still be high enough if the dopands (for example oxygen vacancies) are located at different layers from those in which the conduction occurs, for example if the oxygen vacancies are mainly on the $LaAlO_3$ layer, whereas the conduction occurs in the $SrTiO_3$ layer due to the mixed valence of Ti ions.

An important question, from our perspective, is whether TM compounds are in any respect different from other materials, such as $LaAlO_3$. One can argue that indeed they are. This is connected with the fact that very often TM ions can accept different valences, as already mentioned above. Thus for example in $LaAlO_3$ all constituting elements have well-defined valence: La^{3+}, Al^{3+}, which cannot change. In contrast, $SrTiO_3$ contains Ti^{4+}, which can easily accept one electron and become Ti^{3+}. Apparently it is this flexibility of TM ions which strongly facilitates charge redistribution in the STO/LAO interface, driven by the polarization catastrophe. In this sense the contacts of ordinary band insulators such as $LaAlO_3$ with materials based on TM elements, which typically are Mott insulators, present special interest.[14] The properties on such interfaces, as well as those between different Mott insulators, show very specific behavior and definitely deserve special attention.

There may occur other different phenomena at such interfaces. Some of them have already been mentioned above, such as orbital reconstruction, appearance of new magnetic

---

to create, instead, an oxygen hole, $\frac{1}{2}$ hole per unit cell of AlO_2. In reality this would mean that there would be oxygen vacancies created in this layer, with nominal concentration $\frac{1}{8}$, that is the composition of this layer should become $AlO_{2-\delta}$, $\delta = \frac{1}{8}$.

[14] One should not call $SrTiO_3$ a Mott insulator, as is often done, as it contains formally Ti^{4+} ions with the empty d-shell d^0, that is it can be classified as a band insulator – but still with the possibility of changing the valence of Ti by changing the occupation of its d levels.

states different from the bulk, etc. Yet another effect was observed by Reyren *et al.* (2007) on the same STO/LAO interface. It turned out that not only does this interface become metallic, but in certain situations there appears superconductivity in it, with $T_c \sim 0.2\,\mathrm{K}$. Most probably this phenomenon occurs due to the effective electron doping of the surface layers of $SrTiO_3$: it is known that the bulk $SrTiO_3$ becomes superconducting, with comparable values of $T_c \sim 0.3$–$0.4\,\mathrm{K}$, when slightly doped by Nb (Koonce *et al.*, 1967) (Nb is usually 5+, i.e. it serves as an electron dopand for $SrTiO_3$). Interestingly enough, in some situations on such STO/LAO interfaces there appear simultaneously both superconductivity and ferromagnetism (Dikin *et al.*, 2011; Li *et al.*, 2011). One could argue that in this case phase separation occurs, see Section 9.7, so that the superconducting and ferromagnetic regions become spatially separated, with such phase separation most probably facilitated by the creation of oxygen vacancies. At the moment of writing this question is still unresolved.

Working with multilayers can be viewed as a type of "material engineering": it allows one to create artificial materials not existing under natural conditions. Thus, for example, by growing consecutive [111] layers of perovskites ABO_3 and $AB'O_3$, with different TM elements B and B', we can create a double perovskite $A_2BB'O_6$ with the TM ions B and B' ordered in two sublattices in a checkerboard fashion. Such double perovskites are known to exist in the bulk, but in many cases the ions B and B' in them remain disordered, if the material is prepared using conventional techniques.[15] However, layer-by-layer growth can create a perfectly ordered system. Such material engineering can be used to prepare artificial materials with novel and unusual properties.

The study of films and multilayers of transition metal compounds presents considerable interest not only from the physical point of view, because of the very interesting physical phenomena observed or expected in these situations. These systems also attract significant attention from the practical perspective: they promise very important applications, some of which are already being realized. One should also note that the phenomena at the surfaces of transition metals and compounds, which often rely on intrinsic strong electron correlations, are very important for a quite different field – the field of catalysis, see for example Schrieffer and Gomer (1971). All this makes the study of these systems very important.

---

[15] Usually, ordered structures are obtained in double perovskites, and in many other mixed TM systems, if the ions B and B' have very different ionic radii and strongly different valences, as for example in Sr_2FeMoO_6 (Fe^{3+}, Mo^{5+}). This ordered double perovskite is ferromagnetic with a rather high critical temperature $T_c = 415\,\mathrm{K}$ and with large negative magnetoresistance (Kobayashi *et al.*, 1998). But if the two respective ions are rather similar – for example Co^{2+} and Ni^{2+} – one usually obtains disordered systems.

S.9 Summary of Chapter 9

As discussed in the general scheme of the Introduction, when considering systems with correlated electrons, in particular TM compounds, one usually starts with the simplest situation of integer number of electrons per site, for example $n = N_{el}/N = 1$, and with a strong interaction, for example when the on-site Coulomb (Hubbard) interaction U is much larger than the electron hopping t or the corresponding bandwidth $W \sim t$. In this case we deal with Mott insulators with localized electrons. This is the situation considered in Chapter 1. From here we can go in several directions. The first is "to make the description more realistic": remaining in the regime of integer n and $U \gg t$, include in the description orbital degrees of freedom, Hund's rule coupling, spin–orbit interaction, multiplet effects, etc. The next step is to relax the restrictions gradually: first consider the situations with noninteger electron density n, or with partially filled bands, and then relax the condition $U \gg t$ and consider the crossover to the weakly interacting regime. In this chapter we treat the first situation, that of systems which are predominantly still strongly correlated, but which have a fractional number of electrons per site. The next step – looking at what happens if the interaction becomes weaker, $U \lesssim t$ – is considered mainly in Chapter 10.

When talking about the systems in which *crystallographically equivalent* TM ions have a noninteger number of d-electrons, or intermediate valence, the first question that arises is whether the system would behave as insulating or metallic, and then what kind of insulator or metal it would be. Some of the possibilities have already been discussed previously, in Chapter 7 and Section 5.2. In these situations there may appear a charge ordering, with the electrons localized at particular sites in some regular fashion, after which typically the material becomes insulating. But the system may also remain homogeneous, in which case the electrons (or holes) can hop from the occupied to the empty sites, which would lead to metallic behavior. But in many cases such metals are different from normal metals such as Cu or Al, which are described by the standard band theory and which, in a more formal sense, are Fermi liquids. And even if the ground state is formally of a Fermi-liquid type, correlations can still be very important. Thus the magnetic susceptibility at elevated temperatures can still be Curie-like, $\chi \sim 1/T$, instead of the Pauli susceptibility $\chi \sim$ const. (only at very low temperatures does χ become constant, although possibly very large). The resistivity in metals with strong correlations but with a Fermi-liquid ground state at low temperatures behaves as $\rho \sim \rho_0 + AT^2$, with $A \sim m^{*2} \sim 1/W^2$, where m^* is the effective mass and W is the bandwidth.

But often the systems with correlated electrons, even if metallic, behave as non-Fermi liquids. Such are the high-T_c cuprates such as $La_{2-x}Sr_xCuO_4$ close to optimal doping, at which the superconducting order temperature is maximal. In such cases for example the resistivity can behave as $\rho \sim T$, or $\sim T^\alpha$ with different exponents α, say $\alpha = 1.5$. Often there is no saturation of resistivity at high temperatures, and the resistivity continues to increase, or the conductivity decreases below the value of the *Mott's minimum metallic conductivity* which is obtained, crudely speaking, when the electron mean free path becomes of the order of lattice spacing, $l \lesssim a$ (the Yoffe–Regel limit). Thus the normal state properties of metallic systems with strong correlations can often be quite different from those of ordinary metals such as Cu.

Sometimes this special behavior can be connected with the presence of *quantum critical points* – points at which, with the change of some external parameter (pressure, magnetic field, doping), some type of ordering present in the system (often magnetic ordering) disappears, for example the Néel temperature is gradually suppressed to $T = 0$. Quantum fluctuations, definitely important in this case, can strongly change the properties of such a system, in particular making it a non-Fermi liquid. This is more common in rare earth compounds, but can also occur in some TM compounds. There are some suggestions that the anomalous normal state properties of high-T_c cuprates can be connected with the presence of a "hidden" quantum critical point lying under the superconducting dome, close to the doping concentration giving the maximal T_c. But in some other TM compounds, for example in MnSi or $PrNiO_3$, such a non-Fermi-liquid phase exists not just close to a certain point in the phase diagram, for example at the critical pressure P_c, but in a broad region of pressures above P_c. The detailed nature of such a state is not yet clear.

A very interesting feature of strongly correlated metals was discovered in underdoped and optimally doped cuprates. It was found that in photoemission one does not see the whole expected Fermi surface, but only some parts of it – the so-called Fermi arcs in the directions [11] and [$\bar{1}$1] in a tetragonal setting, that is close to the $(\pm\frac{\pi}{2}, \pm\frac{\pi}{2})$ points of the Brillouin zone. One possible explanation is the role of the background antiferromagnetic order present in undoped cuprates, for example in La_2CuO_4, or of the corresponding antiferromagnetic fluctuations preserved at certain doping. As discussed in Chapter 1, antiferromagnetism prevents the motion of electrons or holes on such a background, that is the hopping between antiferromagnetic sublattices. But charge carriers can move "along the diagonals," on the same sublattice. Thus hole motion can be coherent in these directions, but strongly damped in other directions. This could explain the presence of well-defined coherent quasiparticles along the diagonals of the Brillouin zone, and their absence in x- or y-directions. If there should exist in this case a real two-sublattice antiferromagnetic ordering, then in a mean-field approximation it would truncate the large original Fermi surface and leave only small hole pockets around the points $(\pm\frac{\pi}{2}, \pm\frac{\pi}{2})$; this could explain the observation of Fermi arcs.

This picture of small hole pockets due to the underlying antiferromagnetic ordering typical for strongly correlated systems can also help to explain yet another special feature observed in doped Mott insulators such as cuprates. It is qualitatively clear that when we remove some electrons from the normal Mott insulator with one electron per site, the real current will be carried by these $\delta = |1 - n|$ holes thus created, so for example the Hall coefficient $R_H \sim 1/\delta ec$ would be large and positive. This is indeed observed at low doping, for example in $La_{2-x}Sr_xCuO_4$ (with $\delta = x$). But the total number of electrons n is large, and in the conventional metallic phase these electrons should form a large Fermi surface; this follows from the *Luttinger theorem*, which is valid for normal metals. But such a large Fermi surface would rather give a small negative Hall coefficient. How to reconcile these two features, large positive Hall coefficient scaling as $1/\delta$ ($= 1/x$ in $La_{2-x}Sr_xCuO_4$) and large Fermi surface, is a well-known problem in the physics of high-T_c superconductors and other doped Mott insulators. Its solution is not really clear. One option could be that such strongly correlated metals are actually not Fermi liquids, in which case the Luttinger theorem may not work. Another option is that mentioned above, namely that due to the underlying magnetic correlations we may end up not with the large Fermi surface but with one having only small hole pockets. Once again, this second scenario is based on many simplifications such as mean-field treatment, which are also questionable. In any case, this again demonstrates that strongly correlated metals may behave quite differently from conventional metals. It also illustrates the point already made above, that in systems with strong correlations, such as doped Mott insulators, there is a very strong interplay of charge and magnetic degrees of freedom. This is one of the most typical features of strongly correlated systems, which comes up again and again in different situations.

There is also the inverse influence of the motion of doped charge carriers on the magnetic structure. We have already seen in Chapters 1 and 5 that antiferromagnetism suppresses coherent motion of electrons. In this situation we can gain more kinetic energy of these carriers if we make the material ferromagnetic. This is seen in the simple Hubbard model, but more clearly in more realistic situations with several d-electrons, for example due to the double exchange, see Section 5.2.[16]

For correlated systems with partially filled bands one can have not only ferromagnetism, but also other types of magnetic ordering. In metallic systems this is determined by the Stoner criterion $I\chi(q) > 1$, where I is the corresponding interaction and $\chi(q)$ is the q-dependent magnetic susceptibility. If this condition is first met at a certain $q = Q$, then there will appear in a system a magnetic ordering with the periodicity determined by this Q, $a \sim \hbar/Q$, for example a helicoidal (spiral) or sinusoidal spin density wave (SDW) with this wave vector. A typical situation leading to that is the situation with *nesting* of the Fermi surface – the presence of overlapping (e.g., flat) parts of the Fermi surface. This is apparently the origin of magnetic ordering in many rare earth metals and compounds, in metallic Cr and presumably in iron-based superconductors.

Doping can also strongly influence many other phenomena in correlated systems. Thus, it can suppress or strongly modify the orbital ordering observed in many TM compounds with Jahn–Teller ions, see Chapter 6. This is for example the case in CMR manganites such as $La_{1-x}Ca_xMnO_3$, in which orbital ordering is suppressed by doping and disappears at $x \sim 0.2$, after which the system becomes a metallic ferromagnet with CMR. Orbital ordering is also suppressed when we go to the itinerant electronic state with weak electronic correlations – but this is rather a matter for the next chapter.

---

[16] Just for the nondegenerate Hubbard model the situation is less clear. It is not really established whether such a model can ever give a ferromagnetic state at any doping, or whether we really need different types of d-electrons for that, in which case the ferromagnetism is related to the "atomic ferromagnetism" due to the Hund's rule. Rigorous results in this direction for the doped nondegenerate Hubbard model exist only in very few limiting cases, and the existing numerical calculations do not give a conclusive answer.

Also, the spin state of some ions such as Co^{3+} can be changed by doping. This happens for example in $La_{1-x}Sr_xCoO_3$. Whereas the undoped system is a nonmagnetic insulator with low-spin $Co^{3+}(t_{2g}^6 e_g^0)$ with $S = 0$, with doping ($x \gtrsim 0.2$) it is transformed to a ferromagnetic metal. And in this state not only the Co^{4+} ions which appear formally are magnetic (usually Co^{4+} is low-spin, $t_{2g}^5 e_g^0$, $S = \frac{1}{2}$), but also a large fraction of the remaining Co^{3+} ions are promoted to a magnetic state, either with high or intermediate spin, see Section 3.3. This again can be connected with the tendency to gain the maximum kinetic energy of doped holes and also with the phenomenon of *spin blockade*. To gain kinetic energy it is favorable to promote some t_{2g}-electrons to the broader e_g band, creating some magnetic states from nonmagnetic $Cr^{3+}(t_{2g}^6)$. But in some cases one cannot move the resulting magnetic states, for example high-spin $Co^{2+}(t_{2g}^4 e_g^2, S = \frac{3}{2})$ on the background of low-spin $Co^{3+}(t_{2g}^6 e_g^0, S = 0)$: one cannot exchange states with $S = \frac{3}{2}$ and $S = 0$ simply by moving one electron. This spin blockade can thus suppress such motion and reduce the kinetic energy gain; spin states would be chosen so as to avoid spin blockade.

Probably one of the most interesting and important phenomena in correlated systems is the appearance of superconductivity in some of them, in particular of high-T_c superconductivity. It was discovered in cuprates by Bednorz and Müller in 1986. In 2008 the second class of such materials was discovered – the iron-based superconductors. And although the maximum values of T_c in these superconductors (~ 55 K) are much lower than for cuprates (~ 130 K), the nature of superconducting pairing in these two classes of superconductors may have much in common, though many of their properties are different.

It is important that the dominant electron–electron interaction in correlated systems is a repulsion, whereas for the pairing in conventional superconductors such as Sn or Al one needs an effective electron–electron attraction. One consequence is that the character of superconducting pairing in these novel systems is different from that in the "old" superconductors: it is not the conventional s-wave pairing with the pairing amplitude $\Delta(\mathbf{k}) \sim$ const., but the unconventional pairing with the pairing amplitude $\Delta(\mathbf{k})$ changing sign at different parts of the Fermi surface. This is the d-wave pairing with $\Delta(\mathbf{k}) \sim \cos k_x - \cos k_y$ in cuprates, and most probably the so-called $s_\pm$-pairing in Fe-type superconductors: it is s-wave, but with different signs of the superconducting order parameter on different pockets of the Fermi surface of Fe pnictides or chalcogenides. This allows us to make use of the repulsion between the electrons: the repulsion is detrimental for the normal s-wave pairing with $\Delta(\mathbf{k}) \sim$ const., but may help d- or $s_\pm$-pairing.

Another way of describing this situation is by saying that the intermediate bosons which provide the interaction between the electrons are not phonons, as in the conventional superconductors, but spin fluctuations, typical for correlated systems and doped Mott insulators. Depending on the dominant fluctuations (characterized by the wave vector $\mathbf{Q}$ at which the susceptibility $\chi(\mathbf{Q})$ is maximal) one can have a tendency toward d-wave or, for iron-based systems, $s_\pm$-pairing when $\mathbf{Q} \sim (\pi, \pi)$, that is when the dominant fluctuations are antiferromagnetic, and it can be for example a triplet p-wave pairing if the system is close to ferromagnetism ($\chi(\boldsymbol{Q})$ is maximal for $\boldsymbol{Q} = 0$). This is presumably the nature of superconductivity in Sr_2RuO_4.

There are still many unsolved problems in the physics of cuprates and Fe-based systems. Thus the nature of the so-called *spin gap phase* in low-doped cuprates is still a matter of considerable debate. One point of view is that it is a precursor of superconductivity, a consequence of superconducting pairing, only without the phase coherence needed for real superconductivity. An alternative point of view is that this pseudogap has nothing to do with superconducting pairing, and is due to some other type of ordering, competing with superconductivity.

Yet another open question is what is actually the main energy gain in the superconducting state of cuprates: whether it is the gain in the interaction energy, as in the conventional superconductors such as Sn, or the gain in the kinetic energy of electrons.

One more suggestion already mentioned above is that there exists a quantum critical point of some kind, close to which high-T_c superconductivity appears. Note that if in cuprates this is still a rather controversial suggestion, there exist several other systems, for example $CePd_2Se_2$, $CeIn_3$, UGe_2, in which superconductivity indeed appears just in the vicinity of such a quantum critical point.

Yet another general question appears when we consider doped correlated systems. The standard approach is to treat these systems as homogeneous. However in many such cases the homogeneous state may be unstable with respect to phase separation, with the appearance of inhomogeneous states. Such a tendency is seen clearly in many models: in the Hubbard and in the t–J models, in the model of double exchange, in charge-ordered systems close to commensurate states, etc. Similarly, experimentally there are many systems in which such phase separation has been observed: in manganites with colossal magnetoresistance, in doped cobaltites, in some cuprates.

Very generally the main driving force of such phase separation is connected with the competition of the tendency to form some type of ordering, and the kinetic or band energy of electrons, which may "prefer" other states. Thus we have already seen for example in Chapter 1 or in Section 5.2, and mentioned above, that the motion of charge carriers, electrons or holes, is hindered by the antiferromagnetic background, so that to gain more kinetic energy it may be favorable to make the system ferromagnetic. But for low doping this may not happen in the whole sample – the number of doped carriers is simply not enough to change the state of the whole system from antiferromagnetic to ferromagnetic. In such cases this may happen *in parts of the sample*, so that the material would be split into ferromagnetic (and metallic) regions, in which all the carriers (e.g., holes) would be concentrated, and the remaining undoped, in this case antiferromagnetic, regions. The same mechanism – competition of a particular ordering and the electron kinetic energy – can also lead to similar phase separation for doped orbitally degenerate systems, or for charge-ordered systems when the electron concentration deviates slightly from $n = 0.5$. In such situations the bulk properties, for example conductivity, should be described using the language of percolation.

An important factor not included in the simple models such as the Hubbard or the double exchange model is the long-range Coulomb interaction. It is clear that it can strongly oppose the tendency toward such phase separation. The Coulomb interaction favors states with charge neutrality, and thus it would prevent at least the large-scale phase separation of the type described above, where for example all electrons are collected "in one corner" of the crystal. But still the tendency toward phase separation may remain, and together with the Coulomb interaction it can lead to the formation of an inhomogeneous state, but with the inhomogeneities of a small size, for example of nanoscale.

Experimentally one sometimes sees large-scale inhomogeneities, but this typically occurs in systems which are close to I order phase transitions. In this case there usually exists a hysteresis in the system, there may appear superheating and supercooling, and inside the hysteresis region there exist a mixture of different phases – for example water droplets in a fog, which may have quite large, macroscopic size. One has to be careful not to mix such large-scale phase separation close to I order transitions and the intrinsic phase separation inherent in many doped correlated systems. The large-scale phase separation in the hysteresis region of a I order transition does not require the redistribution of electrons and does not lead to the formation of charged "droplets," as in the case of the phase separation described above. Experimentally both these phenomena, both types of phase separation, have been observed in doped correlated systems.

A special kind of inhomogeneous system are the artificially created materials such as thin films and multilayers; similarly, surfaces of bulk materials present a special case. The properties of surfaces and interfaces of TM compounds may be quite different from those in the bulk, and rather special phenomena can occur in these cases. Thus, both the effective bandwidth and the Coulomb (Hubbard) interaction can change in these cases, as a result of which the surfaces and interfaces of insulating compounds may become metallic, or, vice versa, materials metallic in the bulk may become insulating in thin-film form. Different types of ordering – charge, magnetic, orbital – may change at surfaces and interfaces. All this opens up the possibilities of "material engineering" – artificial creation of materials with novel properties, not found in the bulk.

Special phenomena are encountered in films and multilayers with charged layers, such as in (001) films of $LaMnO_3$, in which the charged layers $Mn^{3+}O_2^{2-}$ and $La^{3+}O^{2-}$ alternate. In these cases we meet the phenomenon of *polarization catastrophe* – a linear growth of electric potential with the thickness of the film, so that for thick enough films the potential drop across the film may exceed the energy gap. This would correspond to an instability of the system. The situation can be "corrected" either extrinsically, for example by creating oxygen vacancies, or by spontaneous electron reconstruction – a transfer of some electrons from one surface of the film to the other. In both cases there may appear charge carriers at the surface or interface, which could then become metallic, even though the respective bulk materials are good insulators. This is observed for example at the interface of $SrTiO_3$ and $LaAlO_3$, and one can even obtain superconductivity or ferromagnetic ordering at this interface. Transition metal compounds play a special role in these phenomena, both because of their eventual ordering (orbital, magnetic), which can change at the interface, but first of all because of the ability of most TM ions to accept different valence states (for example by changing the valence of Ti ions in $SrTiO_3$ from Ti^{4+} to Ti^{3+}), so that their doping, which in this case occurs spontaneously (e.g., due to the polarization catastrophe) can happen quite naturally.

10

Metal–insulator transitions

In analyzing various phenomena in TM compounds in the previous chapter, we have already several times come across the situation when a material, depending on conditions, can be in an insulating or in a metallic state. Such metal–insulator transitions can be caused either by doping (a change in band filling) or by temperature, pressure, magnetic field, etc. The topic of metal–insulator transitions is one of the most interesting in the physics of systems with correlated electrons. Such metal–insulator transitions often lead to dramatic effects and a drastic change in all properties of the system; and the large sensitivity of materials close to such transitions to external perturbations can be used in many practical applications.

In principle, metal–insulator transitions are not restricted to systems with correlated electrons. They are often observed in more conventional solids, well described by the one-electron picture and standard band theory. However the most interesting such transitions, often significantly different from those in "band" systems, are indeed met in systems with strongly correlated electrons, in particular in transition metal compounds – see for example Mott (1990) or Gebhard (1997).

10.1 Different types of metal–insulator transitions

One can divide all metal–insulator transitions into three big groups; these are discussed in the sections below.

10.1.1 Metal–insulator transitions in the band picture

The first group of metal–insulator transitions are transitions which can be understood on the one-electron level in the framework of band theory – although, of course, interactions of some type are always necessary for such transitions. The general explanation of metal–insulator transitions in band theory is that, as is well known, the electron spectrum in such systems consists of energy bands divided by regions of forbidden energy – energy gaps. And if the band filling is such that a band, or several overlapping bands, are partially filled, Fig. 10.1(a), we are dealing with metals. If however some bands (valence bands) are

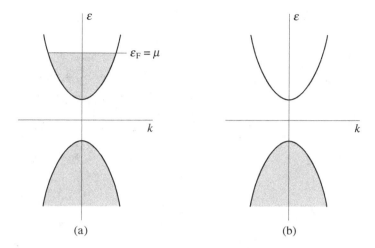

Figure 10.1 The difference between a metal and an insulator in the simple band picture.

filled, but the higher-lying conduction band(s), separated from the filled valence bands by an energy gap, are empty, the material is an insulator (or a semiconductor), Fig. 10.1(b). Note right away that according to such a band picture the systems with odd number of electrons per unit cell (e.g., one electron per site, in simple nondegenerate bands) would always be metallic, with partial band filling: in the conventional band theory every state can be filled, according to the Pauli principle, by two electrons, and we can have completely filled valence band(s) only if we have an even number of electrons (2, 4, etc.) per unit cell.

From this simple picture we see immediately that there are in principle two possibilities of transforming the band insulator of Fig. 10.1(b) into a metal. One is to change the electron concentration, for example by doping, alloying, etc.: that is, to add extra electrons to the system. These electrons will fill the conduction bands, Fig. 10.1(a), which in the original insulator of Fig. 10.1(b) were empty. (One can, of course, also use doping by holes, reducing the number of electrons; this will lead to the valence band being partially filled.) Such phenomena are seen in many cases, for example in heavily doped semiconductors, or in certain alloys (one sometimes refers to *band-filling-controlled* insulator–metal transitions; see Imada *et al.*, 1998). And very often this is the source of the insulator–metal transition in correlated systems as well: we have seen this for example with CMR manganites, doped cobaltites, and high-T_c cuprates, Chapter 9 (of course, in those cases one has to take into account the strong electron correlations).

Another possibility for getting an insulator–metal transition, often met in various systems, is to change the band structure without changing the electron concentration, so that the energy gaps open or close, Fig. 10.2. This can happen for example if one simply moves the valence and conduction bands relative to each other, say by pressure, Fig. 10.2(a,b). But most often this happens when we change the periodic potential which leads to the

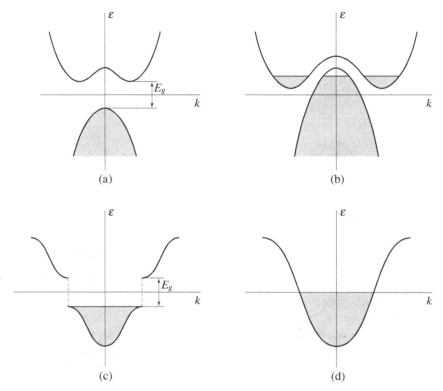

Figure 10.2 Possible transition between a semiconductor (a), and a (semi)metal (b) opening of the gap (c) in a metallic spectrum (d) due to structural distortions, for example the Peierls distortion in the one-dimensional case.

band formation, for example by changing the crystal (and sometimes magnetic) structure, Fig. 10.2(c,d). This happens for example at the Peierls transition in quasi-one-dimensional systems, with one electron per site and with a half-filled band, where the dimerization of the lattice (chain of atoms) leads to opening up of a gap at the Fermi surface, as a result of which the energy spectrum changes from the metallic one of Fig. 10.2(d) to a spectrum with a gap, Fig. 10.2(c).

The examples of such transitions, caused by a change in crystal structure, are numerous. Probably the oldest known such case is the transition between white and gray tin (the famous "tin plague" known since the Middle Ages). White tin, stable at room temperature, is a well-known metal. It has a slightly distorted diamond lattice. At low temperatures, however, the stable phase of Sn is gray tin with an undistorted diamond structure, which is insulating (the energy gap in it is practically zero, but the bulk properties are really those of an insulator). Such a white tin–gray tin transition is accompanied by a large increase in volume, which often leads to the destruction of the crystals, with such manifestations as for example the transformation of a box of tin spoons, or tin buttons on military uniforms, into gray powder.[1]

One can understand this tendency when one looks at the Periodic Table (at the end of the book): when we go down the column containing C, Si, Ge, Sn, Pb, we see that the gap in the diamond structures decreases with increasing atomic number, from C to Sn. Carbon (diamond itself), as well as Si and Ge, are insulators; Sn exists in both metallic and insulating modifications (with the insulating one having zero gap); while Pb is always metallic. And the insulator–metal transition is connected with the change in crystal structure.

We see another example of similar phenomena in the same compounds: whereas solid Ge and Si with the diamond structure are insulators (or, better, semiconductors), they become metallic when they melt. In the liquid state there is no long-range order, but it is mostly the short-range order which determines the metallic or insulating behavior of the system. The short-range order in liquid Ge and Si changes drastically in comparison with the order in the crystal phase, as a result of which molten Ge and Si become metals.

The periodic potential which leads to the formation of bands in a solid and which determines whether the material should be a metal or an insulator in the band picture, is created predominantly by the underlying atomic lattice. But there is also an important contribution to the periodic potential coming from the electrons themselves, in the band theory in the form of a self-consistent electron contribution. If there occurs a periodic change in the *density* of electrons, as in some systems with *charge density waves* (CDW), it acts in the same way as the lattice potential and can open up partial or full gaps at the Fermi surface; this leads to a metal–metal CDW transition, for example in $NbSe_2$ (partial gapping of the Fermi surface), or to a metal–insulator transition in TaS_2 (full gap). Such CDW are of course always accompanied (and may largely be driven) by the corresponding lattice distortions. But there exists also the exchange part of this self-consistent potential, which in principle can lead to an opening of the gap in the spectrum (and which is coupled to the much weaker lattice distortions). This is what sometimes happens in magnetically ordered states, in particular in some systems with itinerant electrons and with spin density waves (SDW). Such SDW also contribute to the periodic potential and can create energy gaps on part of or the whole Fermi surface. In the first case the material would remain metallic; this is what happens apparently in the antiferromagnetic state of metallic Cr. If, however, such a gap covers the whole Fermi surface, the system would become insulating. This is the situation in many interesting organic materials. Such an insulating SDW state may be suppressed for example by pressure, which often leads to quantum critical points, often with the appearance of superconductivity in their vicinity – see Section 9.5.3.

---

[1] As mentioned in some reports, this phenomenon could have played a crucial role in the tragic fate of the famous polar explorer R. Scott, whose team perished in the Antarctic in 1912. One possible reason for this could have been the fact that they lost their fuel stored in flasks sealed with tin, which started to leak when, in the cold climate of the Antarctic, these tin seals decomposed due to the white tin–gray tin transition.

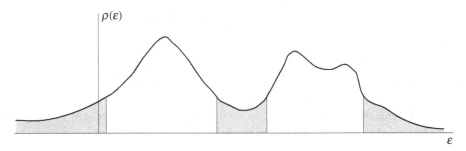

Figure 10.3 Mobility gaps and Anderson localization in disordered systems; shaded regions are the regions with zero mobility.

10.1.2 Anderson transitions in disordered systems

The second class of metal–insulator transitions is met in disordered systems. Such transitions are known as *Anderson transitions*, and they are connected with the *Anderson localization* in disordered systems (Anderson, 1958). This is a major field of study in itself, see for example Brandes and Kettemann (2003); here we will only explain the main ideas. In disordered systems one should consider the motion of noninteracting electrons in an external potential, which is however not periodic but random. The energy spectrum itself can be continuous, without any energy gaps in the density of states $\rho(\varepsilon)$, Fig. 10.3. But, as was shown by Anderson (1958), some of these states, for example in the shaded regions in Fig. 10.3, can in fact be localized: the electrons put into such states cannot move to infinity but remain in a certain vicinity of their original positions. Other states, in the "white" regions in Fig. 10.3, may still be delocalized, and if the number of electrons is such that the uppermost occupied level (the Fermi level) is in this delocalized region, the system should have metallic conductivity. If however the Fermi level ε_F is in the region of localized states (the shaded regions in Fig. 10.3), the material would be insulating – not because there exists an energy gap in the spectrum, but because the *mobility* of the corresponding states is zero. This leads to the notion of a *mobility edge*: the boundaries of shaded and unshaded regions in Fig. 10.3 are such mobility edges, and if the band filling is changed so that ε_F crosses such a mobility edge, there will occur in the system a metal–insulator transition of the Anderson type.

One can understand what happens here qualitatively in the "mountain landscape" model. Suppose our random potential has the form shown in Fig. 10.4 – a "mountain landscape," with some deep valleys, craters and lakes, and mountain ranges. It is clear that if we have a few electrons ("water"), they will fill the bottom of this potential profile, forming isolated lakes, and there will be no current flowing from one side of the mountain range (the sample) to the other. But if we increase the filling, the "water level" will increase until we reach the situation where percolation from the left to the right edge of the system takes place – shown for example by the dashed line in Fig. 10.4. This will be the mobility edge, and for the "water level" (the Fermi level) above this threshold the system would be conducting.

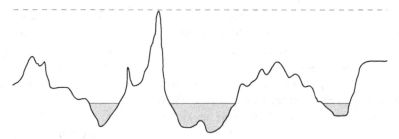

Figure 10.4 Qualitative picture of localization of (electron) liquid in a random potential landscape.

This picture is purely classical, and serves only for illustration. The actual Anderson localization is in fact a quantum phenomenon, and it occurs largely because of quantum interference. What determines whether the state will be delocalized or localized due to this process is the ratio of the typical electron kinetic energy (our familiar electron hopping t) and the amplitude or strength of the disorder potential (the typical "spread" between mountains and valleys in Fig. 10.4). If the disorder is too strong, the electron would be localized in a certain part of the crystal, and would not diffuse from this region to infinity – this is Anderson localization.

A specific feature of this approach is that the standard treatment of the Anderson insulator–metal transition ignores the interaction between the electrons; this phenomenon occurs already for noninteracting electrons (in a random external potential). As we have seen in the previous chapters, electron–electron interactions can also lead to electron localization. A very interesting question arises as to what could be the interplay of these two mechanisms of electron localization: that due to electron correlations (Mott localization) and that due to disorder (Anderson localization), that is what would be the properties of disordered strongly correlated systems? This question is now under active investigation, and the answers obtained until now show that the situation may be far from trivial: these two factors, or two mechanisms of localization, do not simply add, but can even counteract one another. The simplest qualitative picture of what could happen can again be illustrated with a picture similar to that shown in Fig. 10.4. Suppose one electron is localized at one local minimum of the random potential of Fig. 10.4 and is thus immobilized. But in the presence of electron–electron interactions the second electron approaching the first one can "push" it out of its potential well, forcing it to move through the crystal, thus facilitating metallic conductivity. That is, in this picture the electron–electron interaction could in principle counteract the Anderson localization, not enhance it.

10.1.3 Mott transitions

The third, and for us the most interesting mechanism of metal–insulator or insulator–metal transitions is the mechanism connected with electron correlations themselves. We have already seen in Chapter 1 that due to electron–electron interactions the standard band picture may break down, and the electrons may become localized, so that the system may become a novel type of insulator – a Mott insulator. The insulating character of such systems is caused not by the external periodic potential of the lattice, which can lead to the insulating character of some systems with proper filling of one-electron states (see, e.g., Figs 10.1(b), 10.2(a, c)), but by strong enough Coulomb interaction of electrons, for example by large on-site Hubbard repulsion U. As we have seen in Chapter 1, such systems are insulating for integer number of electrons per site, for example $n = N_{el}/N = 1$, and for strong enough interaction $U/t > (U/t)_{crit}$. If we increase the electron hopping t, or the electron bandwidth $W = 2zt$ (where z is the number of nearest neighbors), there should occur sooner or later a transition to a metallic state – the Mott, or Mott–Hubbard transition,[2] see Mott (1990).[3] Mott transitions typically occur under pressure, but often also as a function of temperature; these are the most interesting cases. They are often accompanied by a change in some other properties of the system: by disappearance or modification of magnetic ordering, by a change in orbital ordering, by structural transformations. Therefore it is sometimes difficult to decide whether the observed metal–insulator transition is a Mott transition, or whether it can be explained in the standard band picture, for example as a consequence of a structural transition. The point is that in many real materials both factors are often present: for example lattice distortion, which in itself can lead to a metal–insulator transition, and electronic correlations. Often this leads to a considerable controversy as to which explanation is actually correct. The best known such example is the metal–insulator transition in VO_2; we will discuss this case in detail below.

This also concerns the metal–insulator transition caused by a change of band filling, for instance by doping. As discussed in Section 10.1.1, in principle these transitions can be explained just in the band picture. But in many real materials, such as doped manganites or cuprates, the electron correlation effects definitely play an important role, and one has to incorporate them in the band picture. Indeed, as discussed in Chapter 9, the properties of systems with partially filled correlated bands are often quite different from those with ordinary bands of noninteracting electrons. Whether one should still classify such transitions as Mott transitions is a matter of convention; most often this term is reserved for metal–insulator transitions at fixed electron concentration, caused for example by a change of temperature or pressure.

10.2 Examples of metal–insulator transitions in systems with correlated electrons

There exist several "classical" systems with metal–insulator transitions in correlated systems. Among them are several vanadium oxides (see, e.g., Brückner *et al.*, 1983): V_2O_3, VO_2, the so-called Magnéli phases V_nO_{2n-1} "interpolating" between V_2O_3 and VO_2 (for

---

[2] We use the term "Mott transition" for both the transition from a metal to a Mott insulator and for the reverse insulator–metal transition.

[3] We do not consider here special cases with nesting of the Fermi surface, see for example Khomskii (2010), which can make the system insulating also for large W or small U: usually such nesting, which may exist in some simple lattices such as square or cubic ones for nearest-neighbor hopping, is destroyed and consequently transitions from an insulating to a metallic state restored, when we include further-neighbor hopping.

$n = 2$ this corresponds to V_2O_3; the next ones are V_3O_5, V_4O_7, ..., up to VO_2 which would correspond formally to $n = \infty$). The valence in these systems changes from $V^{3+}(d^2)$ in V_2O_3 to $V^{4+}(d^1)$ in VO_2. Especially interesting is the situation in V_4O_7 which contains formally an equal number of V^{3+} and V^{4+}.

Similar metal–insulator transitions are observed in many Ti oxides, in particular in similar Magnéli phases Ti_nO_{2n-1} ($n = 2$ is Ti_2O_3, ..., $n = \infty$ is TiO_2). However in contrast to the vanadium Magnéli phases, one of the valence states in the case of Ti, Ti^{4+} has an empty d-shell and is nonmagnetic.

As mentioned above, many metal–insulator transitions are accompanied by (or driven by) certain extra types of ordering: antiferromagnetic ordering in V_2O_3, lattice dimerization and charge ordering in V_4O_7, dimerization in VO_2. This is also the case in many other systems: the metal–insulator transitions often occur together with charge ordering, as happens for example in the magnetite Fe_3O_4, or in some manganites; they are accompanied by structural distortions with orbital ordering and with the formation of TM dimers or more complicated "molecules," such as dimers in the spinel $MgTi_2O_4$ (Khomskii and Mizokawa, 2005) or in the layered system $La_4Ru_2O_{10}$ (Khalifah *et al.*, 2002); with the formation of trimers in $LiVS_2$ (Katayama *et al.*, 2009) and heptamers ("molecules" made of seven V ions) in AlV_2O_4 (Horibe *et al.*, 2006). Nevertheless, electron correlations play a very important role in all these phenomena.

10.2.1 V_2O_3

The metal–insulator transition in V_2O_3 occurs with decreasing temperature at $T_c = 150$ K. It is a strong I order transition: the jump in resistivity reaches 10^6 (Fig. 10.5), see McWhan *et al.* (1975); there the series of curves of $V_{2-x}Cr_xO_3$ for different Cr substitutions are shown, see below. This transition is accompanied by a structural change from the corundum structure of V_2O_3 to a monoclinic structure at $T < T_c$; the lattice change is so strong that sometimes single crystals of V_2O_3 crack and practically explode when going through this transition.

The low-temperature states of V_2O_3 are antiferromagnetic, with the magnetic structure shown schematically in Fig. 10.6 (where only the V sublattice is shown; cf. Fig. 3.44). In the honeycomb V lattice in the ab-plane the spins are ordered in two antiferromagnetic sublattices, and the V pairs in the c-direction have parallel spins.

Sometimes one treats V_2O_3 paying main attention to these V pairs in the c-direction. This is quite reasonable for the analogous system Ti_2O_3, where indeed such pairs become strongly bound below T_c; but this is not at all the case in V_2O_3. The c/a ratio in V_2O_3 drops below T_c; however the V–V distance in these c-pairs does not decrease, but actually *increases*. Thus it makes no sense for V_2O_3 to consider these pairs as the natural "building blocks," as is sometimes done.

The insulator–metal transition in V_2O_3 can be influenced strongly by pressure, or by some substitution, as for example in $V_{2-x}Cr_xO_3$, where the increase in Cr content plays

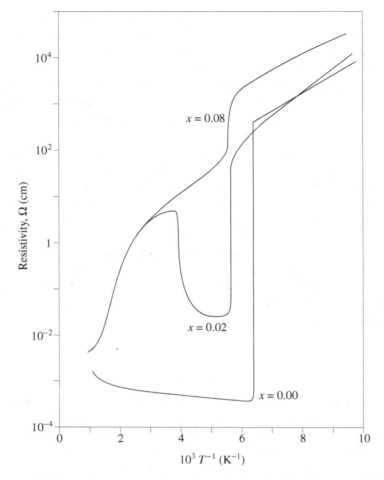

Figure 10.5 Metal–insulator transition in $V_{2-x}Cr_xO_3$, after McWhan *et al.* (1975).

a role of the negative pressure. The resulting phase diagram, Fig. 10.7 is a representative phase diagram for many similar insulator–metal transitions; one can use it to illustrate several general features of such transitions.

At ambient pressure there exists one sharp transition from the metallic to the antiferromagnetic insulating phase, which is suppressed by pressure. But for negative pressures, or for a certain Cr concentration, there are actually two transitions: with increasing temperature the system goes from an antiferromagnetic insulator to a paramagnetic metal, but then again to an insulator, which does not have any long-range magnetic ordering; this is seen for example in the curve for $x = 0.02$ in Fig. 10.5. The first transition we can in principle explain by a change in band structure caused by the structural transition and magnetic ordering; these could in principle be sufficient to open up a gap in the spectrum and cause

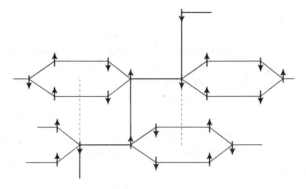

Figure 10.6 Schematic picture of crystal and magnetic structure of V_2O_3.

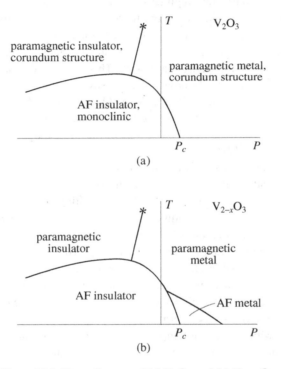

Figure 10.7 Phase diagram of (a) V_2O_3 and (b) $V_{2-x}O_3$.

the insulator–metal transition. But this is definitely not the case for the high-temperature transition. This transition is isomorphous, that is it occurs without any change in the lattice symmetry (only the lattice parameters change), and it is a paramagnetic–paramagnetic transition. Thus there occurs no symmetry change at this transition, which is also seen from the fact that it ends at a critical point (marked by ∗ in Fig. 10.7), above which all properties change continuously. This transition can be nothing but a Mott transition in a pure sense, not accompanied by any other ordering. This proves the importance of electron correlations in V_2O_3, so that the lower-temperature metal–insulator transition to an anti-ferromagnetic insulating phase should definitely also be classified as a Mott transition (but occurring together with structural and magnetic transitions).

Several other typical features of Mott transitions can be seen in Fig. 10.7. One is that dT_N/dP is positive at negative pressures, where the electrons are more localized and where the transition is from an antiferromagnetic insulator to a still-insulating but para-magnetic phase. This is simply the regime of localized electrons, with the low-temperature phase antiferromagnetically ordered, with $T_N \sim J \sim t^2/U$, and with the high-temperature phase which is a standard Mott insulator. With increasing pressure the hopping integral t increases, and with it T_N, as we see in Fig. 10.7.

But in approaching the itinerant regime with increasing pressure, the Néel temperature passes through a maximum and then starts to decrease: in this regime, when the band-width W or electron hopping t become comparable to the Coulomb repulsion U, T_N should start to go down, because the localized magnetic moments themselves start to decrease and finally disappear in the metallic phase for $P > P_c$. This is a typical behavior of the ordering of localized electrons when they approach the crossover to the itinerant regime, which is also seen for example in the series CdV_2O_4–MnV_2O_4–ZnV_2O_4–MgV_2O_4 (Blanco-Canosa *et al.*, 2007). This should also be the case for orbital ordering, see Fig. 7.19.

Another, at first glance counterintuitive feature of the phase diagram of V_2O_3 (Fig. 10.7) is the positive slope $dT_c/dP > 0$ of the high-temperature insulator–metal transition. Naively we expect that this transition, which we have classified as a real Mott tran-sition, that is a transition from localized electrons ("electron crystal") to delocalized ones ("electron liquid"), should occur similarly to melting of ordinary crystals and the high-temperature phase should be "liquid," that is metallic. Experimentally, at the high-temperature transition the situation is just the opposite: with increasing temperature there occurs a transition from a metal to an insulator, see Figs 10.5 and 10.7.

We have already discussed briefly a similar situation and its possible physical explana-tion in Section 7.2. Essentially, the physical picture is the following: as follows from the general statistical mechanics, the choice of one or other phase is determined by the con-dition of the minimum of the free energy $F = E - TS$, see eq. (7.1), and the sequence of different phases with increasing temperature is always such that the entropy S of the high-temperature phase is larger than that of the low-temperature one. With increasing tem-perature the second term in the expression for free energy, $-TS$, becomes more important and we need to make it larger in order to decrease the total free energy.

In V_2O_3 close to the higher-temperature insulator–metal transition two phases compete: a paramagnetic metal and a paramagnetic insulator. The metallic phase, though paramag-netic, is described by the filled Fermi sea and at least at low temperatures it is a unique state with zero entropy. Of course at finite temperatures this is no longer the case, but the entropy remains low, $S \sim \gamma T \sim T/\varepsilon_F$, where γ is the coefficient in the linear specific heat, $c = \gamma T$. In the paramagnetic insulating state, however, we have disordered localized spins, in V_2O_3 nominally spins $s = 1$ (V^{3+} has configuration d^2; here we denote spin by s, so as not to mix it with the entropy S). Consequently, if we ignore short-range spin correlations, the entropy of this state would be $S = k_B \ln(2s + 1) = k_B \ln 3$ – much larger than the entropy of the metallic state. Therefore with increasing temperature we go here from the metallic to the paramagnetic insulator state. Of course if we take into account spin correlations in the insulating state, as well as the effects of electron correlations in the metallic state, the difference in entropies of these two states would be reduced. But apparently the tendency is still preserved, and the magnetic entropy of the paramagnetic insulator is bigger than that of the metallic state, which determines the sequence of phases with increasing temperature. When spins order, in the antiferromagnetic state, this magnetic entropy of the insulating state is reduced, and the order of phases is reversed: with increasing temperatures there occurs a transition from an antiferromagnetic insulator to a metal.

There are a few other interesting features of the insulator–metal transition in V_2O_3. One can show that this transition is accompanied by a certain orbital repopulation, or orbital ordering (Park *et al.*, 2000), although the detailed picture of this ordering is very different from that proposed in Castellani *et al.* (1978): the latter was based on the picture of V dimers in the c-direction, which is definitely not applicable, and this picture led to a wrong prediction of the spin state of V pairs. Nevertheless, the very idea of Castellani *et al.* (1978) that orbital degrees of freedom play some role in the metal–insulator transition in V_2O_3 is definitely correct, although the orbital ordering does not seem to be crucial here.

Yet one more interesting aspect of the behavior of V_2O_3 was discovered by Bao *et al.* (1993): it was found that in slightly nonstoichiometric samples $V_{2-x}O_3$ the transition under pressure occurs not from an antiferromagnetic insulator to a paramagnetic metal but, first, close to the transition a metallic state with some magnetic ordering (of a spiral type) appears below 9 K, and only at higher pressures does the material go over to a normal paramagnetic metal.

10.2.2 VO$_2$

The situation in VO_2 is also very interesting, and quite different from that in V_2O_3. In VO_2 there also occurs a metal–insulator transition from a high-temperature metallic to a low-temperature insulating state at temperatures slightly above room temperature, $T_c = 68°C$. The behavior of resistivity resembles that of V_2O_3 (Fig. 10.5, curve $x = 0$), with a jump in resistivity of up to 10^5. But in contrast to V_2O_3, the low-temperature insulating phase of VO_2 is a nonmagnetic insulator. Apparently this is caused by the different electronic configuration of V^{4+} ions with one d-electron compared with $V^{3+}(d^2)$ in V_2O_3, and with different types of structural distortions accompanying or causing the metal–insulator transition. Notably, in VO_2, which has the rutile structure, one can talk about V chains running in the c-direction, and the lattice distortions occurring below T_c are connected with this one-dimensionality – although the high-temperature phase of VO_2 is rather isotropic, that is in this phase such one-dimensional chains are electronically not yet well formed. We will discuss this point later.

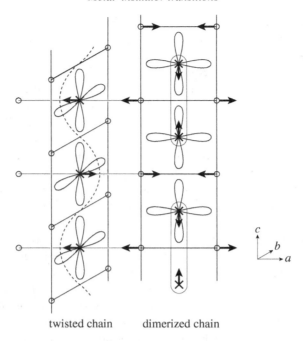

twisted chain   dimerized chain

Figure 10.8 Orbital occupation and lattice distortion (arrows) in the M_2 phase of VO_2. Crosses are V^{4+} ions, circles are oxygens.

The structural distortions occurring in VO_2 below T_c are of two types: first, there appears V dimerization in these chains, and simultaneously these V pairs tilt somewhat in the crystal ("twisting"), that is in fact the V ions shift to the left and to the right from the centers of VO_6 octahedra (analogously to antiferroelectric distortions), Fig. 10.8. The resulting structure is shown schematically in Fig. 10.9. The formation of short V–V dimers leads to singlet pairings of two electrons on such V pairs, of the type $\frac{1}{\sqrt{2}}(1\uparrow 2\downarrow - 1\downarrow 2\uparrow)$; this effectively removes the localized spins, leading to a reduction of magnetic susceptibility below T_c and to the formation of a nonmagnetic insulating state.

The distortions observed in VO_2 below T_c strongly resemble those at the Peierls transitions in one-dimensional systems. This dimerization itself could lead to an opening of the gap in the system, and that is why one of the first explanations proposed for the metal–insulator transition in VO_2 connected it to just this structural transition, that is explained it using the conventional band picture (Adler and Brooks, 1967; Wentzcovitch *et al.*, 1994). However several experiments pointed to the importance of correlation effects in VO_2. One of the most convincing arguments for this was the modification of the properties of VO_2 when V is replaced by Cr (Marezio *et al.*, 1972), and also under uniaxial stress (Pouget *et al.*, 1975; Rice *et al.*, 1994). The resulting phase diagram of $V_{1-x}Cr_xO_2$, shown in Fig. 10.10, has not only a high-temperature metallic rutile phase and a low-temperature insulating phase (M_1) with a monoclinic structure, shown schematically in Fig. 10.9, but

Figure 10.9 The structural distortions of VO_2 below the metal–insulator transition.

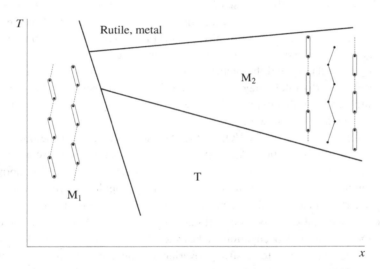

Figure 10.10 Phase diagram of VO_2 doped by Cr or under axial stress, which shows, besides the standard M_1 phase in which there exist dimerization and twisting in all V^{4+} chains, also the M_2 phase in which half of the V chains are dimerized but not twisted, and the other half are twisted but not dimerized.

also another monoclinic phase M_2, in which the distortion pattern is different: half of the V chains have only dimerization but no twisting, and the other half have only twisting but no dimerization, see Figs 10.8–10.10. There exists also a triclinic phase T, in which the distortions change gradually from those of the M_2 phase to those of M_1 (or this phase T is just a fine mixture of regions of M_1 and M_2 phases). It is important that the M_2 phase, in which only half of the V ions form singlet pairs, is already insulating; this is a strong argument in favor of the importance of electron correlations (Mott–Hubbard physics) in the metal–insulator transition in VO_2.

The distortion pattern in the phase M_2 (and M_1) can be understood from the following arguments, see Fig. 10.8. In this figure we show schematically two neighboring chains in the rutile structure of VO_2, with V marked by crosses and oxygens by circles. When dimerization occurs in one, say the right chain, the shift of V ions in one pair toward one another pushes the oxygen ions sitting between the chains out of the first chain, and the corresponding V ions in the neighboring chains move away from the shortened V bonds in dimers and toward the elongated V–V bonds between dimers. That is, the dimerization in one chain, say the one on the right, necessarily causes twisting in the neighboring chain (the one on the left). These distortions and the corresponding order parameters are linearly coupled, and it is in fact one common distortion, say η_1. Similarly, dimerization of the chain on the left would cause twisting in the chain on the right; let us call this distortion η_2. The complex distortion observed in the M_1 phase of VO_2 consists of both dimerization and twisting in every chain, $\eta_1 = \eta_2$. In the M_2 phase it "splits" into either only dimerization or only twisting in the neighboring chains, that is $\eta_1 \neq 0$, $\eta_2 = 0$ or $\eta_1 = 0$, $\eta_2 \neq 0$.[4]

As mentioned above, it is important that the M_2 phase, in which half of the V^{4+} ions remain undimerized and retain their localized spins (as seen by magnetic susceptibility), is already insulating. This shows that one cannot explain the entire metal–insulator transition in VO_2 using band theory and the picture of Peierls dimerization (Rice *et al.*, 1994); the electron correlations definitely play an essential role in this transition – although structural distortions are certainly also important. There exist other experimental indications of the importance of electronic correlations in VO_2.

Orbital ordering also occurs in VO_2 in the insulating state, and it definitely plays an important role in this transition. As is shown by high-energy spectroscopy (Haverkort *et al.*, 2005), at T_c a significant orbital repopulation occurs in VO_2: if above T_c all three t_{2g} orbitals of $V^{4+}(t_{2g}^1)$ are more or less equally populated, which leads to rather isotropic properties, below T_c the electrons occupy predominantly the orbital which has strong overlap with the corresponding orbitals of neighboring V ions in the c-direction, see Fig. 10.8. This orbital occupation makes the electronic structure of VO_2 below T_c more one-dimensional (due to a strong overlap of the corresponding orbitals in the c-direction) and the resulting one-dimensional chain with a half-filled nondegenerate band made of these orbitals (one electron per site in $V^{4+}(d^1)$) becomes unstable toward Peierls dimerization. (This formation of an effectively one-dimensional electronic structure from

---

[4] Interestingly enough, distortion very similar to that in the M_2 phase of VO_2 is observed in another V oxide, $K_2V_8O_{16}$ (with hollandite structure), which also undergoes a metal–insulator transition (Isobe *et al.*, 2006). This material has valence of V close to that of VO_2 ($V^{3.75+}$), and its crystal structure resembles that of VO_2. It also contains V chains (rotated 90°) running in the c-direction; however these are not single chains as in rutile, but double chains, with hollow channels between them in which the K ions are located. Detailed structural studies (Komarek *et al.*, 2011) have demonstrated that there occurs in these double chains of $K_2V_8O_{16}$ a distortion very similar to that in the M_2 phase of $V_{1-x}Cr_xO_2$: one half of the chain shows dimerization, and the other half twisting.

three-dimensional ones above T_c is yet another example of the phenomenon of reduced dimensionality due to orbital ordering, discussed in Section 6.2.)

In fact of course all these phenomena – orbital ordering, dimerization, transition to an insulating state – occur simultaneously, and all of them contribute to the corresponding energy gain and make the transition possible. Thus we can see that the metal–insulator transition in VO_2 has a "combined" nature: it has both the features of a Mott transition and of a band-like one. Electron correlations and modifications of energy bands by distortion both play a role here. The detailed theoretical calculations using *ab initio* band structure calculations combined with dynamical mean field theory (DMFT) indeed confirm this picture (Biermann *et al.*, 2005).

The discussion of VO_2 gives us a good opportunity to present one rather interesting aspect of the behavior of this system, which should actually be observed (and sometimes indeed is) in other materials with metal–insulator transition as well. Notably, it was discovered (see, e.g., Stefanovich *et al.*, 2000) that one can induce a transition from an insulating to a conducting state in VO_2 not only by increasing the temperature, but also by running a strong enough current through the VO_2 samples, or by applying a large voltage to it. It was found that the $I–V$ characteristic of VO_2 is strongly nonlinear and has an S-shaped form, Fig. 10.11. Notably, if we apply a large enough voltage to VO_2 (in the insulating phase, e.g. at room temperature), at a certain critical voltage there occurs a *switching* to a low-resistivity state. When we reduce the voltage afterwards, the system may remain in the conducting state down to $V = 0$, or it may switch back to the high-resistivity state, Fig. 10.11 (usually these experiments are performed in current-controlling mode).

There has been significant discussion in the literature whether this transition is caused just by the heating of the sample, or whether there is a direct influence of the electric field or carrier injection on the electronic state of the system. The first mechanism is of course the most natural: when a current runs through a sample there is Joule heating, and this is enough to heat the sample by 40–50 degrees, inducing the transition to the metallic state

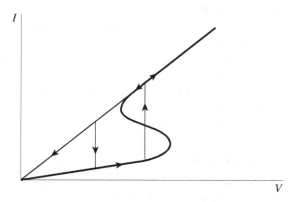

Figure 10.11 Schematic picture of nonlinear (S-shape) $I–V$ characteristic of VO_2 films, showing the possibility of switching.

(remember that these experiments were done at room temperature, and the insulator–metal transition in VO_2 occurs at $\sim 68°C$).[5] However there also exist arguments that this is not the full explanation, and that there may be a direct influence of nonthermal effects such as carrier injection on such switching.

It was also shown that this switching does not occur homogeneously, that is simultaneously in the whole sample: there appear conducting channels in the sample in which the insulator–metal transition occurs, with a local change of crystal and electronic structure. Such recrystallized channels can even be seen under the microscope. When the current or voltage was reduced, sometimes these conducting channels remained, in which case the resistivity remained low down to $V = 0$; in other cases they were broken, and the system returned to a high-resistivity state. If the metallic channels survived at $V = 0$, one could return the sample to the high-resistivity state by applying a strong short voltage pulse, which apparently "melted" such metallic channels, as happens in a fuse. Similar phenomena are now studied actively not just in this but also other classes of materials, with the aim of using them for switchable memory media.

10.2.3 Magnéli phases Ti_nO_{2n-1} and V_nO_{2n-1}

Very interesting information can be obtained from the study of the Magnéli phases Ti_nO_{2n-1} and V_nO_{2n-1}, which "interpolate" between Ti^{3+} and Ti^{4+}, and V^{3+} and V^{4+} ionic states. For Ti_nO_{2n-1} the corresponding electronic configurations are d^1 for Ti^{3+} and d^0 for Ti^{4+}, and for V_nO_{2n-1} they are respectively d^2 for V^{3+} and d^1 for V^{4+}.

Many of these compounds show metal–insulator transitions. The simplest is the situation in Ti_2O_3 – the $n = 2$ member of the corresponding family, and a close analog of V_2O_3. Similar to V_2O_3, Ti_2O_3 also shows a metal–insulator transition but its features are very different compared with that in V_2O_3. The transition in Ti_2O_3 between 400 K and 500 K is rather broad, and the low-temperature insulating phase is nonmagnetic. Structural data show that in Ti_2O_3 the Ti pairs in the c-direction (see the structure in Fig. 10.6) indeed form pairs below the transition: the Ti–Ti distance in these pairs decreases in the insulating phase. Apparently this leads to the formation of singlet pairs (each Ti^{3+} has one d-electron), which makes the material effectively nonmagnetic. Accordingly, in contrast to V_2O_3, for Ti_2O_3 the theoretical treatment proceeding from such pairs is really justified and gives quite a good description of its properties.

The same tendency of d^1 ions with spin $S = \frac{1}{2}$ to form singlet pairs is seen even more clearly in Ti_4O_7, another member of the Ti Magnéli family. These materials, which can be presented as $M_2O_3 + (n-2)MO_2$, consist of rutile-type slabs as in TiO_2 or VO_2, extending

---

[5] There exists a beautiful effect connected with such heating. Suppose we start with an insulating state at room temperature, at which the resistivity of VO_2 is large, and run a current through the sample. The Joule heating is then also large, $\sim IV = I^2R$ (the experiment is done in the current-controlling mode). Owing to strong heating the temperature of the sample rises, eventually becoming higher than T_c, and the material switches to the low-resistivity (metallic) state. But in this state the Joule heating becomes smaller and the sample may cool down below T_c, after which the resistivity again jumps to a high value and the process is repeated. In effect there appear oscillations in the material kept at a d.c. voltage, which are really observed experimentally.

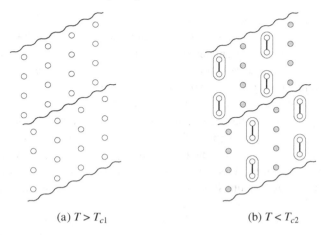

(a) $T > T_{c1}$ (b) $T < T_{c2}$

Figure 10.12 The crystal structure of the Magnéli phase Ti_4O_7 at (a) high and (b) low temperatures. Gray circles are $Ti^{4+}(d^0, S = 0)$ ions, white circles are $Ti^{3+}(d^1, S = \frac{1}{2})$ ions forming singlet pairs.

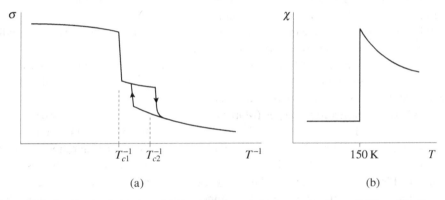

(a)   (b)

Figure 10.13 Qualitative behavior of (a) resistivity and (b) magnetic susceptibility of V_4O_7, after Lakkis *et al.* (1976).

in two directions, separated by shear planes, at which the MO_6 octahedra share faces, as in the c-direction in the corundum structure of Ti_2O_3 and V_2O_3 (Fig. 10.12). In Ti_4O_7 there exist two consecutive transitions with decreasing temperatures, at $T_{c1} = 150\,\text{K}$ and at $T_{c2} = 120\,\text{K}$, at which the resistivity increases in a jump-like manner, see Fig. 10.13 (Lakkis *et al.*, 1976) (the upper transition is the I order transition with a rather broad hysteresis). That is, the upper transition is an insulator–metal transition, and the lower one is an insulator–insulator transition. The magnetic susceptibility is Curie-like in the high-temperature metallic phase, which is thus a "Curie metal"; it drops at the higher transition temperature T_{c1}, and below it the material is practically nonmagnetic. Structural studies (Marezio *et al.*, 1973a) have shown that below T_{c2} there occurs, first, a charge ordering: Ti^{3+} and Ti^{4+} are segregated in consecutive chains in the c-direction, Fig. 10.12(b), in which we denote Ti^{3+} by empty circles and Ti^{4+} by filled ones. Besides that, Ti^{3+} ions are grouped in pairs with much shorter Ti–Ti distance inside the pair than between pairs. This again leads to the formation of singlet states at such pairs, as in Ti_2O_3; this tendency is even stronger in Ti_4O_7. Apparently this dimerization, together with charge ordering, plays an important role in the very phenomenon of metal–insulator transition in Ti_4O_7. One can also see a strong similarity of this phenomenon to what happens below the metal–insulator transition in VO_2, which also has a rutile structure, as the "middle" block in Ti_4O_7, and in which there also occurs singlet pairing of d^1 ions with spin $\frac{1}{2}$, in that case $V^{4+}(d^1)$.

Very interesting is the situation in the intermediate phase of Ti_4O_7 between T_{c1} and T_{c2}. The magnetic susceptibility in this phase is small; apparently the pairing of Ti^{3+} ions has already taken place. But in contrast to the low-temperature phase in which these pairs are ordered as shown in Fig. 10.12(b), in the intermediate phase such pairs are disordered, and can fluctuate and move through the crystal, forming something like a liquid of such pairs. This picture has much in common with the picture of resonating valence bonds (RVB) of Anderson, discussed in Section 5.7.2, but with definite dynamics, determined not so much by quantum but rather by thermal fluctuations. This gave rise to considerable theoretical activity: the intermediate phase of Ti_4O_7 could be an example of mobile bipolarons – bound states of two electrons which can move through the crystal together with the corresponding distortion, which keeps these two electrons together.

The situation in another "half-filled" material of this class, V_4O_7, seems similar in some respects, but also different in others. This material also experiences an insulator–metal transition at $T_c = 237$ K, but this is followed by magnetic ordering at $T_N = 34$ K. Below T_c there also occur structural changes (Marezio *et al.*, 1973b; Hodeau and Marezio, 1978), similar to those in Ti_4O_7, Fig. 10.14: V^{3+} and V^{4+} ions also order in alternating chains, and there occurs some dimerization of V^{3+} ions. V^{4+} ions, with configuration d^1, behave in a more complicated way: those inside the rutile slabs form tightly bound dimers (shown by ovals in Fig. 10.14), but half of the V^{4+} ions, those which lie close to shear planes, remain unpaired, and apparently keep their spin-$\frac{1}{2}$ moments. At lower temperatures those unpaired V^{4+}, and probably partially also the weakly paired V^{3+}, order antiferromagnetically at 34 K (Botana *et al.*, 2011). In fact the details of the low-temperature magnetic structure of V_4O_7, in particular the question of to what extent both V^{4+} and V^{3+} ions participate in this ordering, still remain unclear.

In any case, one feature common to VO_2, Ti_2O_3, and Ti_4O_7 is also seen in V_4O_7: in all these systems the d^1 ions with spin $\frac{1}{2}$ have a strong tendency to form singlet dimers (though for V^{4+} sites close to shear planes this does not happen).

Vanadium Magnéli phases are interesting in yet another respect: all of them, except V_7O_{13} (V_3O_5, V_4O_7, V_5O_9, V_6O_{11}, V_8O_{15}) show metal–insulator transitions, see Fig. 10.15 (Kosuge *et al.*, 1972). In all systems with metal–insulator transitions, except V_3O_5, there occurs some pairing in the insulating phase, which actually starts already above T_c. The remaining unpaired electrons apparently contribute to the magnetic ordering.

The only Magnéli phase system which remains metallic down to the lowest temperatures is V_7O_{13}; the reason for this different behavior of the material is not clear.

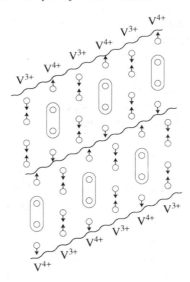

Figure 10.14 Low-temperature structure of V_4O_7.

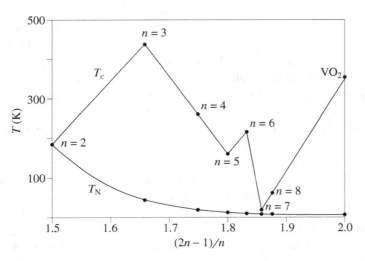

Figure 10.15 Metal–insulator transitions and Néel points in the vanadium Magnéli phases V_nO_{2n-1}, $n = 2, 3, \ldots, 8$ and $n = \infty$ (VO_2), after Kosuge *et al.* (1972).

What general lessons can we draw from studying these different Ti and V compounds, many of which show metal–insulator transitions? One has already been mentioned above: it seems that whenever we have TM ions with d^1 configuration with $S = \frac{1}{2}$, they show a tendency to form singlet dimers. We have seen the same tendency in other compounds, for example in the spinel $MgTi_2O_4$. This tendency is especially clear in systems which are close to the localized–itinerant crossover, for example those showing metal–insulator transitions, and having more complicated, not too symmetric lattices, such as those containing MO_6 octahedra with common edges or faces. For more symmetric cases such as that of perovskites this trend is much less pronounced: thus for example $LaTiO_3$ with $Ti^{3+}(d^1)$ remains undistorted and shows antiferromagnetic ordering of Ti ions at low temperatures. Thus this dimerization with the formation of singlets does not have to occur always; but it is a very plausible scenario for many such systems.

Yet another feature is seen when we compare the materials with integer number of d-electrons, such as V_2O_3 or VO_2, with Magnéli phases with $n > 2$, which nominally contain TM ions with different valences, that is have partially filled (Hubbard) bands. In all vanadium Magnéli phases in the metallic phase above T_c we have magnetic susceptibility following the Curie law, and not Pauli susceptibility as is the case for normal metals. Thus these materials present examples of "Curie metals." These are not the only systems showing this property. Thus the well-known layered material Na_xCoO_2 has the same property for $x \gtrsim 0.5$ (the famous Li_xCoO_2, which is the basis for the best rechargeable batteries in all laptops and mobile phones, belongs to the same class of materials). Apparently many metals with narrow bands and strong correlations can exhibit such behavior.

Interestingly enough, V_2O_3 and VO_2, with integer d-shell occupation (d^1, d^2), show a much more "normal" behavior of susceptibility in the metallic phase, not Curie-like. Thus it seems that in materials of this type the systems with partially filled Hubbard bands behave *magnetically* in a more correlated way than those with exactly one or two electrons per site above their Mott transition. One can understand this qualitatively as follows: for Hubbard systems with $n = 1$ the electrons can move only by "hopping on top of each other," that is creating doubly occupied and empty sites. In the simple Hubbard model such states are nonmagnetic ($|0\rangle$ or $|\uparrow\downarrow\rangle$). Consequently they do not contribute to (Curie) susceptibility. However if we have for example relatively few electrons in the partially filled Hubbard band with strong interaction (large U/t), these electrons can easily move through the crystal – and gain the corresponding kinetic energy – avoiding each other, that is without the necessity of forming doubly occupied sites. Consequently all such electrons would, on the one hand, contribute to conductivity and on the other hand, retain their spins, that is they would contribute to the magnetic response in the same way as localized spins, thus giving Curie susceptibility. This picture, in fact, follows from actual calculations (Khomskii, 1977).

10.2.4 Metal–insulator transitions due to charge ordering

One of the mechanisms of insulator–metal transitions in TM compounds may be charge ordering. We have already met several such examples in Chapters 7 and 9. As discussed in those chapters, many TM compounds with partially filled bands, that is with noninteger number of d-electrons per site, may be metallic even in the presence of strong correlations. Metallic conductivity in such cases may be connected with a rapid exchange of different valence states of TM ions, for example of Fe^{2+} and Fe^{3+} in Fe_3O_4 (Section 7.4), or Mn^{3+} and Mn^{4+} in doped manganites (Section 9.2.1). This picture can be used if the energy bands are narrow, with the bandwidth W smaller than the on-site Hubbard repulsion U. In these cases the metallic character of the corresponding materials may be different from that of normal metals, although the ground state at $T = 0$ may still be of Fermi-liquid type, albeit with strongly renormalized parameters (but it can also be of non-Fermi-liquid type). Very often the strong interactions present in this regime will cause some ordering at low temperatures, most often charge ordering. And the electron localization occurring at such ordering can render the material insulating.

As discussed in Chapter 7, most often such charge ordering occurs for commensurate electron concentration, for example for one electron per two sites, etc. This is the situation in half-doped manganites (Section 7.1) or in the Magnéli phases V_4O_7 and Ti_4O_7 (Section 10.2.3). However, as discussed in Section 7.4 using the example of magnetite Fe_3O_4, in frustrated lattices the situation can be rather complicated even in this case.

The charge-ordered state with localized d-electrons should also show some magnetic and sometimes orbital ordering, and it is usually accompanied by a structural transition. However, often it is quite difficult to observe such charge ordering directly. Thus for example most probably insulator–metal transitions observed in several vanadium systems with the hollandite[6] structure, such as $K_2V_8O_{16}$, $Ba_{1.2}V_8O_{16}$, $Bi_xV_2O_{16}$ (Isobe *et al.*, 2006) should be accompanied by charge ordering (V ions in these systems have valence between $3+$ and $4+$), but apparently the signatures of charge ordering are relatively weak (Komarek *et al.*, 2011).

A very interesting class of insulator–metal transitions is met in materials with spontaneous charge disproportionation, discussed in Section 7.5. In particular, sharp insulator–metal transitions were observed in $PrNiO_3$ and $NdNiO_3$ (Medarde *et al.*, 1992; Torrance *et al.*, 1992). In these systems this transition is accompanied by the appearance of antiferromagnetic order in the low-temperature insulating phase, of a rather special type – with spin alternation of $\uparrow\uparrow\downarrow\downarrow$ type along x-, y-, and z-directions. The initial explanation of this transition and of the magnetic structure appearing in it was in terms of a Mott transition with orbital ordering: low-spin Ni^{3+} ions in these materials have configuration $t_{2g}^6 e_g^1$, and they should be strong Jahn–Teller ions. However none of the possible types of orbital ordering can explain the observed magnetic structure. The resolution of this problem, obtained theoretically (Mizokawa *et al.*, 2000) and confirmed experimentally (Alonso *et al.*, 1999), is the phenomenon of charge disproportionation discussed in Section 7.5: Ni^{3+} ions in these systems disproportionate, formally, into Ni^{2+} and Ni^{4+}, and these ionic states order in the perovskite lattice in a checkerboard fashion. This charge disproportionation occurs at the insulator–metal transition and apparently drives the transition. Interestingly enough, this transition occurs in nickelates $RNiO_3$ simultaneously with magnetic ordering for larger

---

[6] Hollandites form a rather large class of materials (belonging to the so-called tunnel compounds), and in some of them metal–insulator transitions are observed. Besides the V hollandites mentioned above, one can also mention $K_2Cr_8O_{16}$, in which there occurs a metal–insulator transition at $\sim 90\,K$ in the ferromagnetic phase ($T_c = 160\,K$) (Hasegawa *et al.*, 2009; Toriyama *et al.*, 2011), so that this system at low temperatures is a rather rare case of a ferromagnetic insulator.

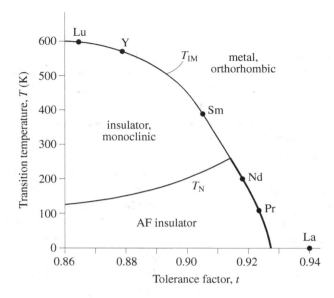

Figure 10.16 Metal–insulator and antiferromagnetic transitions in nickelates $RNiO_3$.

rare earth ions R = Pr, Nd, but for smaller rare earth ions these transitions are decoupled, see the phase diagram in Fig. 10.16 (already presented in a different context in Section 3.5), where thin lines show the II order transitions and the thick line shows the I order transition.

10.2.5 Other examples of metal–insulator transitions in transition metal compounds

There are many other examples of TM compounds showing metal–insulator transitions – some as a function of temperature, which was the main focus of the discussion above, and many more under pressure. One can say that formally all such materials, and even many simpler ones, should become metallic under pressure: the overlap of electronic wavefunctions (and with it the bandwidth) increases with compression, and one may expect that sooner or later, at high enough pressure, all materials would become metallic. However in some cases one may need enormous pressures for this to happen, which are difficult to reach in a laboratory. But such phenomena may definitely occur for example deep inside the Earth, and one indeed believes that many usually insulating materials, including TM compounds, may become metallic in the Earth's interior.

Going back to temperature- or composition-induced metal–insulator transitions, one should mention some sulfides and selenides. The best known case is that of NiS. It has a (metastable) NiAs structure shown in Fig. 10.17, consisting of chains of NiS_6 octahedra in the c-direction sharing common faces, with hexagonal ordering of these chains in the

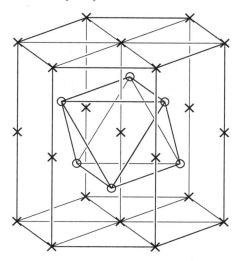

Figure 10.17 The structure of NiS. Here crosses are Ni ions and circles are sulfurs.

ab-plane, cf. the discussion in Section 3.5. NiS experiences an insulator–metal transition at $T_c \sim 240$ K (the exact value depends on the sample), from the antiferromagnetic nonmetallic (possibly semimetallic) low-temperature phase to a paramagnetic metal above T_c; this transition is suppressed under pressure and with deviations from stoichiometry, Fig. 10.18. The situation in NiS is sometimes considered a bona fide example of a Mott transition, because the lattice symmetry does not change at this transition (Imada *et al.*, 1998); only the volume of the crystal decreases above T_c by $\sim 1.9\%$, which could lead to an increase of the bandwidth and to a Mott transition. But apparently the electron correlations are less important in NiS than for example in V oxides with metal–insulator transitions: in contrast to those the high-temperature or high-pressure metallic phase of NiS is almost a normal metal, with a rather small Pauli susceptibility and with the coefficient γ of linear specific heat $c = \gamma T$ having the value $\gamma \sim 6$ mJ/mole $\cdot$ K^2 – typical for normal metals but an order of magnitude smaller than for example in V$_2$O$_3$.[7] Also the resistivity of the metallic phase of NiS, reached under pressure, is similar to that of weakly correlated metals: $\rho \sim T^2$ in the small temperature interval up to ~ 10 K, whereas in V$_2$O$_3$ $\rho \sim T^2$ up to ~ 100 K. The low-temperature properties of NiS resemble more those of semimetals or very narrow-gap semiconductors, not those of real Mott insulators. Nevertheless the presence of antiferromagnetic ordering below T_c, and the strong jump in volume at T_c, may signal some importance of electronic correlations.

---

[7] Remember that the coefficient γ is $\gamma \sim N(0) \sim m^*$, that is, it is a measure of the corresponding bandwidth W (it is larger for narrow bands and for high density of states at the Fermi level $N(0) \sim 1/W$), and it is enhanced by electronic correlations and by electron–phonon interaction. For normal metals with broad bands, such as Cu or Ag, the values of γ are ~ 1 mJ/mole $\cdot$ K^2; for metallic TM compounds γ may be $\sim 10^1$–10^2 mJ/mole $\cdot$ K^2; while it reaches $\sim 10^3$ mJ/mole $\cdot$ K^2 for heavy-fermion systems, mostly on the basis of rare earths and actinides ($4f$ and $5f$ elements) – see Chapter 11.

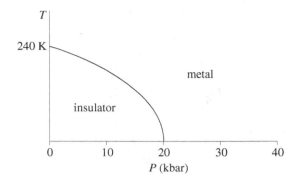

Figure 10.18 Pressure dependence of the insulator–metal transition in NiS.

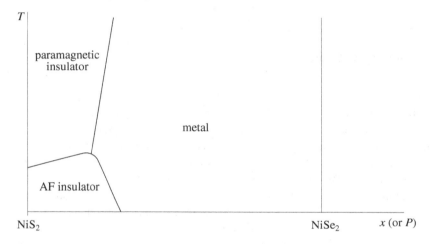

Figure 10.19 Schematic phase diagram of the system $NiS_{2-x}Se_x$.

Yet another system often cited in the context of metal–insulator transitions in TM compounds is the system $NiS_{2-x}Se_x$, see for example Mott (1990). This system has a pyrite structure described in Section 3.5: it can be represented as a NaCl-type structure made of Ni^{2+} ions and $(S_2)^{2-}$ or $(Se_2)^{2-}$ molecules. NiS_2 is an insulator, and $NiSe_2$ is a metal. The insulator–metal crossover occurs in $NiS_{2-x}Se_x$ at low temperatures at $x \sim 0.3$. The insulating phase is antiferromagnetic, and apparently the metallic phase for larger x (or with increased pressure) is also antiferromagnetic. This resembles the situation in non-stoichiometric V_2O_3, see Section 10.2.1. The whole phase diagram of this system, see Fig. 10.19, also resembles the phase diagram of V_2O_3, Fig. 10.7. Apparently the explanation of the positive slope dT_c/dx or dT_c/dP for the high-temperature metal–nonmetal transition is the same as for V_2O_3: the spin entropy in the paramagnetic insulating phase is larger than the entropy of the metallic phase, which leads to a metal–insulator transition with increasing temperature.

Several interesting systems with insulator–metal transitions demonstrate the tendency, already noticed in Chapter 1, that in strongly correlated systems Mott insulators are most often antiferromagnetic, whereas ferromagnetism is usually stabilized by the kinetic energy of electrons and typically coexists with metallic conductivity (although there are exceptions to this general rule – ferromagnetic insulators and antiferromagnetic metals). Some insulator–metal transitions can be explained by this general tendency: the change in the magnetic structure can sometimes be a cause of such transitions. Thus for example in $Pr_{1-x}Ca_xMnO_3$ the application of a magnetic field can lead to a very sharp transition from an antiferromagnetic insulator (with charge ordering) to a ferromagnetic metallic state, see Fig. 7.10, with a jump in resistivity of up to 10^8 (Tomioka *et al.*, 1996; Tokunaga *et al.*, 1997) – larger than for example in V_2O_3. In some such cases the metallic (and ferromagnetic) state appears with decreasing temperature, in contrast to most metal–insulator transitions in TM systems. This happens for example in CMR manganites such as $La_{1-x}Ca_xMnO_3$ ($0.25 \lesssim x \lesssim 0.5$). These phenomena have already been discussed in Section 9.2.1.

Possibly the strongest insulator–metal transition, with the largest drop in resistivity, of $\sim 10^{10}$, occurs with decreasing temperature in nonstoichiometric EuO during the transition to a ferromagnetic state at $T_c = 70\,K$ (Shapira *et al.*, 1973). A simple explanation of this transition is illustrated in Fig. 10.20. Owing to nonstoichiometry there are impurity levels (levels of oxygen vacancies) below the bottom of the conduction band, Fig. 10.20(a). Below T_c, when EuO becomes ferromagnetic, the conduction band experiences exchange splitting, and the subband with spins parallel to the magnetization (caused by localized $4f$ spins of Eu) crosses the vacancy level, Fig. 10.20(b). Then the electrons from the impurity levels "spill off" into the conduction band, and the material becomes metallic. A strong shift of the bottom of the conduction (sub)band to lower energies at low temperatures is seen clearly in optical data – the famous "red shift," which in EuO reaches $\sim 0.4\,eV$.

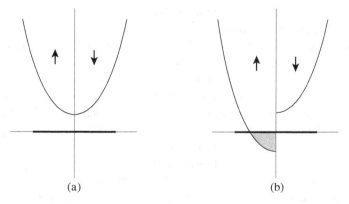

(a)   (b)

Figure 10.20 Qualitative explanation of the insulator–metal transition in EuO (slightly nonstoichiometric). (a) $T > T_c$, a paramagnetic insulating phase; (b) $T < T_c$, a ferromagnetic metallic phase. The thick horizontal line is the vacancy level.

10.3 Theoretical description of Mott transitions

A theoretical description of metal–insulator transitions is a very difficult task. Of course it depends on the detailed type of transition. Those insulator–metal transitions which are connected mainly with a change in the crystal structure, and which can be explained within the conventional band theory, present a simpler problem: one can use the well-developed apparatus of band structure calculations for these systems and employ the conventional methods of treatment of interactions, which in these cases can often be considered as weak perturbations (which nevertheless lead to an opening of the gap in the spectrum, i.e. to a metal–insulator transition). Such is for example the situation for the Peierls transition or for charge or spin density waves, see for example Khomskii (2010). Likewise, the metal–insulator transition caused by charge ordering can often be explained in a similar way. In all these cases the metal–insulator transition is connected with the appearance in the system of a certain (long-range) ordering – Peierls dimerization in quasi-one-dimensional systems, charge ordering in half-doped manganites, etc., and it is this ordering which leads to an opening of the gap in the electron spectrum and to a metal–insulator transition (though actually the energy gain due to the opening of this gap is often the driving force of this very ordering). In any case, in such situations we can describe the metal–insulator transition by introducing corresponding order parameters, we know which symmetry changes occur at these transitions, etc.

The situation is completely different for real Mott transitions in systems with strongly correlated electrons. By their very definition, Mott transitions occur just in the situation where the electron–electron interaction becomes of the order of the bandwidth, that is when the interaction energy of the system becomes comparable with the kinetic energy. For a theoretical description this is the worst case: all quantities become of the same order, there is no small parameter, etc. But most well-developed theoretical methods usually use one or the other form of perturbation theory: most often in the weak interaction, as for example in the Feynman diagram technique, or vice versa, treating the hopping as a perturbation, as we did for the strongly interacting Hubbard model with $t/U \ll 1$ (Chapter 1). Mott transitions are bona fide examples of situations where no perturbation theory can work. This makes their theoretical description extremely difficult – maybe one of the most difficult problems in the whole theory of condensed matter.

Yet another complication is that we do not know what the order parameter of Mott transitions is and which symmetry, if any, is broken at these transitions (when they occur in a pure form, without any conventional ordering such as magnetic or structural). Usually we describe II order phase transitions, such as many magnetic transitions, by pointing out which symmetry is broken at this transition and by introducing the corresponding order parameter, for example the magnetization of a ferromagnet. For Mott transitions these characteristics are not known; it is quite possible that in this sense Mott transitions present a novel class of phase transitions, different from the conventional ones (or perhaps Mott transitions should always be I order, like liquid–gas phase transitions, for which the symmetry is the same in both phases).

In any case, in the absence of small parameters we have to devise approximate schemes for the description of Mott transitions. Until now two such schemes have proved to be the most successful. The first is the Gutzwiller variational treatment (Gutzwiller, 1965), applied specifically to Mott transitions by Brinkman and Rice (1970). The other method is the dynamical mean field method (Georges *et al.*, 1996; Kotliar and Vollhardt, 2009), which is widely used nowadays both for the description of general features of Mott transitions and, especially in combination with band-structure calculations using density functional methods such as local density approximation (LDA), for the description of insulator–metal transitions in specific materials.

Both of these are rather sophisticated theoretical methods, and we will not be able to consider them in detail in this book. We will only describe their main ideas and present some general results.

10.3.1 Brinkman–Rice treatment of metal–insulator transitions

One of the main features of systems with strongly correlated electrons, for example described by the Hubbard model (1.6) with strong on-site repulsion $U \gg t$, is that due to this interaction the probability of finding two electrons at the same site is strongly suppressed. For noninteracting electrons filling free-electron bands, for example in the case of one electron per site, $n = N_{el}/N = 1$, the wavefunction takes the form of plane waves, and one can easily see that the probabilities of finding, at a given site, no electrons, one electron with spin $\uparrow$, one electron with spin $\downarrow$, or two electrons $\uparrow\downarrow$ are all equal – namely $\frac{1}{4}$. Thus the chance of having two electrons at a given site, which is measured by the average value of the on-site correlation function $\langle n_{i\uparrow}n_{i\downarrow}\rangle$, is in this case $\frac{1}{4}$. For the Hubbard model with $U \gg t$ this probability will be strongly reduced. Gutzwiller (1965) proposed taking this factor into account in a variational scheme, reducing the contribution of the state with two electrons at the same site to the total wavefunction by a factor $g < 1$, with g being the variational parameter, so that if there are D such doubly occupied sites in the system, the corresponding part of the wavefunction would be suppressed by the factor g^D. If we then write down the expression for the average band (kinetic) energy, and for the average interaction energy, the resulting expressions would depend on the parameter g, and optimizing the total energy in g we obtain a solution which takes into account strong electron correlations, which suppress double occupancy for large U/t.

This general method was applied to Mott transitions in the nondegenerate Hubbard model with one electron per site, for which we expect a metal–insulator transition with increasing U/t, by Brinkman and Rice (1970). In their approximation the double occupation $\langle n_{i\uparrow}n_{i\downarrow}\rangle$, or the variational parameter g, was nonzero in the metallic state for $U < U_c = 8|\bar{\varepsilon}_0|$, as it should be in metals (here $\bar{\varepsilon}_0$ is the average kinetic energy of electrons, of the order of the hopping t or bandwidth $W = 2zt$), but this on-site correlation function decreased for $U \to U_c$ and became zero for $U > U_c$, which in this description corresponded to a Mott insulator. Strictly speaking, though, this is not completely true – we have seen in Chapter 1 that even in Mott insulators there is some virtual hopping of electrons from site to site, giving nonzero probability of finding two electrons at a site, $\langle n_{i\uparrow} n_{i\downarrow} \rangle \sim (t/U)^2$. Nevertheless the Brinkman–Rice treatment gives quite a good description of such systems on the metallic side of Mott transitions, that is it describes very reasonably correlated metals. In particular Brinkman and Rice have shown that on approaching the Mott transition from the metallic side, that is for $U \leqslant U_c$, the effective mass of the electron diverges as

$$m^* = \frac{m_0}{1 - (U/U_c)^2} \, , \tag{10.1}$$

and the magnetic susceptibility is

$$\chi \to \frac{\mu_B \, \rho(\varepsilon_F)}{1 - (U/U_c)^2} \sim \frac{m^*}{m_0} \, . \tag{10.2}$$

The approximation of Brinkman and Rice has been applied to many systems with strongly correlated electrons, and not only to those but also, for example, to another system of fermions with very strong on-site repulsion – ^{3}He. One can find a good description of this approach in the review article of Vollhardt (1984).

10.3.2 Dynamical mean-field method for Mott transitions

Another rather successful approach to the theoretical description of Mott transitions, including real TM systems, is the dynamical mean-field theory (DMFT) (see, e.g., Georges *et al.*, 1996). In this method one treats on-site effects exactly, and the coupling of a site to the rest of the system is taken into account in a mean-field way, as in all mean-field theories. However, in contrast to the standard mean-field treatment, for example in the theory of magnetism, the interactions are not treated statically, but the dynamic effects are taken into account. This means that the corresponding quantities should be time- or frequency-dependent. The resulting self-consistency equations are written for the quantities characterizing the behavior of electrons – notably for the one-electron Green functions or for self-energies, see for example Georges *et al.* (1996) or Kotliar and Vollhardt (2009). Such Green functions "know" about the bare noninteracting spectrum of electrons $\varepsilon_0(k)$, but when treating the interactions the momentum dependence is usually ignored and only the frequency dependence is taken into account. This is the approximation which can be justified in certain specific cases, but which in general can lead to some deficiencies. At present there are many efforts being made to go beyond this approximation, which, hopefully, would finally correct this drawback. Nevertheless even in the simple form, which completely ignores k-dependence of the self-energy, this method gives a very appealing description of Mott transitions.

As we have already mentioned in Section 9.3 (see also Khomskii, 2010), the electron spectrum in the free-electron gas (or for weak interactions) can be described as that of coherent quasiparticles, maybe with a renormalized spectrum $\varepsilon(k)$, and with some

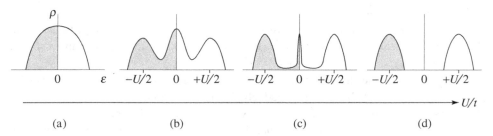

Figure 10.21 Evolution of the density of states in the nondegenerate Hubbard model in single-site dynamical mean-field theory.

weak relaxation, giving a finite lifetime τ of such quasiparticles. Mathematically this quasiparticle spectrum is described by the poles of the one-electron Green function,

$$G(\boldsymbol{k}, \omega) \sim \frac{1}{\omega - \epsilon(\boldsymbol{k}) + i\Gamma} \, , \qquad \Gamma \sim \frac{\hbar}{\tau} \, . \tag{10.3}$$

In systems with strong correlations, however, there appears, besides this coherent part of the spectrum, also an incoherent part, which is not given by the poles of the Green function. In particular, the lower and upper Hubbard bands described in Chapter 1 largely present such an incoherent part of the spectrum.

The DMFT treatment of the simple nondegenerate Hubbard model (in the approximation ignoring the k-dependence of interaction effects) leads to the picture of the approach of the system to Mott transition, illustrated in Fig. 10.21, where we show a typical evolution of the density of states $\rho(\epsilon)$ with increasing electron correlation U/t. For noninteracting electrons we have the ordinary metallic band, filled up to the Fermi level, which for one electron per site, $n = 1$, lies in the middle of the band, see Fig. 10.21(a), where the density of states of excitations obtained from the corresponding Green function is shown. With increasing U/t this quasiparticle band close to zero energy changes its shape, the total number of states in it (the area of this peak) decreases, and there appear satellite peaks at the energies $\sim -U/2$ and $+U/2$, which represent the future lower and upper Hubbard bands, Fig. 10.21(b) (the Fermi energy here is taken as 0). As U/t grows, Fig. 10.21(c), the quasiparticle peak at the Fermi energy narrows, its area decreases still further, and spectral weight is transferred from it to the upper and lower incoherent peaks. At the approach to the Mott transition from the metallic side, $U \to U_c$, the width of the central quasiparticle peak goes to zero (while its height remains constant as a consequence of the so-called Friedel sum rule), and for $U > U_c$ it disappears completely, leaving only the filled lower and empty upper Hubbard bands, with the energy gap $\sim U$ between them, Fig. 10.21(d). That is, we have obtained an insulating state for $U > U_c$ – the Mott insulator.

This picture, especially the description of the metallic state close to the crossover to a Mott insulator as having both coherent Fermi-like quasiparticles and a large incoherent part of the spectrum, is very appealing, and agrees in its main features with the results of for example high-energy spectroscopy such as photoemission and inverse photoemission.

The progressive narrowing of the coherent peak for $U \rightarrow U_c$ also agrees with the results of Brinkman and Rice: the effective mass m^* (10.1) is $\sim 1/W_{\mathrm{coh}}$, where W_{coh} is the width of the coherent peak in Fig. 10.21: as $W_{\mathrm{coh}} \rightarrow 0$ for $U \rightarrow U_c$, the effective mass should go to infinity, as also follows from eq. (10.1).

There are still some drawbacks in the approximation used in the version of DMFT which led to Fig. 10.21. It reduced effectively the treatment to that of a single site, ignoring intersite correlations. But, as we have seen in the case of some Mott transitions, for example in VO_2 (Section 10.2.2), such intersite correlations can sometimes be very important: in the case of VO_2 these are the correlations within singlet dimers. Some generalizations of single-site DMFT now allow us to take such effects into account, at least partially; a good example is the cluster DMFT treatment of the metal–insulator transition in VO_2 (Biermann *et al.*, 2005), which gives quite a reasonable description of this transition.

From the general point of view, the single-site treatment leading to the "sequence of events" shown in Fig. 10.21 used the approximation in which the height of the quasiparticle peak is constant, which follows from the Friedel sum rule valid for a single site such as an isolated impurity in a metal. This approximation can in principle break down in concentrated systems, if we include intersite correlations. And indeed this is the case in some generalizations of DMFT which include many-site effects, in which a pseudogap can form in the center of the quasiparticle peak during the approach to a Mott transition. There is also the question of whether the Mott transition in the DMFT treatment should always be an I order transition (it is I order in the standard treatment described e.g. in Georges *et al.*, 1996, which seems to be a reasonable result). Nevertheless, despite all these problems, the DMFT method, especially with its modern extensions, seems to be at present one of the most promising approaches to the theoretical treatment of Mott transitions.

10.4 Insulator–metal transitions for different electronic configurations

Until now we have discussed mainly metal–insulator transitions proceeding from the basic theoretical models, such as the band model or the nondegenerate Hubbard model. As we have seen, even in these conceptually simple cases there are still many fundamental open questions. When we turn to real materials, the specific features of each can also play a very important role, modifying significantly and often determining largely the detailed features of such transitions. Following our general line, formulated in the Introduction, we will now discuss how metal–insulator transitions depend on the details of the electronic structure of the corresponding compounds, such as orbital state, multiplet effects (spin state), the presence of anion (ligand) p-electrons, etc.

When we want to take into account all these effects, it may be more convenient to start from the insulating state of the material and consider its transition to the metallic state, that is to talk not of metal–insulator, but of insulator–metal transitions; we will follow this approach in this section. We consider how specific details of the electronic structure, discussed for insulators for example in Chapters 2–4, may influence insulator–metal transitions in the corresponding materials.

10.4.1 Multiplet effects, spin crossover, and interaction-controlled insulator–metal transitions

When discussing insulator–metal transitions in correlated systems, one sometimes refers to two main types (Imada *et al.*, 1998): band-filling-controlled transitions and bandwidth-controlled transitions. By band-filling-controlled transitions we mean the situation described in Chapter 9, where the system becomes metallic under doping or under a change of electron concentration, for example by photoexcitation, injection from contacts, etc. By bandwidth-controlled transitions we mean the standard Mott transitions for systems with integer (or at least fixed) number of d-electrons, discussed for example in Sections 10.1.3 and 10.3. In this case the criterion for having the insulating or metallic state is the ratio of the effective electron–electron interaction, that is the on-site Hubbard repulsion U, to the corresponding bandwidth $W \sim 2zt$, where t is the electron–electron hopping and z is the number of nearest neighbors. For $U/W > (U/W)_c$ the material is a Mott insulator, and for smaller U/W it is a metal. Such transitions can be caused for instance by applying pressure: pressure increases the overlap of wavefunctions and consequently the electron hopping t and the bandwidth W. The repulsion U is usually taken as constant.

This simplification, however, may turn out to be insufficient in some cases. First of all, the Coulomb interaction, even the on-site U, can be screened. And when the material approaches the itinerant regime and becomes metallic, this screening increases, which would lead to a progressive decrease of U. This can, in particular, accelerate the insulator–metal transition under pressure: not only would the bandwidth W increase with pressure, but U could decrease, leading to a faster decrease of the ratio U/W. Such a "bootstrap" mechanism can even make such insulator–metal transition I order.

There exists, however, another, even stronger mechanism which can lead to a modification of the effective value of U. The actual definition of U for TM ions is the energy cost of the transfer of a d-electron from one TM ion to another, that is the energy of the transition $d^n + d^n \rightarrow d^{n+1} + d^{n-1}$,

$$U_{\text{eff}} = 2E(d^n) - E(d^{n+1}) - E(d^{n-1}) . \tag{10.4}$$

This energy in general depends both on the detailed states of the corresponding ions and also on the spin correlation of the corresponding ions. As these effects are less well known, we will discuss them here in somewhat greater detail.

We will discuss first the possible role of intersite spin correlations. Consider for example a ferromagnetic Mott insulator, for example YTiO$_3$ (ferromagnetic due to a particular orbital ordering). It contains one t_{2g}-electron. When we transfer an electron from site i to a

$$\Delta E = U - J_{\text{H}} \qquad\qquad\qquad \Delta E = U$$

(a)   (b)

Figure 10.22 Excitation energy for intersite electron transfer which depends on the corresponding spin–spin correlations.

neighboring site j, the lowest excitation energy would be $U_{\text{eff}} = U - J_{\text{H}}$, see Fig. 10.22(a): in the excited state we will have at site j two electrons with parallel spins, which, according to our rule (Section 2.2), gives the energy $U - J_{\text{H}}$ (each pair of parallel spins gives the energy $-J_{\text{H}}$).[8]

If however we transfer an electron between two sites with antiparallel spins, the energy U_{eff} would be larger, $U_{\text{eff}} = U$, see Fig. 10.22(b). Thus above the critical temperature, when YTiO$_3$ becomes paramagnetic, there would be pairs of sites both with parallel and antiparallel spins; thus the average U_{eff} would increase from its low-temperature value $U - J_{\text{H}}$.

If in a similar situation we were to start from the antiferromagnetic state, such as in LaTiO$_3$, then similar arguments show that the effective U_{eff} for hopping *between nearest neighbors* would *decrease* above the magnetic ordering (Néel) temperature. Thus a change of spin correlations may change the value of U_{eff} and, as a result, could modify the conditions for the Mott transition. This factor may play some role for example in the metal–insulator transition in V$_2$O$_3$ (Park *et al.*, 2000): as follows for example from the results of Taylor *et al.* (1999), spin correlations in V$_2$O$_3$ not only change from antiferromagnetic to paramagnetic at the insulator–metal transition, but even become ferromagnetic in the insulating phase. This should lead to a significant reduction of U_{eff} during the process of going from an insulator to a metal in V$_2$O$_3$ (Park *et al.*, 2000).

Yet another mechanism which can lead to the same effect – a change in the effective Hubbard interaction U_{eff} – is the change of the spin state, or the multiplet state of the corresponding ions, see Sections 3.3 and 9.5.2. This phenomenon was described by Ovchinnikov (2008) and Lyubutin *et al.* (2009), mainly considering systems containing Fe^{3+}. The idea of this effect is best explained by this example.

In normal conditions, for TM compounds with not extremely strong ligands, Fe^{3+} is usually in the high-spin state, Fig. 10.23(a). In this case the transfer of an electron from one Fe^{3+} ion to another (which is only possible if these ions have opposite spins), Fig. 10.23(a), would cost the energy

$$U_{\text{eff}}^{\text{HS}}(\text{Fe}^{3+}) = U - \Delta_{\text{CF}} + 4J_{\text{H}} \qquad\qquad (10.5)$$

(the "best" transition, costing minimal energy, is the one shown in Fig. 10.23(a): by transferring an e_g-electron from one site to the t_{2g} level on the other site we gain the

---

[8] For simplicity we take in this section, as in several other places of this book, the Hubbard's repulsion U as constant, ignoring its possible dependence on specific orbital occupation, see the detailed discussion in Section 3.3. In effect the coefficients with which the Hund's coupling J_{H} enters different expressions may change somewhat, but the qualitative conclusions presented below remain valid.

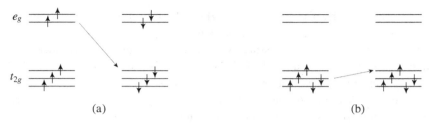

Figure 10.23 Illustration of the dependence of the lowest energy of d–d intersite excitations (i.e., U_{eff}) on the spin state of the corresponding ions, with the example of $Fe^{3+}(d^5)$.

energy of the t_{2g}–e_g splitting Δ_{CF}, but we lose the energy of the Hund's interaction of this electron with four other electrons with parallel spins, which we had in the metallic state).[9]

However if the Fe^{3+} ion should be in the low-spin state of Fig. 10.23(b) (which can happen when, e.g. under pressure, the crystal field splitting Δ_{CF} becomes large enough), the similar "cheapest" process of electron transfer, Fig. 10.23(b), would cost energy

$$U_{\text{eff}}^{\text{LS}}(Fe^{3+}) = U - J_{\text{H}} : \tag{10.6}$$

the electron remains at the t_{2g} level, that is we do not change the crystal field energy, and we gain J_{H} because in the initial state the Hund's energy of down-spin electrons on two sites was $-2J_{\text{H}}$, but in the final state, after electron transfer, this energy is $-3J_{\text{H}}$. In effect, comparing eqs (10.5) and (10.6), we see that if $\Delta_{\text{CF}} < 5J_{\text{H}}$, then $U_{\text{eff}}^{\text{LS}}(Fe^{3+}) < U_{\text{eff}}^{\text{HS}}(Fe^{3+})$. The typical values of Δ_{CF} are for example ~ 2 eV, and $J_{\text{H}} \sim 0.8$–0.9 eV, so this condition is fulfilled.[10]

In contrast, one sees easily that Fe^{3+} on an isolated site would go to the LS state if $\Delta_{\text{CF}} > 3J_{\text{H}}$ (see eq. (3.36)).[11]

Suppose now that we start from the insulating state of some material containing Fe^{3+} ions in the HS state, and apply pressure to this system. Under pressure the value of the crystal field splitting Δ_{CF} increases, and if it becomes bigger than $3J_{\text{H}}$, there should occur a HS $\rightarrow$ LS spin-state transition at each Fe^{3+}. However if the crystal field splitting is still smaller than $5J_{\text{H}}$, then, according to eqs (10.5), (10.6), $U_{\text{eff}}^{\text{LS}} < U_{\text{eff}}^{\text{HS}}$. And there may be situations in which $U_{\text{eff}}^{\text{HS}} > U_{\text{crit.}}$, that is the HS Fe^{3+} is in an insulating state, but the effective Hubbard's U in the low-spin state, U_{eff}^{LS}, is already *smaller* than the critical

[9] Remember that according to our rule, formulated in Chapters 2 and 3, the total Hund's energy of a particular spin configuration is $(-J) \times$ (number of pairs of parallel spins).

[10] We do not discuss here some details of this spin crossover, such as the possibility of having not only both HS states (Fig. 10.22(a)) of both LS states (Fig. 10.22(b)) in the final state, but also the "mixed" state with the resulting d^6 configuration in the LS state (Ovchinnikov, 2008); this could lead to the intermediate situation for $2\Delta_{\text{CF}} < J_{\text{H}} < 3\Delta_{\text{CF}}$, but such excitation would require a change of spins during the process.

[11] The energy of the LS state of Fe^{3+}, Fig. 10.23(b), would be $E^{\text{LS}}(Fe^{3+}) = -4J_{\text{H}}$ (the number of pairs of parallel spins here is 4), and the energy of the high-spin state of Fig. 10.23(b) would be $E^{\text{HS}}(Fe^{3+}) = 2\Delta_{\text{CF}} - 10J_{\text{H}}$ (in going from the LS to the HS state we moved two e_g-electrons from t_{2g} to e_g levels, which costs us the energy $2\Delta_{\text{CF}}$, but the Hund's energy gain in the HS state is bigger, since there are 10 pairs of parallel spins in this state). Comparing these energies we get the condition $\Delta_{\text{CF}} = 3J_{\text{H}}$ for the spin-state crossover for isolated Fe^{3+} ions.

value $U_{crit.}$. Thus, the sharp drop in U_{eff} at the HS $\to$ LS transition can lead to this transition simultaneously becoming a metal–insulator transition.

Such a spin crossover under pressure, with the reduction in U_{eff}, was indeed observed in a number of Fe^{3+} compounds: in $FeBO_3$, in $GdFeO_3$, and in $BiFeO_3$, see Lyubutin *et al.* (2009) and references therein. In $BiFeO_3$ it is simultaneously an insulator–metal transition (in the other two systems the energy gap drops at this transition, but still remains finite). Apparently such spin-state transitions, with the corresponding reduction in U_{eff}, help the insulator–metal transition. Thus this is an example of an *interaction-controlled insulator–metal transition.*

One should note that the reduction in U_{eff} in the LS state, demonstrated for Fe^{3+}, is not a general rule. Depending on the particular electronic configuration, U_{eff} may either decrease or increase at the HS–LS transition, or may remain unchanged. Thus, for ions with configuration d^6, such as Co^{3+} and Fe^{2+} (typical ions in which spin-state transitions are observed at not too extreme conditions), the effect is the opposite, $U_{eff}^{LS}(Co^{3+}) > U_{eff}^{HS}(Co^{3+})$. Indeed, the same considerations as those presented above show that for the HS d^6 ions

$$U_{eff\uparrow\uparrow}^{HS}(d^6) = U - J_H \tag{10.7}$$

for the HS initial states when two ions have parallel spins, Fig. 10.24(a), or

$$U_{eff\uparrow\downarrow}^{HS}(d^6) = U - \Delta_{CF} + 3J_H \tag{10.8}$$

for ions with antiparallel spins, Fig. 10.24(b). As the HS state of such d^6 ions is lower in energy than the LS state for $\Delta_{CF} < 2J_H$, see Section 3.3, eq. (3.38), the value of U_{eff} for parallel spins (10.7) is smaller than that for antiparallel spins (10.8).

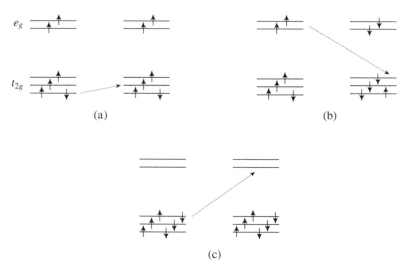

Figure 10.24 Intersite electron transitions for ions with the configuration d^6 (Co^{3+}, Fe^{2+}) for different spin states and different spin orientations, see text.

If however Δ_{CF} becomes larger than $2J_H$, such ions go over to the LS state. In this case, as one can easily calculate by this method from the scheme presented in Fig. 10.24(c), we have

$$U_{eff}^{LS}(d^6) = U + \Delta_{CF} - J_H . \qquad (10.9)$$

Comparing eqs (10.9) and (10.7), we see that if $\Delta_{CF} > 2J_H$ (which is the condition for the existence of the LS state for each d^6 ion), then $U_{eff}^{LS} > U_{eff}^{HS}$.[12] Thus, when, for example under pressure, Δ_{CF} increases and the material transforms from the HS to the LS state, the effective Hubbard repulsion would *increase*, and the system would be "more insulating." This is consistent with the tendency observed in $LaCoO_3$, for which the low-temperature ground state with LS Co^{3+} is insulating, but the high-temperature state ($T \gtrsim 400\,K$), with many Co^{3+} ions promoted to a magnetic state (most probably the HS state), is conducting: according to our considerations, in the high-spin state U_{eff} is reduced, thus facilitating metallic conductivity. We have also seen such an unusual transition from the metallic to the insulating state under pressure in doped $La_{1-x}Sr_xCoO_3$ (Lengsdorf *et al.*, 2004): whereas almost always materials become more conducting under pressure, in this system pressure causes a transition from the HS (or IS) state to the LS state of Co, accompanied by a transition from the conducting to an insulating state.

Once again, because of our simplifications (in particular in taking the same U independent of orbital occupation) some numerical estimates presented above may not be exactly accurate, see also Section 3.3 and footnote 8 on p. 410. But the qualitative conclusions reached remain correct.

In any case, we see that in considering insulator–metal transitions for many-electron TM ions one has to take into account the possible multiplet structure of the corresponding ions, as well as their magnetic correlations. The account of these factors, together with the possible role of screening, can lead to a change in effective Coulomb repulsion during an insulator–metal transition, and it can modify the features of such transitions, leading, in particular, to *interaction-controlled* insulator–metal transitions, in addition to the usually discussed bandwidth- and band-filling-controlled ones (Imada *et al.*, 1998). These effects have to be taken into account in models and in *ab initio* calculations for systems with insulator–metal transitions (the standard approaches almost always use the description with Hubbard's U the same in both the insulating and metallic phases), and in interpreting insulator–metal transitions in real materials.

10.4.2 Orbital-selective Mott transitions

In theoretical treatments of Mott transitions one usually starts from the simple nondegenerate Hubbard model (1.6). However, as we have argued in many places in this book, orbital degrees of freedom play a very important role in determining many properties of TM compounds. And we have already seen in several examples of metal–insulator transitions

---

[12] Again, we do not discuss here some fine details, connected with the possibility of a transition from the initial LS states of d^6 ions to the HS state of the "excited" d^5 and d^7 states, which would also require a change in the total spin of such an excitation.

above, such as those in V_2O_3 and VO_2, that these effects may be very important for Mott transitions as well.

The possible role of orbital effects in Mott transitions has been studied theoretically in many papers, starting from the paper Anisimov *et al.* (2002). The first such effect was investigated by Koch *et al.* (1999). The standard condition for Mott transition is that the Hubbard repulsion U should be larger than the corresponding bandwidth $W = 2zt$ (where t is the electron hopping and z is the number of nearest neighbors),

$$U > U_{c0} = aW = a \cdot 2zt, \qquad (10.10)$$

where a is a numerical constant ~ 1, which depends on the details of the lattice structure, etc. Suppose now that there are several (e.g., N) equivalent orbitals on a site, to which an electron from another site can hop. The arguments presented in Koch *et al.* (1999) show that in this case the effective hopping and the corresponding bandwidth are renormalized: for half-filled bands (when the number of electrons per site is equal to the orbital degeneracy N) the effective hopping is modified $t \rightarrow t\sqrt{N}$ (and there is a similar, but somewhat weaker enhancement for other electron concentrations). That is,

$$W = 2zt \longrightarrow \sqrt{N}\, W = \sqrt{N}\, 2zt . \qquad (10.11)$$

Effectively, the presence of many orbitals gives rise to several channels for electron hopping from site to site, which increases their kinetic energy, that is the effective bandwidth. (The situation here resembles the increase in effective hybridization for Zhang–Rice singlets, Section 4.3, although the detailed physics is not the same.) But then one would need larger values of the Coulomb (Hubbard) interaction U to overcome this kinetic energy and make the system a Mott insulator: from (10.10), (10.11) we see that

$$U_c = \sqrt{N}\, 2zt = \sqrt{N}\, U_{c0} . \qquad (10.12)$$

Thus, in a system with many orbitals a metallic state is more robust than in the single-band case, and it is more difficult to make such a system insulating.

A more detailed treatment of this situation using dynamical mean-field theory shows (Florens *et al.*, 2002) that with decreasing U the (Mott) insulating state indeed becomes unstable at $U_{c1} \sim \sqrt{N}\, W$, but going from the metallic state, such a state exists with increasing U up to $U_{c2} \sim NW$. That is, in this approximation at low temperatures the Mott transition turns out to be an I order transition, and the hysteresis limits U_{c1} and U_{c2} depend differently on the degeneracy.

The assumptions used in obtaining the result (10.12) are not very realistic for actual TM compounds, although the qualitative tendency caught by this treatment is correct, and for some systems, for example fullerenes R_xC_{60} (R = K, Rb, etc.) it may work directly. In typical TM systems the strongly directional character of orbitals makes corresponding hoppings orbital-dependent. Thus, in perovskites the e_g-electrons hop to e_g orbitals of neighboring ions, and the t_{2g}-electrons to t_{2g} orbitals. In effect there will be separate bands of particular orbital character formed, with different bandwidths, determined by the corresponding hopping matrix elements. The effective e_g–e_g hopping in these systems (with $\sim 180°$ M–O–M bonds) is larger than that for t_{2g} electrons, $t_{e_g-e_g} \simeq 2t_{t_{2g}-t_{2g}}$

(Harrison, 1989), that is the e_g bands are broader than the t_{2g} ones. The question arises whether we would in this case have an *orbital-selective Mott transition*. For example, according to the standard arguments, eq. (10.10), one can expect that if $U < W_{e_g}$, but $U > W_{t_{2g}}$, we would have a metallic e_g band with itinerant e_g-electrons, whereas the t_{2g}-electrons would be on the insulating side of the Mott transition. This was the question addressed in many theoretical papers, see for example Anisimov *et al.* (2002), Koga *et al.* (2004), Liebsch (2004), de Medici *et al.* (2009). And although some contradictory claims have been made, it seems that our qualitative expectations described above are indeed largely supported theoretically – notably that localized electrons, which have already undergone Mott transition, may coexist with itinerant electrons in broader bands. Actually we have already used this picture, maybe without explicitly stating this problem, in many parts of this book – for example in treating the double-exchange model of colossal magnetoresistance manganites, Sections 5.2 and 9.2.1, when we considered t_{2g}-electrons as localized, forming localized spins, and treated electrons in the partially filled e_g band as itinerant.

Actually, the notion that different electrons can display different degrees of correlations has been used already long before the very notion of Mott insulators was introduced. Thus for example in treating the transition metals themselves, such as iron or nickel, Mott (see Mott and Jones, 1958) used the picture in which the d-bands with more t_{2g}-like character are more localized than the bands of predominantly e_g-like character (although for real metals these notions are less well-defined). Similarly we use this picture when we are dealing for example with rare earth metals or compounds, see Chapter 11 below, where we always start from the picture of localized $4f$-electrons and itinerant electrons of conduction bands (s-, p-, d-electrons). As we will see in Chapter 11, in some cases the hybridization of these different types of electrons can lead to quite nontrivial states, such as the heavy fermion state. Apparently the same effects were the source of some conflicting statements as to the existence of orbital-selective Mott transitions in d-electron systems. But at least qualitatively the picture of two possible different states of electrons coexisting in the same solid, localized electrons in narrow bands and conduction electrons in broader bands, is a good starting point for discussing the properties of many real materials.

10.5 Insulator–metal transitions in Mott–Hubbard and charge-transfer insulators

As discussed in Chapter 4, one can divide all strongly correlated insulating transition metal compounds into two groups: Mott–Hubbard insulators and charge-transfer insulators, see for example the Zaanen–Sawatzky–Allen phase diagram in Fig. 4.7. In Chapter 4 we considered some differences in the properties of insulating systems in these two regimes, such as the character of the lowest current-carrying excitations or the detailed form of the exchange interaction. In that chapter we mentioned briefly that when we go from the insulating regime with localized d-electrons to the regime with itinerant electrons, we can have two different situations, see Fig. 10.25. The system can evolve either starting from the Mott–Hubbard regime and following the route A $\rightarrow$ A$'$, or starting from the charge-transfer regime along the route B $\rightarrow$ B$'$.

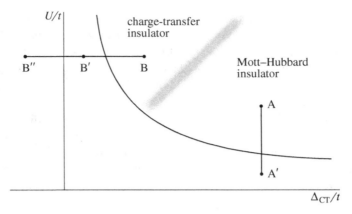

Figure 10.25 Possible transitions from localized to itinerant electrons, starting from the Mott–Hubbard or charge-transfer insulators.

One can easily understand that the features of transitions in these two cases, and, moreover, the very nature of the resulting states at the points A′ and B′ may be very different. The first case (the trajectory A → A′) describes a conventional Mott transition. Using the schematic form of the density of states such as that shown in Fig. 4.12, we can illustrate the corresponding changes in Fig. 10.26, where we show the filled $2p$-band of a ligand (oxygen) and the lower and upper Hubbard bands in the Mott–Hubbard insulator phase (point A), and in the metallic phase (point A′). The lowest charged excitations in this case correspond to d–d transitions $d^n d^n \rightarrow d^{n-1} d^{n+1}$. We see that the insulator–metal transition in this case corresponds to a closure of the Mott–Hubbard gap, with the formation of a normal partially filled d-band in the metallic state. The filled oxygen p-band in this case remains deep below the Fermi energy and does not play a significant role. Note that the electron correlations, characterized by the Hubbard interaction U, or U/t, decreases strongly in going to the metallic regime, so that formally we go over to the itinerant electrons without significant correlations.

On the contrary, when we start from the charge-transfer insulator phase, that is the point B in Fig. 10.25, the possible development is that shown in Fig. 10.27. We see that in going from point B to point B′ in Fig. 10.25 we decrease the charge-transfer gap Δ_{CT}, but the Hubbard's U, and with it the electron correlations, remain strong. The resulting state, illustrated in Fig. 10.27(b), is rather unusual: it corresponds to the closing of the charge-transfer gap – the energy of the transition $d^n p^6 \rightarrow d^{n+1} p^5$. In effect the broad p-band (we consider it as uncorrelated) overlaps with the upper Hubbard band, which corresponds to still strongly correlated d-electrons. The resulting situation can be shown schematically in Fig. 10.28. It strongly resembles that of mixed-valence rare earth compounds, which will be described in more detail in Chapter 11 (see Figs. 11.2, 11.7). That is, in this regime we have all the problems and all the possibilities encountered in rare earth compounds in the analogous regime. The core of the problem lies in the fact that here electronic states

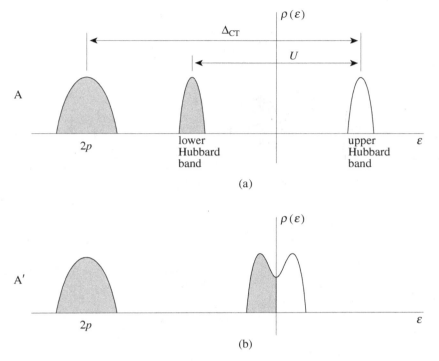

Figure 10.26 Schematic form of the density of states for (a) the Mott–Hubbard insulator, point A in Fig. 10.25 and (b) the metallic phase, point A' in Fig. 10.25.

with very different character lie close in energy: the broad itinerant p-band, which can accommodate two electrons per state, and strongly correlated states of the (upper) Hubbard band, having place for only one electron per state. And these very different states have to hybridize, they have different interactions, etc. The same problem which we meet in $4f$ and $5f$ systems, for example those with mixed valence, appears here also – "how to hybridize correlated bands" (or, more accurately, bands with different degrees of correlation – here the uncorrelated p-bands and still strongly correlated $(U/t \gg 1)$ d-bands). Depending on the situation, different types of resulting state are feasible here. One can end up, as shown in Fig. 10.27(b), in a metallic state, though with strongly renormalized parameters and with strong electron correlations still present. One can then expect, for example, that the resulting metallic state would still have Curie, or Curie–Weiss, and not Pauli magnetic susceptibility. Such states ("Curie–Weiss metals") are often observed in metallic TM compounds, for example in $Na_x CoO_2$ for $x > 0.5$, or in $LaNiO_3$, see Fig. 10.16. The situation described above is the most plausible explanation of the appearance of metals with Curie–Weiss susceptibility. In some cases this state can also develop some kind of magnetic ordering, remaining metallic; such may be the situation with the formation of a helicoidal magnetic structure in $SrFeO_3$, see for example Mostovoy (2006).

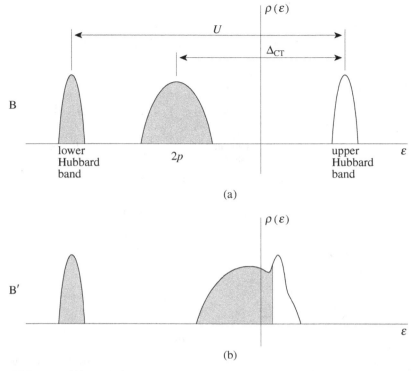

Figure 10.27 A possible evolution of the density of states for an insulator–metal transition occurring with decrease of the charge-transfer energy Δ_{CT}, the trajectory B $\rightarrow$ B$'$ $\rightarrow$ B$''$ in Fig. 10.25. (a) The initial insulating state (charge-transfer insulator, point B in Fig. 10.25). (b) The final metallic state, with still strong correlations for d-electrons (points B$'$, B$''$ in Fig. 10.25).

But this is not the only possibility. Similar to rare earth compounds, in certain cases in this regime the resulting state can become insulating – similar to the so-called Kondo insulators in $4f$ compounds (Aeppli and Fisk, 1992). In the simplest treatment such a gap can appear just due to the p–d hybridization, Fig. 10.29. However this simple explanation may not work, or at least may not be sufficient both here and in $4f$ Kondo insulators. A more reasonable picture of such a state may be based on the fact that after the occupied $2p$-bands start to overlap with the empty upper Hubbard d-band, some former p-electrons would be transferred to the d-states, leaving p-holes behind. Then just "by construction" the number of resulting d-electrons would be equal to the number of p-holes thus formed. In this case they may form bound states of excitonic type due to the electron–hole attraction, which can make the resulting state insulating: every p-hole can find a d-mate. The resulting state would then resemble strongly that of excitonic insulators, see for example the detailed discussion in Khomskii (2010).

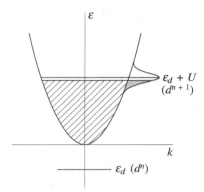

Figure 10.28 Energy spectrum of the metallic state reached after the insulator–metal transition caused by the decrease in charge-transfer energy Δ_{CT} (points B$'$, B$''$ in Fig. 10.25).

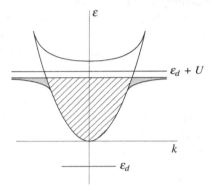

Figure 10.29 Possible energy spectrum of an insulating state which can appear for small or negative charge-transfer energy (points B$'$, B$''$ in Fig. 10.25), cf. the situation with Kondo insulators for $4f$ compounds (Chapter 11).

One can also notice the relation of such states to that of the Zhang–Rice singlets discussed in Section 4.4 (bound states of a d-electron and a p-hole). In certain cases such states with many Zhang–Rice singlets can be insulating. Thus, for example, this is the way to describe the insulating and nonmagnetic state of NCuF$_2$ and KCuF$_2$ (Hesterman and Hoppe, 1969; Mizokawa *et al.*, 1991). The main structural block of these systems is the chain of edge-sharing CuO$_6$ octahedra, Fig. 10.30. Formally in these compounds the valence of Cu is 3+. But, as discussed in Section 4.3, Cu^{3+} has a negative charge-transfer gap, see Fig. 4.8. That is, we are here at the point B$''$ in Fig. 10.25. In this case we should take, instead of the state Cu$^{3+}(d^8)$, rather the state Cu$^{2+}(d^9)\underline{L}$, where $\underline{L}$ denotes the ligand (here oxygen) hole. Such p-holes would form Zhang–Rice singlet bound states with the "extra" d-electrons of Cu^{2+} (d–p excitons), and in this case we would have such Zhang–Rice singlets at every unit cell (shown by dashed circles in Fig. 10.30). Such a

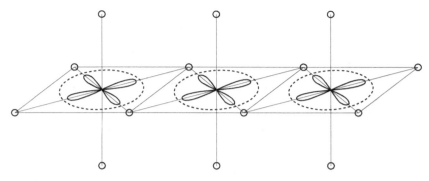

Figure 10.30 The main structural blocks of NaCuO$_2$ and KCuO$_2$, with Zhang–Rice singlets at every unit cell.

state would in effect be insulating and diamagnetic. This is also one of the possible outcomes of the transition from the charge-transfer insulator (point B in Fig. 10.25) to the state with small or negative charge-transfer gap (points B$'$ and B$''$). In this sense one can say that the title of this section is not very accurate: one of the possible outcomes of the transition from localized to itinerant electron states could in fact be not a metal, but still an insulator.

Note that the nature of the resulting state is different from that of the original charge-transfer insulator, which also leads for example to quite different magnetic properties of this state. Thus, whereas the initial charge-transfer insulator in systems such as KCuO$_2$, but with positive charge-transfer gap (point B in Fig. 10.25), with localized d-electrons and the corresponding localized magnetic moments, should have some kind of magnetic ordering, the final state such as that at points B$'$ or B$''$ would be nonmagnetic. (One can say that there occurs here a spin-state transition from the high-spin state typical for the large positive charge-transfer gap, point B in Fig. 10.25, to the low-spin state at points B$'$, B$''$ in that figure – similar e.g. to the transition from the high-spin state of Ni^{2+}(d^8, $S = 1$) ($\Delta_{CT} > 0$, see Fig. 4.8) to the low-spin state of Cu^{3+}(d^8, $S = 0$) = Cu^{2+}(d^9)$\underline{L}$ ($\Delta_{CT} < 0$).)

Similar to Kondo insulators in rare earth compounds, the formation of such an insulating state requires certain specific conditions; that is why we do not meet such states very often. More often we end up, by decreasing Δ_{CT}, in some kind of metallic state – but, as stressed above, of a rather unusual kind, with strong electron correlations still playing an important role.

10.6 Formation of molecular clusters and "partial" Mott transitions

The electron–electron repulsion, for example Hubbard's U, is essentially an atomic property and does not depend on interatomic distance, whereas the electron hopping strongly depends on this distance. This leads to the concept, largely advocated for example by Goodenough (1963), that there should exist a critical interatomic distance R_c (different for different transition metal ions) dividing the regions of localized and itinerant behavior of electrons. When the distance between transition metal ions changes, for example under pressure, there should occur in solids an insulator–metal (Mott) transition (in molecules this would correspond to a crossover from the Heitler–London description to the molecular orbital description).

Usually Mott transition in TM compounds such as TM oxides is supposed to occur homogeneously in the whole sample. But, as gradually became clear, this is not the only possibility. There may appear novel, inhomogeneous states close to a Mott transition, so that we have, in a sense, a "partial" Mott transition: in some parts of the system, in some particular clusters, the interatomic distances may already be smaller than R_c, so that these clusters may be better described using molecular orbitals, whereas the distance *between* such clusters remains large, so there will be no net metallic conductivity. The formation of such "metallic" clusters has much in common with the valence bond solids.

This phenomenon, that is such a "fractional" Mott transition, is most often observed in low-dimensional and in frustrated systems. Indeed, in regular three-dimensional lattices such as perovskites we usually have homogeneous ordered states, such as 3d antiferromagnetic ordering. Molecular-type clusters such as valence bond states are much more plausible in low-dimensional and frustrated systems. In this section we describe this situation using several specific examples of systems with triangular and pyrochlore (spinel) lattices, and also some quasi-one-dimensional systems; see also Khomskii (2011).

10.6.1 Formation of dimers in spinels

Metal–insulator transitions with the formation of unusual structures are observed in many spinels. Some examples are $MgTi_2O_4$ (Schmidt *et al.*, 2004), $CuIr_2S_4$ (Radaelli *et al.*, 2002), or AlV_2O_4 (Matsuno *et al.*, 2001). Structural studies have shown that in all these cases rather unusual structural modifications take place, which can often be described as the formation of molecular clusters. In $MgTi_2O_4$ there appears a "chiral" structure with alternation of long and short Ti–Ti bonds (Schmidt *et al.*, 2004), in $CuIr_2S_4$ Ir octamers are formed (Radaelli *et al.*, 2002), but both these phenomena can in fact be explained by the formation of Ti or Ir *singlet dimers*, which in a frustrated lattice of *B* sites of a spinel finally give rise to these chiral or octamer structures (Khomskii and Mizokawa, 2005). In AlV_2O_4 the molecular clusters formed consist of seven vanadiums – really large "heptamer molecules." As discussed in Section 6.2, orbital ordering usually plays a significant role in these phenomena.

In all these cases the formation of these clusters is caused largely by the specific features of the *B*-site spinel lattice, which may be represented as a collection of 1d chains running in xy (or $x\bar{y}$), xz and yz-directions, see Fig. 6.15. The t_{2g} orbitals of Ti or V ions on this lattice have a strong direct d–d overlap, such that for example the electrons from the xy orbitals can hop to the same orbitals in neighboring ions in the xy-direction, or similarly yz-electrons can hop in the yz-direction, etc., see Fig. 6.15. As a result the electronic structure in these, basically cubic crystals has essentially one-dimensional character.

And in the insulating phase one forms a Peierls state with singlet dimers in these 1d chains (Khomskii and Mizokawa, 2005), which for a spinel lattice give the chiral structure in $MgTi_2O_4$, the heptamers in AlV_2O_4, and octamers (consisting actually of dimers) in $CuIr_2S_4$. Note that the interatomic distances in these clusters (dimers) are rather short: they are shorter than the critical distance R_c for the localized–itinerant crossover (Goodenough, 1963), or maybe even shorter than the metal–metal distance in the corresponding elemental metals, so that one can consider the electronic state in these clusters as forming molecular orbitals, that is in these clusters we are already on the "metallic" side of the Mott transition. But the distance *between* these clusters is larger than R_c and as a result the total compound is insulating.

The "metallic" clusters in the examples discussed above are singlets. But this is not the only possibility. Magnetic (e.g., triplet) clusters can also be formed in some cases. This seems to happen in ZnV_2O_4 (Pardo *et al.*, 2008). Vanadium spinels such as MgV_2O_4, ZnV_2O_4, CdV_2O_4 show cubic–tetragonal structural transitions (Furubayashi *et al.*, 1994) with the formation of a very specific magnetic structure at lower temperatures, with $\uparrow\uparrow\downarrow\downarrow$ pattern in xz- and yz-chains (and with $\uparrow\downarrow\uparrow\downarrow$ ordering in the xy-direction) (Reehuis *et al.*, 2003), see Fig. 10.31. Usually the properties of these systems are explained by some type of orbital ordering: the xy orbitals are always occupied due to tetragonal distortion with $c/a < 1$, and the remaining second t_{2g}-electron of the $V^{3+}(d^2)$ ions is supposed to form an additional orbital ordering of some type (Motome and Tsunetsugu, 2004; Tchernyshyov, 2004; Khomskii and Mizokawa, 2005). But *ab initio* calculations (Pardo *et al.*, 2008) have shown that there should exist in ZnV_2O_4 a dimerization in xz- and yz-chains, see Fig. 10.31, with, surprisingly, ferromagnetic bonds becoming shorter (thick bonds in Fig. 10.31)! One can explain this tendency by noticing that, in contrast to most

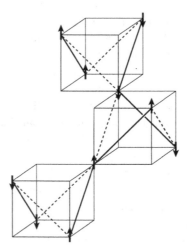

Figure 10.31 Formation of short (bold) and long (dashed) V–V bonds in ZnV_2O_4, following Pardo *et al.* (2008); the same pattern also exists in CdV_2O_4.

examples of valence bond solids with singlet bonds, here we are dealing not with ions with one electron with $S = \frac{1}{2}$, but with d^2 ions with $S = 1$. One can argue that the extra delocalization of one of two electrons in a short bond enhances the ferromagnetic (actually double-exchange) interaction between these d^2, $S = 1$ ions. The *ab initio* calculations of another material of this family, CdV_2O_4, done by a different method, have confirmed the formation of these dimers, and demonstrated, in addition, that their formation can lead to the appearance of ferroelectricity, which was indeed observed in CdV_2O_4 (Giovanetti *et al.*, 2011).

Interestingly enough, the short V–V distances obtained theoretically for ZnV_2O_4 are only 2.92 Å – again shorter than the critical V–V distance of 2.94 Å (Goodenough, 1963). Thus again these V dimers may be considered as "metallic," though the material itself remains insulating, albeit with a rather small energy gap ~ 0.2 eV (Pardo *et al.*, 2008).

10.6.2 *"Metallic" clusters in layered materials*

Some more examples of "metallic" clusters in an overall insulating matrix are found in layered materials, notably with triangular lattices, such as $LiVO_2$ (Pen *et al.*, 1997), $LiVS_2$ (Katayama *et al.*, 2009), TiI_2 (Meyer *et al.*, 2009); these were already mentioned in a different context in Section 6.2.

The triangular lattice is usually considered as frustrated, meaning that it is not bipartite (cannot be subdivided into two sublattices). But in $LiVO_2$, $LiVS_2$, TiI_2 we are dealing with ions Ti^{2+}, V^{3+} with two t_{2g} d-electrons, which, besides spin, also have triple orbital degeneracy. And if one cannot subdivide a triangular lattice into two sublattices, one can naturally subdivide it into three! That is indeed what happens in these materials (Pen *et al.*, 1997), see Fig. 10.32(a) (already presented earlier, Fig. 6.17), in which the occupied orbitals for $V^{3+}(d^2)$ or $Ti^{2+}(d^2)$ are shown.

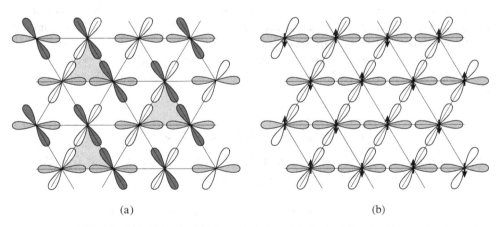

(a)   (b)

Figure 10.32 (a) Orbital ordering in $LiVO_2$ with three orbital sublattices, leading to the formation of spin singlet states on tightly bound V triangles (shaded), following Pen *et al.* (1997). (b) Orbital and magnetic ordering in $NaVO_2$, following McQueen *et al.* (2008a).

After forming this orbital superstructure, we have strong antiferromagnetic exchange in the shaded triangles of Fig. 10.32(a) and in effect these bonds become shorter than the others, and there are singlet states formed *on such trimers* (three $S = 1$ ions with strong antiferromagnetic coupling have a singlet ground state with $S_{tot} = 0$). Note that in this case the short distance in a cluster – here a trimer – is 2.84 Å, which is even shorter than the V–V distance in metallic V, 2.62 Å, so these clusters can indeed be considered as "metallic." (The distances between trimers remain quite long, 3.02 Å.)

If however we increase the total lattice parameter (so that we are further away from the localized–itinerant crossover), which we can do for example by going from $LiVO_2$ to $NaVO_2$, the situation changes drastically. As shown by McQueen *et al.* (2008a), the orbital ordering in $NaVO_2$ is quite different from that in $LiVO_2$: only two types of orbitals are occupied, see Fig. 10.32(b). As a result, after such orbital ordering the system becomes topologically a square lattice, and as such it can develop the ordinary Néel magnetic ordering, which is actually observed. Note also that, in contrast to $LiVO_2$, in $NaVO_2$ the bonds with strong orbital overlap become *not shorter, but longer*!

In effect both short and long bonds here (2.977 Å and 3.015 Å) are above the Goodenough critical value, that is this material remains far on the insulating side of the Mott transition, in contrast to $LiVO_2$ and $LiVS_2$ (in the latter this trimerization is simultaneously a real metal–insulator transition; see Katayama *et al.*, 2009).

One can explain this behavior of $NaVO_2$ as a competition between two effects: stronger covalency tends to make such bonds shorter, but the bond–charge repulsion (Coulomb repulsion of electron clouds on orbitals directed toward each other) would tend to make such bonds longer. Apparently when the system has a relatively small lattice parameter, or small average TM–TM distance, that is when the system is close to the localized–itinerant crossover, the hybridization is rather strong, and the first effect, the tendency to increase it still further, prevails, so that such bonds become even shorter; this is what happens in $LiVO_2$ and $LiVS_2$. If, however, the system has localized d-electrons and is far from the Mott transition, the d–d overlap is exponentially small, and the second effect, the bond–charge repulsion, may become more important, as a result of which it is favorable to make such bonds even longer, to reduce this repulsion further. This is apparently what happens in $NaVO_2$.

10.6.3 Spin-Peierls–Peierls transition in TiOCl

A very clear example of the Peierls phenomenon, which we have met already several times in this book, in particular in connection with the metal–insulator transition in VO_2, is observed in TiOCl.

This quasi-one-dimensional material is insulating with localized d-electrons at ambient pressure, and it has a spin-Peierls transition at ~ 60 K (with an incommensurate phase at 60 K $< T \lesssim 90$ K). Under pressure its energy gap decreases strongly, and it approaches an insulator–metal transition at ~ 10 GPa (Kuntscher *et al.*, 2006, 2007).

However direct d.c. resistivity measurements have shown (Forthaus *et al.*, 2008) that TiOCl actually remains insulating even for $P > 10\,\text{GPa}$, at least up to $30\,\text{GPa}$, albeit with a smaller gap. Surprisingly, *ab initio* calculations with lattice optimization have shown (Blanco-Canosa *et al.*, 2009) that in the high-pressure phase, in which indeed the energy gap strongly decreases, the Ti–Ti dimerization itself does not decrease but *increases* instead.

According to these calculations, the short Ti–Ti distance becomes $2.95\,\text{Å}$ – already close to the Ti–Ti distance $2.90\,\text{Å}$ in Ti metal, whereas the long Ti–Ti distance is much bigger, $3.69\,\text{Å}$. Thus here also the short Ti–Ti dimers are closer to a molecular orbital description than to a Heitler–London one.

A qualitative explanation of this behavior can be obtained, again using the idea of the proximity of this material to the localized–itinerant crossover. At $P = 0$ the d-electrons are localized, and we have a spin-Peierls transition, with T_c and with the degree of dimerization $\delta d/d$ proportional to

$$T_c \sim \frac{\delta d}{d} \sim J\, e^{-1/\lambda}, \tag{10.13}$$

where J is the exchange interaction ($J \sim t^2/U$) and λ is the spin–phonon coupling constant. Note that the energy gap in this regime is still of the Mott–Hubbard type, that is large, $E_g \sim U$. Under pressure TiOCl approaches the insulator–metal transition, as observed in Kuntscher *et al.* (2006, 2007). But after crossing to the itinerant regime, the system still "knows" that it is one-dimensional, and as such it has a Peierls instability, with T_c ($\sim$ degree of dimerization $\sim$ energy gap)

$$T_c \sim E_g \sim \frac{\delta d}{d} \sim t\, e^{-1/\lambda'}, \tag{10.14}$$

with λ' being the electron–electron coupling constant in this regime. That is, the very nature of the insulating state in this regime is *due to Peierls dimerization*, and is not of the Mott insulator type. But the dimerization in this regime is proportional to the bandwidth, or to the hopping matrix element t, and not to $J \sim t^2/U$ as in (10.13). Thus apparently we have here, simultaneously with the localized–itinerant crossover, also a spin-Peierls–Peierls transition.

Summarizing this section, we stress once again that the systems close to Mott transition, or to a localized–itinerant crossover, can show a variety of specific properties, different from those of states deep in the Mott insulator regime or those of conventional metals. Especially interesting is the situation of a "partial Mott transition," in which first certain finite clusters go over to the regime with uncorrelated ("itinerant") electrons and only later does there occur a real Mott transition in the whole sample. As we saw with several examples, the most favorable conditions for these phenomena are met in low-dimensional and in frustrated systems; they are less probable in simple lattices such as perovskites. Thus this crossover region between localized and itinerant electronic states presents a rich playground for interesting physics, with the eventual appearance of some novel states.

10.7 Mott transition: a normal phase transition?

One interesting feature of the DMFT treatment of Mott transitions is that this transition at low temperatures turns out to be a I order transition, see Section 10.3. This, in particular, is connected with the quite open questions of the general description of Mott transitions. In the standard treatment of phase transitions in condensed matter, such as magnetic or superconducting transitions, the main notions are the notions of symmetry and of an order parameter. The order parameters are for example the average magnetization in a ferromagnet or polarization in a ferroelectric material. The order parameter is the *ground state average* of a certain operator, $\eta = \langle 0|\hat{A}|0\rangle$, for example of the local magnetization, and such an order parameter is zero in the disordered and nonzero in an ordered phase. Correspondingly, we usually have a change of symmetry in the transition: the disordered phase usually has higher symmetry, and the symmetry is broken in the ordered phase. Thus for example in a ferromagnet the time-reversal and rotational symmetries are broken, in a ferroelectric the inversion symmetry is broken, etc. (These are at least the necessary features of the II order phase transitions: the I order phase transitions such as liquid–gas transitions can occur without a change in symmetry, and in some of them one cannot define the conventional order parameter.)

In this sense it is not clear how to describe Mott transitions: it is not known which symmetries are broken in such transitions, if any, and how to introduce the appropriate order parameter, differentiating between the insulating and the metallic state. Indeed, a Mott insulator differs from a metal not so much by the properties of the *ground state* itself, but rather by the characteristics of the *lowest current-carrying excited states*. Whereas in a metal such excitations have no gaps, in insulators there is a finite gap for such excitations. That is maybe why one cannot differentiate between a Mott insulator and a metal by the *ground state average* $\langle 0|\hat{A}|0\rangle$ of any operator, so that in this sense Mott transitions could indeed be different from the standard II order phase transitions in solids. If true, we can then conclude that a Mott transition could be either a I order transition or just a gradual crossover. There are different suggestions in the literature of possible order parameters for Mott transitions, but none of them are satisfactory.

It is not even clear which state, metallic or insulating, should be treated as ordered and which as disordered. One could give arguments that it is the metallic state which should be considered as the ordered state, and the Mott insulator without magnetic ordering as the disordered one. Indeed, as we have already mentioned above, a metal at $T = 0$ has a filled Fermi surface – a unique quantum state with zero entropy, whereas the state with localized electrons with disordered spins has a finite spin entropy $k_B N \ln 2$. Indeed, the "pure" Mott transition in doped V_2O_3 (the high-temperature transition line in the phase diagram of Fig. 10.7) has a positive slope, which, as explained there, implies that the entropy of the Mott insulating state is higher than that of the metallic state, so the metallic state is more ordered.

In reality the transition to a Mott insulator is almost always accompanied by some other, more conventional ordering, for example antiferromagnetic. In this case one could associate the transition just with this ordering. But in this case we would in fact describe this transition in the effective one-electron picture, considering energy bands for the motion of one electron in a self-consistent exchange field created by magnetic sublattices. This is not the Mott transition *per se* – the transition from itinerant to localized electrons, not necessarily accompanied by any usual long-range, say magnetic, ordering. For example, many TM oxides are good insulators also above the Néel temperature. Similarly, antiferromagnetic ordering may be suppressed for example in frustrated systems, but with the system remaining insulating. Once again, it is not clear how to describe such a "pure" Mott transition in terms of the conventional notions of symmetry change and order parameter. And it could well be that this standard description is not really applicable to Mott transitions. As mentioned above, in this case the transition should then either be I order, or just a gradual crossover.

It is still not excluded that one could introduce something like an order parameter for Mott transitions, but it could be a highly nontrivial one – for example a nonlocal quantity, operating with such notions as entanglement entropy, etc. (Kitaev and Preskill, 2006; Levin and Wen, 2006). Such attempts, however, are at a very preliminary stage.

S.10 Summary of Chapter 10

Probably one of the most interesting phenomena, but also one of the most difficult to describe theoretically, in the physics of systems with strongly correlated electrons is that of metal–insulator transitions observed in some of these systems with a change in temperature, pressure, composition, etc. From the point of view of the general scheme presented in the Introduction, we now relax the condition of having strong correlations $U \gg t$ and consider the limit $U \lesssim t$. The transition from the metallic state for $U < t$ to the Mott insulator state for $U > t$ is called a *Mott transition*.

In principle there can exist different types of metal–insulator transitions in solids. One can subdivide them into three main categories. The first is transitions which can be explained in the framework of the standard band theory, considering the motion of independent noninteracting electrons in a periodic potential. Depending on the specific form of the energy spectrum and on the electron concentration, we can have here either insulators or semiconductors, in which the filled valence bands are separated from the empty conduction band(s) by an energy gap, or metals with partially filled conduction band or bands. Insulators of this type are *band insulators*. If, for example, in such a system at a structural transition the form of the electronic spectrum changes and the energy gap disappears, this will give a transition from an insulator to a metal, or vice versa if such distortion leads to an opening of the energy gap at the Fermi surface of the original metallic state. Examples of such transitions are the transition of white tin (a good metal) to gray tin (a zero-gap semiconductor), or Peierls dimerization in quasi-one-dimensional systems, or melting of the classical semiconductors Ge and Si, which become metals after melting because the short-range order in the liquid state is different from the diamond structure of solid Ge and Si. The periodic potential leading to an opening of the gap may be the lattice potential due to electron–ion interactions, but it may also be a self-consistent potential due to electron–electron interactions. That is, in some cases the interaction between electrons can also be important for this mechanism of insulator–metal transitions, but the nature and origin of the insulating state here is still the existence of an energy gap in a one-electron spectrum, treated in the band picture. One of the consequences of this is that for such systems the materials with odd number of valence electrons per unit cell should always be metallic: as, according to the Pauli principle, in the band picture every state should be filled by two electrons with opposite spins, one can have filled valence bands separated from the conduction bands by an energy gap only for an even number of electrons per unit cell; for an odd number of electrons in this picture some band or bands should necessarily be partially filled and the system should be a metal. (But the even electron concentration does not yet guarantee the insulating character of the system: different energy bands can overlap, so they may then be partially filled for an even number of electrons per unit cell.)

There is yet another type of metal–insulator transition in systems with noninteracting electrons – the *Anderson transition* in disordered systems. For strong enough disorder, crudely speaking when the mean-square fluctuations of the random potential $\langle \Delta V^2 \rangle^{1/2}$ are larger than the electron hopping t or the bandwidth $W \sim zt$, the electron diffusion would be suppressed, and an electron put at some place in the system would remain in the vicinity of its original position and never go to infinity, that is the conductivity of such a system would be zero and the system would be insulating. But the one-electron energy spectrum in this case may still be continuous, so there may be no gap in the energy spectrum. What suppresses conductivity in this case is that *the mobility* μ of the electrons is zero. This situation can be illustrated with a classical picture of a "mountain landscape": if a potential relief has deep wells ("craters" or "valleys"), the electrons ("liquid") can fill these deep wells and there will be no possibility of a "flow" from one well to the other across the high barriers ("mountain ridges").

For a weaker disorder for some states, in some parts of the energy spectrum, the mobility can be nonzero, and if the electron filling is such that the Fermi energy would fall into these regions with nonzero mobility, the system would be a metal. But in other parts of the spectrum, for example in its tails, the mobility can still be zero, and there may exist a *mobility edge*, dividing the regions of insulators ($\mu = 0$) and metals ($\mu \neq 0$). If by changing the electron concentration or by reducing the strength of disorder we cross the mobility edge, we have an insulator–metal transition.

For us, in dealing with TM compounds having strong electron correlations, the main interest is presented by the third type of metal–insulator transitions – Mott transitions between a metal and a Mott insulator. As discussed in Chapter 1, in strongly correlated systems the material can become insulating not because of the external (or self-consistent) potential, but because of strong electron correlations: in the simplest case of the Hubbard model – for one electron per site $n = 1$ and for strong on-site repulsion U larger than the electron hopping t. As we have discussed in several places, in particular in Chapter 9, the material can become metallic even for $U \gg t$ if we dope the system, $n \neq 1$. But there may also occur transitions to a metallic state for a fixed electron concentration, say $n = 1$, if we reduce the interaction strength U, or rather increase the electron hopping t, which we can do for example by applying pressure. The transitions between metals and Mott insulators are Mott transitions.

Such transitions are observed, for example as a function of temperature, in quite a few TM compounds. But in specific cases it is sometimes not easy to decide whether the particular metal–insulator transition is indeed a Mott transition, or whether it can be connected for example to lattice distortions which often occur during such transitions, so that it can be explained in the band picture.

The "classical" examples of metal–insulator transitions in TM oxides are the transitions in V_2O_3 and in VO_2. In the first of these systems the metal–insulator transition at ambient pressure is a transition between a paramagnetic metal and an antiferromagnetic insulator, which has all the features of a Mott insulator. Thus the consensus is that this is really a Mott transition – although in this case there is also a strong change in the crystal structure, from corundum at high temperatures to a monoclinic structure at $T < T_c$. In the phase diagram of V_2O_3, shown in Fig. 10.7, there is a separate line of metal–insulator transition at negative pressures and elevated temperatures, at which neither magnetic nor crystal structures change (there is only a jump in lattice parameters, but the lattice symmetry remains the same). This transition is definitely a Mott transition – it cannot be anything else.

The situation in VO_2 is more controversial. The metal–insulator transition in this material, occurring at $T_c \sim 70°C$, is accompanied by the dimerization of V chains, running in the c-direction in the rutile structure of VO_2, and many have explained the very metal–insulator transition in this system in the band picture as a consequence of this dimerization (i.e., treating this transition as a Peierls transition). Indeed, this lattice distortion definitely plays an important role in the metal–insulator transition in VO_2. However the electron correlations are also very important here, which can be shown for example by replacing V by Ti and Cr. Thus the metal–insulator transition in VO_2 has a combined nature, having the features of both Mott and Peierls transitions.

Similarly, one should analyze the metal–insulator transition observed in many other TM compounds. Thus, the metal–insulator transitions in the Magnéli phases V_nO_{2n-1} and Ti_nO_{2n-1} are definitely connected with electronic correlations, but with the structural details also playing an important role. In contrast, the features of the metal–insulator transitions in NiS point to its more band-like character, although the presence of antiferromagnetism in the low-temperature insulating phase may signal an important role of correlations.

An interesting feature of some systems with one d-electron and with $S = \frac{1}{2}$, showing metal–insulator transitions, is that often in the insulating phase there occurs a formation of singlet dimers of transition metals. This is the case in VO_2, in the vanadium hollandite $K_2V_8O_{16}$, in some Magnéli phases such as V_4O_7, in Ti_2O_3, and in the pyroxene $NaTiSi_2O_6$. But the ions with bigger spins, say with $S = 1$, can also form pairs, which, however, can be both singlet and triplet. Similarly, more complicated clusters can accompany the metal–insulator transition, for example singlet trimers formed in $LiVS_2$ (see also Chapter 6).

The theoretical description of Mott transitions presents a very challenging problem. The main difficulty is connected with the fact that these transitions occur just in the situation where all the parameters are of the same order, the interaction U being of the same order as the kinetic or band energy t. That is, there is no small parameter in this case, which one could use to build a regular theoretical description, and one has to rely on some approximate schemes.

There are many such theoretical approaches proposed in the literature. Probably the most successful are two of them: the approach using the *Gutzwiller method*, and the *dynamical mean-field theory*. In the Gutzwiller method one uses the variational approach, suppressing the contribution of the polar states (states with double occupancy of sites) in the total wave function by some parameter. This parameter is then found variationally, by minimizing the total energy of the system. This approach is used widely not only for considering Mott transitions, but also for describing the properties of correlated systems themselves, in particular of high-T_c superconductors. For Mott transitions this method, mainly due to Brinkman and Rice, gives a reasonable description of the approach to this transition from the metallic side, showing for example the behavior of some characteristics of the system when the correlation strength increases. This approach is now used widely not only for solids, but also for example for ^{3}He, in which the correlations are strong.

Another successful approach to describe metal–insulator transitions, in particular in some real materials, is the dynamical mean-field theory (DMFT), combined with real *ab initio* calculations for the given material. The general scheme used in the DMFT approach is the mean-field description, which takes into account dynamical effects and which is formulated in the language of the electron Green functions or of the corresponding self-energies. This allows us to obtain the effective electron spectral function or the density of states $\rho(\varepsilon)$, which includes both the contribution of coherent quasiparticles and an incoherent contribution. In this approach the crossover from a normal metal to a Mott insulator with increasing electron correlations (Hubbard's U) occurs by first creating incoherent "wings" in $\rho(\varepsilon)$, which develop into the lower and upper Hubbard bands, with the coherent peak at the Fermi level reducing in intensity, until it completely disappears at $U > U_c$ ($\sim 2zt$) in the Mott insulator phase. There are some limitations to this approach, mainly connected with the neglect of spatial correlations, that is of the k-dependence of the Green function and the self-energy. This approximation can influence the results. Nevertheless, the general description of metal–insulator transitions in this method seems to be very appealing. And there are at present many attempts to develop this method still further, to correct for its possible deficiencies, in particular by including intersite correlations.

When discussing Mott transitions in real materials, one has to take into account not only the details of the crystal structure but also the possible role of magnetic correlations and multiplet effects. They can lead to the consequence that the effective value of the Hubbard repulsion U_{eff} could change, for example with the spin state of the corresponding ions. Thus, the high spin–low spin transition in some Fe^{3+} compounds, which can be induced by pressure, would be accompanied by a decrease in U_{eff}, so that such an HS–LS transition could simultaneously be an insulator–metal transition. Thus, besides occupation-controlled and bandwidth-controlled insulator–metal transitions, there may also exist interaction-controlled transitions.

Mott transitions in systems with several bands of different character can have specific features. In this case there can in principle occur orbital-selective Mott transitions: more narrow bands can already be on the insulating side of the Mott transition, whereas electrons in broader bands could still be treated as itinerant. We have, in fact, already used such a picture in previous chapters, for example in the description of colossal magnetoresistance manganites: the t_{2g}-electrons were treated as localized, producing localized spins, while the e_g-electrons were treated as itinerant (though with certain correlations still present), leading to double-exchange ferromagnetism. Theoretically the problem of orbital-selective Mott transitions is not a trivial one, as there always exist different processes (hybridization, interactions) which couple different bands. Still, the consensus seems to be that indeed Mott transitions in different bands can largely occur separately.

Close to Mott transition there may appear inhomogeneous states, connected with the formation of tightly bound metal clusters (dimers, trimers, bigger clusters). Within such clusters the electron hopping may already be sufficiently large, larger than Hubbard's repulsion U, so that such clusters can be described in the molecular orbital picture, and the electrons in such clusters would already be on the itinerant side of the localized–itinerant crossover. At the same time, the distance between such clusters may still be large, and the hopping between them small, so the whole system would still be insulating. And the real insulator–metal transition in the whole sample may occur only later, for example at higher pressure. Such *partial* or *stepwise Mott transitions* are more likely in low-dimensional or frustrated materials. Examples of such behavior are met in $MgTi_2O_4$, $LiVO_2$, $LiVS_2$, and $TiOCl$.

As to the conceptual points, a very specific feature of Mott transitions in the pure sense (localized–itinerant transition without any accompanying long-range order, e.g. magnetic) is that they are very different from the standard phase transitions considered for example in Landau theory (see, e.g., Appendix C). It is actually not clear whether there exists some symmetry which distinguishes the metallic and the insulating phase, and what the actual order parameter of Mott transitions is. It could be the case that the pure Mott transition is either a I order transition or a continuous one – similar to liquid–gas transitions. It is still not excluded that one could introduce some notion of an order parameter for Mott transition – however it could be a highly nontrivial one, for example nonlocal.

11

Kondo effect, mixed valence, and heavy fermions

The topic of this book is the physics of transition metal compounds. In all their properties strong electron correlations play a crucial role. However TM compounds are not the only materials in which electron correlations are extremely important. Other such systems are substances containing rare earth elements with partially filled $4f$ shells or actinide compounds with $5f$-electrons. These systems show a lot of very interesting special properties such as mixed valence and heavy fermion behavior. And though these phenomena were discovered and are mostly studied in $4f$ and $5f$ systems, similar effects, maybe less pronounced, are also observed in some TM compounds. The main concepts, and also the main problems in the physics of rare earth (and actinide) compounds are very similar to those in TM systems. Therefore we also include in this book, formally devoted to TM materials, this short chapter in which we summarize the main phenomena discovered in $4f$ and $5f$ systems, and compare them and their description with that of TM systems. Some of these phenomena were even discovered first in materials with TM ions, but later proved to be essential in treating rare earth systems; while other notions were introduced for rare earth compounds and later transferred to the study of transition metal systems.

There exists a significant body of literature devoted specifically to some of the topics discussed briefly below. One can find detailed descriptions for example in Hewson (1993) or Coleman (2007). The last chapter of Khomskii (2010) gives a somewhat more detailed and more theoretical treatment of the material presented below.

11.1 Basic features of f-electron systems

Partially filled d-shells (orbital moment $l = 2$) in TM compounds ($3d$, $4d$, $5d$) determine their main properties, described in this book. Similarly to d-shells, in lanthanides (rare earths, $4f$ series) and in actinides ($5f$ series) the f-shells (orbital moment $l = 3$) are partially occupied. Therefore most of the descriptions presented above apply to these elements as well. The f-electrons, especially $4f$ ones, have an even smaller radius of the corresponding wavefunction, that is they are more strongly localized close to the atomic core. This has several important consequences. On the atomic level the classification of electronic states similar to that of TM elements applies, but due to this tight localization, and due to the heavier atomic mass, the spin–orbit coupling in f-elements is much stronger than in TM elements. Therefore the multiplet scheme used for d-elements, especially for

$3d$ ones – the LS or Russell–Saunders scheme, in which one first sums the spins and orbital moments of all atomic electrons into the total spin S_{tot} and total orbital moment L_{tot} and then combines them into $J = S_{\text{tot}} + L_{\text{tot}}$ – is not applicable for rare earth elements. In the latter case one has to use the so-called jj scheme: first the orbital moment l_i and spin s_i for each electron combine into $j_i = l_i + s_i$, and then one builds the total moment of the whole atom or ion as $J = \sum_i j_i$. One very important consequence of strong spin–orbit coupling is that in f-electron ions the g-factor may be rather large, and also sometimes very anisotropic, so that the corresponding ions may behave practically as Ising ions. Thus for example for Tb, $g_{\|} \sim 18$ and $g_{\perp} \sim 0$, that is Tb^{3+} can be treated as an Ising ion. Such very strong single-site anisotropy makes some such systems very interesting from the physical point of view, and also makes some rare earth compounds very useful for practical applications, for example in strong permanent magnets with large coercive force.

The second special feature of f-ions is that, again due to the strong localization of f-electrons,[1] when f-ions are put in a crystal, the crystal field splitting is usually much smaller than that of TM ions. Instead of the crystal field splitting for TM ions, which is of order up to $\sim$ 2–3 eV, in rare earth ions it is typically less than 0.1 eV.

The third consequence of the same fact – small radius of f-states – is that in concentrated systems the f-electrons, especially $4f$ ones, almost always behave as localized, with magnetic moments practically coinciding with those of isolated ions. In our language, we can say that the f-electrons are almost always deep on the insulating side of the Mott transition. Therefore most of these compounds are strongly magnetic, often with quite large magnetic moments. At the same time, the magnetic ordering in them often occurs at relatively low temperatures – again for the same reason, strong localization of f-electrons practically inside the ionic core, as a result of which their exchange interaction is rather small.

On the contrary, in many rare earth and actinide materials there exist, besides these very strongly localized f-electrons, conduction electrons forming broad bands. Thus, despite the f-electrons being in a "Mott insulator" state, the materials can be, and often are, metals. Usually it is the interaction with these conduction electrons, or the interaction between f-electrons via conduction bands, which determines the main physical properties of these systems. Thus the most interesting phenomena are met in metallic systems containing rare earth ions.

The insulating materials with f-electrons present somewhat less interest, except possibly for two cases. One is systems containing, besides rare earth elements, also TM elements, such as CMR manganites $(LaSr)MnO_3$ or $(PrCa)MnO_3$, or mixed rare earth–TM orthoferrites and garnets. The dominant magnetic ordering of such systems is due to TM spins, but the coupling with rare earth ions may determine their detailed magnetic behavior. If however we have only magnetic rare earth ions, magnetism usually appears at rather low temperatures.

---

[1] We will talk mostly about rare earth systems, to which all the features described above apply. The $5f$-electrons in actinides have bigger radius, and they resemble somewhat the $3d$ series, with $5f$-electrons less localized than in rare earth systems – although for example the spin–orbit coupling in them is again very strong.

Some such systems still attract considerable attention, especially systems with frustrations. As mentioned above, due to strong single-site anisotropy some rare earth ions behave practically as Ising ions, and they present a good example of Ising systems on frustrated lattices. Such are for example the pyrochlores RTi_2O_7 with $R = $ Tb, Dy, which give classical examples of *spin ice*, see Section 5.7. The very fact that all the energy scales (such as exchange interaction) are here much smaller than in TM systems makes them quite appealing for experimental studies, which, though usually requiring low temperatures, also give the possibility of influencing them by relatively small and accessible magnetic fields (the "rule of thumb" is that a field of 1 T is equivalent to energy ~ 1 K for the g-factor ~ 2, so if the exchange interaction or the eventual ordering temperature was ~ 5–10 K one could influence these systems by fields ~ 5–10 T). Such insulating frustrated rare earth systems are studied very actively nowadays and provide quite interesting information.

Some interesting effects are also observed in other insulating or semiconducting rare earth compounds, for example the insulator–metal transition in EuO, discussed in Chapter 10. Also the cooperative Jahn–Teller effect and orbital ordering are observed in some materials of this class, for example in zircons such as $TmVO_4$ or $DyVO_4$ (Gehring and Gehring, 1975; Kaplan and Vekhter, 1995). But the corresponding transition temperatures are again much smaller than in similar TM compounds: thus in $TmVO_4$, $T_c \sim 2.4$ K. In rare earth compounds one often talks not about the Jahn–Teller effect, but about quadrupolar ordering, because in some of these materials the mechanism of such ordering is really the classical quadrupole–quadrupole interaction, which in TM compounds is much weaker than electron–phonon or exchange interactions leading to orbital ordering.

For the most interesting, metallic rare earth compounds, the models describing these systems should contain both strongly correlated f-electrons and weakly correlated band electrons. The basic such model for f-systems, which in a sense replaces the Hubbard model that we use as the prototype model for TM systems, is the Anderson model describing isolated f-ions embedded in a metallic host, or its generalization to the concentrated f-electron system – the Anderson lattice. The Hamiltonian of the Anderson model for a single magnetic impurity is (Anderson, 1961)

$$\mathcal{H}_A = \varepsilon_f \sum_\sigma f_\sigma^\dagger f_\sigma + \sum_{k,\sigma} \varepsilon(k) c_{k\sigma}^\dagger c_{k\sigma} + U n_{f\uparrow} n_{f\downarrow} + \sum_{k,\sigma} (V_k f_\sigma^\dagger c_{k\sigma} + \text{h.c.}) . \quad (11.1)$$

Here $f_\sigma^\dagger$, f_σ are creation and annihilation operators of f-electrons with spin σ, and $c_{k\sigma}^\dagger$, $c_{k\sigma}$ describe conduction electrons with band spectrum $\varepsilon(k)$. The third term in (11.1) is the on-site Coulomb interaction of f-electrons, and the last term is the f–c hybridization.

The corresponding model for a concentrated system – the Anderson lattice – looks exactly like the model (11.1), with many f-electron sites:

$$\mathcal{H}_{AL} = \varepsilon_f \sum_{i,\sigma} f_{i\sigma}^\dagger f_{i\sigma} + \sum_{k,\sigma} \varepsilon(k) c_{k\sigma}^\dagger c_{k\sigma} + U \sum_i n_{fi\uparrow} n_{fi\downarrow} + \sum_{i,k,\sigma} (V_{ik} f_{i\sigma}^\dagger c_{k\sigma} + \text{h.c.})$$

$$(11.2)$$

where, due to the periodicity of the system,

$$V_{ik} = V_k \, e^{i k \cdot R_i} \, . \tag{11.3}$$

This model is of course simplified; it ignores complications connected with the orbital quantum numbers of f-electrons (for $l = 3$ there are $(2l + 1) = 7$ different f orbitals), as well as the multiplet structure, etc. But it captures the main physical effects: the presence of strongly localized f-electrons (with large U) hybridized with the electrons of the broad conduction band.

When comparing these basic models, especially the Anderson lattice model (11.2), with the basic models describing TM systems, we see that, similar to the Hubbard model (1.6), it contains strong on-site Coulomb (Hubbard) interaction U, which, when large, would prevent on-site charge fluctuations, that is double occupancy of the f level. But in contrast to d-electrons, for which the overlap of d-functions on neighboring TM ions is not negligible, there is no direct f–f hopping term in the Hamiltonian (11.2) similar to the hopping term $t c_{i\sigma}^{\dagger} c_{j\sigma}$ in the Hubbard model (1.6). Instead, there is here the hybridization of localized f-electrons with the conduction band, due to which in principle the f-electrons could also acquire some kinetic energy, moving from site to site by hopping via the conduction band. In this sense the Anderson lattice resembles more the d–p model (4.1), which describes TM compounds, such as perovskite oxides, by including the d–p hybridization. The difference is that here the conduction electrons form their own band with the dispersion $\varepsilon(\mathbf{k})$, whereas in the d–p model (4.1) we did not include the intrinsic width of the p band – although in the more realistic description of TM compounds in the d–p model one could (and sometimes should) include also the p–p hopping t_{pp} which would lead to a dispersion of p bands. If we add to the model (4.1) a direct p–p hopping between different oxygen sites $\sim t_{pp} p_{j\sigma}^{\dagger} p_{j\sigma}$, these models would actually become equivalent.

11.2 Localized magnetic moments in metals

The single-site Anderson model (11.1) was used widely to describe the properties of magnetic impurities in nonmagnetic metals – not only rare earth impurities, but even more the TM impurities such as Mn in Cu. In fact this model was first proposed (Anderson, 1961) not for rare earth, but for TM impurities. The first question which Anderson investigated in that paper was the condition for the formation, or rather preservation, of localized magnetic moments when magnetic impurities are dissolved in nonmagnetic metals. In simple terms what can happen in this case is that due to the f–c hybridization[2] a localized electron of the impurity, say with spin up, can hop to the conduction band, and an electron from the conduction band with the opposite spin can then hop back to the impurity level. In effect the impurity moment will start to fluctuate, so that the average magnetic moment can disappear.

---

[2] We will continue to talk here about f-electrons, although instead of f-electrons these could be the d-electrons of a TM impurity.

The theoretical treatment carried out by Anderson (1961) (see also a more detailed presentation in Chapter 13 of Khomskii, 2010) has clarified the conditions for this to happen, or, vice versa, the conditions for the localized magnetic moments to survive. Without describing the technical details, we present here the main conclusions, which were obtained in the mean-field approximation. The conditions for the existence of a localized magnetic moment on the magnetic impurity depend on the position of the f level with respect to the Fermi level of the conduction electrons, on the on-site interaction U, and on the $f-c$ hybridization V_k. For f level lying in the conduction band, such a level would acquire a finite width

$$\Gamma = \pi |V|^2 \rho(\varepsilon_{\mathrm{F}}) , \qquad (11.4)$$

where $\rho(\varepsilon_{\mathrm{F}})$ is the density of states of conduction electrons at the Fermi energy ε_{F}, and where we have neglected the k-dependence of the hybridization matrix element V_k in (11.1). The quantity Γ in some sense plays the role of effective "bandwidth" (level width) of f-electrons, and the real dimensionless parameter characterizing the tendency of electron localization is U/Γ, just like U/W in the Hubbard model.

The region of the existence of localized moments obtained in Anderson (1961) is the shaded region in Fig. 11.1. We see that such localized states are more probable for large U and small Γ, and also for the most symmetric position of the f level relative to the Fermi level, see Fig. 11.2. Indeed, if in this case there is one electron at this level, it will be at energy ε_f. If we put another electron at the same level, which would make the total spin of the impurity zero, the energy of such a state would be $\varepsilon_f + U$. The most symmetric location of these levels, ε_f and $\varepsilon_f + U$, relative to the Fermi level ε_{F}, is the most favorable for the preservation of the localized magnetic moment (LMM) at this impurity. As already mentioned above, the processes which could lead to the disappearance of such moments would be transitions of an electron from the f level to the conduction band (to the Fermi level), or the trapping of a second electron from the conduction band by the impurity; as is clear from Fig. 11.2, these processes cost respectively energy $(\varepsilon_{\mathrm{F}} - \varepsilon_f)$ and $(\varepsilon_f + U - \varepsilon_{\mathrm{F}})$. However they are promoted by the $f-c$ hybridization V, so the bigger V, or Γ (11.4),

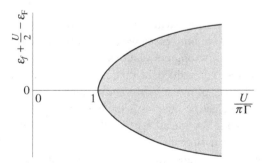

Figure 11.1 Region of existence of localized magnetic moments for magnetic impurity in a metal, after Anderson (1961).

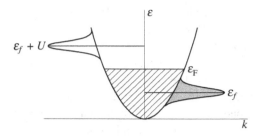

Figure 11.2 Virtual bound states giving localized magnetic moments in a metal.

the more probable such processes become. Nevertheless, in a large part of the parameter space the magnetic impurities, such as TM of rare earth atoms in nonmagnetic metals, can preserve their localized spins.

11.3 Kondo effect

This is, however, not the end of the story. Suppose that we are indeed in the regime in which, according to the mean-field solution of Anderson (1961), the impurity has a localized spin. Such a spin will still have an exchange interaction with the conduction electrons, which will have an antiferromagnetic character. This can be described by the effective Hamiltonian

$$\mathcal{H} = \sum_{kk'} J_{kk'} c^{\dagger}_{k\sigma} \sigma_{\sigma\sigma'} c_{k'\sigma'} \cdot S \sim J\sigma \cdot S, \tag{11.5}$$

where σ is the spin density of conduction electrons at the position of the impurity, and S is the impurity spin. The exchange constant J can be obtained in perturbation theory in the hybridization V, and it has the form

$$J = 2V_{k_F}^2 \left(\frac{1}{\varepsilon_F - \varepsilon_f} + \frac{1}{\varepsilon_f + U - \varepsilon_F} \right) = 2V^2 \frac{U}{(\varepsilon_F - \varepsilon_f)(\varepsilon_f + U - \varepsilon_F)}. \tag{11.6}$$

One can notice here a close analogy with the treatment of the exchange interaction in the Hubbard model in Chapter 1: in both cases we proceed from the electronic Hamiltonians, (1.6) or (11.1); in the limit of strong correlations the electrons (or the impurity electron) become localized, so that we can reduce the model to the effective models (1.12) or (11.5) dealing only with spin degrees of freedom; the antiferromagnetic exchange (1.12) or (11.6) appears in second order in electron hopping – the hopping t between sites in the Hubbard model, and the f–c hopping (hybridization) V in the Anderson model.

The presence of such an antiferromagnetic exchange (11.5) can lead to drastic consequences. It turns out that at high temperatures the magnetic impurity indeed behaves as a localized spin, with rapid spin fluctuations, but still giving a Curie-like susceptibility $\chi \sim 1/T$. But with decreasing temperature the effective coupling of the impurity

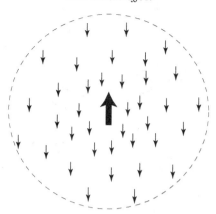

Figure 11.3 Qualitative picture of Kondo screening of localized magnetic moments in a metal.

spin with the spins of the conduction electrons becomes stronger and stronger, and as it is antiferromagnetic (J in (1.16) is > 0), there will be conduction electrons with spin $\downarrow$ accumulated close to an impurity with spin $\uparrow$. In effect this magnetization of conduction electrons will *screen out* the localized spin (see a qualitative picture in Fig. 11.3) so that, seen from outside, it would seem as though the impurity spin has disappeared. This phenomenon is known as the *Kondo effect*, and the model with interaction (11.5) is called the Kondo model (Kondo, 1964). A generalization of this model to the case of a periodic system with localized spins at each lattice site is called the Kondo lattice. Theoretical treatment of the Kondo effect presents quite a difficult problem, and several approximate methods have been proposed; the exact solution of this model was obtained later, by N. Andrei, and by P. B. Wiegmann and A. M. Tsvelik, see for example Andrei *et al.* (1983) and Tsvelik and Wiegmann (1983). One can find some extra details in Khomskii (2010).

Thus it seems that we have first struggled hard to get a localized magnetic moment at the impurity, only to see it finally disappear anyway. But the resulting picture is still much richer and much more interesting than what we would have if the localized magnetic moment had not appeared in the first place, that is if the system were to lie outside the shaded region of Fig. 11.1. In the Kondo case the strong effective interaction increases with decreasing temperature, and the corresponding progressive screening of the impurity spin leads to strong modifications of most properties of such systems. One can show that the energy, or temperature scale at which such a screening of localized magnetic moments occurs is given by the expression

$$T_K \sim \varepsilon_F \, e^{-1/J\rho(\varepsilon_F)} \qquad (11.7)$$

(for weak coupling $J\rho(\varepsilon_F) \sim J/\varepsilon_F \ll 1$, which is practically always the case). For $T > T_K$, as we have said above, the impurity spins behave as free localized magnetic moments. But when the temperature approaches the Kondo temperature T_K, the exchange scattering of the conduction electrons on the magnetic impurity become stronger and

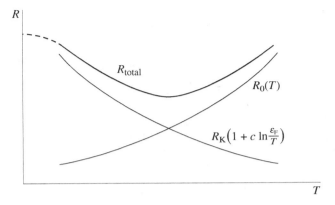

Figure 11.4 Resistance minimum for metals with magnetic impurities, explained by the Kondo effect.

stronger, and one can say that for example $\downarrow$ spins of conduction electrons will "stick" to the impurity spin $\uparrow$. This will, in particular, lead to an *increase* in resistivity with decreasing temperature, see Fig. 11.4 – a behavior very untypical for normal metals. This effect, with the appearance of a resistivity minimum at a certain temperature, was noticed in simple metals such as Cu or Au long ago, and for a long time its origin was a mystery. Now we know that this effect appears due to an uncontrolled presence of some TM impurities such as Fe or Mn in these metals.

There are many other consequences of the Kondo effect. They may simply be understood in the picture of *Kondo resonance* (sometimes also called Abrikosov–Suhl resonance): as discussed above, as $T \to 0$ the localized spin of the impurity is screened and effectively disappears, and the system again behaves as a standard metal, or a Fermi liquid, but with strongly renormalized parameters: it looks as if there appears a narrow peak in the density of states, Fig. 11.5, and the Fermi level will lie at this peak. The width of this peak is $\sim T_K$, that is the effective density of states at the Fermi level would increase strongly, $\rho^*(\varepsilon_F) \sim 1/T_K$ (of course in the case of dilute impurities the effects will be proportional to the impurity concentration v, but, as we will see in the next section, in concentrated systems described by the Anderson lattice model (11.2) all sites can contribute). In any case, as we know, in normal metals the density of states at the Fermi level determines most low-temperature properties, such as the linear specific heat

$$c = \gamma T , \qquad \gamma \sim \rho^*(\varepsilon_F) \tag{11.8}$$

or the magnetic susceptibility at low temperature

$$\chi(T \to 0) \sim \rho^*(\varepsilon_F) . \tag{11.9}$$

As we have here $\rho^*(\varepsilon_F) \sim 1/T_K$ (or rather v/T_K, where v is the concentration of impurities), both specific heat and magnetic susceptibility in the Kondo systems would be strongly

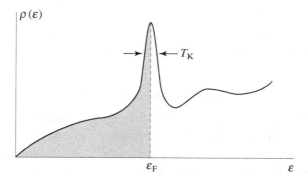

Figure 11.5 Effective density of states of a metal with magnetic impurity, showing the formation of the Kondo (or Abrikosov–Suhl) resonance at the Fermi energy at low temperatures.

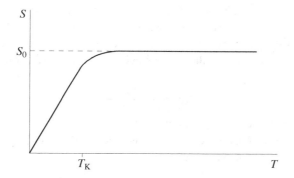

Figure 11.6 Illustration of the appearance of large linear specific heat in Kondo systems due to the release of large spin entropy $\sim k_B \ln(2s + 1)$ in a narrow temperature interval $\sim T_K$.

enhanced for $T < T_K$ (above T_K we have $\rho^* \sim \rho \sim 1/W$, where W is the bandwidth of the conduction electrons). And, whereas typically $W \sim 2$–5 eV, the Kondo temperature (11.7) is usually much smaller, ~ 10–100 K $\sim 10^{-3}$–10^{-2} eV, so the Kondo renormalization of many properties can be extremely strong.

One can give a simple qualitative explanation of these results. The picture of the Kondo effect is that for $T > T_K$ the impurity spins behave as free spins s, and in the paramagnetic state they would give the entropy $S_0 = k_B \ln(2s + 1)$. But at low temperatures, due to the antiferromagnetic coupling (11.5), these spins are screened, the ground state is effectively a nonmagnetic singlet, and, in agreement with the Nernst theorem, the entropy goes to zero, in Fermi systems linearly. Thus the total behavior of the entropy should look as shown in Fig. 11.6. That is, the entire magnetic entropy is released in the narrow temperature interval $\sim T_K$, so that the slope of the entropy in this region is large, $S(T) \sim T/T_K$ (or $S_0 T/T_K$). If we recall that the specific heat is $c = T \, \partial S/\partial T$, we see that the specific heat would also behave as $c \sim T/T_K$, as we have written in (11.8) (with $\rho^*(\varepsilon_F) \sim 1/T_K$).

All these effects would be even more pronounced if instead of dilute magnetic impurities we were to have such Kondo-like spins at every site of the lattice. This is the situation which gives rise to the state known as a *heavy fermion state*, to be discussed in the next section.

11.4 Heavy fermions and mixed valence

When we consider concentrated systems with coexisting localized and itinerant electrons, such as rare earth metals or intermetallic compounds, we have to take into account a very important factor – the interaction between localized electrons on different sites. It will act simultaneously with local effects such as the Kondo effect, and may counteract the latter. And this interplay of local and intersite correlations can lead to very rich and nontrivial properties.

Physically one could expect two possible outcomes. One is that the localized magnetic moments, for example those of $4f$ shells of rare earths, would have an exchange interaction, and due to that some type of magnetic ordering would appear at low temperatures. The mechanism of this exchange in most cases is exchange via the conduction electrons: due to local $c-f$ exchange (11.5) the f spins induce a spin polarization of the conduction electrons, which behaves as

$$M(r) \sim \frac{\cos(2k_F r)}{r^3} \,, \tag{11.10}$$

and another localized f-electron, at another site, "feels" this polarization and orients its spin accordingly. This mechanism of exchange is called Ruderman–Kittel–Kasuya–Yosida, or RKKY exchange, and has the form

$$\mathcal{H}_{RKKY} = \sum_{ij} I_{ij} \, \mathbf{S}_i \cdot \mathbf{S}_j \,, \qquad I_{ij} \sim \frac{J^2}{\varepsilon_F} \frac{\cos(2k_F r_{ij})}{|r_{ij}|^3} \,, \tag{11.11}$$

where J is the Kondo coupling (11.6). This is the main mechanism of exchange interaction in rare earth metals or intermetallic compounds.[3]

Note that the RKKY interaction (11.11) is oscillating in space. Thus in principle it can lead to different types of magnetic ordering (see also Chapter 5). For regular, periodic systems this can be ferro- or antiferromagnetic ordering or some type of helicoidal or spiral ordering (with the wave vector of the spiral determined by the structure of the Fermi surface, i.e. by the values of k_F). But in disordered systems the same interaction can lead for example to the formation of a *spin-glass state*, see for example Mydosh (1993).

However, as we have seen in the previous section, there exists also an opposing tendency: due to the Kondo effect each localized magnetic moment may be screened by conduction

---

[3] There exists also an analog of the RKKY interaction at work in insulating rare earth compounds such as EuO, EuS, etc.: it is connected with the virtual transfer of an f-electron to the empty conduction band (or the transfer of an electron from the valence band to the f level). These excited electrons or holes then move through the conduction or valence band, and thus promote coupling to other sites. This leads to an exchange which behaves as $I_{ij} \sim e^{-\sqrt{2m\Delta}|r_{ij}|}$, where Δ is the energy of the f–c excitation (playing the role of the energy gap). This mechanism of exchange is sometimes called the Bloembergen–Rowland mechanism (Bloembergen and Rowland, 1955).

electrons and may eventually disappear, so that there would remain no spins to order. If this tendency dominates and such screening occurs before the spins have a chance to order, the outcome, and the type of the ground state would be completely different: instead of a metal with long-range magnetic ordering of localized (e.g., $4f$) spins the ground state would be a nonmagnetic metal of Fermi-liquid type, but with strongly renormalized parameters. This state is known as a *heavy fermion state*.

Which tendency prevails is determined by which energy, or which temperature scale, is larger. The tendency to magnetic ordering is characterized by the strength of the RKKY interaction (11.11), that is it is $T_c \sim I \sim J^2/\varepsilon_F$. On the contrary, the tendency to "demagnetization" has energy scale $T_K \sim \varepsilon_F e^{-1/J\varepsilon_F}$, see (11.7). If we remember that the Kondo f–c exchange J is given by the expression (11.6), which for very large U is simply

$$J = \frac{2V^2}{\varepsilon_F - \varepsilon_f} \, , \tag{11.12}$$

we see that for the f level ε_f lying deep below the Fermi level ε_F we have $J \to 0$. But, as the RKKY interaction is quadratic in J, $I \sim J^2/\rho(\varepsilon_F)$, and the Kondo temperature is exponentially small, $T_K \sim e^{-J\rho(\varepsilon_F)}$, in this regime $T_c \sim I_{RKKY} \gg T_K$, so in this case we expect that a long-range magnetic order would be established. This is the case in most rare earth metals or compounds, such as the metals Gd, Tb, etc.

But there are special situations in which the f level lies not very deep, and in which it can approach the Fermi level. Of course if $\varepsilon_F - \varepsilon_f$ becomes small, we cannot, strictly speaking, use the expressions (11.6), (11.12) obtained in perturbation theory in $J/|\varepsilon_F - \varepsilon_f|$. But the tendency is clear: in this case the Kondo temperature increases strongly and may exceed $T_c \sim I_{RKKY}$. Thus in this case the Kondo effect could prevail, and a nonmagnetic ground state would be formed.

When are such situations possible? To understand that we have to recall the well-known chemical properties of rare earth elements. When they form different compounds, almost always they accept valence 3+: Gd^{3+}, Dy^{3+}, etc. That is, they give two s-electrons and one $4f$-electron to chemical bonds or to metallic bands.

Most rare earth elements are indeed very stable in the valence state 3+. But there are some exceptions. These are met at the very beginning (Ce, Pr), at the very end (Tm, Yb), and close to the middle of the $4f$ series (Eu, Sm). These elements, besides the common valence state R^{3+}, can exist in other ionization states: Ce exists in the states $Ce^{3+}(4f^1)$ and $Ce^{4+}(4f^0)$; Yb as $Yb^{3+}(4f^{13})$ and $Yb^{2+}(4f^{14})$; Eu as $Eu^{3+}(4f^6)$ and $Eu^{2+}(4f^7)$. The stability of these unusual valence states (4+; 2+) is caused by the extra stability of empty (here $4f^0$), completely full ($4f^{14}$), or exactly half-full ($4f^7$) f shells, well known in atomic physics. This is the reason why the states $Ce^{4+}(4f^0)$, $Yb^{2+}(4f^{14})$, and $Eu^{2+}(4f^7)$ compete with the usual valence 3+. This tendency is also seen in the "neighboring" elements such as Sm and Tm, although it is less pronounced in these.

Now, from our point of view this tendency, the proximity to a different valence state, means that the corresponding f level lies close to ε_F and may in principle even cross

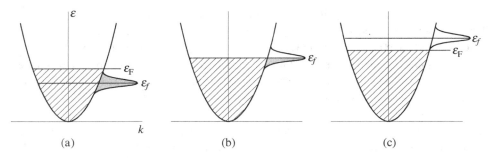

Figure 11.7 Evolution of the energy spectrum when the localized f level ε_f moves from below the Fermi level ε_F to above it.

it. Consider for example the case of Ce. The "normal" valence Ce^{3+} in the Ce metal at ambient pressure and at room temperature means that this system, being a metal, has one localized f-electron per Ce. That is, the corresponding energy diagram looks as shown in Fig. 11.7(a), with the f level ε_f below the Fermi-energy ε_F (we also show here the broadening of this f level due to its hybridization with the conduction band, cf. Fig. 11.5).

But the alternative valence state, $Ce^{4+}(4f^0)$, if it is realized in a "clean" form, would correspond to the situation of Fig. 11.7(c), with the f level *above* ε_F and with all former f-electrons now "spilled off" into the conduction band. And indeed such a transition occurs in Ce: under pressure it goes from the so-called γ phase, which is magnetic, to the α phase, with the same crystal structure (but smaller volume) which is nonmagnetic, that is does not contain localized $4f$-electrons. A more detailed picture of this transition will be described below; here we want first to stress that the very fact that Ce can have both valences 3+ and 4+ tells us that these states should not differ very much in energy, which means that even when Ce is Ce^{3+}, its f level should lie relatively close to the Fermi level, so $\varepsilon_F - \varepsilon_f$ should be small which, as we have seen above, is the condition for $T_K \gtrsim T_{RKKY}$ and for the realization of the heavy fermion state. Indeed, it is just in many compounds containing Ce that we find this heavy fermion state. Alternatively, we can talk not about one f-electron, but about an f-hole, and then Yb would be the analog of Ce ($Yb^{3+}(4f^{13})$ has one $4f$-hole and $Yb^{2+}(4f^{14})$ with filled f shell has no f-holes). Consequently, one can expect heavy fermion behavior in some Yb compounds, which is indeed the case. Similarly, heavy fermion behavior is observed in some actinide $5f$ compounds, with more or less the same tendency: it is met mostly at the beginning of the $5f$ series (U, Np), although due to larger radius and stronger hybridization of $5f$-electrons the identification of respective valence states in metallic compounds is more difficult.

But Fig. 11.7 tells us yet more: when (e.g., under pressure) the f level ε_f moves up, approaching and then crossing the Fermi level ε_F, we do not always go directly from the state of Fig. 11.7(a) with $n_f \simeq 1$ to the state of Fig. 11.7(c) with an empty f level, $n_f \simeq 0$; rather the system can become "stuck" in the intermediate regime with $\varepsilon_f \simeq \varepsilon_F$, Fig. 11.7(b). This should be the case if such a transition is continuous. In such a state we would have a *partial occupation* of the f level, $0 < n_f < 1$, or, in other words, the valence

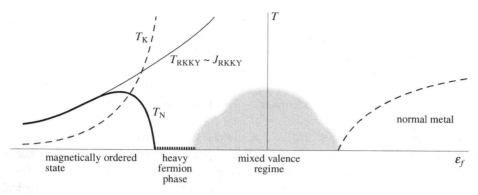

Figure 11.8 Different states depending on the position of the f level ε_f relative to the Fermi energy ε_F (here taken as zero).

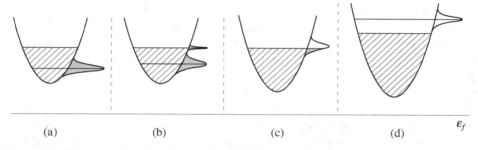

Figure 11.9 Change in densities of state: from magnetic metal (phase (a)) through heavy fermion state (phase (b)) to the mixed valence state (phase (c)), and finally to the nonmagnetic metal with the empty f level (phase (d)).

of the ion would be intermediate between 3+ and 4+. Such a state is called an *intermediate valence*, or *mixed valence* (MV) state. In Ce the γ–α transition is I order, but still the high-pressure nonmagnetic α phase does not have a completely empty f level, rather it is of mixed valence type, $\sim Ce^{3.6+}$.

Thus the whole "sequence of events," the sequence of different phases which occur when we gradually move the f level from deep below the Fermi level toward and eventually above it would look as shown in Fig. 11.8.

The effective form of the energy spectrum in different regions is shown in Fig. 11.9. We start from the state (a) with a very deep f level, in which the RKKY interaction dominates, and in this state the long-range magnetic ordering is established because of it. When the f level moves up, the Kondo effect starts to compete with the RKKY interaction, and when it wins, the system would be in the heavy fermion regime. The f level itself in this phase is still below the Fermi level, so that n_f is almost 1, but there is a strong renormalization of all properties, which can be described by the appearance of a collective Kondo resonance – a narrow peak, with width $\sim T_K$, in the density of states at the Fermi level, Fig. 11.9(b). It is in this phase, in which the effective density of states $\rho(\varepsilon_F) \sim 1/T_K$

becomes very large, that we have the strongest renormalization of all low-temperature properties such as the specific heat (11.8) and the magnetic susceptibility (11.9). And, in contrast to the case of Kondo impurities, here we have such "impurities" at every site, or in every unit cell, so that the experimentally observed effects are really huge. Thus, whereas the coefficient γ in the low-temperature specific heat $c = \gamma T$ in the conventional metals is $\gamma \sim 1\text{--}5\,\text{mJ/mole} \cdot \text{K}^2$, in heavy fermion systems it is $\sim 10^3$ times larger. For example, in CeAl$_3$ $\gamma \sim 1700\,\text{mJ/mole} \cdot \text{K}^2$.

As in metals $\gamma \sim \rho(\varepsilon_F) \sim m^*$, one can say that the effective mass of fermions at the Fermi level becomes very large, $\sim 10^2\text{--}10^3 m_0$. That is why one talks here about heavy electrons, or *heavy fermions*. This strong renormalization can be explained by the "dressing" of electrons by magnetic fluctuations, and it is large because here $T_K \ll \varepsilon_F$: in typical heavy fermion compounds $T_K \sim 1\text{--}10\,\text{K}$, whereas the typical values of the Fermi energy in normal metals are $\sim 2\text{--}5\,\text{eV}$, that is $\sim 10^4\text{--}10^5\,\text{K}$.

When the f level moves even closer to ε_F, the system can go to the state with real mixed valence, $0 < n_f < 1$. In this case the f level itself is close to the Fermi level, Fig. 11.9(c). The width of this level is $\Gamma = \pi \rho(\varepsilon_F)|V|^2$, see eq. (11.4). This width is also much smaller than ε_F, but is usually larger than T_K. Therefore in this state the electron effective mass is also enhanced, but not as strongly as in the heavy fermion regime. The typical values of the coefficient γ in the linear specific heat are in this case $\gamma \sim 10^2\,\text{mJ/mole} \cdot \text{K}^2$ (the γ values are usually used as a measure of how "heavy" the electrons are).

There are many very interesting aspects in the behavior of the heavy fermion and mixed valence compounds on the basis of rare earth and actinide elements. Thus, some of these materials, remaining strongly renormalized, still develop small energy gaps. These systems are sometimes called mixed valence or Kondo insulators. Examples are SmS and SmB$_6$. The detailed nature of the energy gap in these materials is still not clarified, see for example Khomskii (2010).

There is also a number of very interesting superconductors among heavy fermion systems, such as CeCu$_2$Si$_2$ or UPt$_3$. Most probably the type of pairing in them is unconventional – not the standard s-wave pairing, but p-wave or d-wave. The mechanism of pairing in these systems is presumably connected with their proximity to a magnetic state and is caused by spin fluctuations. In this respect these superconductors can be analogous to the high-T_c superconductors on the basis of TM, such as high-T_c cuprates or iron-based superconductors, see Section 9.7. Very interesting is also the discovery of superconductivity in some heavy fermion compounds which do not have an inversion symmetry. In this situation the classification of possible types of pairing would be different from the conventional one, for example the s and p pairings would mix. These materials present definite interest.[4]

---

[4] In a sense, heavy fermion superconductors are more "high-T_c superconductors" than cuprates or Fe-based systems, although on the absolute scale the transition temperatures in them are small, typically $T_c \sim 1\,\text{K}$: whereas in the standard high-T_c superconductors the values of T_c are still much smaller than the Fermi energy, which is the characteristic energy of normal states, $T_c \ll \varepsilon_F$, in heavy fermion superconductors the values of T_c have to be compared with the normal state energy scale, which in this case is defined by the Kondo temperature, and in heavy fermion superconductors it may be that already $T_c \sim \varepsilon_F^* = T_K$, which is the real definition of "high-T_c superconductivity."

Yet another aspect specific for many rare earth and actinide heavy fermion compounds is that due to relatively small energy scales in them one can rather easily change their properties by accessible pressure, magnetic fields, etc., in particular one can tune some of them to the quantum critical regime, see Section 9.5.3. The study of quantum critical phenomena is done nowadays mostly on these compounds, although some TM systems also show these properties. Thus these $4f$ and $5f$ heavy fermion compounds present a "paradise for experimentalists" (although simultaneously a "hell for theoreticians").

Going back to our main topic – TM compounds – we can say that the experience obtained in the study of rare earth heavy fermion systems, and the notions developed in this field, are starting to play a more and more important role also in the study of TM compounds. Owing to the much larger typical energy scales in TM compounds compared with rare earth ones, the observed effects are usually weaker in the former. But still some of the TM systems show behavior similar to that of rare earth systems. The first such example was the spinel LiV_2O_4, which was presented as the first heavy fermion TM compound, see Kondo *et al.* (1997). As shown there, in this system the γ-value in the linear specific heat is $\gamma \sim 450\,mJ/mole \cdot K^2$ – comparable with that in rare earth heavy fermion systems. The first explanation proposed for this behavior was that it is due to the same physics as in rare earth systems, that is the large value of γ is due to the release of spin entropy, cf. Fig. 11.6. But this is not the only possible explanation. If we look at the chemical formula of this compound, we realize that it has V ions with formal valence $V^{3.5+}$, that is we have here one electron per two sites on the B sublattice of a spinel. This situation resembles strongly that of Fe_3O_4, see Section 7.4, and one could expect here some type of charge ordering. However, as discussed in Section 7.4, here we have a strongly frustrated situation. In combination with the larger hopping of vanadium d-electrons compared with that on Fe sites in Fe_3O_4, this can suppress the eventual charge ordering, but still the system may be in close proximity to it. And this tendency to charge ordering, with the corresponding charge localization, can also lead to a large increase in electron effective mass (Jönsson *et al.*, 2007), cf. for example (10.1). Thus it could be not the spin, but charge degrees of freedom which lead to the heavy fermion behavior of LiV_2O_4, in contrast to the conventional heavy fermion $4f$ and $5f$ systems. Still, as we have tried to show in this chapter, there are many analogies (but also many differences) in the behavior of unconventional $4f$ and $5f$ systems, especially those with mixed valence and heavy fermions, and in transition metal compounds: the strong electron–electron interaction finally determines the properties of both these classes of materials.

S.11 Summary of Chapter 11

Electron correlations, crucial for TM compounds, also play a very important role in many other materials, in particular in compounds containing rare earth and actinide elements. These materials are typically also magnetic, by the same reasons (electron localization due to strong electron–electron interaction) as TM compounds. But they also display very specific phenomena, which are in principle similar to the corresponding phenomena in TM systems, but often lead to much stronger effects. The main topic of this book is the physics of TM compounds; but, as many effects in rare earth systems are similar and, in contrast, as some of the concepts developed for rare earth systems are now also used in studying TM compounds, this chapter summarizes the properties of rare earth systems and describes briefly main specific effects in these systems.

Most rare earth and actinide compounds contain partially filled $4f$ or $5f$ shells. The $4f$ states are even more localized than the $3d$ ones, therefore the on-site Hubbard repulsion for them is always much larger than the corresponding hopping, which means that these electrons are always in the localized limit, and they have localized magnetic moments. The $5f$-electrons in actinides are often also localized, although in this sense they are more similar to $3d$-electrons.

One more specific factor is that for $4f$- and $5f$-electrons the relativistic spin–orbit interaction is much more important than for TM elements, because of stronger localization of f-electrons and because of larger mass of ions. Because of that, in rare earth and actinide atoms and ions one cannot use the Russell–Saunders scheme for forming atomic multiplets, but the jj scheme is applicable instead, in which first the spin and orbital moment of each electron are combined into the total moment j, and then these electron moments are summed to give the total moment of an atom or an ion. Strong spin–orbit interaction gives, in particular, g-factors much different from the electronic g-factor 2. These may also become very anisotropic. Thus for Tb^{3+} ions usually $g_{\parallel} \simeq 18$ and $g_{\perp} \simeq 0$, that is Tb^{3+} is practically an Ising ion.

Because of strong localization especially of $4f$ states their interaction with the crystal surrounding is much weaker than for d-electrons, and in effect the crystal field splitting of f levels is much smaller than for d states: typically for $4f$ states it is 10^{-2}–10^{-3} eV, instead of ~ 1 eV for TM ions. Because of this strong localization the $4f$ exchange interaction in insulators is usually much weaker than in TM compounds, so that magnetic ordering occurs in such systems at lower temperatures, often ~ 1–10 K.

But many $4f$ and $5f$ systems are in fact metallic, that is there exist in them, besides localized f-electrons, also itinerant s-, p-, or d-electrons forming broad bands. The interaction of localized f-electrons with delocalized metallic electrons is the most important factor, and it is this interaction which leads to the most interesting, often quite spectacular effects. Such effects appear already in the case of one localized electron in a metal, for example for a magnetic impurity in a nonmagnetic metal. Such magnetic impurity can be a TM impurity, say Mn in Cu, or a rare earth impurity. In this case, first of all, the magnetic moment existing in an isolated atom or an ion can disappear, "dissolve" in the sea of conduction electrons. The simplest process leading to that could be the transition of a localized electron, say with spin ↑, to the conduction band, after which the electron with the opposite spin ↓ can be trapped by the ion. In effect the magnetic moment of an impurity starts to fluctuate, which can lead to the disappearance of the average moment of the site. The conditions for the appearance, or rather for preservation of such a moment on a magnetic impurity in nonmagnetic metal, are usually derived in the Anderson model for magnetic impurities as described in Section 11.2.

But even if at this level the localized magnetic moment survives, there is still an exchange interaction between such a moment of the impurity and the spins of conduction electrons, which is usually antiferromagnetic. And this interaction leads to drastic changes in the properties of such systems at low temperatures. Notably, due to such antiferromagnetic interaction there occurs a screening of the impurity spin by the spins of the conduction electrons. In effect at $T \to 0$, "looking from outside" we would not see a localized spin: it would be surrounded by a cloud of oppositely polarized conduction electrons, which effectively screen it. As a result, for example, the susceptibility, which at high temperatures is that of localized spins and which follows the Curie law, $\chi \sim c/T$, for $T \to 0$ saturates and tends to a constant, $\chi \to$ const. But this behavior is typical for normal metals, Fermi liquids. The temperature at which there occurs such a (gradual) crossover from localized moments with Curie-like susceptibility to screened moments is called the *Kondo temperature* T_K, and the very effect of such screening is the *Kondo effect*.

Metals with magnetic impurities with Kondo effect show many other anomalies, both in thermodynamic and in transport properties. Thus there exists a resistivity minimum at a certain temperature: the scattering of conduction electrons on such impurities becomes stronger at low temperatures, which leads to an increase in resistivity at temperatures $\lesssim T_K$. It is this resistivity minimum which was the first experimental indication of the phenomenon which later became known as the Kondo effect.

If we have not just isolated magnetic impurities, but a concentrated, regular system such as a rare earth metal or intermetallic compound, the situation changes. In this case we have such "impurities" at every site, and one has to take into account an interaction between such f centers. This interaction occurs not via the direct f–f hopping or exchange, which is usually small because the f wavefunctions are strongly localized, but rather with and through the conduction electrons. There are two consequences of this interaction. One is the so-called RKKY (Ruderman–Kittel–Kasuya–Yosida) exchange interaction between different sites, $J_{\mathrm{RKKY}}(r) \sim (J^2/\varepsilon_F) \cos(2k_F r)/r^3$, where J is the exchange interaction of localized (f) and conduction (c) electrons, and k_F and ε_F are the Fermi momentum and Fermi energy of conduction electrons. As we see, this interaction is oscillating in space. It can lead to a formation of one or the other type of magnetic ordering, in particular helicoidal ordering typical for many rare earth metals.

But there exists here an alternative possibility – the screening of magnetic moments of every site due to the Kondo effect. And if this effect is stronger than the RKKY interaction, that is if $T_K > J_{\mathrm{RKKY}}$, then with decreasing temperature the moments would disappear at

$T \lesssim T_K$ – before they have a chance to order. In this case the ground state would again be that of a nonmagnetic metal, a normal Fermi liquid, but with strongly renormalized parameters: the presence of the original localized spins, which disappear only below T_K, still plays a role and leads to the formation of a huge peak in the density of states at the Fermi level (the Kondo or Abrikosov–Suhl resonance), with width T_K. Correspondingly, the renormalized density of states at the Fermi energy $\rho^*(\varepsilon_F) \sim 1/T_K$, and with it for example the susceptibility $\chi(T=0) \sim \rho^*(\varepsilon_F) \sim 1/T_K$, or the coefficient γ in the linear specific heat $c \sim \gamma T$ would be $\gamma \sim 1/T_K$, that is the effective mass of fermions at the Fermi level would be $m^* \sim m_0 \varepsilon_F / T_K$. This state is the state of *heavy fermions*: as for $4f$ and $5f$ systems usually $T_K \ll \varepsilon_F$, the effective mass becomes $m^* \gg m_0$, and all low-temperature properties of the system would be extremely strongly renormalized. For heavy fermion rare earth systems m^* can be $\sim 10^3 m_0$.

One can show that such states appear in metallic systems with $4f$ or $5f$ ions in situations in which the f level ε_f lies relatively close to the Fermi level ε_F. The RKKY interaction $J_{RKKY} \sim J^2/\varepsilon_F$, where the f–c exchange $J \sim V^2/|\varepsilon_f - \varepsilon_F|$ (here V is the f–c hybridization). On the contrary, the Kondo temperature $T_K \sim \varepsilon_F \exp(-J/\varepsilon_F)$, that is for a deep f level and small J it is exponentially small. Therefore for very deep f levels the RKKY interaction wins and a magnetic state will be realized. Such are most rare earth metals and compounds.

If, however, $\varepsilon_f - \varepsilon_F$ becomes small, the Kondo effect may become stronger, in which case we can end up in a heavy fermion state. This may happen for rare earth elements at the beginning (Ce), the end (Tm, Yb), and in the very middle (Sm, Eu) of the $4f$ series. And if, for example under pressure or with some substitution, we move ε_f directly to the Fermi level, the material would go from the heavy fermion state to a *mixed valence* state, with partial occupation of the f level. Thus for example at the γ–α transition in the metal Ce, occurring under pressure, $Ce^{3+}(4f^1)$ goes over to the mixed, or intermediate valence state $Ce^{\sim 3.7+}$. Mixed valence systems are typically also metals, but with renormalization "in between" that in heavy fermion and normal metals, with $m^* \sim 10^1$–$10^2 m_0$.

Thus the sequence of phases when we move the f level from deep below the Fermi energy toward and finally above ε_F is the following: RKKY-dominated magnetically ordered state $\longrightarrow$ heavy fermion state $\longrightarrow$ mixed valence state $\longrightarrow$ completely nonmagnetic (spd) metal without any renormalization and with an empty f level (an example of such a state is the metal La).

In some specific situations mixed valence of heavy fermion systems may still develop a small gap at the Fermi level and become insulators; they are called mixed valence or Kondo insulators. Examples are SmB_6, YbB_{12}, and the "gold" phase of SmS. In other situations there may appear quantum critical points in heavy fermion systems: such quantum critical phenomena are studied experimentally mostly in these systems. Yet in some of them superconductivity appears ($CeCu_2Si_2$, UPt_3, etc.). Most probably superconductivity in them is also unconventional, as for example in high-T_c cuprates: not the standard s-wave, but for example d-wave. And often superconductivity in these systems appears close to quantum critical points.

As mentioned above, one sees the most pronounced effects, such as heavy fermion renormalization of all properties, in rare earth ($4f$) and actinide ($5f$) compounds. But many of these phenomena are met, albeit in a weaker form, in TM compounds as well (but at higher temperatures). Also, in metallic systems on the basis of TM one sees a renormalization of the normal state properties such as the effective mass, specific heat, etc. The Kondo effect is seen also for TM impurities (and it was actually discovered in TM systems). The RKKY interaction is also important for these systems. Recently some TM compounds were discovered which really behave as heavy fermion systems. The first such system is LiV_2O_4 – a metallic spinel with the γ-value in specific heat $c = \gamma T$ being $\gamma \sim 450 \, mJ/mol \, K^2$ (for normal metals $\gamma \sim 1 \, mJ/mol \, K^2$; for real rare earth heavy fermion systems $\gamma \sim 1000 \, mJ/mol \, K^2$). But the mechanism of heavy fermion behavior in LiV_2O_4 may be different from that in rare earth heavy fermion compounds. Instead of the magnetic (Kondo) mechanism, here it may be due to charge degrees of freedom: mixed valent $V^{3.5+}$ would like to order into V^{3+} and V^{4+}, but in a frustrated spinel lattice such a charge ordering may be suppressed, still leaving strong charge fluctuations which can strongly increase the effective mass of carriers.

Thus a lot of effects which are met in rare earth and actinide materials have close parallels in TM compounds. Fundamentally they are all due to strong electron correlations, although specific manifestations of course depend on the specific details of a particular system.

Appendix A

Some historical notes

The history of the development of some of the key concepts discussed in this book is quite interesting and has some rather unexpected twists and turns. In this section we discuss briefly the history of the concepts of Mott insulators, the Jahn–Teller effect, and the Peierls transition.

A.1 Mott insulators and Mott transitions

The notion of a Mott insulator as a state conceptually different from the standard band-like insulators and metals can be introduced using two approaches. In the main text, for example in Chapter 1 we described the approach that uses the Hubbard model (1.6) with short-range (on-site) electron–electron repulsion and attributes the insulating nature for strong interaction to the fact that an electron transferred to an already occupied site experiences repulsion from the electron already sitting on that site. This is the picture most often used nowadays to explain the idea of Mott insulators.

But historically these ideas first appeared in a different picture, presented in a paper by Mott published in 1949 (Mott, 1949) – although it already contained some hints about the picture mostly used nowadays, formalized in the Hubbard model. But the main arguments of Mott in this paper rely rather on the long-range character of Coulomb interaction, and the main statement is that, starting from an insulator, one cannot get a metal by exciting a small number of electrons and holes. These electrons and holes would attract each other by the (screened) Coulomb interaction with potential

$$V(r) = -\frac{e^2}{r}e^{-\kappa r} = -\frac{e^2}{r}e^{-r/r_D},$$ (A.1)

where the Debye screening length r_D is given by the expression

$$\kappa^2 = \frac{1}{r_D^2} = \frac{4me^2 n^{1/3}}{\hbar^2} = \frac{4n^{1/3}}{a_0}$$ (A.2)

with Bohr radius

$$a_0 = \frac{\hbar^2}{me^2}.$$ (A.3)

The concentration of free carriers, electrons and holes, is here n.

One can also include here the static dielectric constant; this does not change the final conclusions.

Now, the arguments presented in Mott (1949) were that at small concentration of charge carriers n and, consequently, weak screening of Coulomb interaction, there would always be an electron–hole bound state (an exciton) for the interaction (A.1), that is the excited electrons and holes thus created (with small concentration) would be bound to neutral excitons and would not produce metallic conductivity. Thus, from these arguments it is clear that because of such Coulomb interaction the insulating and metallic states should be different states of a solid, and continuous transition between them is not possible. Only when we create simultaneously a substantial number of electrons and holes would the Debye screening be so strong that the exciton bound state would disappear, and the metallic state with the finite (large) concentration of free carriers would be formed.

The condition for the disappearance of the bound state in the potential (A.1) is

$$\kappa > \frac{me^2}{\hbar^2} \tag{A.4}$$

from which, using also eqs (A.2) and (A.3), we can get the condition for the existence of the metallic state

$$a_0 n^{1/3} > 0.25. \tag{A.5}$$

In the first paper by Mott (1949) there were only qualitative arguments, without any formulae and numerical estimates presented above, but they were contained in his later publications, for example Mott (1961).

This explanation of the difference between insulators and metals, and of the transition between them, was the first picture of Mott insulators and Mott transition. In this picture the transition should necessarily be jump-like, that is I order, with the immediate appearance in a metal of a finite electron concentration $n^{1/3} > 0.25/a_0$, see (A.5), the state with smaller concentration remaining insulating not because of the standard band effects but because of the Coulomb interaction (here in the form of attraction of electrons and holes). It was picked up by the community, and reproduced for example in Section 5.9 of a very good book by Ziman on the theory of solids (Ziman, 1964) – one of the best books in this field even now. And it was only later, especially after the papers by Anderson (1959) and Hubbard (1963), that another language became more popular (although, as already mentioned, these two pictures are in fact closely related, and already in the first paper by Mott (1949) there were arguments pointing in this direction, e.g. the comparison of the Heitler–London picture to that of molecular orbitals, which is essentially equivalent to the band picture, see Chapter 1).

Now, let us take one step back in time. During the Second World War, Lev Landau published a short paper with a then rather young colleague, Yakov Zeldovich (Landau and Zeldovich, 1944). Published in the Soviet Union, this paper remained largely unnoticed in the West, but now it is easily available since it was reproduced in the Collected

Papers of L. D. Landau (Landau, 1965, p. 380). In this paper the authors were interested mostly in the relation between liquid–gas and metal–insulator transitions in liquid metals such as mercury or molten sodium. At the start of this paper there is the following passage:

A dielectric differs from a metal by the presence of an energy gap in the electronic spectrum. Can however this gap tend to zero when the transition point into a metal is approached (on the side of the dielectric)? In this case we should have to do with a transition without latent heat, without change of volume and of other properties. Peierls has pointed out that a continuous transition—in this sense— is impossible. Let us consider the excited state of the dielectric in which it is capable of conducting an electric current: an electron has left its place, leaving a positive charge in a certain place of the lattice and is moving throughout the latter. At large distances from the positive charge, the electron must certainly suffer a Coulomb attraction tending to bring it back. In a Coulomb attraction field there always exist discrete levels of negative energy, corresponding to a binding of the electron; the excited conducting state of the dielectric must therefore always be separated from the fundamental one, in which the electron is bound, by a gap of a finite width.

As we see, essentially the same arguments as those appearing in the first paper by Mott are already contained here, although without the numerical estimate (A.5). Landau and Zeldovich refer here to yet another famous 20th-century physicist, Rudolf Peierls, as one of the proponents of the main idea.

For a long time I thought that this was just a remark of Peierls made in informal discussion, or in private communication, never published. However recently I found that these ideas actually appeared in a short paper published in the *Proceedings of the Physical Society* in 1937, as a discussion after the talk of Dutch physicists de Boer and Verwey, who had reported that NiO, which according to band theory should be a metal, is actually a very good insulator. This short note was actually written by Mott, using, as mentioned there, notes of Peierls: see Mott and Peierls (1937). Most probably, in this short paper Mott also included some of his earlier thoughts on the subject, although he attributes the main ideas to Peierls. This text is so interesting that I reproduce a large part of it below:

Prof. Peierls agreed with Dr. de Boer that the existence of semi-conductors with incomplete d-bands could not be understood merely by considering the low transparency of the potential barrier. This transparency could be perhaps 10^{-2} or 10^{-3}, less than for ordinary metal but not 10^{-10}. He suggested that a rather drastic modification of the present electron theory of metals would be necessary in order to take these facts into account. The solution of the problem would probably be as follows: if the transparency of the potential barriers is low, it is quite possible that the electrostatic interaction between the electrons prevents them from moving at all. At low temperatures the majority of the electrons are in their proper places in the ions. The minority which have happened to cross the potential barrier find therefore all the other atoms occupied, and in order to get through the lattice have to spend a long time in ions already occupied by other electrons. This needs a considerable addition of energy and so is extremely improbable at low temperatures.

It appears, therefore, that if the transparency is at all small, the electrostatic interaction would reduce the conductivity still further and thus that at low temperatures the conductivity is proportional to a high power of the initial transparency. No mathematical development of these ideas has however yet been given.

Prof. Peierls made some further remarks about semi-conductors containing impurities. Suppose the impurity is capable of giving off an electron; it is then not sufficient to remove the electron into one of the adjoining lattice ions, in order to make it free for electronic conduction. This is because the impurity is then left in an ionized state and will attract the electron, which therefore moves in the periodic field of the lattice with a Coulomb field superimposed. It is known that in such a field the electron is capable of discrete bound states of negative energy (as in the field of a positive ion). Only if the energy of the electron is still further increased so that it can overcome this attraction and move right away from the impurity centre will it become a conduction electron in the proper sense. This suggests that, even if, as might conceivably happen, an electron would gain energy in moving from the impurity atom to an adjacent lattice ion, nevertheless the lowest state of the electron would always be a bound state in the neighbourhood of the impurity centre, in which it, so to speak, revolves round the centre, even though it is not definitely attached. Hence a finite activation energy should always be necessary to produce conduction electrons. This conclusion no longer holds when the number of impurity centres becomes large because then the electron need not overcome the attraction of its 'home' centre completely; already at a small distance from it, it will come under the simultaneous attraction of other centres and will thus be released.

From these arguments one may draw the conclusion that one should never find a substance in which a small concentration of impurity can produce a conductivity without activation energy.

It may be possible to explain on these lines the dependence of activation energy on concentration shown by most semi-conductors.

Similarly it should be true that pure substances of stoichiometric composition should never show a very small number of conduction electrons with zero activation energy.

As we see, in this text, in the remarks attributed to Peierls, already *both* pictures of Mott insulators used nowadays to explain this phenomenon are already expressed in a very clear form. The second part of this text presents the idea that the unscreened Coulomb attraction of electrons and holes (Peierls mentions here not holes, but positively charged impurities) would always lead to the formation of nonconducting bound states, and that for a metallic state one needs finite concentration of charge carriers – the idea used later by Landau and Zeldovich, and which apparently influenced the first significant publications of Mott in 1949 and later. But the first part of these remarks by Peierls formulates the picture we now use referring to the Hubbard model: that the electrons would be localized at their sites, and it would cost large energy to put extra electrons on the already occupied sites, which could suppress conductivity completely. And this does not rely on the long-range character of Coulomb interaction. This is exactly what we now call the Hubbard model and Mott or Mott–Hubbard insulators.

Thus, although the great role of Mott in this field cannot be overemphasized and we call these phenomena by his name deservedly, two other giants of theoretical physics 20th-century, Peierls and Landau, also stood at the cradle of these concepts, which now play such an important role in modern solid-state physics, in particular that of transition metal compounds.

A.2 Jahn–Teller effect

Interestingly enough, it seems that Landau also played an important role in the development of the ideas which are known as the Jahn–Teller effect. This is what Teller himself wrote in the preface to a book on the Jahn–Teller effect (Englman, 1972). Here is an excerpt from this preface (which Teller himself called "An historical note"):

In the year 1934 both Landau and I were in the Institute of Niels Bohr at Copenhagen. I had many discussions. I told Landau of the work of one of my students, R. Renner, on degenerate electronic states in the linear CO_2 molecule.... He said that I have to be very careful. In a degenerate electronic state the symmetry on which this degeneracy is based ... will in general be destroyed. ...

I proceeded to discuss the problem with H. A. Jahn who, as I, was a refugee from the German university. We went through all possible symmetries and found that the linear molecules constitute the only exception. In all other cases Landau's suspicion was verified. ...

This is the reason why the effect should carry the name of Landau. He suspected the effect, and no one has given a proof that mathematicians would enjoy. Jahn and I merely did a bit of a spade work.

Of course, Teller definitely underestimated his contribution: it was not just "spade work," indeed it played a crucial role in opening up the large and very important field of the Jahn–Teller effect. Still, it is interesting that this field, as so many others, bears an imprint of Landau.

A.3 Peierls transition

The name of Peierls, which has appeared in connection with the history of Mott transitions, see Section A.1 above, is mentioned in this book in another context. In many places we refer to Peierls instability, or Peierls transition, or Peierls dimerization – especially in connection with orbital physics (Chapter 6) and metal–insulator transitions (Chapter 10). This is a phenomenon initially proposed for one-dimensional systems. Its essence is that if in the band picture (without electron–electron interaction taken into account) a (tight-binding) band is half-filled, as for example in Fig. 1.8, this system is unstable with respect to lattice dimerization. Indeed, if we introduce a lattice distortion by shifting every second atom, making alternating short and long bonds, the resulting unit cell will be doubled, that is the boundaries of the Brillouin zone would be not at $\pm\pi$, but at $\pm\frac{\pi}{2}$; so, a gap will open at the Fermi surface, see Fig. A.1. Thus the resulting state would be insulating.

Interestingly enough, this idea, which is now so important for many systems, not only quasi-one-dimensional, was introduced by Peierls not even in a special publication; rather, it was first mentioned in a short remark in his book *Quantum Theory of Solids* (Peierls, 1955). As Peierls himself later remarked in an interesting collection of

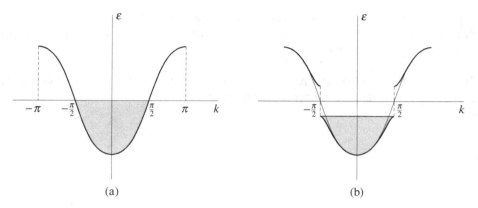

Figure A.1 Change in energy spectrum of one-dimensional metals with a half-filled band (a), after dimerization (b).

essays (Peierls, 1991), he thought the result so obvious that he did not even include any mathematics proving the instability of the homogeneous one-dimensional metal with respect to such distortions. But, in fact, this result needs some mathematical justification. Indeed, when we distort the lattice as explained above, the energy of all occupied states goes down and thus we gain some energy, but simultaneously there is an elastic energy loss due to the distortion, $E_{\text{elastic}} \sim Bu^2/2$, where u is the lattice distortion (dimerization). One has to show that the electron energy gain exceeds the elastic energy loss; only then would the homogeneous system be unstable and Peierls dimerization occur. In fact the calculations presented by Peierls much later (Peierls, 1991; see also, e.g., Khomskii, 2010) have shown that the electron energy gain is always bigger than the elastic energy loss: the electronic energy is $E_{\text{el}} \sim -u^2 \ln u$, and a one-dimensional system is indeed always unstable with respect to such distortion. (Actually this holds true not only for half-filled one-dimensional bands and the corresponding dimerization, but for arbitrary concentration of electrons in one-dimensional systems with the band filled up to a certain Fermi energy ε_F and Fermi momentum k_F: one can always make a distortion with period $l = a\hbar/2k_F$ so that all electronic states below the Fermi energy would go down, and produce the same energy gain $\sim u^2 \ln u$. The same arguments also work in some two-dimensional and three-dimensional systems with the nesting of the Fermi surface, see Sections 7.3, 9.1.)

As already mentioned, in the original publication (Peierls, 1955) Peierls had not included these calculations, thinking the conclusion self-evident, but reproduced them in detail in Peierls (1991). There he wrote:

This instability came to me as a complete surprise when I was tidying material for my book [Peierls 1955], and it took me a considerable time to convince myself that the argument was sound. It seemed of only academic significance, however, since there are no strictly one-dimensional systems in nature (and if there were, they would become disordered at any finite temperature ...) I therefore did not think it worth publishing the argument, beyond brief remark in the book, which did not even mention logarithmic behaviour.

But this short remark and the physics contained in it turned out later to be extremely important in many fields of condensed matter physics.

Appendix B

A layman's guide to second quantization

From the very beginning, from Chapter 1 in this book, when introducing and discussing models relevant to our field we have used the notation and technique known as the method of second quantization. It is a very convenient method, used widely and described in detail in many books on quantum theory of solids (and in other fields). It is indeed a special technique, with its own detail and subtle points, but in essence it is very simple; and just some basic notions are sufficient for our purposes.

The method of second quantization works mainly with operators such as $c^\dagger$ and c, known as creation and annihilation operators. If we start from a certain state of the system, described by some wavefunction, say $|\Psi\rangle$, the operators $c^\dagger$ and c describe the processes of adding or removing a particle – it may be an electron, a phonon, etc. For instance to create an extra electron at site i with spin σ, we apply the corresponding creation operator $c^\dagger_{i\sigma}$, that is the new state with one extra electron will be $c^\dagger_{i\sigma}|\Psi\rangle$. Similarly, we can describe the process by which we remove an electron, say from site j with spin σ'; this process is described by applying the annihilation operator $c_{j\sigma'}$, and the resulting state will be $c_{j\sigma'}|\Psi\rangle$. One immediately sees that with the help of this notation we can very conveniently describe for example the process of electron hopping from site to site: if we apply the combined operator $c^\dagger_{i\sigma}c_{j\sigma}$ to the initial state $|\Psi\rangle$ (which, we suppose, already contains an electron with spin σ on site j), we remove the electron from site j (operator $c_{j\sigma}$) but then (re)create it at site i. In other words, the term

$$c^\dagger_{i\sigma}c_{j\sigma}|\Psi\rangle \tag{B.1}$$

describes the process of hopping of an electron from site j to site i (with the spin σ of course conserved during such process), see for example the first term in the Hamiltonian of the Hubbard model (1.6).

One can also describe in this language the occupation of a particular site. It is given by the occupation number operator $\hat{n}_{i\sigma} = c^\dagger_{i\sigma}c_{i\sigma}$. When we apply this operator to a state in which there is no such electron at site i, the operator $c_{i\sigma}$ would give zero (it has to annihilate an electron from site i, but if there is no such electron there, it gives 0). Assuming there is an electron in this state, then when we apply the operator $\hat{n}_{i\sigma}$ to such a state we recover the same state: $\hat{n}_{i\sigma}|\Psi\rangle = c^\dagger_{i\sigma}c_{i\sigma}|\Psi\rangle = |\Psi\rangle$ (the annihilation operator destroys the electron $\{j, \sigma\}$, but then the creation operator recreates it in the same state). In effect we have

$$\hat{n}_{i\sigma}|\Psi\rangle = c_{i\sigma}^{\dagger}c_{i\sigma}|\Psi\rangle = \begin{cases} 0 & \text{if there is no electron } \{i, \sigma\}, \\ \\ |\Psi\rangle & \text{if there is one such electron.} \end{cases} \quad (\text{B.2})$$

Thus, we may write

$$\hat{n}_{i\sigma}|\Psi\rangle = c_{i\sigma}^{\dagger}c_{i\sigma}|\Psi\rangle = n_{i\sigma}|\Psi\rangle, \quad (\text{B.3})$$

where $n_{i\sigma}$ is the number of electrons in the corresponding state (which for electrons, i.e. fermions, can be only 0 or 1). Thus the occupation number n is an eigenvalue of the number operator $\hat{n} = c^{\dagger}c$. Using these number operators, we can describe, for instance, the Hubbard on-site repulsion of electrons – the second term in the Hamiltonian of the Hubbard model (1.6).

We can introduce such operators for electrons, or for phonons, or for magnons, etc. As we know, electrons are fermions, obeying the Pauli exclusion principle, according to which there should be no more than one fermion at each state (with the same spin). This means that they satisfy the Fermi statistics, which tells us that the total wavefunction should be antisymmetric with respect to interchange of two particles (fermions). In the second quantization formalism this is reflected in the condition that the fermion creation and annihilation operators should obey anticommutation relations

$$\{c_{i\sigma}^{\dagger}, c_{j\sigma'}^{\dagger}\} = c_{i\sigma}^{\dagger}c_{j\sigma'}^{\dagger} + c_{j\sigma'}^{\dagger}c_{i\sigma}^{\dagger} = 0, \quad (\text{B.4})$$

$$\{c_{i\sigma}, c_{j\sigma'}\} = c_{i\sigma}c_{j\sigma'} + c_{j\sigma'}c_{i\sigma} = 0, \quad (\text{B.5})$$

$$\{c_{i\sigma}, c_{j\sigma'}^{\dagger}\} = c_{i\sigma}c_{j\sigma'}^{\dagger} + c_{j\sigma'}^{\dagger}c_{i\sigma} = \delta_{ij}\delta_{\sigma\sigma'}, \quad (\text{B.6})$$

where we use the notation $\{\ ,\ \}$ for the anticommutator of two operators, $\{a, b\} = ab + ba$; the delta-symbol δ_{ij} is equal to 1 for $i = j$ and 0 for $i \neq j$, and the same for $\delta_{\sigma\sigma'}$. One can show that the Pauli principle is naturally incorporated in these (anti)commutation relations, and that the result (B.3) is perfectly consistent with it.

Similarly to electrons (fermions) we can also introduce the second quantization operators (creation and annihilation, number operator, etc.) for bosons (particles with integer spin, e.g. phonons). Such operators, $b_{\alpha}^{\dagger}, b_{\alpha}$, where the index α may be, for example, the phonon momentum and index of a phonon branch (longitudinal, transverse), obey not anticommutation but commutation relations

$$[b_{\alpha}, b_{\beta}^{\dagger}] = b_{\alpha}b_{\beta}^{\dagger} - b_{\beta}^{\dagger}b_{\alpha} = \delta_{\alpha\beta}. \quad (\text{B.7})$$

Because of the different statistics, reflected in the different commutation relations for bosons (B.7) compared with those for fermions (B.4)–(B.6), the boson wavefunctions should be symmetric under interchange of two bosons, and there may be many bosons in one state. This, in particular, can lead to the phenomenon of Bose-condensation, see for example chapter 5 in Khomskii (2010). This, however, is not important for our present topics (although sometimes one uses the notions of Bose-condensation in application to certain magnetic systems based on transition metals, see e.g. Sonin, 2010). We only mention that, similar to the expression (B.3), one can introduce the number operator for bosons, $\hat{n}_\alpha = b_\alpha^\dagger b_\alpha$, so that again

$$\hat{n}_\alpha|\Phi\rangle = n_\alpha|\Phi\rangle, \tag{B.8}$$

where n_α is the number of bosons in state α (e.g., with momentum k) in the total many-particle function $|\Phi\rangle$ (which for bosons can be any number).

In fact this information, especially that contained in the expression (B.1), is sufficient for understanding the notations and notions used in the main text. Those who want to get a deeper understanding of this technique, its virtues and possibilities, can find more material in the following books for example: Mattuck (1992); Mahan (2000); Kittel (2004b); Khomskii (2010).

Appendix C

Phase transitions and free energy expansion: Landau theory in a nutshell

C.1 General theory

In several places in this book we have used the language and notions first developed by Landau to describe second-order phase transition, but which are used nowadays in a much broader context. Here we summarize the basics of this theory and illustrate different situations in which it is used. One can find a more detailed description for example in the brilliant original presentation of Landau and Lifshitz (1969), or in Khomskii (2010) (which is more or less followed below).

The original aim of Landau was to describe II order phase transitions – transitions in which a certain ordering, for example ferromagnetic, appears with decreasing temperature at some critical temperature T_c in a continuous manner. But it turned out later that the approach developed has much broader applicability than originally planned.

In thermodynamics and in statistical physics the optimal equilibrium state of a many-particle system is determined by the condition of the minimum of the Helmholtz free energy

$$F(V, T) = E - TS \qquad (C.1)$$

or of the Gibbs free energy

$$\Phi(P, T) = E - TS + PV \qquad (C.2)$$

at given temperature and either fixed volume (C.1) or fixed pressure (C.2); more often in reality we are dealing with the second situation. When a certain ordering appears in the system – it may be magnetic ordering, for example ferro- or antiferromagnetic; or ferroelectricity; or an ordering in a structural phase transition – one can introduce a measure of such ordering, different for specific situations, which is called the *order parameter*; let us denote it η. It may be a scalar, for example electron density at charge ordering; or a vector, such as polarization for ferroelectrics; or a pseudovector such as magnetization in a ferromagnet; or a tensor for quadrupole ordering; and so on. The free energy of the system should depend on the degree of this ordering, that is on the order parameter η, for example $\Phi(P, T, \eta)$. The equilibrium value of the corresponding order parameter is again determined by the condition of the minimum of the free energy, that is by the condition

$$\frac{\partial \Phi(P, T, \eta)}{\partial \eta} = 0 \tag{C.3}$$

(of course one has to check that the solution of this equation really corresponds to a minimum of the free energy, and not to a maximum or a saddle point).

For phase transitions the order parameter η is zero in the disordered phase at $T > T_c$, and becomes nonzero below T_c. Close to T_c it should be small, otherwise it would not be a continuous II order transition. In this situation, according to the idea of Landau, one can make an expansion of the free energy in small η, which for a scalar order parameter η can be written

$$\Phi(P, T, \eta) = \Phi_0 + \alpha\eta + A\eta^2 + C\eta^3 + B\eta^4 + \cdots \tag{C.4}$$

For more complicated order parameters (vectors, tensors) one can write similar expressions, including terms allowed by symmetry, see below.

Which terms may exist in the general expression (C.4) is determined by some general requirements. Thus, we suppose that in the absence of an external field the order parameter at $T > T_c$ should be zero. This implies that the coefficient α in the linear term $\alpha\eta$ in the expansion (C.4) should be zero: otherwise the minimum of the free energy given by equation (C.3) would always correspond to $\eta \neq 0$. Similarly, in most cases the free energy should not depend on the sign of the order parameter, that is it should be an even function of η. Such is for example the case for ferromagnetic ordering, for which in the isotropic system the states with magnetization M and $-M$ should be equivalent. In this case we can conclude that the cubic term in (C.4), $C\eta^3$, should also be absent, $C = 0$. Then we are left with the expression

$$\Phi = \Phi_0 + A\eta^2 + B\eta^4 + \cdots \tag{C.5}$$

(This is the reason why in writing the expression (C.4) we have used a somewhat unnatural notation for coefficients, with C entering before B – in most cases this term drops out anyway. But there are also interesting situations in which such cubic invariants are present in the expansion of the free energy; these would give rise to jump-like I order transitions, see e.g. chapter 2 in Khomskii (2010).)

To describe the II order phase transitions, a natural assumption is that the coefficient A in the expansion (C.5) is positive for $T > T_c$, and negative for $T < T_c$. In this case indeed we would get the solutions with $\eta = 0$ for $T > T_c$, and $\eta \neq 0$ for $T < T_c$. The free energy (C.5) in this phase has the form shown in Fig. C.1 (we assume here that the coefficient B in (C.5) is positive, $B > 0$).

Making the simplest assumption, $A = a(T - T_c)$, consistent with the assumed behavior of $A(T)$, we obtain from the minimization (C.3):

$$\eta^2 = -\frac{A}{2B} = \frac{a}{2B}(T_c - T) \tag{C.6}$$

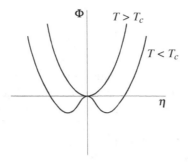

Figure C.1 Free energy as a function of the order parameter for $T > T_c$ and $T < T_c$.

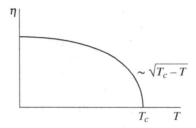

Figure C.2 The behavior of the order parameter $\eta(T)$ in Landau theory of second-order phase transitions.

for $T < T_c$, and $\eta = 0$ for $T > T_c$. Thus, η behaves close to T_c as $\eta \sim \sqrt{T_c - T}$, see Fig. C.2. The free energy itself behaves as

$$\Phi_{\min} = \Phi_0 - \frac{A^2}{4B} = \Phi_0 - \frac{a}{2B}(T_c - T). \tag{C.7}$$

Using these results we can also describe many other properties of the system close to such phase transitions, such as the behavior of the specific heat, susceptibility, etc. These results are not presented here; one can find the details in Landau and Lifshitz (1969) or in Khomskii (2010). Rather, we concentrate on the different uses of the Landau expansion for free energy.

C.2 Dealing with the Landau free energy functional

As we have already said above, in writing the Landau expansion for the free energy we have to include only terms allowed by the symmetry of the corresponding system. Thus, for a scalar order parameter we should keep terms included in the expansion (C.5). For more complicated situations, and for other order parameters, there may be other terms which should enter such an expansion. Thus for example for the isotropic ferromagnet with order parameter M (average magnetization) the free energy has the form

$$\Phi = AM^2 + BM^4 - H \cdot M, \tag{C.8}$$

where we include also the interaction with the external magnetic field H. For the usual two-sublattice antiferromagnetic ordering the order parameter is not the total magnetization, but the sublattice magnetization, or the difference between the moments of two sublattices:

$$L = M_1 - M_2. \tag{C.9}$$

It is clear that this order parameter does not couple linearly to the magnetic field, thus we have to write this coupling in a different form, so that the total expansion for an antiferromagnet would be

$$\Phi = \Phi_0 + AL^2 + BL^4 + K(H \cdot L)^2 - \tfrac{1}{2}\kappa_p H^2. \tag{C.10}$$

For the coupling to a magnetic field we have used here the lowest-order term allowed by symmetry: it should be of second order in L, and it should be invariant with respect to rotations and time inversion (for which the magnetic vectors H, M, L change sign). We have also included here the last term, describing the energy of the paramagnetic phase in the external field. From the expression (C.10) we can, for example, obtain the well-known behavior of magnetic susceptibility of an antiferromagnet, shown in Fig. C.3, which, in particular, shows that below the critical temperature (Néel temperature T_N) the susceptibility is anisotropic (which actually follows from the form of the coupling to the magnetic field in the expression (C.10)).

Similarly, using general symmetry requirements, one can also write expressions for the free energy for systems in which there is a coupling of a particular ordering, for example magnetic, to some other degrees of freedom, say lattice distortions. One can also have different orderings coupled to each other, such as ferroelectric and magnetic ordering in multiferroics, discussed in Chapter 8. In all these cases one should keep in the free energy only terms (invariants) allowed by symmetry. Furthermore, if some such terms are allowed by symmetry, they *have to be included* in the corresponding free energy; that is, if certain

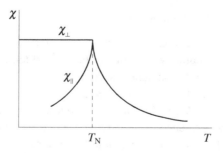

Figure C.3 The behavior of parallel and perpendicular magnetic susceptibility in an isotropic antiferromagnet.

terms are allowed by symmetry, they *will appear*. Maybe they will enter with small coefficients, so that their effect on the behavior of the system would be minor – but in any case all such terms should be included and their possible consequences investigated.

C.3 Some examples

1. *Phase transitions in a compressible lattice*

When we consider phase transitions, characterized by order parameter η, in a compressible lattice, the simplest form of free energy including the coupling to the lattice is

$$\Phi = A\eta^2 + B\eta^4 + \lambda\eta^2 u + \frac{bu^2}{2}. \tag{C.11}$$

Here we have included the coupling $\lambda\eta^2 u$ of the order parameter η with the deformation (homogeneous compression) u. It again should be quadratic in η, because in most cases the free energy should not change when $\eta \rightarrow -\eta$. We have also added the last term describing the elastic energy of the lattice with the bulk modulus b.

Minimizing this free energy in the distortion u, we obtain

$$\frac{\partial\Phi}{\partial u} = bu + \lambda\eta^2 = 0 , \qquad u = -\frac{\lambda\eta^2}{b} \tag{C.12}$$

and, putting this expression back into the free energy (C.11), we get

$$\Phi = A\eta^2 + B\eta^4 + \frac{\lambda^2\eta^4}{2b} - \frac{\lambda^2\eta^4}{b} = A\eta^2 + \left(B - \frac{\lambda^2}{2b}\right)\eta^4. \tag{C.13}$$

We see from the expression (C.13) that for strong enough coupling with the lattice (large coupling constant λ) and for soft lattice (small bulk modulus b) the renormalized coefficient at the quadratic term in (C.13), $B' = B - \lambda^2/2b$, may become negative. But one can show that in this case the phase transition would become jump-like I order, see for example Khomskii (2010). This is, in particular, the mechanism which can make some magnetic phase transitions first order (the Bean–Rodbell mechanism mentioned in Section 5.4).

2. *Finding T_c in a ferromagnet; the Arrott plot*

One can use the Landau expansion of type (C.5), or, more specifically, (C.8), to devise a method of determining accurately the value of the critical temperature of ferromagnetic ordering. As is well known, in ferromagnets without or in a weak external field there always appear magnetic domains, and this hinders significantly the accurate determination of the critical temperature. One can use the expansion (C.8) to overcome this obstacle. Minimizing this expression in the presence of an external field we obtain (for magnetization M parallel to magnetic field H, so that we can omit vector notation)

$$\frac{\partial\Phi}{\partial M} = 2AM + 4BM^3 - H = 0. \tag{C.14}$$

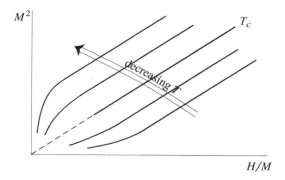

Figure C.4 The Arrott plot allowing one to determine accurately the critical temperature of a ferromagnet.

We can rewrite eq. (C.14) in the form

$$2BM^2 = H/2M - A. \tag{C.15}$$

Remember that in Landau theory the coefficient A is positive for $T > T_c$, and negative for $T < T_c$, and close to T_c it can be approximated by the expression $A = a(T - T_c)$. Then eq. (C.15) becomes

$$2BM^2 = H/2M - a(T - T_c). \tag{C.16}$$

Thus, if we measure the magnetization experimentally at different temperatures and fields – pretty routine measurements – and plot M^2 vs H/M at different temperatures, we would get a series of curves as shown in Fig. C.4. This is called an Arrott plot. At strong enough fields, when all the domain physics is already gone (there are no domains left, all spins point in the same direction) the magnetization is still not equal to the maximum possible value (the order parameter in Fig. C.2 at nonzero temperatures is still not maximal), and M still increases with the field, as follows from the relation (C.16) (this is sometimes called a paraprocess). This gives the linear parts of the curves in Fig. C.4. According to (C.16) these straight lines in coordinates $(M^2, H/M)$ extrapolate for $T > T_c$ to negative values of M^2, and for $T > T_c$ to positive ones. Exactly at T_c one such straight line extrapolates to zero – and this line gives the value of T_c. (Close to the origin these lines become curved due to domain effects, etc. But this region is not essential for this method, one uses the part of the plot where the dependence of M^2 on H/M is linear.)

References

Aarborgh, H. M. *et al.* (2006), *Phys. Rev.* **B 74**, 134408.

Abbate, M. *et al.* (1963), *Phys. Rev.* **B 47**, 16124.

Abragam, A. and Bleaney, B. (1970), *Electron Paramagnetic Resonance of Transition Ions.* Oxford: Clarendon Press.

Adler, D. and Brooks, H. (1967), *Phys. Rev.* **155**, 826.

Aeppli, G. and Fisk, Z. (1992), *Comments Cond. Mat. Phys.* **16**, 155.

Alonso, J. A. *et al.* (1999), *Phys. Rev. Lett.* **82**, 3871.

Anderson, P. W. and Hasegawa, H. (1955), *Phys. Rev.* **100**, 675.

Anderson, P. W. (1956), *Phys. Rev.* **102**, 1008.

Anderson, P. W. (1958), *Phys. Rev.* **109**, 1492.

Anderson, P. W. (1959), *Phys. Rev.* **115**, 2.

Anderson, P. W. (1961), *Phys. Rev.* **123**, 41.

Anderson, P. W. and Blount, E. I. (1965), *Phys. Rev. Lett.* **14**, 217.

Anderson, P. W. (1973), *Mat. Res. Bull.* **8**, 153.

Anderson, P. W. (1997), *The Theory of Superconductivity in the High-T_c Cuprate Superconductors.* Princeton, NJ: Princeton University Press.

Andrei, N., Furuya, K. and Löwenstein, H. H. (1983), *Rev. Mod. Phys.* **55**, 331.

Andres, K., Bucher, E., Darack, S. and Maita, J. P. (1972), *Phys. Rev. Lett.* **6**, 2716.

Anisimov, V. I., Elfimov, I. S., Korotin, M. A. and Terakura, K. (1997), *Phys. Rev.* **B 55**, 15494.

Anisimov, V. I. *et al.* (2002), *Eur. Phys. J.* **B 25**, 191.

Aoki, H. and Kamimura, H. (1987), *Solid State Comm.* **63**, 665.

Arima, T. (2007), *J. Phys. Soc. Japan* **76**, 073702.

Ascher, E., Schmid, H. and Tar, D. (1964), *Solid State Comm.* **2**, 45.

Ascher, E., Rieder, H., Schmid, H. and Stossel, H. (1966), *J. Appl. Phys.* **37**, 1404.

Ashcroft, N. W. and Mermin, N. D. (1976), *Solid State Physics.* Philadelphia: Saunders/Harcourt.

Astrov, D. N. (1960), *Sov. Phys.—JETP* **11**, 708.

Astrov, D. N. (1961), *Sov. Phys.—JETP* **13**, 729.

Attfield, J. P. *et al.* (1998), *Nature* **396**, 655.

Attfield, J. P. (2006), *Solid State Sci.* **8**, 861.

Babushkina, N. A. *et al.* (1998), *Nature* **391**, 159.

Ballhausen, C. J. (1962), *Introduction to Ligand Field Theory.* New York: McGraw-Hill.

Bao, W. *et al.* (1993), *Phys. Rev. Lett.* **71**, 766.

Bean, C. P. and Rodbell, D. S. (1962), *Phys. Rev.* **126**, 104.

Bednorz, J. G. and Müller, K. A. (1986), *Zeitschr. für Physik B: Condens. Matter.* **64**, 189.

Belik, A. A. *et al.* (2005), *Chem. Mater.* **17**, 269.

Belik, A. A. *et al.* (2006), *Chem. Mater.* **18**, 798.

Berry, M. V. (1984), *Proc. Roy. Soc. London* **A 392**, 45.

Bersuker, I. B. (1962), *Zh. Exp. Teor. Fiz.* **43**, 1315 [*Sov. Phys.—JETP* **16**, 933 (1963)].

Bersuker, I. B. (2006), *The Jahn–Teller Effect*. Cambridge: Cambridge University Press.

Bersuker, I. B. (2010), *Electronic Structure and Properties of Transition Metal Compounds*. Hoboken, NJ: Wiley.

Bertaut, E. F. (1965), in G. T. Rado and H. Suhl (eds), *Magnetism*, vol. 3, p. 149. New York: Academic Press.

Bhattacharjee, S., Bosquet, E. and Ghoses, P. (2009), *Phys. Rev. Lett.* **102**, 117602.

Biermann, S. *et al.* (2005), *Phys. Rev. Lett.* **94**, 026404.

Birgenau, R. J., Als-Nielsen, J. and Bucher, F. (1972), *Phys. Rev. Lett.* **6**, 2724.

Blanco-Canosa, S. *et al.* (2007), *Phys. Rev. Lett.* **99**, 187201.

Blanco-Canosa, S. *et al.* (2009), *Phys. Rev. Lett.* **102**, 056406.

Blinc, R. (2011), *Advanced Ferroelectricity*. Oxford: Oxford University Press.

Bloembergen, N. and Rowland, T. J. (1955), *Phys. Rev.* **97**, 1679.

Bocquet, A. E. *et al.* (1992), *Phys. Rev.* **B 45**, 3771.

Bode, M. *et al.* (2007), *Nature* **447**, 190.

Bogdanov, A. N. and Yablonskii, D. A. (1989), *Sov. Phys.—JETP* **68**, 101.

Bos, J.-W. G. *et al.* (1998), *Phys. Rev.* **B 78**, 094416.

Botana, A. S. *et al.* (2011), *Phys. Rev.* **B 84**, 115138.

Brandes, T. and Kettemann, S. (2003), *The Anderson Transition and its Ramifications — Localization, Quantum Interference and Interactions*. Berlin: Springer-Verlag.

Brinkman, W. F. and Rice, T. M. (1970), *Phys. Rev.* **B 2**, 4302.

Brown, N. F., Hornreich, R. M. and Shtrikman, S. (1968), *Phys. Rev.* **168**, 574.

Brückner, W. *et al.* (1983), *Vanadiumoxide: Darstellung, Eigenshaften, Einwendungen*. Berlin: Akademie-Verlag (in German).

Bulaevskii, L. N., Nagaev, E. L. and Khomskii, D. I. (1968), *Zh. Exp. Teor. Fiz.* **54**, 1562 [*Sov. Phys.—JETP* **27**, 836].

Bulaevskii, L. N. *et al.* (2008), *Phys. Rev.* **B 78**, 024402.

Buyers, W. J. L. *et al.* (1975), *Phys. Rev.* **B 11**, 266.

Campbell, B. J. *et al.* (2001), *Phys. Rev.* **B 65**, 014427.

Carlson, E. W. *et al.* (1998), *Phys. Rev.* **B 57**, 14704.

Castellani, C., Natoli, C. R. and Ranninger, J. (1978), *Phys. Rev.* **B 18**, 4945, 4967, 5001.

Castelnovo, C., Moessner, R. and Sondhi, S. L. (2008), *Nature* **451**, 42.

Cava, R. J. *et al.* (1988), *Nature* **332**, 814.

Chakhalian, J. *et al.* (2006), *Nature Phys.* **2**, 244.

Chakhalian, J. *et al.* (2007), *Science* **318**, 1114.

Chappel, E. *et al.* (2000), *Eur. Phys. J.* B **17**, 609, 615.

Cheong, S.-W. and Mostovoy, M. V. (2007), *Nature Mater.* **6**, 13.

Choi, Y. J. *et al.* (2008), *Phys. Rev. Lett.* **100**, 047601.

Coey, J. M. D. (2010), *Magnetism and Magnetic Materials*. Cambridge: Cambridge University Press.

Cohen, R. E. (1992), *Nature* **358**, 136.

Coleman, P. (2007), in H. Kronmüller and S. Parkin (eds), *Handbook of Magnetism and Advanced Magnetic Materials*. New York: Wiley.

Cooper, B. R. (1972), in R. J. Elliott (ed.), *Magnetic Properties of Rare Earth Metals*. London: Plenum Press.

Cotton, F. A., Wilkinson, G. and Gaus, P. L. (1995), *Basic Inorganic Chemistry*. New York: Wiley.

Cox, P. A. (1992), *Transition Metal Compounds: An Introduction to their Electronic Structure and Properties*. Oxford: Clarendon Press.

Curie, P. (1894), *J. Physique* **3**, 393.

Cwik, M. *et al.* (2009), *Phys. Rev. Lett.* **102**, 057201.

Dagotto, E. (2003), *Nanoscale Phase Separation and Colossal Magnetoresistance: The Physics of Manganites and Related Compounds*. Berlin: Springer-Verlag.

Daoud-Aladine, A. *et al.* (2002), *Phys. Rev. Lett.* **89**, 097205.

de Boer, J. H. and Verwey, E. J. W. (1937), *Proc. Phys. Soc.* **A49**, 59.

de Gennes, P. G. (1960), *Phys. Rev.* **118**, 141.

de Groot, J. *et al.* (2012), *Phys. Rev. Lett.* **108**, 187601.

Delaney, K. T., Mostovoy, M. and Spaldin, N. A. (2009), *Phys. Rev. Lett.* **102**, 157203.

de Medici, L. *et al.* (2009), *Phys. Rev. Lett.* **102**, 126401.

Diep, H. T. (ed.) (2004), *Frustrated Spin Systems*. Singapore: World Scientific.

Dikin, D. A. *et al.* (2011), *Phys. Rev. Lett.* **107**, 056802.

Di Matteo, S., Jackeli, G. and Perkins, N. B. (2005), *Phys. Rev.* **B 72**, 024431.

Doumerc, J.-P. *et al.* (1999), *J. Solid State Chem.* **147**, 211.

Doumerc, J.-P. *et al.* (2001), *J. Mater. Chem.* **11**, 78.

Dzyaloshinskii, I. E. (1958), *J. Phys. Chem. Solids* **4**, 241.

Dzyaloshinskii, I. E. (1959), *Sov. Phys.—JETP* **10**, 628.

Efremov, D. V., van den Brink, J. and Khomskii, D. I. (2004), *Nature Mater.* **3**, 853.

Ehrenstein, W., Mazur, N. and Scott, J. F. (2006), *Nature* **442**, 759.

Emery, V. J. and Reiter, G. (1988), *Phys. Rev.* **B 38**, 4547.

Emery, V. J., Kivelson, S. A. and Lin, H. Q. (1991), *Phys. Rev. Lett.* **64** 475.

Englman, R. (1972), *The Jahn–Teller Effect in Molecules and Crystals*. New York: Wiley.

Eshelby, J. D. (1956), in F. Seitz and D. Turnbull (eds), *Solid State Physics*, vol. 3, p. 79, New York: Academic Press.

Ezhov, S. Yu. *et al.* (1999), *Phys. Rev. Lett.* **83**, 4136.

Fauth, F. *et al.* (2002), *Phys. Rev.* **B 66**, 184421.

Fernandez-Diaz, M. T. (2001), *Phys. Rev.* **B 34**, 144417.

Ferriani, P. *et al.* (2008), *Phys. Rev. Lett.* **101**, 027201.

Fiebig, M. (2005), *J. Phys. D Appl. Phys.* **38**, 12123.

Florens, S., George, A., Kotliar, G. and Parcollet, O. (2002), *Phys. Rev.* **B 66**, 205102.

Forthaus, M. K. *et al.* (2008), *Phys. Rev.* **B 77**, 165121.

Freund, H.-J., Kuhlenbeck, H. and Staemmler, V. (1996), *Rep. Progr. Phys.* **59**, 283.

Fukazawa, H., Hoshikawa, A., Ishii, Y, Chakoumakos, B. C. and Fernandez-Baca, J. A. (2006), *Astrophys. J.* **652**, L57.

Furubayashi, T. *et al.* (1994), *J. Phys. Soc. Japan* **66**, 778.

Galasso, F. (1970), *Structures and Properties of Inorganic Solids*. Oxford: Pergamon Press.

Garcia, J. and Subias, G. (2004), *J. Phys. Condens. Matter* **16**, 145.

Garcia-Fernandez, P. *et al.* (2010), *J. Phys. Chem. Lett.* **1**, 647.

Gebhard, F. (1997), *The Mott Metal–Insulator Transitions: Models and Methods*. Berlin: Springer-Verlag.

Gehring, G. A. and Gehring, K. A. (1975), *Rep. Progr. Phys.* **38**, 1.

Georges, A. *et al.* (1996), *Rev. Mod. Phys.* **68**, 13.

Ghoses, P., Michenaud, J.-P. and Gonze, X. (1998), *Phys. Rev.* **B 58**, 6239.

Gianmarchi, T., Ruegg, C. and Tchernyschyov, O. (2008), *Nature Phys.* **4**, 198.

Giovanetti, G. *et al.* (2009), *Phys. Rev. Lett.* **103**, 156401.

Giovanetti, G. *et al.* (2011), *Phys. Rev.* **B 83**, 060402(R).

Glazer, A. M. (1972), *Acta Cryst.* **B 28**, 3384.

Goniakowski, J., Finocchi, F. and Noguera, C. (2008), *Rep. Progr. Phys.* **71**, 016501.

Gonzalo, J. A. (2006), *Effective Field Approach to Phase Transitions and Some Applications to Ferroelectrics.* Singapore: World Scientific.

Goodenough, J. B. (1963), *Magnetism and the Chemical Bond.* New York: Interscience.

Goodenough, J. B. and Longo, J. M. (1970), *Magnetic and Other Properties of Oxides and Related Compounds.* In *Numerical Data and Functional Relations in Science and Technology, New Series*, vol. III. 4. Berlin: Springer-Verlag.

Goodenough, J. B. (1971), *J. Solid State Chem.* **3**, 490.

Goodenough, J. B. (2004), *Rep. Progr. Phys.* **67**, 1915.

Griffith, J. S. (1971), *The Theory of Transition-Metal Ions.* Cambridge: Cambridge University Press.

Grohol, D. *et al.* (2005), *Nature Mater.* **4**, 323.

Grüninger, M. *et al.* (2002), *Nature* **418**, 39.

Gschneider, K. A. and Eyring, L. (eds) (1978), *Handbook of the Physics and Chemistry of Rare Earths*, vol. 1: *Metals.* Amsterdam: North Holland.

Güttlich, P. and Goodwin, H. A. (eds) (2004), *Spin Crossover in Transition Metal Compounds*, I–III. Berlin: Springer-Verlag.

Gutzwiller, M. C. (1965), *Phys. Rev.* **137**, A1726.

Halcrow, M. A. (2013), *Spin-crossover Materials: Properties and Applications.* New York: Wiley.

Haldane, D. (1983), *Phys. Rev. Lett.* **50**, 1153.

Ham, F. S. (1972), in S. Geshwind (ed.), *Electron Paramagnetic Resonance.* New York: Plenum Press.

Harris, A. B., Yildirim, T., Aharoni, A., Entin-Wohlman, O. and Korenblit, I. Ya. (2003), *Phys. Rev. Lett.* **91**, 987206.

Harris, A. B., Aharony, A. and Entin-Wohlman, O. (2008), *J. Phys. Condens. Matter* **20**, 434202.

Harrison, W. A. *et al.* (1978) *Phys. Rev.* **B 18**, 4402.

Harrison, W. A. (1989), *Electronic Structure and the Properties of Solids.* New York: Dover.

Hasegawa, K. *et al.* (2009), *Phys. Rev. Lett.* **103**, 146403.

Haule, K. and Kotliar, G. (2009), *Nature Phys.* **5**, 796.

Haverkort, M. W. *et al.* (2005), *Phys. Rev. Lett.* **95**, 196404.

Haverkort, M. W. *et al.* (2006), *Phys. Rev. Lett.* **97**, 176405.

Heinrich, V. E. and Cox, P. A. (1994) *The Surface Science of Metal Oxides.* Cambridge: Cambridge University Press.

Heinze, S. *et al.* (2012), *Nature Phys.* **7**, 713.

Hellberg, C. S. and Manousakis, E. (1997), *Phys. Rev. Lett.* **78**, 4609.

Hemberger, J. *et al.* (2002), *Phys. Rev.* **B 66**, 094410.

Hennion, M. and Moussa, F. (2005), *New J. Phys.* **7**, 84.

Hesper, R., Tjeng, L. H., Heeres, A. and Sawatzky, G. A. (2000), *Phys. Rev.* **B 62**, 16046.

Hesterman, K. and Hoppe, R. (1969), *Z. Anorg. Allg. Chem.* **367**, 249, 261.

Hewson, A. C. (1993), *The Kondo Problem to Heavy Fermions.* Cambridge: Cambridge University Press.

Hill, N. A. (2000), *J. Phys. Chem.* B **104**, 6694.

Hirschfeld, P. J., Korshunov, M. M. and Mazin, I. I. (2011), *Rep. Progr. Phys.* **74**, 124508.

Hoch, M. J. R. *et al.* (2004), *Phys. Rev.* **B 70**, 174443.

Hodeau, J.-L. and Marezio, M. (1978), *J. Solid State Chem.* **23**, 253.

Horibe, Y. *et al.* (2006), *Phys. Rev. Lett.* **96**, 086406.

Horiuchi, S. and Tokura, Y. (2008), *Nature Mater.* **7**, 357.

Huang, H.-Y. *et al.* (2011), *Phys Rev.* **B 84**, 235125.

Hubbard, J. (1963), *Proc. Roy. Soc.* **A 276**, 238.

Hur, N. *et al.* (2004), *Nature* **429**, 392.

Hwang, H. *et al.* (2012), *Nature Mater.* **11**, 103.

Hyde, B. G. and Andersson, S. (1989), *Inorganic Crystal Structures*. New York: Wiley.

Ikeda, N. *et al.* (2005), *Nature* **436**, 2005.

Imada, M., Fujimori, A. and Tokura, Y. (1998), *Rev. Mod. Phys.* **70**, 1039.

Ishikawa, A., Nohara, J. and Sugai, S. (2004), *Phys. Rev. Lett.* **93**, 136401.

Isobe, M. *et al.* (2002), *J. Phys. Soc. Japan* **71**, 1423.

Isobe, M. *et al.* (2006), *J. Phys. Soc. Japan* **75**, 73801.

Ito, Y. and Akimitsu, J. (1976), *J. Phys. Soc. Japan* **40**, 1333.

Ivanov, M. A. *et al.* (1983), *J. Magn. Magn. Mater.* **36**, 26.

Jackeli, G. and Khaliullin, G. (2009), *Phys. Rev. Lett.* **103**, 067205.

Jahn, H. A. and Teller, E. (1937), *Proc. Roy. Soc.* **A 161**, 220.

Jia, C. *et al.* (2007), *Phys. Rev.* **B 76**, 144424.

Johnson, R. D. *et al.* (2011), *Phys. Rev. Lett.* **107**, 137205.

Johnson, R. D. *et al.* (2012), *Phys. Rev. Lett.* **108**, 067201.

Jönsson, P. E. *et al.* (2007), *Phys. Rev. Lett.* **99**, 167402.

Jorgensen, C. K. (1962), *Orbitals in Atoms and Molecules*. London: Academic Press.

Jorgensen, C. K. (1971), *Modern Aspects of Ligand Field Theory*. Amsterdam: North Holland.

Kagan, M. Yu., Khomskii, D. I. and Mostovoy, M. V. (1999), *Eur. Phys. J.* **12**, 217.

Kagan, M. Yu., Kugel, K. I. and Khomskii, D. I. (2001), *JETP* **93**, 415.

Kagoshima, S., Kanoda, K. and Mori, T. (2006), *Organic Conductors*. Tokyo: Physics Society of Japan.

Kanamori, J. (1960), *J. Appl. Phys. Suppl.* **31**, 14S.

Kamihara, Y. *et al.* (2008), *J. Am. Chem. Soc.* **130**, 3296.

Kaplan, M. D. and Vekhter, B. G. (1995), *Cooperative Phenomena in Jahn–Teller Crystals*. New York: Plenum Press.

Katayama, N. *et al.* (2009), *Phys. Rev. Lett.* **103**, 146405.

Katsura, H., Nagaosa, N. and Balatsky, A. V. (2005), *Phys. Rev. Lett.* **95**, 057205.

Keldysh, L. V. (1979), *JETP Lett.* **29**, 658.

Kenzelmann, M. *et al.* (2007), *Phys. Rev. Lett.* **98**, 267205.

Khalifah, P. *et al.* (2002), *Science* **297**, 2237.

Khaliullin, G. and Maekawa, S. (2000), *Phys. Rev. Lett.* **85**, 3950.

Khaliullin, G. (2005), *Progr. Theor. Phys. Suppl.* **160**, 155.

Khaliullin, G. (2013), arXiv:1310.0767.

Khomskii, D. I. (1970), *Sov. Phys.—Physics of Metals and Metallography* **29**, 31.

Khomskii, D. I. and Kugel, K. I. (1973), *Solid State Comm.* **13**, 763.

Khomskii, D. I. (1977), *Sov. Phys.—Solid State* **19**, 1850.

Khomskii, D. I. (1997), *Lithuanian J. Phys.* **37**, 65 (also arXiv: cond-mat/0101164).

Khomskii, D. I. (2000), *Physica* B **280**, 325.

Khomskii, D. I. and van den Brink, J. (2000), *Phys. Rev. Lett.* **85**, 3329.

Khomskii, D. I. (2001), *Bull. Am. Phys. Soc.* C 21.002.

Khomskii, D. I. and Kugel, K. I. (2001), *Europhys. Lett.* **55**, 208.

Khomskii, D. I. and Kugel, K. I. (2003), *Phys. Rev.* **B 67**, 134401.

Khomskii, D. I. and Löw, U. (2004), *Phys. Rev.* **B 69**, 184401.

Khomskii, D. I. (2005), *Physica Scripta* **72**, cc8.

Khomskii, D. I. and Mizokawa, T. (2005), *Phys. Rev. Lett.* **94**, 156402.

Khomskii, D. I. (2006), *J. Magn. Magn. Mater.* **306**, 1.

Khomskii, D. I. (2009), *Physics (Trends)* **2**, 20.

Khomskii, D. I. (2010), *Basic Aspects of the Quantum Theory of Solids: Order and Elementary Excitations.* Cambridge: Cambridge University Press.

Khomskii, D. I. (2011), *J. Phys.: Conf. Ser.* **320**, 012055.

Kim, B. J. *et al.* (2008), *Phys. Rev. Lett.* **101**, 076402.

Kimura, T. *et al.* (2003a), *Nature* **426**, 55.

Kimura, T. *et al.* (2003b), *Phys. Rev.* **B 68**, 060403.

Kiss, A. and Fazekas, P. (2005), *Phys. Rev.* **B 71**, 054415.

Kitaev, A. (2006), *Ann. Phys.* (N.Y.) **321**, 2.

Kitaev, A. and Preskill, J. (2006), *Phys. Rev. Lett.* **96**, 110404.

Kittel, Ch. (2004a), *Introduction to Solid State Physics*, 8th edn. New York: Wiley.

Kittel, Ch. (2004b), *Quantum Theory of Solids*. New York: Wiley.

Kobayashi, K.-I. *et al.* (1998), *Nature* **395**, 677.

Koch, E., Gunnarsson, O. and Martin, R. M. (1999), *Phys. Rev.* **B 60**, 15718.

Koga, A., Kawakami, N., Rice, T. M. and Sigrist, M. (2004), *Phys. Rev. Lett.* **92**, 216402.

Koizumi, H. and Bersuker, I. B. (1999), *Phys. Rev. Lett.* **83**, 3009.

Kojima, N. *et al.* (1994), *J. Am. Chem. Soc.* **116**, 11368.

Komarek, A. C. *et al.* (2011), *Phys. Rev. Lett.* **107**, 027201.

Kondo, S. (1964), *Progr. Theor. Phys.* **3**, 37.

Kondo, S. *et al.* (1997), *Phys. Rev. Lett.* **78**, 3729.

König, E. (1991), *Struct. Bonding* **1976**, 51.

Konishi, Y. *et al.* (1999), *J. Phys. Soc. Japan* **68**, 3790.

Koonce, S. C. *et al.* (1967), *Phys. Rev.* **163**, 380.

Kopaev, Yu. V. (2009), *Physics–Uspekhi* **52**, 1111.

Korotin, M. A. *et al.* (1996), *Phys. Rev.* **B 54**, 5309.

Korotin, M. A., Anisimov, V. I., Khomskii, D. I. and Sawatzky, G. A. (1997), *Phys. Rev. Lett.* **80**, 4305.

Kosuge, K., Okinaka, H. and Kashi, S. (1972), *IEEE–Trans. Mag.*, Mag-8, N3, p. 581.

Kotliar, G. and Vollhardt, D. (2009), *Physics Today* **57**, 53.

Kubetska, A. *et al.* (2002), *Phys. Rev. Lett.* **88**, 057201.

Kugel, K. I. and Khomskii, D. I. (1973), *Zh. Exp. Teor. Fiz.* **64**, 1429 [*Sov. Phys.—JETP* **37**, 725].

Kugel, K. I. and Khomskii, D. I. (1982), *Usp. Fiz. Nauk* **136**, 621 [*Sov. Phys.—Uspekhi* **25**, 231].

Kugel, K. I. *et al.* (2008), *Phys. Rev.* **B 78**, 155113.

Kuntscher, C. A. *et al.* (2006), *Phys. Rev.* **B 74**, 184402.

Kuntscher, C. A. *et al.* (2007), *Phys. Rev.* **B 76**, 241101.

Kuramoto, Y., Kusunose, H. and Kiss, A. (2009), *J. Phys. Soc. Japan* **78**, 072001.

Lacroix, C., Mendels, P. and Mila, F. (eds) (2011), *Introduction to Frustrated Magnetism.* Springer Series in Solid State Sciences, vol. 164. Berlin: Springer-Verlag.

Lakkis, S. *et al.* (1976), *Phys. Rev.* **B 14**, 1429.

Landau, L. D. (1965), *Collected Papers of L. D. Landau*, edited and with an introduction by D. ter Haar. Oxford: Pergamon Press.

Landau, L. D. and Lifshitz, E. M. (1960), *Electrodynamics of Continuous Media*. Oxford: Pergamon Press.

Landau, L. D. and Lifshitz, E. M. (1965), *Quantum Mechanics*. Oxford: Pergamon Press.

Landau, L. D. and Lifshitz, E. M. (1969), *Statistical Physics*. Reading, MA: Addison-Wesley.

Landau, L. D. and Zeldovich, Ya. (1944), *Acta Phys.-Chim. USSR* **18**, 194; also *ZhETF* **14**, 32 (1944).

Lee, S. *et al.* (2006), *Nature Mater.* **5**, 471.

Leggett, A. J. (2006), *Quantum Liquids: Bose Condensation and Cooper Pairing in Condensed Matter Systems*. Oxford: Oxford University Press.

Lengsdorf, R. *et al.* (2004), *Phys. Rev.* **B 69**, 140403.

Leonov, I., Yaresko, A. N., Antonov, V. N., Korotin, M. A. and Anisimov, V. I. (2004), *Phys. Rev. Lett.* **93**, 146404.

Levanyuk, A. P. and Sannikov, D. G. (1974), *Sov. Phys.—Uspekhi* **17**, 199.

Levin, M. and Wen, X.-G. (2006), *Phys. Rev. Lett.* **96**, 110405.

Li, L. *et al.* (2011), *Nature Physics* **7**, 762.

Liebsch, A. (2004), *Phys. Rev.* **B 70**, 165103.

Lines, M. E. and Glass, A. M. (1977), *Principles and Applications of Ferroelectrics and Related Materials*. Oxford: Oxford University Press.

Logginov, A. S. *et al.* (2007), *JETP Lett.* **86**, 115.

Logginov, A. S. *et al.* (2008), *Appl. Phys. Lett.* **93**, 182510.

Longuet-Higgins, H. C., Öpik, U., Price, M. H. L. and Sack, R. A. (1958), *Proc. Roy. Soc. London* **A 244**, 1.

Lorenz, B., Wang, Y. Q. and Chu, C. W. (2006), *Phys. Rev.* **B 76**, 104405.

Loudon, J. C. *et al.* (2005), *Phys. Rev. Lett.* **94**, 097202.

Lummen, T. T. A. *et al.* (2008), *J. Phys. Chem.* **C 112**, 14158.

Lyubutin, I. S., Ovchinnikov, S. G., Gavrilyuk, A. G. and Struzhkin, V. V. (2009), *Phys. Rev.* **B 79**, 085125.

Mahan, G. D. (2000), *Many Particle Physics*. Berlin: Springer-Verlag.

Maignan, A. *et al.* (2004), *Phys. Rev. Lett.* **93**, 026401.

Majumdar, C. K. and Ghosh, D. K. (1969), *J. Math. Phys.* **10**, 1399.

Malashevich, A. and Vanderbilt, D. (2008), *Phys. Rev. Lett.* **101**, 037210.

Mannhart, J., Blank, D. H. A., Hwang, H. Y., Millis, A. J. and Triscone, J.-P. (2008), *MRS Bull.* **38**, 1027.

Mannhart, J. and Schlom, D. G. (2010), *Science* **327**, 1607.

Marezio, M., McWhan, D. B., Remeika, J. P. and Dernier, P. D. (1972), *Phys. Rev.* **B 5**, 2541.

Marezio, M., McWhan, D. B., Dernier, P. D. and Remeika, J. P. (1973a), *J. Solid State Chem.* **6**, 213.

Marezio, M., McWhan, D. B., Dernier, P. D. and Remeika, J. P. (1973b), *J. Solid State Chem.* **6**, 419.

Mathur, N. D. *et al.* (1998), *Nature* **394**, 39.

Matsuno, K. *et al.* (2001), *J. Phys. Soc. Japan* **70**, 1456.

Mattuck, R. D. (1992), *A Guide to Feynman Diagrams in Many-Body Problems*. New York: Dover.

Mazin, I. I. *et al.* (2008a), *Phys. Rev. Lett.* **98**, 176406.

Mazin, I. I. *et al.* (2008b), *Phys. Rev. Lett.* **101**, 057003.

Mazin, I. I. and Schmalian, J. (2009), *Phys. C—Supercond. Appl.* **469**, 614.

Mazin, I. I. (2011), *Physics (Trends)* **4**, 26.

McWhan, D. B. *et al.* (1973), *Phys. Rev.* **B 7**, 1920.

McWhan, D. B., Jayaraman, A., Remeika, J. P. and Rice, T. M. (1975), *Phys. Rev. Lett.* **34**, 547.

McQueen, T. M. *et al.* (2008a), *Phys. Rev. Lett.* **101**, 166402.

McQueen, T. M. *et al.* (2008b), *J. Phys. Condens. Matter* **20**, 235210.

Medarde, M. *et al.* (1992), *Phys. Rev.* **B 45**, 14974.

Megaw, H. D. (1957), *Ferroelectricity in Crystals*. London: Methuen.

Methfessel, S. and Mattis, D. C. (1968), *Magnetic Semiconductors*. In *Handbuch der Physik*, vol. XVII, Part 1, p. 389. Berlin: Springer-Verlag.

Meyer, G., Gloger, T. and Beekhuizen, J. (2009), *Z. Anorg. Allg. Chemie* **635**, 1497.

Mielke, A. and Tasaki, H. (1993), *Comm. Math. Physics* **158**, 341.

Mitsui, T. *et al.* (1981), *Ferroelectrics and Related Substances*. In *Numerical Data and Functional Relations in Science and Technology, New Series*, vol. 16(1). Berlin: Springer-Verlag.

Mizokawa, T. *et al.* (1991), *Phys. Rev. Lett.* **67**, 1638.

Mizokawa, T., Khomskii, D. I. and Sawatzky, G. A. (2000), *Phys. Rev.* **B 61**, 11263.

Molegraaf, H. J. A. *et al.* (2002), *Science* **295**, 2239.

Monceaux, P., Nad', F. Y. and Brazovskii, S. (2001), *Phys. Rev. Lett.* **86**, 4080.

Moon, R. M. (1970), *Phys. Rev. Lett.* **25**, 527.

Moriya, T. (1960), *Phys. Rev.* **120**, 91.

Mostovoy, M. V. and Khomskii, D. I. (2002), *Phys. Rev. Lett.* **89**, 227203.

Mostovoy, M. V. and Khomskii, D. I. (2003), *J. Phys. A: Mat. and General* **36**, 9197.

Mostovoy, M. V. and Khomskii, D. I. (2004), *Phys. Rev. Lett.* **92**, 167201.

Mostovoy, M. V. (2006), *Phys. Rev. Lett.* **86**, 067601.

Mostovoy, M. V. (2008), *Phys. Rev. Lett.* **100**, 089702.

Motome, Y. and Tsunetsugu, H. (2004), *Phys. Rev.* **B 70**, 184427.

Mott, N. F. and Peierls, R. (1937), *Proc. Phys. Soc.* **A49**, 72.

Mott, N. F. (1949), *Proc. Phys. Soc.* **A62**, 416.

Mott, N. F. and Jones, H. (1958), *The Theory of the Properties of Metals and Alloys*. New York: Dover.

Mott, N. F. (1961), *Phil. Mag.* **6**, 287.

Mott, N. F. (1990), *Metal–Insulator Transitions*. London: Taylor & Francis.

Mülbauer, S. *et al.* (2009), *Science* **323**, 915.

Mydosh, J. A. (1993), *Spin Glasses*. Philadelphia: Taylor & Francis.

Nagaev, E. L. (1969), *Zh. Exp. Teor. Fiz.* **57**, 1274 [*Sov. Phys.—JETP* **30**, 693 (1970)].

Nagaev, E. L. (1983), *Physics of Magnetic Semiconductors*. Moscow: MIR.

Nagaoka, J. (1966), *Phys. Rev.* **147**, 392.

Noguchi, S., Kawamata, S., Okuda, K., Nojiri, H. and Motokawa, M. (2002), *Phys. Rev.* **B 66**, 094404.

Noguera, C. (1996), *Physics and Chemistry at Oxide Surfaces*. Cambridge: Cambridge University Press.

Norman, M. R. and Pepin, C. (2003), *Rep. Progr. Phys.* **66**, 1547.

Nussinov, Z. and van den Brink, J. (2013), arXiv:1303.5922.

Ohtomo, A. and Hwang, H. Y. (2004), *Nature* **427**, 423.

Okamoto, Y., Nohara, M., Aruga-Katori, H. and Takagi, H. (2007), *Phys. Rev. Lett.* **99**, 137207.

Oles, A. (2012), *J. Phys. Condens. Matter* **24**, 313201.

Ovchinnikov, S. G. (2008), *Zh. Exp. Teor. Fiz.* **134**, 172 [JETP **107**].

Pardo, V. *et al.* (2008), *Phys. Rev. Lett.* **101**, 256403.

Park, J.-H. *et al.* (2000), *Phys. Rev.* **B 61**, 11506.

Pauling, L. (1998), *General Chemistry*. New York: Dover.

Pavarini, E., Koch, E. and Lichtenstein, A. I. (2008), *Phys. Rev. Lett.* **101**, 266405.

Pavarini, E. and Koch, E. (2010), *Phys. Rev. Lett.* **104**, 086402.

Peierls, R. (1955), *Quantum Theory of Solids*. Oxford: Oxford University Press.

Peierls, R. (1991), *More Surprises in Theoretical Physics*. Princeton, NJ: Princeton University Press.

Pen, H. *et al.* (1997), *Phys. Rev. Lett.* **78**, 1323.

Penn, D. R. (1966), *Phys. Rev.* **142**, 350.

Pfleiderer, C. *et al.* (2004), *Nature* **427**, 227.

Phelan, D. *et al.* (2006), *Phys. Rev. Lett.* **96**, 027201; *Phys. Rev. Lett.* **97**, 235501.

Phillips, P. (2009), *Rep. Progr. Phys.* **72**, 03601.

Picozzi, S. *et al.* (2006), *Phys. Rev.* **B 74**, 094402.

Pimenov, A. *et al.* (2006), *Nature Phys.* **2**, 97.

Pimenov, A. *et al.* (2008), *J. Phys. Condens. Matter* **20**, 434209.

Plakhty, V. P. (2005), *Phys. Rev.* **B 71**, 214407.

Plakida, N. (2010), *High-Temperature Cuprate Superconductors: Experiment, Theory and Applications*. Berlin: Springer-Verlag.

Podlesnyak, A. *et al.* (2006), *Phys. Rev. Lett.* **97**, 247208.

Podlesnyak, A. *et al.* (2008), *Phys. Rev. Lett.* **101**, 247603.

Pomyakushin, V. Yu. *et al.* (2002), *Phys. Rev.* **B 66**, 184412.

Pouget, J. P. *et al.* (1975), *Phys. Rev. Lett.* **35**, 873.

Racah, P. M. and Goodenough, J. B. (1967), *Phys. Rev.* **155**, 932.

Radaelli, P. G. *et al.* (2002), *Nature* **416**, 155.

Radaelli, P. G. and Chapon, L. C. (2008), *J. Phys. Condens. Matter* **20**, 434212.

Ramesh, R. and Spaldin, N. A. (2007), *Nature Mater.* **6**, 21.

Rao, C. N. R. and Raveau, B. (1998), *Transition Metal Oxides: Structure, Properties and Synthesis of Ceramic Oxides*. New York: Wiley.

Reehuis, M. *et al.* (2003), *Eur. Phys. J.* **B35**, 311.

Ren, Y. *et al.* (1998), *Nature* **396**, 441.

Ren, Y. *et al.* (2000), *Phys. Rev.* **B 62**, 6571.

Reynaud, F. *et al.* (2001), *Phys. Rev. Lett.* **86**, 3638.

Reyren, N. *et al.* (2007), *Science* **317**, 1196.

Rice, T. M. and Sneddon, L. (1981), *Phys. Rev. Lett.* **47**, 689.

Rice, M. J. and Choi, H. Y. (1992), *Phys. Rev.* **B 45**, 10173.

Rice, T. M., Launois, H. and Pouget, J. P. (1994), *Phys. Rev. Lett.* **73**, 3042.

Rodriguez-Carvajal, J. *et al.* (1998), *Phys. Rev.* **B 57**, R3189.

Ropka, Z. and Radwanski, R. J. (2003), *Phys. Rev.* **B 67**, 172401.

Rosenberg, M. J. *et al.* (1995), *Phys. Rev. Lett.* **75**, 105.

Sachdev, S. (2011), *Quantum Phase Transitions*. Cambridge: Cambridge University Press.

Sadovskii, M. V. (2008), *Physics–Uspekhi* **51**, 1201.

Saitoh, E. *et al.* (2001), *Nature* **410**, 180.

Sakai, H. *et al.* (2011), *Phys. Rev. Lett.* **107**, 137601.

Sato, O. *et al.* (1996), *Science* **272**, 704.

Saxena, S. S. *et al.* (2000), *Nature* **406**, 587.

Sboychakov, A. O. *et al.* (2009), *Phys. Rev.* **B 80**, 024429.

Schlapp, R. and Penney, W. G. (1932), *Phys. Rev.* **42**, 666.

Schmid, H. (1994), *Ferroelectrics* **162**, 317.

Schmid, H. (2008), *J. Phys. Condens. Matter* **20**, 434201.

Schmidt, M. *et al.* (2004), *Phys. Rev. Lett.* **92**, 056402.

Schrieffer, J. R. (1964), *Theory of Superconductivity*. New York: W. A. Benjamin.

Schrieffer, J. R. and Gomer, R. (1971), *Surface Sci.* **25**, 315.

Schrieffer, J. R. (ed.) (2007), *Handbook of High-Temperature Superconductors*. New York: Springer-Verlag.

Schweika, W., Valldor, M. and Lemmens, P. (2007), *Phys. Rev. Lett.* **98**, 067201.

Scott, J. F. (2000), *Ferroelectric Memories*. Berlin: Springer-Verlag.

Seki, S., Onose, Y. and Tokura, Y. (2008), *Phys. Rev. Lett.* **101**, 067204.

Seki, S. *et al.* (2012), *Science* **336**, 198.

Senn, M. S., Write, J. P. and Attfield, J. P. (2011), *Nature* **481**, 137.

Sergienko, I. A. and Dagotto, E. (2006), *Phys. Rev.* **B 73**, 094434.

Sergienko, I. A., Sen, C. and Dagotto, E. (2006), *Phys. Rev. Lett.* **97**, 227204.

Seshadri, R. (2006), *Solid State Sci.* **8**, 259.

Shannon, R. D. (1976), *Acta Crystallogr. Sect. A*, **A 32**, 751.

Shapira, Y., Foner, S. and Reed, T. B. (1973), *Phys. Rev.* **B 8**, 2299.

Sharma, N. *et al.* (2008), *J. Phys. Condens. Matter* **20**, 025215.

Shekhtman, L., Entin-Wohlman, O. and Aharony, A. (1992), *Phys. Rev. Lett.* **69**, 836.

Shitade, A. *et al.* (2009), *Phys. Rev. Lett.* **102**, 256403.

Slater, J. C. (1963), *Quantum Theory of Molecules and Solids*, vol. 1. New York: McGraw-Hill.

Slater, J. C. (1968), *Quantum Theory of Matter*. New York: McGraw-Hill.

Slichter, C. P. and Drickamer, H. G. (1972), *J. Chem. Phys.* **56**, 21242.

Smolenskii, G. A. and Chupis, I. E. (1982), *Sov. Phys.—Uspekhi* **25**, 475.

Sonin, E. B. (2010), *Adv. Phys.* **59**, 181.

Spaldin, N. A., Fiebig, M. and Mostovoy, M. (2008), *J. Phys. Condens. Matter* **20**, 434203.

Spaldin, N. A. (2012), *J. Solid State Chem.* **195**, special issue S1, p. 2.

Starykh, O. A. *et al.* (1996), *Phys. Rev. Lett.* **77**, 2558.

Stauffer, D. and Aharony, A. (1994), *Introduction to Percolation*. London: Taylor & Francis.

Stefanovich, G., Pergament, A. and Stefanovich, D. (2000), *J. Phys. Condens. Matter* **12**, 8833.

Stewart, G. R. (2011), *Rev. Mod. Phys.* **83**, 1589.

Streltsov, S., Popova, O. V. and Khomskii, D. I. (2006), *Phys. Rev. Lett.* **96**, 249701.

Sturge, M. D. (1967), in H. Ehrenreich, F. Seitz and D. Turnbull (eds), *Solid State Physics*, v. 20, p. 91. New York: Academic Press.

Sudayama, T. *et al.* (2011), *Phys. Rev.* **B 83**, 235105.

Sugano, S., Tanabe, Y. and Kamimura, H. (1970), *Multiplets of Transition Metal Ions in Crystals*. New York: Academic Press.

Sushkov, A. B. *et al.* (2008), *J. Phys. Condens. Matter* **20**, 434210.

Takano, M., Kawachi, J., Nakanish, N. and Takeda, Y. (1981), *J. Solid State Chem.* **39**, 75.

Tanabe, Y. and Sugano, S. (1954), *J. Phys. Soc. Japan* **9**, 753.

Tanabe, Y. and Sugano, S. (1956), *J. Phys. Soc. Japan* **11**, 864.

Taniguchi, S. *et al.* (1995), *J. Phys. Soc. Japan* **64**, 2758.

Taylor, J. W. *et al.* (1999), *Eur. Phys. J.* **B 12**, 199.

Tchernyshyov, D. (2004), *Phys. Rev. Lett.* **93**, 157206.

Tchernyshyov, D., Moessner, R. and Sondhi, S. L. (2002), *Phys. Rev.* **B 66**, 064403.

Thiel, S. *et al.* (2006), *Science* **313**, 1942.

Tjeng, L. H. *et al.* (1989), in H. Fukuyama, S. Maekawa and A. P. Malozemoff (eds), *Strong Correlations and Superconductivity*. Springer Series in Solid State Sciences, vol. 89, p. 85. Berlin: Springer-Verlag.

Tokunaga, M. *et al.* (1997), *Phys. Rev.* **B 57**, 5259.

Tokura, Y. and Nagaosa, N. (2000), *Science* **288**, 462.

Tomioka, Y. *et al.* (1996), *Phys. Rev.* **B 53**, R1689.

Toriyama, T. *et al.* (2011), *Phys. Rev. Lett.* **107**, 266402.

Torrance, J. B. *et al.* (1992), *Phys. Rev.* **B 45**, 8209.

Tsuda, N., Nasu, K., Yanase, A. and Siratori, K. (1991), *Electronic Conduction in Oxides.* Berlin: Springer-Verlag.

Tsvelik, A. M. and Wiegmann, P. B. (1983), *Adv. Phys.* **32**, 453.

Turov, E. A. (1994), *Usp. Fiz. Nauk* **164**, 325.

Ulrich, C. *et al.* (2002), *Phys. Rev. Lett.* **89**, 167202.

Ushakov, A. V., Streltsov, S. V. and Khomskii, D. I. (2011), *J. Phys. Condens. Matter* **23**, 445601.

Valdes Aguilar, R. *et al.* (2009), *Phys. Rev. Lett.* **102**, 047203.

van Aken, B. B. *et al.* (2004), *Nature Mater.* **3**, 164.

van den Brink, J. and Khomskii, D. I. (1999), *Phys. Rev. Lett.* **82**, 1016.

van den Brink, J. and Khomskii, D. I. (2001), *Phys. Rev.* **B 63**, 140416(R).

van den Brink, J. and Khomskii, D. I. (2008), *J. Phys. Condens. Matter* **20**, 434217.

van der Marel, D. and Sawatzky, G. A. (1988), *Phys. Rev.* **B 37**, 10674.

Varma, C. M. (1988), *Phys. Rev. Lett.* **61**, 2713.

Varma, C. M. (1999), *Phys. Rev. Lett.* **83**, 3538.

Vasiliu-Doloc, L. *et al.* (1999), *Phys. Rev. Lett.* **83**, 4393.

Vedmedenko, E. Y. *et al.* (2004), *Phys. Rev. Lett.* **92**, 077207.

Verwey, E. J. W. (1939), *Nature* (London) **144**, 327.

Verwey, E. J. W. and Haaymann, P. W. (1941), *Physica* (Amsterdam) **8**, 979.

Verwey, E. J. W., Haaymann, P. W. and Romeijn, F. C. (1947), *J. Chem. Phys.* **15**, 174, 181.

Visscher, P. B. (1974), *Phys. Rev.* **B10**, 943.

Vollhardt, D. (1984), *Rev. Mod. Phys.* **56**, 99.

von der Linden, W. and Edwards, D. M. (1991), *J. Phys. Condens. Matter* **3**, 4917.

von Löhneisen, H., Rosch, A., Vojta, M. and Wölfle P. (2007), *Rev. Mod. Phys.* **79**, 1015.

Wang, J. *et al.* (2003), *Science* **299**, 1719.

Wang, K. F., Liu, J.-M. and Ren, Z. F. (2009), *Adv. Physics* **58**, 321.

Wang, Y. (2011), *Advanced Mater.* **23**, 4111.

Wannier, G. H. (1960), *Phys. Rev.* **79**, 357.

Weng, Z. Y., Sheng, D. N., Chen, Y. C. and Ting, S. C. (1997), *Phys. Rev.* **B 55**, 3894.

Wentzcovitch, R. M., Shulz, W. W. and Allen, P. B. (1994), *Phys. Rev. Lett.* **72**, 3389.

White, S. R. and Scalapino, D. J. (2000), *Phys. Rev.* **B 61**, 6320.

Wilson, J. A. (1972), *Adv. Phys.* **21**, 143.

Wright, J. P., Attfield, J. P. and Radaelli, P. G. (2002), *Phys. Rev.* **B 66**, 214422.

Wu, H. *et al.* (2006), *Phys. Rev. Lett.* **96**, 256402.

Wu, H. *et al.* (2009), *Phys. Rev. Lett.* **102**, 026404.

Wu, H. *et al.* (2011), *Phys. Rev.* **B 84**, 155126.

Wu, J. and Leighton, C. (2003), *Phys. Rev.* **B 67**, 174408.

Yamaguchi, S. *et al.* (1996), *Phys. Rev.* **B 53**, 122926.

Yamaguchi, S., Okimoto, Y. and Tokura, Y. (1997), *Phys. Rev.* **B 55**, 128666.

Yamasake, Y. *et al.* (2006), *Phys. Rev. Lett.* **96**, 207204.

Yarkony, D. R. (1996), *Rev. Mod. Phys.* **68**, 985.

Zaanen, J., Sawatzky, G. A. and Allen, J. W. (1985), *Phys. Rev. Lett.* **55**, 418.

Zaanen, J. and Gunnarsson, O. (1989), *Phys. Rev.* **B 40**, 7391.

Zaliznyak, I. A. *et al.* (2000), *Phys. Rev. Lett.* **85**, 4353.

Zaliznyak, I. A. *et al.* (2001), *Phys. Rev.* **B 64**, 195117.

Zapf, V. Z. *et al.* (2006), *Phys. Rev. Lett.* **96**, 077204.

Zapf, V. Z., Jaime, M. and Batista, C. D. (2014), *Rev. Mod. Phys.* **86**, 563.

Zener, C. (1951), *Phys. Rev.* **82**, 403.

Zhang, F. C. and Rice, T. M. (1988), *Phys. Rev.* **B 37**, 3759.

Zheng, H., Qing An Li, Gray, K. and Mitchell, J. F. (2008), *Phys. Rev.* **B 78**, 155103.

Zhitomirsky, M. E. and Tsunetsugu, H. (2005), *Progr. Theor. Phys. Suppl.* **160**, 461.

Zhou, J.-S., and Goodenough, J. B. (2005a), *Phys. Rev. Lett.* **94**, 065501.

Zhou, J.-S., Goodenough, J. B. and Dabrowski, B. (2005b), *Phys. Rev. Lett.* **94**, 226602.

Ziese, M. and Thornton, M. J. (eds.) (2001), *Spin Electronics*. Berlin: Springer-Verlag.

Ziman, J. M. (1964), *Principles of the Theory of Solids*. Cambridge: Cambridge University Press.

Ziman, J. M. (2000), *Electrons and Phonons*. Oxford: Oxford University Press.

Zobel, C., Kriener, M., Bruns, D., Baier, J., Grueninger, M. and Lorenz, T. (2002), *Phys. Rev.* **B 66**, 020402(R).

Index

Periodic Table of the Elements

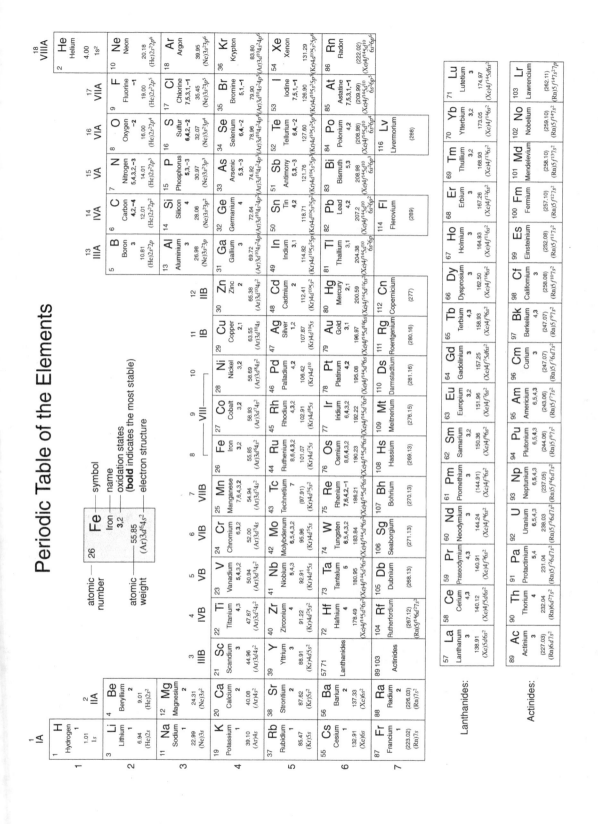